Cyclic Peptides

From Bioorganic Synthesis to Applications

Chemical Biology

Editor-in-chief:
Tom Brown, *University of Oxford, UK*

Series editors:
Kira J. Weissman, *Lorraine University, France*
Sabine Flitsch, *University of Manchester, UK*
Nick J. Westwood, *University of St Andrews, UK*

Titles in the series:
1: High Throughput Screening Methods: Evolution and Refinement
2: Chemical Biology of Glycoproteins
3: Computational Tools for Chemical Biology
4: Mass Spectrometry in Chemical Biology: Evolving Applications
5: Mechanisms of Primary Energy Transduction in Biology
6: Cyclic Peptides: From Bioorganic Synthesis to Applications

How to obtain future titles on publication:
A standing order plan is available for this series. A standing order will bring delivery of each new volume immediately on publication.

For further information please contact:
Book Sales Department, Royal Society of Chemistry, Thomas Graham House, Science Park, Milton Road, Cambridge, CB4 0WF, UK
Telephone: +44 (0)1223 420066, Fax: +44 (0)1223 420247,
Email: booksales@rsc.org
Visit our website at www.rsc.org/books

Cyclic Peptides
From Bioorganic Synthesis to Applications

Edited by

Jesko Koehnke
Helmholtz Centre for Infection Research
Email: jeskokoehnke@gmail.com

James Naismith
University of St Andrews, UK
Email: jhn@st-andrews.ac.uk

and

Wilfred A. van der Donk
University of Illinois, USA
Email: vddonk@illinois.edu

ROYAL SOCIETY OF CHEMISTRY

THE QUEEN'S AWARDS FOR ENTERPRISE: INTERNATIONAL TRADE 2013

Chemical Biology No. 6

Print ISBN: 978-1-78262-528-5
PDF ISBN: 978-1-78801-015-3
EPUB ISBN: 978-1-78801-377-2
ISSN: 2055-1975

A catalogue record for this book is available from the British Library

The Royal Society of Chemistry is a charity, registered in England and Wales, Number 207890, and a company incorporated in England by Royal Charter (Registered No. RC000524), registered office: Burlington House, Piccadilly, London W1J 0BA, UK, Telephone: +44 (0) 207 4378 6556.

For further information see our web site at www.rsc.org

Printed in the United Kingdom by CPI Group (UK) Ltd, Croydon, CR0 4YY, UK

Contents

Chemical Biology No. 6
Cyclic Peptides: From Bioorganic Synthesis to Applications
Edited by Jesko Koehnke, James Naismith and Wilfred A. van der Donk

Published by the Royal Society of Chemistry, www.rsc.org

CHAPTER 1

An Introduction to Cyclic Peptides

MARTIN EMPTING

Helmholtz-Institute for Pharmaceutical Research Saarland, Department Drug Design & Optimization, Campus E81, Saarbrücken 66123, Germany
*E-mail: martin.empting@helmholtz-hzi.de

1.1 Of Peptides and Proteins (and Small Molecules)

When we compare the Ancient Greek etymologies of 'protein' and 'peptide' it becomes evident that the former of these very closely related biomolecules is associated with rather positive attributes, as its name derives from *proteios* (πρωτεῖος) meaning "the first quality".[1] Peptides, on the other hand, seem to be considered a mere rudiment of their bigger ancestors, as they are referred to with a terminus derived from *peptós* (πεπτός), meaning "digested" or "cooked".[2] But what is the decisive characteristic that defines the 'quality' of proteins, which peptides supposedly lack? Both biopolymers usually consist of a linear sequence of amide-linked building blocks, which in most cases are a selection from the standard repertoire of the 20 proteinogenic amino acids. Usually, amino acid sequences shorter than 50 residues are considered as peptides and, thus, these oligomers reside in the so-called 'middle space' (see Figure 1.1).[3,4]

This term has been coined to refer to molecules with a molecular weight between 500 and 5000 Da (or maybe only approximately 3000 Da). Noteworthily, most of the active principles in pharmaceutical drugs belong either to

Chemical Biology No. 6
Cyclic Peptides: From Bioorganic Synthesis to Applications
Edited by Jesko Koehnke, James Naismith and Wilfred A. van der Donk

Published by the Royal Society of Chemistry, www.rsc.org

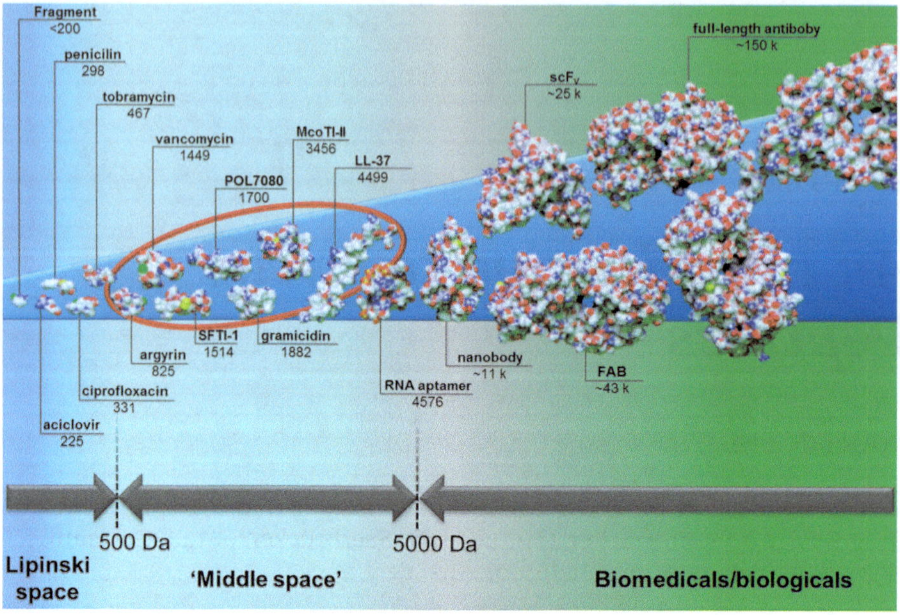

Figure 1.1 The scale of pharmaceutical agents divided into small molecules (Lipinski space), the 'middle space', and the biomedicals/biologicals illustrated by selected anti-infective agents and other examples. The range commonly occupied by (cyclic) peptides is highlighted with a red elliptical shape. From small to large: a fragment-sized compound,[5] acyclovir, penicillin, ciprofloxacin, tobramycin, argyrin,[6] vancomycin,[7] sunflower-trypsin inhibitor-1 (SFTI-1),[8] POL7080,[9] gramicidin,[10] *Momordica cochinchinensis* trypsin inhibitor-II (McoTI-II),[11] human cathelicidin LL-37,[12] an RNA aptamer,[13] a vNAR antibody fragment as a representative for 'nanobodies',[14] a single-chain variable fragment (scF_v),[15] a FAB-fragment,[16] and a full-length IgG antibody.[17] The corresponding molecular weights are given in Da.

small molecules or biomedicals.[18] The former usually obey the well-known rules-of-five set up by Lipinski, while the most prominent members of the latter are immunoglobulins or derivatives thereof.[19] Members of the 'middle space' are still comparably rarely found in pharmaceuticals that are in clinical use and small molecular drugs make up about 90% of the pharmaceutical market.[20,21] However, their share is continuously growing.[20]

So, size as a molecular descriptor might give us a hint toward the 'special quality' of proteins. However, we know of so-called 'mini-proteins' (*e.g.* McoTI, see Figure 1.1), which can be smaller than 40 residues and exert fascinating biological functions, nonetheless.[22,23] Hence, certainly, the line between peptides and proteins is fuzzy and not so clear cut as one might think. Which brings us back to the initial question: 'What is the unique quality of proteins, then?'

One or maybe THE extraordinary characteristic of protein chains is their ability to adopt precise spatial arrangements of each individual rotatable bond within their backbone. Only when the huge array of so-called Φ, Ψ, and ω dihedral angles is assembled in the right way (see Figure 1.2A), is the protein folded into its correct three-dimensional structure.[24] Partial motifs of protein folds (secondary structures) can also be present in peptides (Figure 1.2B).[25] The most prominent secondary structures are of course α-helices and β-sheets. Aside from larger loops, which are usually not well defined, β-turns should be highlighted as important structural motifs connecting, for example, the separate strands of a β-sheet.[26] However, full

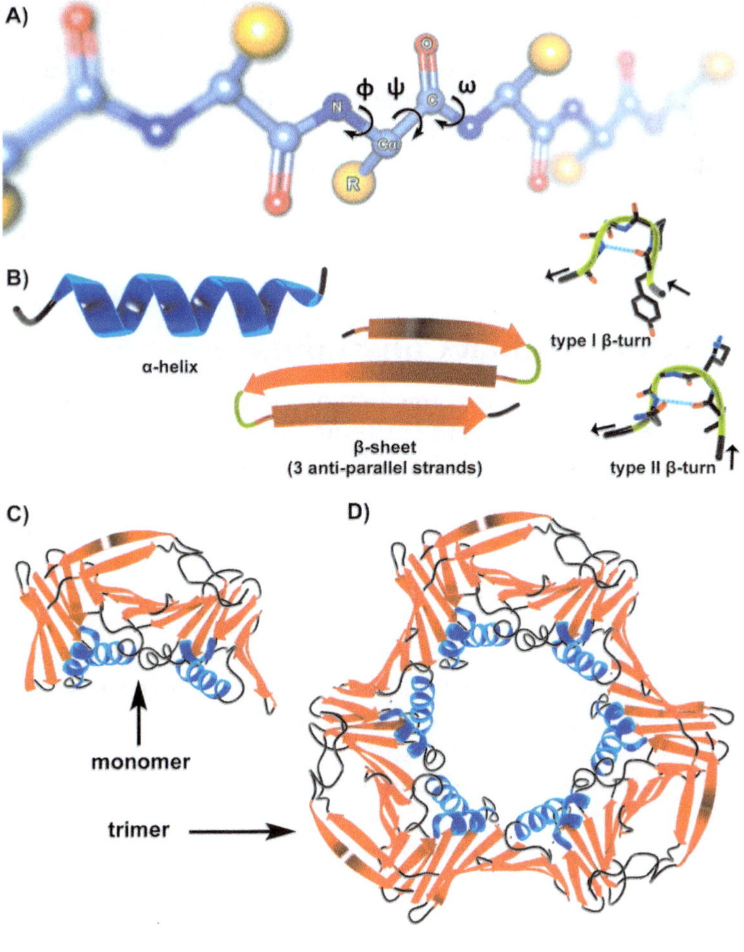

Figure 1.2 (A) A generic peptide chain illustrating the Φ, Ψ, and ω dihedral angles at one amino acid residue. The consecutive order of amino acids in the polypeptide chain is called the primary structure. (B) Schematic depictions of the most common peptide secondary structures: α-helix, β-sheet (3 strands), and type I as well as type II β-turns. (C) and (D) Tertiary and quaternary structure illustrated for the example of proliferating-cell-nuclear-antigen (PCNA).[31]

tertiary or even quaternary assemblies are usually not found for peptides (Figure 1.2C and D). The reason for this is that a short peptide chain offers fewer opportunities for structure-defining intramolecular interactions, which are needed to render a desired three-dimensional structure thermodynamically favorable. Instead, linear peptides are generally quite flexible and do not adopt an unambiguous geometry.[27] As the structure dictates the function of a biomolecule, we may now have approximated the answer to the question raised above: linear peptides often lack a well-defined structure, while proteins exert their various activities through folding into stable three-dimensional assemblies.[28]

This short repetition of textbook knowledge above instantly raises another question: 'Is there a way to fix a well-defined structure within a peptidic biomolecule?' Well, the title of the book you hold in your hands readily reveals the answer: Yes! – and it can be achieved through introducing cyclic or macrocyclic motifs! Nature and scientists alike make use of these conformational constraints, which drastically reduce the degree of rotational freedom within the backbone, thereby orienting the side chains in favorable directions and, thus, tethering fascinating biological activities and other favorable properties into otherwise inactive peptide sequences.[29,30]

1.2 Conformational Constraints

Restricting the conformational freedom of a peptide can be achieved by different means.[29] In addition to the just-mentioned macrocyclization strategy, available torsional angles can be reduced through bulky amino acid side chains, where only some of the possible dihedral combinations are allowed at the corresponding and neighboring residues due to steric hindrance.[32] Furthermore, the presence of a proline readily fixes the Φ dihedral at this residue as a consequence of the covalent linkage of the alkyl side chain to the α-amine. The ω dihedral, which is commonly referred to as the peptide bond, can be cemented into each of the two possible conformations through biomimetic exchange, for example by the use of 1,2,3-triazoles or other heteropentacycles.[33] However, introducing a cycle/macrocycle is the strategy that has the most prominent impact on the overall conformational freedom effecting a multitude of residues and not only some selected positions within the peptide chain.[29]

There are three straightforward concepts to achieve a looped structure within a peptide:

Head-to-tail-, side chain-to-side chain-, and side chain-to-terminus-cyclization.[34,35] Multiple loops are also regularly found in peptides, *e.g.* a combination of a head-to-tail- and a side chain-to-side chain-macrocycle. If the looped amino acids are only connected *via* amide bonds, this compound is called 'homodetic'.[35] In cases where any other linkage is involved (*e.g.* disulfide or depsipeptide bonds), it can be referred to as a 'heterodetic' peptide.[35] According to the IUPAC, a macrocycle consists of at least twelve atoms. Hence the simplest homodetic macrocyclic peptide is a covalent circuit built up of at least four residues. However, smaller (non-macro-)cyclic motifs also occur containing only a limited number of constrained bonds. Figure 1.3 depicts some of the most simple and common motifs.[34] Comparing these structures, it immediately becomes clear that each individual mode of cyclization can result in a completely different geometry of the looped compound.

Noteworthily, far more complex patterns have been observed in peptidic natural products including, for example, thioether linkages commonly found

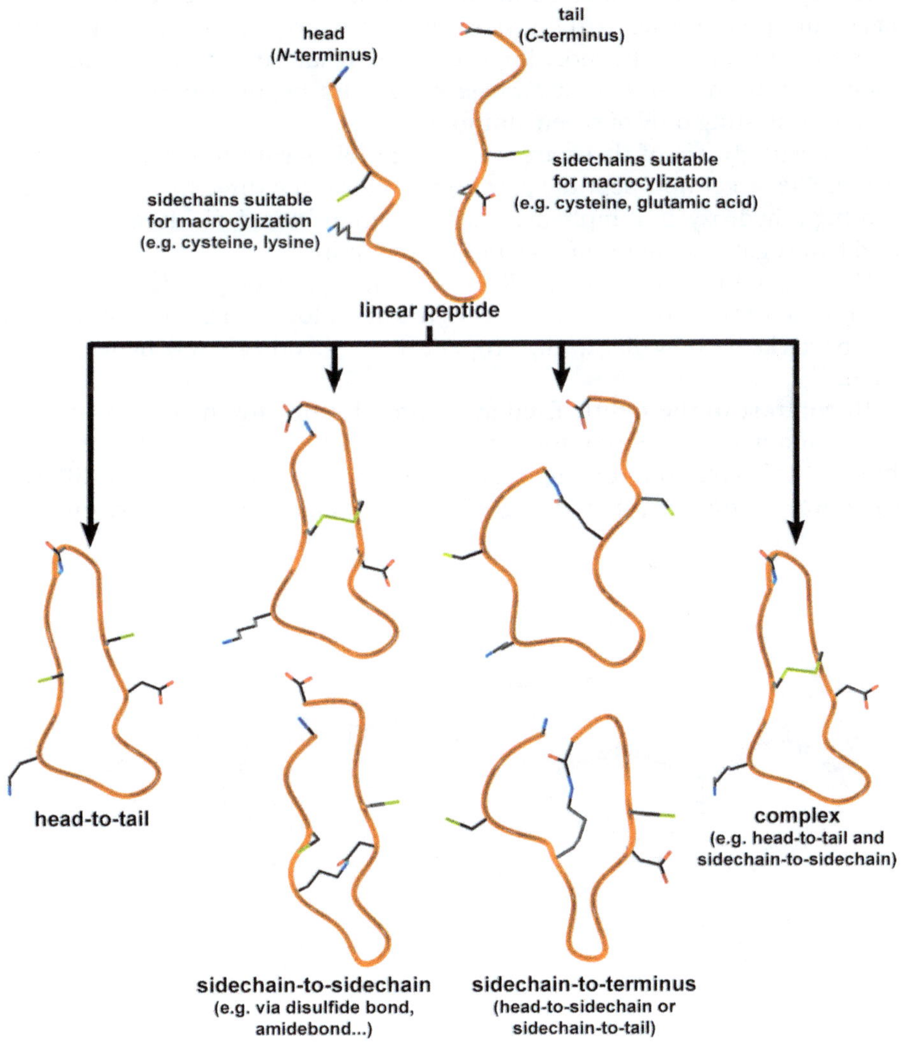

Figure 1.3 Common modes of macrocyclization shown for the example of a generic peptide chain containing two cysteines, a lysine, and a glutamic acid. All other residues are not shown. The peptide backbone is given in orange, carbon in black, nitrogen in blue, oxygen in red, and sulfur in yellow.

in lanthipeptides.[36] As described for proline, the backbone amide nitrogen can also be used as an anchor point for building a covalent ring structure.[37] Considering all these possibilities for the generation of cyclic peptides, the issue of choosing the appropriate one for each situation is difficult. High-throughput screening methodologies employing *e.g.* phage display technology can provide access to interesting cyclized peptidic structures.[40] However, researchers often rely on structures that have been provided by nature and can readily be exploited for various applications.

A very prominent showcase for the complexity and finesse of naturally occurring macrocycles is the glycopeptide vancomycin (Figure 1.4).[38] This compound is one of the most important last resort antibiotics for the treatment of multidrug-resistant *Staphylococci*.[39] Its peptidic portion is rather small, consisting only of seven amino acids.

Interestingly, six of these are non-natural ones with four being D-amino acids. The larger non-peptidic part helps to form the three macrocycles containing a hydroxylated biphenyl motif, an ether-linked dichloroaryl-system and two sugar moieties (glucose and vancosamine).

With its tightly fixed three-dimensional shape, vancomycin is able to selectively bind to D-alanine residues within the peptidoglycan layer of Gram-positive bacteria, thereby disrupting proper cell-wall synthesis and, hence, killing them.[41]

In contrast to the tightly fixed structure of vancomycin, a linear peptide has to adopt a bioactive conformation upon binding to its target. If the flexibility of this compound is high and it can exist in a multitude of different (inactive) geometries, its binding affinity will suffer from a severe entropic

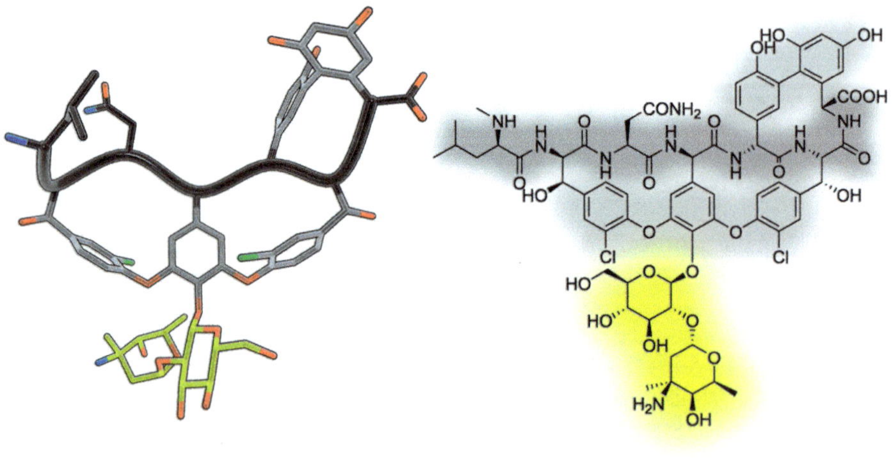

Figure 1.4 Three-dimensional (left) and chemical (right) structure of vancomycin.[7] The peptidic portion is highlighted in dark grey, the hydroxylated biphenyl- and ether-linked dichloroaryl-systems are shown in grey, and the two sugar-derived moieties are given in yellow.

penalty significantly affecting the Gibb's free energy of binding (ΔG) and, hence, its binding/dissociation constant (K_D) (eqn (1.1) and (1.2)).

$$\Delta G = \Delta H - T\Delta S. \tag{1.1}$$

with ΔH being the binding enthalpy, T being the temperature, and ΔS being the binding entropy;

$$K_D = e^{\frac{\Delta G}{RT}} \tag{1.2}$$

with R being the universal gas constant.

The establishment of one or maybe multiple constrained cyclic motifs helps to reduce this unfavorable thermodynamic profile of a peptide. Of course, this benefit can only become evident when the applied mode of rigidification does not interfere with binding (*e.g.* through steric clashes), rendering the rational design of suitable loop structures a difficult, yet, worthwhile endeavor.

1.3 Cyclic Peptides as Pharmaceutical Agents

The number of examples of cyclic peptides from the 'middle space' (*vide supra*) as active principles in clinically used drugs is rather sparse compared to the 'classical' approaches relying on small molecular entities or biologicals. To date, around 40 cyclic peptide-based drugs are in clinical use, while the majority of the market is still governed by New Chemical Entities (NCEs, ~90% market share).[20,45] But the interest of the pharmaceutical industry in developing novel therapeutics based on cyclic peptides is currently growing.[42,43] For sure, this is a notoriously underexploited structural space with high potential for scientific as well as application-driven exploration.[44]

But first of all, let's have a look at the downsides. In general, peptides are commonly considered as not drug-like. From the perspective of small molecule-based drugs, this holds perfectly true: peptides usually violate the well-known 'Lipinski rule-of-five' used to qualitatively estimate the possible oral bioavailability of a given compound.[46] Even a short sequence of only ten amino acids readily has a molecular weight of over 500 Da, more than five hydrogen bond donors, and more than ten hydrogen bond acceptors, while the high number of rotatable bonds is also supposed to negatively affect oral uptake.[47] Importantly, the amide bonds connecting each of the amino acid monomers are susceptible to enzymatic hydrolysis by proteases and peptidases abundantly present in the intestines, leading to the fast degradation of most peptides.[48,49] As a consequence, peptides usually possess no or only a very poor oral bioavailability.[50] Hence, the most common routes for administration of peptidic drugs are intravenous, subcutaneous, or intramuscular. However, even if a peptide has entered the bloodstream, it is usually rapidly eliminated from the system by renal clearance.[51] Although from the perspective of small molecular entities peptides are quite large, they are rather small as compared to other biomedicals like immunoglobulin-based drugs. Hence, they fit through the pores of the glomeruli of the kidney, which have a diameter of around 8 nm and effectively remove hydrophilic substances from the plasma. Only larger polypeptides and proteins with a molecular weight cut-off of around 50 to 70 kDa are less prone to elimination by this glomerular ultrafiltration step. Taken together, all these handicaps usually result in a poor pharmacokinetic profile of linear and natural peptides with *in vivo* half-lives of only a few minutes.[52]

Noteworthily, some of these shortcomings can be overcome by switching the regime from linear to cyclic peptides. First and foremost, it has been well documented that the looped congeners are much more resistant towards hydrolytic degradation.[53] Many peptidolytic enzymes require that the residues at the cleavage site within the peptide substrate accommodate a certain backbone conformation to facilitate proper recognition by the P and P' sites of the protease to promote cleavage of the amide bond.[56] The circular constraint can prevent the compound from adopting this substrate-like conformation and, thus, protect it against the enzymatic attack of exo- and endopeptidases. Of course, the observant reader will notice that the number of rotatable bonds is significantly reduced due to the cyclic constraint as laid out earlier (*vide supra*), which is an additional benefit. In the case of head-to-tail cyclization, two of the ionizable groups, namely the N- and C-terminus, are removed.[53] This results in a lowered polarity of the compound. Additionally, unfavorable free hydrogen donors and acceptors can be masked due to intramolecular hydrogen bonds made possible by fixing an appropriate geometry and in turn further rigidifying the overall structure. In several cases, it has been demonstrated that cyclic peptides possess improved membrane permeability as compared to their linear variants, which is an important parameter for passive absorption in the gastrointestinal tract.[53] As a consequence, cyclic peptides *CAN* indeed demonstrate reasonable oral bioavailability.[54] A prime example is cyclosporin A (Figure 1.5), which is obviously 'not drug-like', but has a reasonable bioavailability of about 30%–40% depending on the formulation.[55] It is in clinical use as an immunosuppressive agent for the treatment of *e.g.* graft-*versus*-host disease in bone marrow recipients.[55]

Having a closer look at the structure of cyclosporin A, it becomes clear that in addition to the 33-atom-spanning macrocycle, several non-natural amino acids as well as *N*-methylation at seven of the eleven amide bonds are present. All of these intriguing motifs contribute to the stability of cyclosporin A against enzymatic degradation. In addition to the macrocycle, the extensive *N*-methylation of this compound has been discussed as another important feature for achieving the unexpectedly good oral bioavailability.[58] The methyl groups reduce the number of unpaired hydrogen bond donors in the molecule, which, as we have learned, should be kept to a minimum. Interestingly, the remaining hydrogen bond donors sum up to a number of five. So at least this Lipinski criterion is fulfilled. As a consequence, masking

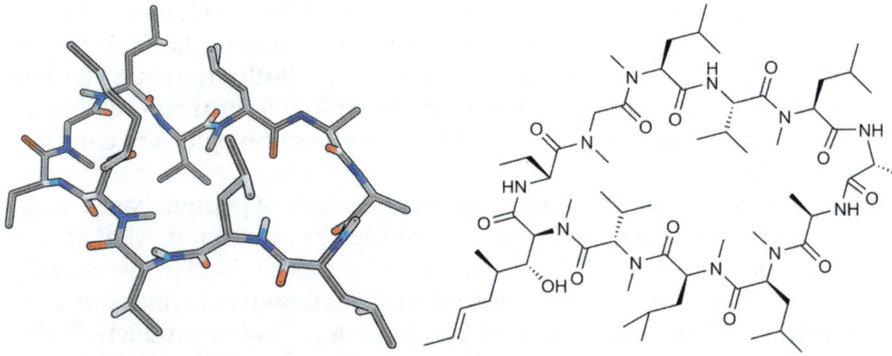

Figure 1.5 Three-dimensional (left) and chemical (right) structure of cyclosporin A.[57] Color code on the left side: carbon = grey, nitrogen = blue, oxygen = red. Hydrogens left out for clarity.

the most exposed amides through *N*-methylation has been proposed as a rational design principle to enhance the membrane permeability and, thus, oral bioavailability of a given peptide.[58] Noteworthily, due to the large number of hydrophobic residues and methylated amide nitrogens, cyclosporin A has an undesirable poor solubility. This highlights the necessity to balance the described measures (reduction of hydrogen bond donors and acceptors) against physicochemical parameters during medicinal chemistry optimization.

This has been an encouraging example for a compound with oral bioavailability far beyond the Lipinski space, which fostered further investigations of the possibilities to develop peptide-based drugs for oral intake. From a clinical standpoint, this is highly desirable as the parenteral routes of administration are usually associated with reduced patient compliance due to their invasiveness and inconvenience.[50] Hence, several methodologies, including the generation of pro-drug conjugates or nanoparticular delivery systems, have been studied to improve oral uptake.[54] But a generic approach has not yet been developed and still remains a challenging task. However, applications where oral administration is not the preferred route, like *e.g.* topical treatment of wounds or pulmonary delivery into infected lungs, can readily make use of cyclic peptides as therapeutic agents.

As pointed out above, clearance is another pharmacokinetic issue of (cyclic) peptides. A straightforward approach to escape glomerular ultrafiltration is to increase the molecular weight.[51] To this end, the peptide drug can be covalently linked to large biocompatible polymers like polyethylene glycol (PEG) or plasma proteins, which inherently provide long circulation times.[51] The latter approach makes use of native substances like albumin or the constant regions of immunoglobulins (Fc), and results in circulation half-lives of up to several days.[51] This strategy results in a permanent increase in the molecular weight of the active principle, which is detrimental for tissue penetration and intracellular delivery. In order to overcome this drawback, reversible attachment can be achieved by conjugation to albumin-binding molecules, resulting in an enhancement of the circulation half-life for the corresponding peptide from minutes to hours.[51] Both strategies, permanent and non-permanent, have already been exploited for the development of peptide-based drugs in clinical use and novel technologies are currently under investigation.[51]

So far, we have debated the pharmacokinetic issues of peptide-based drugs and ways to circumvent them to a greater or a lesser extent. But, what are the advantages of this type of compound? As they are assembled from naturally occurring building blocks (amino acids), drug metabolism is rather predictable resulting in biocompatible and non-toxic degradation products.[44] This holds especially true for solely homodetic compounds, which usually results in good safety and tolerability profiles.[44] Due to their increased size compared to small molecular entities, peptides provide the potential to establish more favorable directed interactions with their target.[59] This has several benefits. First, this leads to a commonly high potency and, thus, efficacy of peptides.[44] Second, increased selectivity can be expected as the specific interactions are very likely to fit only to the target of choice.[44] And last but not least, targets that are notoriously difficult to address with small molecules, like macromolecule–macromolecule interactions, can become druggable when using peptides.[59]

1.4 End of the Prologue

Without doubt, natural product-derived and synthetic cyclic peptides have received growing interest from the scientific community in recent years. This has been accompanied by a noticeable renaissance of activities by the pharmaceutical industry in the field of peptide-based drugs. In this regard, the macrocyclization motif offers a number of advantages over linear analogs, like improved stability, potency, and membrane permeability. These favorable chemical and physical properties can be exploited for various applications, which are certainly not limited to pharmaceutical exercises.

To shed light on the fascinating aspects associated with cyclic peptides, this book will provide the reader with a detailed discussion of the current knowledge on their biosynthesis (Chapters 2 and 3) and information on very recent work looking to incorporate novel structural units (Chapter 4). A subsequent review section on the use of classical chemistry approaches for cyclic peptides will include an introductory part (Chapter 5) accompanied by selected specific examples (Chapters 6–9). Then, an example of how synthesis and biological methods can be combined will be described (Chapter 10) followed by a section focusing on the analysis of cyclic peptides (Chapters 11–13). The last four chapters (Chapters 14–17) of the book will highlight the use of cyclic peptides in biology and in drug discovery.

Finally, the authors of this book hope to generate a spark of appreciation for the potential and beauty of cyclic peptides, contrasting the rather disparaging undertone resonating along with their original word stem *peptós*.

References

1. H. B. Vickery, The origin of the word protein, *Yale J. Biol. Med.*, 1950, **22**, 387–393.
2. Peptone, Dictonary.com Unabridged, 2016, http://www.dictionary.com/browse/peptone.
3. N. Terrett, Drugs in middle space, *MedChemComm*, 2013, **4**, 474.
4. A. T. Bockus, C. M. McEwen and R. S. Lokey, Form and function in cyclic peptide natural products: a pharmacokinetic perspective, *Curr. Top. Med. Chem.*, 2013, **13**, 821–836.
5. M. Zender, T. Klein, C. Henn, B. Kirsch, C. K. Maurer, D. Kail, C. Ritter, O. Dolezal, A. Steinbach and R. W. Hartmann, Discovery and biophysical characterization of 2-amino-oxadiazoles as novel antagonists of PqsR, an important regulator of Pseudomonas aeruginosa virulence, *J. Med. Chem.*, 2013, **56**, 6761–6774.
6. B. Nyfeler, D. Hoepfner, D. Palestrant, C. A. Kirby, L. Whitehead, R. Yu, G. Deng, R. E. Caughlan, A. L. Woods and A. K. Jones, *et al.* Identification of elongation factor G as the conserved cellular target of argyrin B, *PLoS One*, 2012, **7**, e42657.
7. G. M. Sheldrick, *Crystal Structure of Vancomycin at Atomic Resolution*, 1997.
8. M. L. Korsinczky, H. J. Schirra, K. J. Rosengren, J. West, B. A. Condie, L. Otvos, M. A. Anderson and D. J. Craik, Solution structures by 1H NMR of the novel cyclic trypsin inhibitor SFTI-1 from sunflower seeds and an acyclic permutant, *J. Mol. Biol.*, 2001, **311**, 579–591.
9. J. Schmidt, K. Patora-Komisarska, K. Moehle, D. Obrecht and J. A. Robinson, Structural studies of beta-hairpin peptidomimetic antibiotics that target LptD in Pseudomonas sp, *Bioorg. Med. Chem.*, 2013, **21**, 5806–5810.
10. L. E. Townsley, W. A. Tucker, S. Sham and J. F. Hinton, Structures of gramicidins A, B, and C incorporated into sodium dodecyl sulfate micelles, *Biochemistry*, 2001, **40**, 11676–11686.
11. A. Heitz, J. F. Hernandez, J. Gagnon, T. T. Hong, T. T. Pham, T. M. Nguyen, D. Le-Nguyen and L. Chiche, Solution structure of the squash trypsin inhibitor MCoTI-II. A new family for cyclic knottins, *Biochemistry*, 2001, **40**, 7973–7983.
12. P. B. Stathopulos, L. Zheng, G.-Y. Li, M. J. Plevin and M. Ikura, Structural and mechanistic insights into STIM1-mediated initiation of store-operatded calcium entry, *Cell*, 2008, **135**, 110–122.
13. I. Russo Krauss, A. Merlino, A. Randazzo, E. Novellino, L. Mazzarella and F. Sica, High-resolution structures of two complexes between thrombin and thrombin-binding aptamer shed light on the role of cations in the aptamer inhibitory activity, *Nucleic Acids Res.*, 2012, **40**, 8119–8128.
14. O. V. Kovalenko, A. Olland, N. Piche-Nicholas, A. Godbole, D. King, K. Svenson, V. Calabro, M. R. Muller, C. J. Barelle and W. Somers, *et al.*, Atypical antigen recognition mode of a shark immunoglobulin new antigen receptor (IgNAR) variable domain characterized by humanization and structural analysis, *J. Biol. Chem.*, 2013, **288**, 17408–17419.

15. C. Tu, J. Bard and L. Mosyak, anti-CXCL13 scFv - E10, 2015, DOI: 10.2210/pdb5c6w/pdb.
16. G. R. Andersen and E. Spillner, *HHH1 Fab Fragment*, 2016.
17. L. J. Harris, S. B. Larson, K. W. Hasel and A. McPherson, Refined structure of an intact IgG2a monoclonal antibody, *Biochemistry*, 1997, **36**, 1581–1597.
18. P. J. Declerck, Biologicals and biosimilars: A review of the science and its implications, *GaBI J*, 2012, **1**, 13–16.
19. C. Liu, P. P. Constantinides and Y. Li, Research and development in drug innovation: reflections from the 2013 bioeconomy conference in China, lessons learned and future perspectives, *Acta Pharm. Sin. B*, 2014, **4**, 112–119.
20. D. J. Craik, D. P. Fairlie, S. Liras and D. Price, The future of peptide-based drugs, *Chem. Biol. Drug Des.*, 2013, **81**, 136–147.
21. A. A. Kaspar and J. M. Reichert, Future directions for peptide therapeutics development, *Drug discovery today*, 2013, **18**, 807–817.
22. V. Lavergne, R. J. Taft and P. F. Alewood, Cysteine-rich mini-proteins in human biology, *Curr. Top. Med. Chem.*, 2012, **12**, 1514–1533.
23. H. Kolmar, Alternative binding proteins: biological activity and therapeutic potential of cystine-knot miniproteins, *FEBS J.*, 2008, **275**, 2684–2690.
24. S. Perticaroli, J. D. Nickels, G. Ehlers, H. O'Neill, Q. Zhang and A. P. Sokolov, Secondary structure and rigidity in model proteins, *Soft Matter*, 2013, **9**, 9548.
25. K. Usui, T. Ojima, M. Takahashi, K. Nokihara and H. Mihara, Peptide arrays with designed secondary structures for protein characterization using fluorescent fingerprint patterns, *Biopolymers*, 2004, **76**, 129–139.
26. L. A. Moran, *Principles of Biochemistry*, Pearson, Boston, 5th edn, 2012.
27. C. El Amri, F. Bruston, P. Joanne, C. Lacombe and P. Nicolas, Intrinsic flexibility and structural adaptability of Plasticins membrane-damaging peptides as a strategy for functional versatility, *Eur. Biophys. J.*, 2007, **36**, 901–909.
28. L. Mirny and E. Shakhnovich, Protein folding theory: from lattice to all-atom models, *Annu. Rev. Biophys. Biomol. Struct.*, 2001, **30**, 361–396.
29. T. A. Hill, N. E. Shepherd, F. Diness and D. P. Fairlie, Constraining cyclic peptides to mimic protein structure motifs, *Angew. Chem., Int. Ed. Engl.*, 2014, **53**, 13020–13041.
30. S. H. Joo, Cyclic peptides as therapeutic agents and biochemical tools, *Biomol. Ther.*, 2012, **20**, 19–26.
31. J. M. Gulbis, Z. Kelman, J. Hurwitz, M. O'Donnell and J. Kuriyan, Structure of the C-terminal region of p21(WAF1/CIP1) complexed with human PCNA, *Cell*, 1996, **87**, 297–306.
32. D. K. Chang and C. C. Liang, Influence of bulky side chains of amino acids on the solution conformation of peptide fragment (81-92) derivatives of CD4, TYICEVEDQKEE, as studied by NMR spectroscopy and molecular modeling, *Biochim. Biophys. Acta*, 1994, **1205**, 262–267.

33. M. Tischler, D. Nasu, M. Empting, S. Schmelz, D. W. Heinz, P. Rottmann, H. Kolmar, G. Buntkowsky, D. Tietze and O. Avrutina, Braces for the peptide backbone: insights into structure-activity relationships of protease inhibitor mimics with locked amide conformations, *Angew. Chem., Int. Ed. Engl.*, 2012, **51**, 3708–3712.
34. A. Tapeinou, M.-T. Matsoukas, C. Simal and T. Tselios, Review cyclic peptides on a merry-go-round; towards drug design, *Biopolymers*, 2015, **104**, 453–461.
35. J. Spengler, J.-C. Jimenez, K. Burger, E. Giralt and F. Albericio, Abbreviated nomenclature for cyclic and branched homo- and hetero-detic peptides, *J. Pept. Res.*, 2005, **65**, 550–555.
36. E. L. Ongey and P. Neubauer, Lanthipeptides: chemical synthesis *versus in vivo* biosynthesis as tools for pharmaceutical production, *Microb. Cell Fact.*, 2016, **15**, 97.
37. C. Schumann, L. Seyfarth, G. Greiner, S. Reissmann and I. Paegelow, Synthesis and biological activities of new side chain and backbone cyclic bradykinin analogues, *J. Pept. Res.*, 2002, **60**, 128–140.
38. M. Schäfer, T. R. Schneider and G. M. Sheldrick, Crystal structure of vancomycin, *Structure*, 1996, **4**, 1509–1515.
39. K. Chua and B. P. Howden, Treating Gram-positive infections: vancomycin update and the whys, wherefores and evidence base for continuous infusion of anti-Gram-positive antibiotics, *Curr. Opin. Infect. Dis.*, 2009, **22**, 525–534.
40. C. Heinis and G. Winter, Encoded libraries of chemically modified peptides, *Curr. Opin. Chem. Biol.*, 2015, **26**, 89–98.
41. P. J. Loll, R. Miller, C. M. Weeks and P. H. Axelsen, A ligand-mediated dimerization mode for vancomycin, *Chem. Biol.*, 1998, **5**, 293–298.
42. T. Uhlig, T. Kyprianou, F. G. Martinelli, C. A. Oppici, D. Heiligers, D. Hills, X. R. Calvo and P. Verhaert, The emergence of peptides in the pharmaceutical business: From exploration to exploitation, *EuPa Open Proteomics*, 2014, **4**, 58–69.
43. P. Thapa, M. J. Espiritu, C. Cabalteja and J.-P. Bingham, The Emergence of Cyclic Peptides: The Potential of Bioengineered Peptide Drugs, *Int. J. Pept. Res. Ther.*, 2014, **20**, 545–551.
44. K. Fosgerau and T. Hoffmann, Peptide therapeutics: current status and future directions, *Drug Discovery Today*, 2015, **20**, 122–128.
45. A. Zorzi, K. Deyle and C. Heinis, Cyclic peptide therapeutics: past, present and future, *Curr. Opin. Chem. Biol.*, 2017, **38**, 24–29.
46. C. A. Lipinski, F. Lombardo, B. W. Dominy and P. J. Feeney, Experimental and computational approaches to estimate solubility and permeability in drug discovery and development settings1PII of original article: S0169-409X(96)00423-1. The article was originally published in Advanced Drug Delivery Reviews 23 (1997) 3–25.1, *Adv. Drug Delivery Rev.*, 2001, **46**, 3–26.
47. D. F. Veber, S. R. Johnson, H.-Y. Cheng, B. R. Smith, K. W. Ward and K. D. Kopple, Molecular properties that influence the oral bioavailability of drug candidates, *Journal of medicinal chemistry*, 2002, **45**, 2615–2623.

48. D. J. Haines, C. H. Swan, J. R. Green and J. F. Woodley, Mucosal peptide hydrolase and brush-border marker enzyme activities in three regions of the small intestine of rats with experimental uraemia, *Clin. Sci. (Lond.)*, 1990, **79**, 663–668.
49. J. F. Woodley, Enzymatic barriers for GI peptide and protein delivery, *Crit. Rev. Ther. Drug Carrier Syst.*, 1994, **11**, 61–95.
50. J. H. Lin, Pharmacokinetics of biotech drugs: peptides, proteins and monoclonal antibodies, *Curr. Drug Metab.*, 2009, **10**, 661–691.
51. L. Pollaro and C. Heinis, Strategies to prolong the plasma residence time of peptide drugs, *MedChemComm*, 2010, **1**, 319.
52. M. Werle and A. Bernkop-Schnurch, Strategies to improve plasma half life time of peptide and protein drugs, *Amino Acids*, 2006, **30**, 351–367.
53. A. T. Bockus, C. M. McEwen and R. S. Lokey, Form and function in cyclic peptide natural products: a pharmacokinetic perspective: A Pharmacokinetic Perspective, *Curr. Top. Med. Chem.*, 2013, **13**, 821–836.
54. N. Yin, Enhancing the Oral Bioavailability of Peptide Drugs by using Chemical Modification and Other Approaches, *Med. Chem.*, 2014, **4**, 763–769.
55. N. Parquet, O. Reigneau, H. Humbert, M. Guignard, P. Ribaud, G. Socie, A. Devergie, H. Esperou and E. Gluckman, New oral formulation of cyclosporin A (Neoral) pharmacokinetics in allogeneic bone marrow transplant recipients, *Bone Marrow Transplant.*, 2000, **25**, 965–968.
56. M. Prabu-Jeyabalan, E. Nalivaika and C. A. Schiffer, Substrate shape determines specificity of recognition for HIV-1 protease: analysis of crystal structures of six substrate complexes, *Structure*, 2002, **10**, 369–381.
57. H. Ke, Y. Zhao, F. Luo, I. Weissman and J. Friedman, Crystal structure of murine cyclophilin C complexed with immunosuppressive drug cyclosporin A, *Proc. Natl. Acad. Sci. U. S. A.*, 1993, **90**, 11850–11854.
58. C. K. Wang, S. E. Northfield, B. Colless, S. Chaousis, I. Hamernig, R.-J. Lohman, D. S. Nielsen, C. I. Schroeder, S. Liras and D. A. Price, *et al.*, Rational design and synthesis of an orally bioavailable peptide guided by NMR amide temperature coefficients, *Proc. Natl. Acad. Sci. U. S. A.*, 2014, **111**, 17504–17509.
59. E. Marsault and M. L. Peterson, Macrocycles are great cycles: applications, opportunities, and challenges of synthetic macrocycles in drug discovery, *J. Med. Chem.*, 2011, **54**, 1961–2004.

CHAPTER 2

The Biosynthesis of Cyclic Peptides – RiPPs – An Overview

CLARISSA M. CZEKSTER[a] AND JAMES H. NAISMITH*[a,b,c,d]

[a]School of Chemistry, Biomedical Sciences Research Complex, University of St Andrews, North Haugh, St Andrews KY16 9ST, UK; [b]Wellcome Trust Centre for Human Genetics, University of Oxford, Roosevelt Drive, Oxford OX3 7BN, UK; [c]RCaH, Harwell Campus, Oxford OX11 0FA, UK; [d]Biotherapy Centre, Sichuan University, Chengdu, China
*E-mail: naismith@strubi.ox.ac.uk

2.1 Introduction

A survey of sources of drugs covering the years 1981 to 2014 revealed that natural products accounted for approximately 1/3 of total molecules approved for clinical use. Interestingly, for antimicrobial and anticancer agents, 1/4 were natural products and over half were derived from natural products.[1] Perhaps unsurprisingly, compounds that have evolved to regulate biological processes turn out to have useful medicinal properties; something humanity has relied on for many years. In addition, even if not fit to be used as drugs, natural products can disclose new targets for the treatment of disease.[2] Natural products have thus dominated medicine for most of the 20th century.

Advances in high-throughput screening methodology, chemical synthesis and automation have made it possible to screen hundreds of thousands of compounds against a given target or cell line (phenotype) in a relatively short time. Since natural products are known to be useful, it might have

Chemical Biology No. 6
Cyclic Peptides: From Bioorganic Synthesis to Applications
Edited by Jesko Koehnke, James Naismith and Wilfred A. van der Donk

Published by the Royal Society of Chemistry, www.rsc.org

been expected that compound libraries would be dominated by them. However, the high molecular weight (often above the 500 Da threshold set by Lipinski's "rule-of-five"[3]), lack of availability (because they are either hard to synthesize or hard to isolate) and molecular complexity of natural products has made them under-represented in compound libraries.[4] Macrocyclic molecules with their greater information content are uniquely placed to target historically difficult problems such as protein–protein interactions.[5]

Ribosomally synthesized and post-translationally modified peptides (RiPPs) are currently an active area of research. A RiPP is defined as a molecule that is made on the ribosome and has undergone at least one enzyme-catalyzed post-translational modification.[6] These modifications encompass a wide range of chemical transformations, and in this review we focus on the formation of macrocyclic RiPPs. Cyclic peptides have desirable properties, such as resistance to protease degradation and structural rigidity, which make them desirable therapeutic molecules.[7] Since RiPPs are genetically encoded, structural and chemical diversity can be programmed. The enzymes that catalyze post-translational modifications combine exquisite chemical specificity and substrate promiscuity. This apparent contradiction arises from the use of substrate leader or tail sequences outside of the core peptide (which becomes the natural product) to direct the actions of the post-translationally modifying enzymes. Conceptually, RiPPs can be thought of as hybrids between natural products and biologics since they can be genetically encoded, tailored and produced in culture on a large scale.

Additionally, although several strategies for the chemical synthesis of peptide macrocycles exist, they face drawbacks limiting their ring size and composition.[8] For the synthesis of small to medium sized macrocycles, the peptide bond geometry hinders the peptides from adopting a favorable conformation for macrocyclization. In the case of larger rings, although the peptide bond geometry does not present an obstacle, side reactions and other intermolecular reactions must be avoided.[8,9] Enzymes have evolved to overcome these obstacles, performing macrocyclization by a plethora of distinct strategies, which will be discussed here.

In this chapter, major classes of cyclic RiPPs are discussed in terms of their sources, structural diversity, including distinct types of post-translational modifications, and biological activities. We highlight some attempts to increase the scope of the biosynthetic machinery to further expand natural product diversity. The nomenclature and characterization of RiPPs follows the designation proposed for this class of natural products by Arnison *et al.*[6]

2.2 Cyanobactin Biosynthesis

Cyanobactins were first isolated from ascidian marine organisms in the 1980s.[10] They are amongst the most abundant cyclic peptides in nature, and 10-30% of all cyanobacteria on earth produce such molecules.[11] Many cyanobactins have been found to possess biological activity. Examples are kawaguchipeptin B, which has antibacterial activity, phakellistatin 1 and patellamide A, which inhibit cancer cell growth, and patellamides B and C, which reverse multi-drug resistance in cancer cells.[12]

The biosynthetic enzymes involved in cyanobactin biosynthesis were identified in a landmark paper[13] and are shown in Figure 2.1 along with the biosynthetic scheme for patellamide C and trunkamide A. The biosynthetic machinery carrying out cyanobactin production is arranged in a gene cluster in which the sequence for the precursor peptide is one of the genes. Cyanobactin gene clusters always encode two protease genes (A for a protease, G for a protease operating as a macrocyclase), one precursor peptide gene (E) and two genes, B and C, which have unknown function and are non-essential for cyanobactin production *in vitro*. More variable are the presence of genes encoding heterocyclases (D, forms thiazolines and/or oxazolines, also known as cyclodehydratase), prenylases (F) and thiazoline/oxazoline oxidases (often a domain of the G protein).[12]

The precursor peptide encompasses the core sequence that encodes the final highly modified cyclic peptide (typically 6–20 amino acids[14]), and this core is flanked by additional elements required for recognition by the modifying enzymes.[15,16] In many E proteins, there are multiple core peptides within the same precursor. In several cyanobactins, some or all of the cysteines are converted into thiazolines, whilst serine and threonine remain unmodified, whilst in others the serines and/or threonines are also heterocyclized (to oxazolines). The heterocyclase reaction is ATP-dependent and requires the presence of the leader (Leader 1) and must precede cleavage, macrocyclization and oxidation.[17]

Wild-type heterocyclases catalyze their reactions in a distributive ordered fashion. The C-terminal cysteine is heterocyclized first, followed by internal cysteines, and finally threonines and serines.[18,19] The heterocyclase enzyme from patellamide biosynthesis has been engineered to abolish the requirement for a leader, improving the biocatalytic properties of the enzyme.[20]

For macrocyclization to occur, the leader sequence must first be removed, a reaction catalyzed by a protease (A protein).[22] After leader removal, the macrocyclase enzyme catalyzes peptide bond formation between the new N-terminus (an amine) and the acyl enzyme intermediate of the terminal core sequence (in doing so the flanking tail is cleaved from the core).[23,24] The presence of a proline or thiazoline/thiazole at the terminus of the core peptide and immediately preceding C-terminal recognition sequence 1 is required for macrocyclization.[25] This residue kinks the chain such that the peptide does not adopt a β-strand conformation, which would clash with the protein. The macrocyclase recognizes the C-recognition tail and acts *via* an acyl enzyme intermediate. The cyclic peptide product can be further modified by oxidation of the thiazolines/oxazolines to generate thiazoles/oxazoles or O-prenylation of serine, threonine and tyrosine.[13,26]

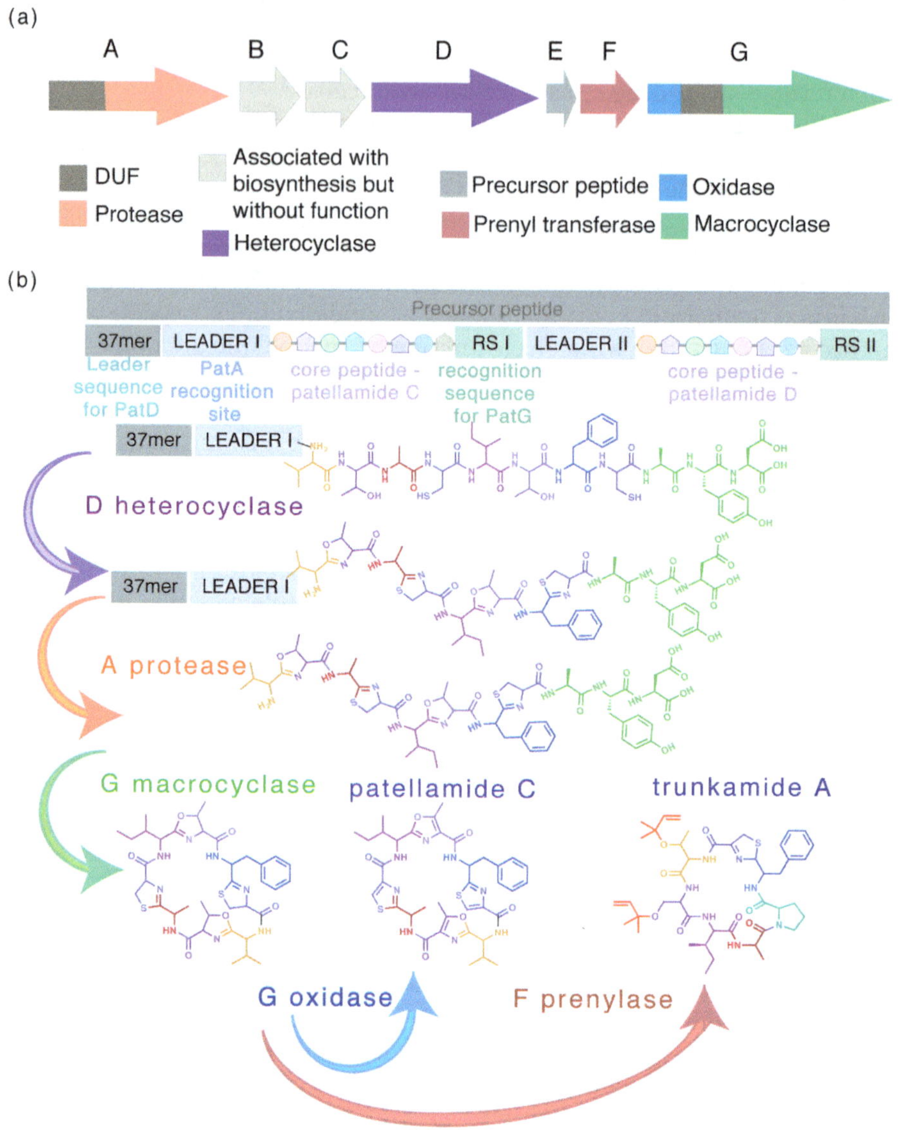

Figure 2.1 Biosynthesis of cyanobactins. As an example, the biosynthetic pathway for patellamides (the most well-studied) is shown. (a) General organization of the gene cluster for cyanobactin biosynthesis. The precursor gene sequence encodes two copies of the precursor peptide. Therefore, Leader I and RS I are for the first precursor peptide and Leader II and RS II are for the second. On panel b, only one copy is shown for simplicity. (b) Biosynthesis of patellamide C and trunkamide A. Modified from *Current Opinion in Chemical Biology*. **35**, C. M. Czekster, Y. Ge and J. H. Naismith, Mechanisms of cyanobactin biosynthesis, 80–88,[21] Copyright (2016), with permission from Elsevier.

2.3 Lanthipeptides

Gram-positive bacteria such as *Streptococcus* and *Streptomyces*,[27] as well as Gram-negative bacteria such as Cyanobacteria, Bacteroidetes and Proteobacteria,[28] produce toxins called lanthipeptides, which inhibit the growth of competing bacterial species. Nisin A (Figure 2.2) is produced by some strains of *Lactococcus lactis* and possesses bactericidal activity caused by its interaction with the precursor of cell wall peptidoglycan lipid II, forming membrane pores.[29] Because of its bactericidal action, nisin A has been used as a food-preservant for decades.[29] Duramycin, another lanthipeptide, is currently being evaluated for the treatment of cystic fibrosis.[30]

Lanthipeptides possess polycyclic thioether amino acids (lanthionine or 3-methyllanthionine – Figure 2.2), derived from dehydroalanine and dehydrobutyrine residues, respectively.[31]

Analogous to cyanobactins, a leader sequence is required for substrate recognition by lantibiotic synthetase, although an engineered variant that does not require the leader sequence has also been developed.[32] The lanthionine or 3-methyllanthionine residues are introduced by a two-step process. The precursor peptide is modified both by a dehydratase and a cyclase enzyme, and these steps can be catalyzed by distinct proteins or by a single protein with different domains. The cross-linking reaction to form the macrocyclic ring proceeds by Michael addition and is shown in Figure 2.2c.

Further modifications can be introduced by other post-translationally modifying enzymes, such as the oxidation of thioether bonds to a sulfoxide,

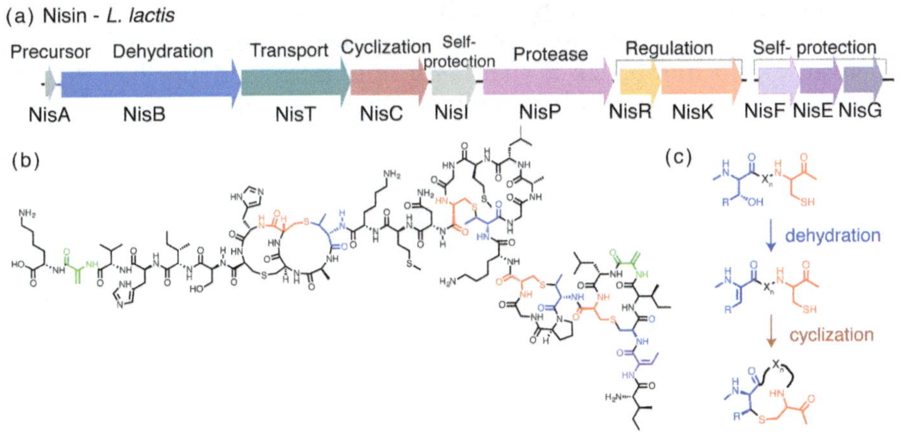

Figure 2.2 Biosynthesis of lanthipeptides. As an example, the biosynthetic pathway for nisin A is shown. (a) General organization of the gene cluster for nisin A biosynthesis. (b) Structure of nisin A. Dehydroalanine residues are in green, covalent bonds to form lanthionine or methyllanthionine are in purple, amino acids covalently bound by a cysteine are in red and blue. (c) Details of the reactions that form the cyclic peptide in this pathway.

hydrogenations of dehydroalanine and dehydrobutyrine to yield D-amino acids, chlorination, hydroxylation, *N*-glycosylation, and others.[33]

The substrate permissiveness of lanthipeptide biosynthetic enzymes has been exploited to produce novel molecules with potential biological activity.[34,35]

2.4 Thiopeptides

Thiostrepton A1 was first isolated from *Streptomyces azureus* ATCC 14921 in the 1950s,[36] but it was only in 2009 that it was classified as a RiPP and genes from its biosynthetic cluster were identified (Figure 2.3a).[37] Several other members of this family of natural products have been identified since, showing remarkable similarity in the gene composition of their gene clusters. Peptides from this class possess a pyridine, piperidine or dehydropiperidine ring in their core structure. Formation of this central six-membered ring structure occurs through an intra-molecular Diels–Alder-like cycloaddition between dehydroalanine residues, followed by dehydration (and in some cases elimination).[38] In some thiopeptides, such as thiomuracin I[39] and thiocillin I,[40] the six-membered ring forms the cyclic scaffold of the peptide final natural product, and there is only one macrocycle present. In others, such as thiostrepton (Figure 2.3b) and nosiheptide, the final natural product possesses two macrocycles, connected by a central six-membered ring.[38] Several other modifications such as dehydration and heterocyclization are also present. Some thiopeptides possess genes involved in transcription regulation,[41] host resistance and natural product efflux, although these are lacking in the gene cluster for thiostrepton A2 (Figure 2.3b).

Several thiopeptides possess biological activities, and two are used commercially: thiostrepton as a veterinary antibiotic, and nosiheptide as a feed additive for animals.[42] Several molecules of this class show potent

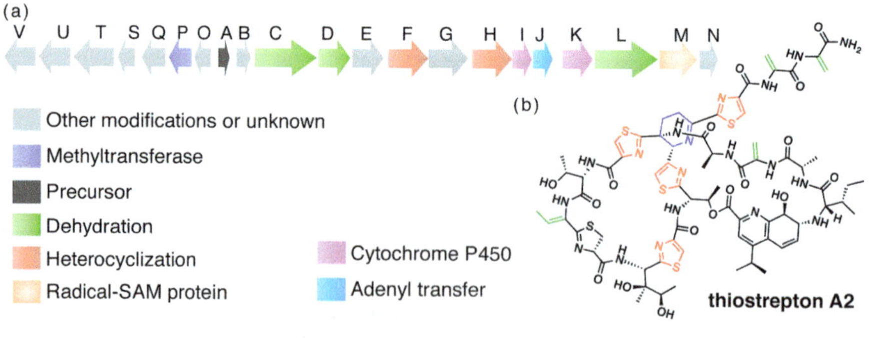

Figure 2.3 Thiostrepton A2. (a) Biosynthetic cluster showing the functions of genes crucial for biosynthesis. The precursor peptide is encoded by gene A (black). (b) Structure of thiostrepton A2. Dehydroalanine- and dehydrobutyrine-modified residues are shown in green, thiazoles are in red, and dehydropiperidine is in purple.

antibacterial activity and clinical trials using thiopeptides are ongoing to treat bacterial infections caused by *Clostridium difficile*.[43]

The precursor peptide contains both an N-terminal leader peptide (between 34–55 amino acids long) and a C-terminal sequence (12 to 17 amino acids) that contains the amino acids that will become the thiopeptide itself and in some cases one or two extra residues that are posteriorly removed.[44] The identity of the proteases required to remove the leader and recognition sequences is unknown, but these steps are required to yield the final thiopeptide product. The backbone modifications present in thiopeptides are introduced by enzymes similar to the ones from cyanobactin (thiazoles), lanthipeptides (dehydrated residues) and linear azole-containing peptides. Cyclization of the linear peptide occurs by the cycloaddition of two dehydroalanine residues to form the central six-membered ring – (piperidine, dehydropiperidine (in Figure 2.3b) or pyridine) – in the peptide structure.[45–47]

Modified thiostrepton variants have been generated.[48–50] The modularity and complexity of the biosynthesis of thiopeptides represents an opportunity for further pathway and natural product engineering.[50]

2.5 Bottromycin

Bottromycins possess antimicrobial activity against multi-drug resistant *Staphylococcus* and enterococci.[51] They contain several distinctive structural features such as a decarboxylated C-terminal thiazole and an amidine ring (Figure 2.4). These molecules also contain several carbon-methylated amino acids.

In bottromycins, a leader peptide is not present at the N-terminus of the precursor peptide, but in the C-terminus (called a "follower" peptide).

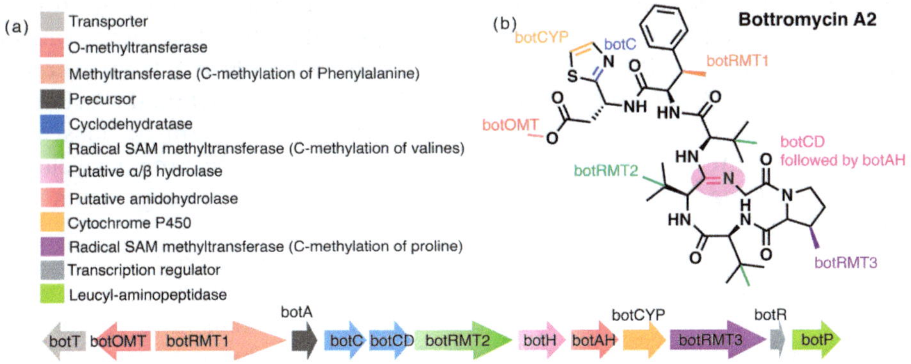

Figure 2.4 Biosynthesis of bottromycin A2. (a) Gene cluster from *Streptomyces* sp. BC16019.[52] (b) Structure of the final natural product, with post-translational modifications color-coded according to the enzymes performing each reaction.[53–55]

This follower peptide is thought to perform a similar function as the leader peptide in other RiPPs.

Recently the order of reactions in bottromycin biosynthesis has been elucidated.[55] In *Streptomyces scabies*, macrocyclization was shown to require both BotCD and BotAH enzymes in a two-step reaction in which the amide is first activated by BotCD in an ATP-dependent manner and then cyclized by BotAH. Oxidation of the thiazoline to thiazole is catalyzed by a P450 enzyme in Bottromycin biosynthesis. The precise mechanism by which these reactions occur remains to be determined, but the order of post-translational modifications has been resolved. The N-terminal methionine of the precursor peptide is cleaved by BotP (BtmM in *S. scabies*), then formation of thiazolines (BotC) and β-methylation occurs (BotMT1, BotMT2 and BotMT3), followed by macrocyclization (BotAH and BotCD), hydrolysis of the follower peptide (BotH), decarboxylation (BotCYP) and O-methylation (BotOMT).[55,56]

2.6 Cyclic RiPPs from Plants: Cyclotides and Orbitides

Flowering plants produce a range of cyclic peptides[57] including cyclotides which typically contain 28–37 amino acids. Their biological function is related to the plant host defenses against pests, such as insects.[58] Several cyclotides have potentially useful biological activities,[59] including hemolytic, cytotoxic and antimicrobial.[60] The enzymes catalyzing cyclization possess remarkable substrate promiscuity and have been extensively utilized for the production of cyclic peptides containing unnatural amino acids,[61] as a strategy for protein ligation[62,63] and labelling.[63] The precursor peptide contains an endoplasmic reticulum signal sequence, an N-terminal domain analogous to the Leader 1 sequence, and the core cyclotide sequence (in some cases multiple cores) followed by a disposable C-terminal recognition sequence. The enzymes responsible for macrocyclization are asparagine-specific peptide ligases such as Butelase 1 from *Clitoria ternatea* and *Oa*AEP1 from *Oldenlandia affinis*. Although detailed mechanistic studies on either enzyme are lacking, they are thought to possess a modified serine peptidase-type mechanism akin to other macrocyclases such as PatGmac and prolyl oligopeptidases (see below on segetalins and amanitins).[64]

The most well-studied cyclotide is kalata B1 (Figure 2.5a and 2.5b) but other cyclotides have been described, with various structures and biological activities. A cell penetrating cyclotide scaffold from *Momordica cochinchinensis* (trypsin inhibitor-II, or MCoTI-II) has been employed as the basis for novel drug molecules that bind to the oncogenic protein BCR-ABL,[65] angiotensin-(1-7)[66] and HIV gp120 protein.[67]

Orbitides are short RiPPs from plants that lack disulfide bonds. They are composed mostly of unmodified L-amino acids, with a few exceptions such

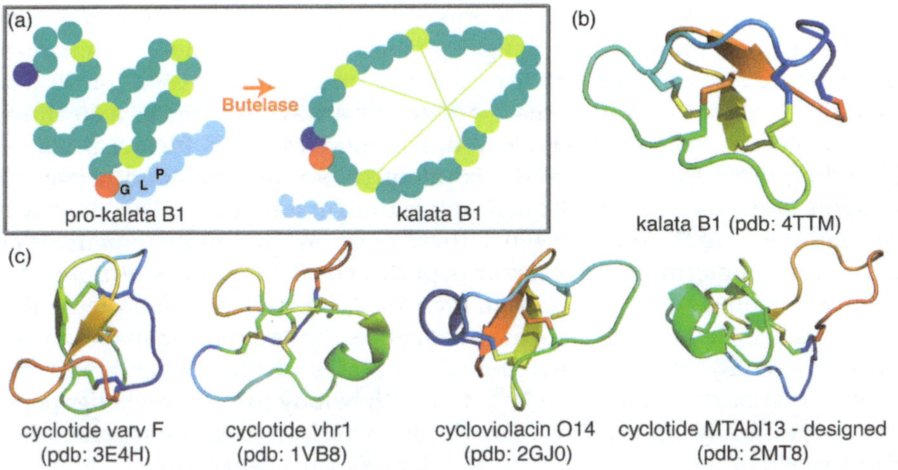

Figure 2.5 Biosynthesis of plant cyclotides. (a) Schematic of the macrocyclization reaction in the biosynthesis of kalata B1, were a pro-peptide is converted into a cyclotide possessing three disulfide bonds. The connectivity between cysteine residues (cysteines are shown as light green spheres) is represented by the light green lines in the scheme. The N-terminal amino acid is colored blue, while the C-terminal residue from the core sequence is in red. (b) Crystal structure of Kalata B1 from *Oldenlandia affinis*.[68] (c) Examples of other cyclotides: cyclotide varv F from *Viola arvensis*,[69] vhr1 from *Viola hederacea*,[70] cycloviolacin O14 from *Viola odorata*[71] and designed cyclotide MTAbl13.[65] For panels b and c, the peptides are colored from N- (blue) to C-terminus (red).

as hydroxylation,[72] an unusual methylation,[73] and a D-tryptophan residue.[6] Some orbitides have been shown to possess biological activity, albeit by undetermined mechanisms.[74]

Segetalins are orbitides consisting of 5–9 amino acids, with some possessing estrogen-like[75] and/or vasorelaxant[76] biological activities. These molecules start from precursor peptides of 30–40 amino acids, in which the core peptide is flanked by a leader sequence and by a recognition sequence. The segetalins described to date comprise only natural amino acids, and the only enzymes involved in their biosynthesis are an unidentified protease (oligopeptidase 1, or OLP1) and a macrocyclase (PCY1, or peptide cyclase 1).[77]

2.7 Cyclic RiPPs from Mushrooms: Amanitins and Dikaritins

Amanitins and dikaritins are derived from a core peptide sequence flanked by disposable recognition sequences. The enzymes in the pathway catalyze peptide bond cleavage, macrocyclization, hydroxylation, epimerization and cysteine-tryptophan cross-linking.[78,79]

Amatoxins are found in mushrooms of the genera *Amanita*, *Galerina*, *Lepiota*, and *Conocybe* from the fungal phylum Basidiomycetes.[78] α-amanitin poisoning causes irreversible liver damage and is responsible for 90% of fatal incidents with mushrooms.[80] Amatoxins are synthesized as a 35 amino acid precursor, which is then cleaved, macrocyclized and further modified including the introduction of the covalent tryptathionine bond between a cysteine and a tryptophan, the defining feature of amatoxins. The identity of the enzymes and the order in which these reactions take place remain experimentally undetermined. Precursor peptides containing the N-terminal signature sequence MSDIN have been identified,[81] as well as the protease that catalyzes the cleavage of the leader sequence (POPA) and the macrocyclase enzyme (POPB).[82,83] These are the only enzymes involved in amanitin biosynthesis characterized to date[82,83], and both belong to the prolyl oligopeptidase superfamily. An extensive kinetic characterization of the macrocyclase from *Galerina marginata* revealed similarities with other prolyl oligopeptidases, as well as pronounced product inhibition caused by the long recognition sequence.[84] Figure 2.6a provides an overview of the biosynthesis of amanitins.

Dikaritins are toxins from filamentous mushrooms, recently identified as RiPPs.[85] They include the tetrapeptidic ustiloxins[86] and phomopsins.[87] Phomopsins are cyclic toxins with six amino acids produced by the Ascomycetes mushroom *Phomopsis leptostromiformis*. Ustiloxins are produced by *Ustilaginoidea virens* and *Aspergillus flavus*. Dikaritins target tubulin, inhibiting microtubule assembly, a valid strategy in the development of antitumor drugs.[88] The biosynthesis of dikaritins is summarized in Figure 2.6b, and the structure and biosynthesis of ustiloxin B is shown in Figure 2.6c. The gene cluster encoding the proteins participating in the biosynthesis of ustiloxin B from *Aspergillus flavus* contains a precursor peptide containing 16 repeats of the core peptide region. These sequences must be cleaved both at the N- and C-terminus by as yet uncharacterized protease(s). The biosynthesis of ustiloxins begins with macrocyclization and oxidation catalyzed by three proteins (UstQYaYb), followed by methylation and oxidation of amino acid side chains (UstM, UstF1 and UstF2). The final reaction is carried out by a PLP-dependent enzyme that catalyzes decarboxylation and condensation (UstD).[89] The biosynthesis of phomopsins shows several similarities to ustiloxins, but several enzymes involved in modifying the cyclic peptide remain undetermined.

2.8 Conclusion and Outlook

RiPPs are of great interest to the pharmaceutical industry. Cyclic RiPPs possess important properties such as cell permeability, resistance to protease degradation and conformational rigidity. Structural and chemical diversity can easily be programmed into RiPPs and they can be extensively

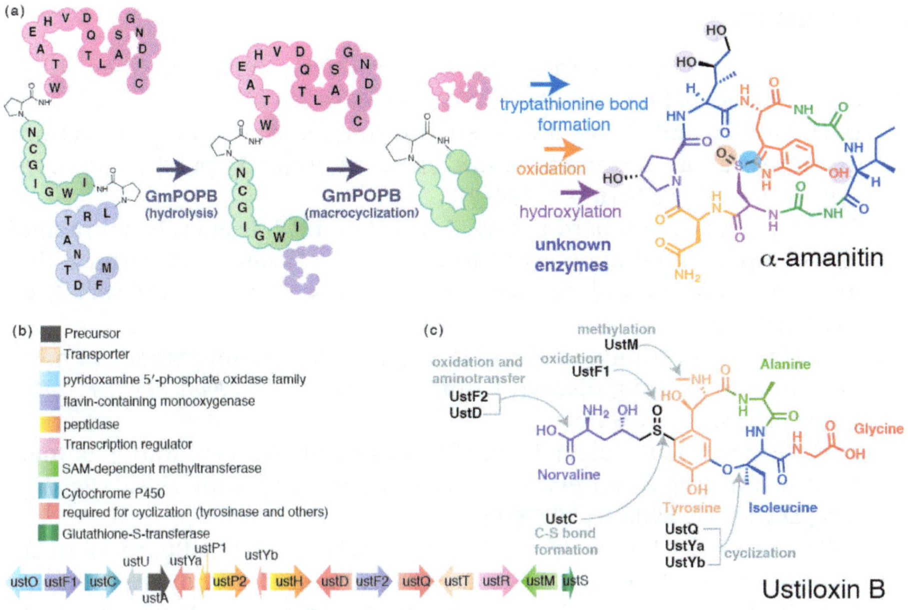

Figure 2.6 Cyclic peptides from mushrooms. (a) Biosynthesis of α-amanitin from the mushroom *Galerina marginata*. The precursor peptide is 35 amino acids long, and a single enzyme from the prolyl oligopeptidase family (GmPOPB) catalyzes peptide bond hydrolysis to release a 10 amino acid leader peptide, followed by macrocyclization, releasing a cyclic peptide product and a 17 amino acid long recognition sequence. The cyclic peptide is further modified by uncharacterized enzymes to yield the final toxic amanitin.[84,90] (b) The gene cluster required to produce ustiloxin toxins from *Aspergillus flavus*.[91] The complex nature of this short cyclic peptide requires several modifying enzymes, as well as regulatory proteins and transporters. (c) The final structure of Ustiloxin B, showing the amino acid scaffold, the post-translational modifications and the enzymes involved.[92,93]

post-translationally modified enzymatically. A particular attraction of these molecules is that they can be made both *in vivo* and *in vitro*. The former approach allows the creation of cell factories, and by using error prone PCR approaches, large numbers of variant core sequences. The later approach holds potential in being able to combine chemical synthesis (thus allowing huge diversity and fine tuning) with enzymatic transformation to produce novel molecules at scale.

Macrocyclic peptides seem poised to become part of the solution to the major problem of targeting protein–protein interactions inside the cell. Their greater structural complexity allows them to encode much more information than traditional small molecules, yet unlike biologics, macrocyclic peptides can passively diffuse into cells.

References

1. D. J. Newman and G. M. Cragg, Natural products as sources of new drugs from 1981 to 2014, *J. Nat. Prod.*, 2016, **79**, 629–661.
2. D. H. Drewry and R. Macarron, Enhancements of screening collections to address areas of unmet medical need: an industry perspective, *Curr. Opin. Chem. Biol.*, 2010, **14**, 289–298.
3. C. A. Lipinski, F. Lombardo, B. W. Dominy and P. J. Feeney, Experimental and computational approaches to estimate solubility and permeability in drug discovery and development settings, *Adv. Drug Delivery Rev.*, 1997, **23**, 3–25.
4. A. L. Harvey, R. Edrada-Ebel and R. J. Quinn, The re-emergence of natural products for drug discovery in the genomics era, *Nat. Rev. Drug Discovery*, 2015, **14**, 111–129.
5. E. M. Driggers, S. P. Hale, J. Lee and N. K. Terrett, The exploration of macrocycles for drug discovery–an underexploited structural class, *Nat. Rev. Drug Discovery*, 2008, **7**, 608–624.
6. P. G. Arnison, M. J. Bibb, G. Bierbaum, A. A. Bowers, T. S. Bugni, G. Bulaj, J. A. Camarero, D. J. Campopiano, G. L. Challis, J. Clardy, P. D. Cotter, D. J. Craik, M. Dawson, E. Dittmann, S. Donadio, P. C. Dorrestein, K. D. Entian, M. A. Fischbach, J. S. Garavelli, U. Goransson, C. W. Gruber, D. H. Haft, T. K. Hemscheidt, C. Hertweck, C. Hill, A. R. Horswill, M. Jaspars, W. L. Kelly, J. P. Klinman, O. P. Kuipers, A. J. Link, W. Liu, M. A. Marahiel, D. A. Mitchell, G. N. Moll, B. S. Moore, R. Muller, S. K. Nair, I. F. Nes, G. E. Norris, B. M. Olivera, H. Onaka, M. L. Patchett, J. Piel, M. J. T. Reaney, S. Rebuffat, R. P. Ross, H. G. Sahl, E. W. Schmidt, M. E. Selsted, K. Severinov, B. Shen, K. Sivonen, L. Smith, T. Stein, R. D. Sussmuth, J. R. Tagg, G. L. Tang, A. W. Truman, J. C. Vederas, C. T. Walsh, J. D. Walton, S. C. Wenzel, J. M. Willey and W. A. van der Donk, Ribosomally synthesized and post-translationally modified peptide natural products: overview and recommendations for a universal nomenclature, *Nat. Prod. Rep.*, 2013, **30**, 108–160.
7. A. Zorzi, K. Deyle and C. Heinis, Cyclic peptide therapeutics: past, present and future, *Curr. Opin. Chem. Biol.*, 2017, **38**, 24–29.
8. C. J. White and A. K. Yudin, Contemporary strategies for peptide macrocyclization, *Nat. Chem.*, 2011, **3**, 509–524.
9. X. Yu and D. Sun, Macrocyclic drugs and synthetic methodologies toward macrocycles, *Molecules*, 2013, **18**, 6230–6268.
10. C. Ireland and P. J. Scheuer, Ulicyclamide and ulithiacyclamide, two new small peptides from a marine tunicate, *J. Am. Chem. Soc.*, 1980, **102**, 5688–5691.
11. N. Leikoski, L. Liu, J. Jokela, M. Wahlsten, M. Gugger, A. Calteau, P. Permi, C. A. Kerfeld, K. Sivonen and D. P. Fewer, Genome mining expands the chemical diversity of the cyanobactin family to include highly modified linear peptides, *Chem. Biol.*, 2013, **20**, 1033–1043.

12. K. Sivonen, N. Leikoski, D. P. Fewer and J. Jokela, Cyanobactins-ribosomal cyclic peptides produced by cyanobacteria, *Appl. Microbiol. Biotechnol.*, 2010, **86**, 1213–1225.
13. E. W. Schmidt, J. T. Nelson, D. A. Rasko, S. Sudek, J. A. Eisen, M. G. Haygood and J. Ravel, Patellamide A and C biosynthesis by a microcin-like pathway in Prochloron didemni, the cyanobacterial symbiont of Lissoclinum patella, *Proc. Natl. Acad. Sci. U. S. A.*, 2005, **102**, 7315–7320.
14. J. A. McIntosh, M. S. Donia and E. W. Schmidt, Ribosomal peptide natural products: bridging the ribosomal and nonribosomal worlds, *Nat. Prod. Rep.*, 2009, **26**, 537–559.
15. T. J. Oman and W. A. van der Donk, Follow the leader: the use of leader peptides to guide natural product biosynthesis, *Nat. Chem. Biol.*, 2010, **6**, 9–18.
16. D. Sardar, E. Pierce, J. A. McIntosh and E. W. Schmidt, Recognition sequences and substrate evolution in cyanobactin biosynthesis, *ACS Synth. Biol.*, 2015, **4**, 167–176.
17. J. A. McIntosh and E. W. Schmidt, Marine molecular machines: heterocyclization in cyanobactin biosynthesis, *ChemBioChem*, 2010, **11**, 1413–1421.
18. J. Koehnke, A. F. Bent, D. Zollman, K. Smith, W. E. Houssen, X. Zhu, G. Mann, T. Lebl, R. Scharff, S. Shirran, C. H. Botting, M. Jaspars, U. Schwarz-Linek and J. H. Naismith, The cyanobactin heterocyclase enzyme: a processive adenylase that operates with a defined order of reaction, *Angew. Chem., Int. Ed. Engl.*, 2013, **52**, 13991–13996.
19. J. O. Melby, K. L. Dunbar, N. Q. Trinh and D. A. Mitchell, Selectivity, directionality, and promiscuity in peptide processing from a *Bacillus* sp. Al Hakam cyclodehydratase, *J. Am. Chem. Soc.*, 2012, **134**, 5309–5316.
20. J. Koehnke, G. Mann, A. F. Bent, H. Ludewig, S. Shirran, C. Botting, T. Lebl, W. E. Houssen, M. Jaspars and J. H. Naismith, Structural analysis of leader peptide binding enables leader-free cyanobactin processing, *Nat. Chem. Biol.*, 2015, **11**, 558–563.
21. C. M. Czekster, Y. Ge and J. H. Naismith, Mechanisms of cyanobactin biosynthesis, *Curr. Opin. Chem. Biol.*, 2016, **35**, 80–88.
22. J. Lee, J. McIntosh, B. J. Hathaway and E. W. Schmidt, Using marine natural products to discover a protease that catalyzes peptide macrocyclization of diverse substrates, *J. Am. Chem. Soc.*, 2009, **131**, 2122–2124.
23. V. Agarwal, E. Pierce, J. McIntosh, E. W. Schmidt and S. K. Nair, Structures of cyanobactin maturation enzymes define a family of transamidating proteases, *Chem. Biol.*, 2012, **19**, 1411–1422.
24. J. Koehnke, A. Bent, W. E. Houssen, D. Zollman, F. Morawitz, S. Shirran, J. Vendome, A. F. Nneoyiegbe, L. Trembleau, C. H. Botting, M. C. Smith, M. Jaspars and J. H. Naismith, The mechanism of patellamide macrocyclization revealed by the characterization of the PatG macrocyclase domain, *Nat. Struct. Mol. Biol.*, 2012, **19**, 767–772.

25. K. Sivonen, N. Leikoski, D. P. Fewer and J. Jokela, Cyanobactins-ribosomal cyclic peptides produced by cyanobacteria, *Appl. Microbiol. Biotechnol.*, 2010, **86**, 1213–1225.
26. J. A. McIntosh, M. S. Donia, S. K. Nair and E. W. Schmidt, Enzymatic basis of ribosomal peptide prenylation in cyanobacteria, *J. Am. Chem. Soc.*, 2011, **133**, 13698–13705.
27. P. D. Cotter, R. P. Ross and C. Hill, Bacteriocins - a viable alternative to antibiotics?, *Nat. Rev. Microbiol.*, 2013, **11**, 95–105.
28. L. M. Repka, J. R. Chekan, S. K. Nair and W. A. van der Donk, Mechanistic understanding of lanthipeptide biosynthetic enzymes, *Chem. Rev.*, 2017, **117**, 5457–5520.
29. E. Breukink, I. Wiedemann, C. van Kraaij, O. P. Kuipers, H. G. Sahl and B. de Kruijff, Use of the cell wall precursor lipid II by a pore-forming peptide antibiotic, *Science*, 1999, **286**, 2361–2364.
30. I. Oliynyk, G. Varelogianni, G. M. Roomans and M. Johannesson, Effect of duramycin on chloride transport and intracellular calcium concentration in cystic fibrosis and non-cystic fibrosis epithelia, *APMIS*, 2010, **118**, 982–990.
31. W. A. van der Donk and S. K. Nair, Structure and mechanism of lanthipeptide biosynthetic enzymes, *Curr. Opin. Struct. Biol.*, 2014, **29**, 58–66.
32. T. J. Oman, P. J. Knerr, N. A. Bindman, J. E. Velasquez and W. A. van der Donk, An engineered lantibiotic synthetase that does not require a leader peptide on its substrate, *J. Am. Chem. Soc.*, 2012, **134**, 6952–6955.
33. W. A. van der Donk and S. K. Nair, Structure and mechanism of lanthipeptide biosynthetic enzymes, *Curr. Opin. Struct. Biol.*, 2014, 58–66.
34. M. Montalban-Lopez, A. J. van Heel and O. P. Kuipers, Employing the promiscuity of lantibiotic biosynthetic machineries to produce novel antimicrobials, *FEMS Microbiol. Rev.*, 2017, **41**, 5–18.
35. E. M. Molloy, R. P. Ross and C. Hill, 'Bac' to the future: bioengineering lantibiotics for designer purposes, *Biochem. Soc. Trans.*, 2012, **40**, 1492.
36. J. D. Dutcher and J. Vandeputte, Thiostrepton, a new antibiotic. II. Isolation and chemical characterization, *Antibiot. Annu.*, 1955, **3**, 560–561.
37. W. L. Kelly, L. Pan and C. Li, Thiostrepton biosynthesis: prototype for a new family of bacteriocins, *J. Am. Chem. Soc.*, 2009, **131**, 4327–4334.
38. M. A. Ortega and W. A. van der Donk, New insights into the biosynthetic logic of ribosomally synthesized and post-translationally modified peptide natural products, *Cell Chem. Biol.*, 2016, **23**, 31–44.
39. G. A. Hudson, Z. Zhang, J. I. Tietz, D. A. Mitchell and W. A. van der Donk, *In vitro* Biosynthesis of the core scaffold of the thiopeptide thiomuracin, *J. Am. Chem. Soc.*, 2015, **137**, 16012–16015.
40. V. S. Aulakh and M. A. Ciufolini, Total synthesis and complete structural assignment of thiocillin I, *J. Am. Chem. Soc.*, 2011, **133**, 5900–5904.
41. R. P. Morris, J. A. Leeds, H. U. Naegeli, L. Oberer, K. Memmert, E. Weber, M. J. LaMarche, C. N. Parker, N. Burrer, S. Esterow, A. E. Hein, E. K. Schmitt and P. Krastel, Ribosomally synthesized thiopeptide antibiotics targeting elongation factor Tu, *J. Am. Chem. Soc.*, 2009, **131**, 5946–5955.

42. X. Just-Baringo, F. Albericio and M. Alvarez, Thiopeptide antibiotics: retrospective and recent advances, *Mar. Drugs*, 2014, **12**, 317–351.
43. J. A. Leeds, M. Sachdeva, S. Mullin, J. Dzink-Fox and M. J. Lamarche, Mechanism of action of and mechanism of reduced susceptibility to the novel anti-Clostridium difficile compound LFF571, *Antimicrob. Agents Chemother.*, 2012, **56**, 4463–4465.
44. L. C. W. Brown, M. G. Acker, J. Clardy, C. T. Walsh and M. A. Fischbach, Thirteen posttranslational modifications convert a 14-residue peptide into the antibiotic thiocillin, *Proc. Natl. Acad. Sci. U. S. A.*, 2009, **106**, 2549–2553.
45. B. W. Bycroft and M. S. Gowland, The structures of the highly modified peptide antibiotics micrococcin P1 and P2, *J. Chem. Soc., Chem. Commun.*, 1978, 256–258.
46. W. J. Wever, J. W. Bogart, J. A. Baccile, A. N. Chan, F. C. Schroeder and A. A. Bowers, Chemoenzymatic synthesis of thiazolyl peptide natural products featuring an enzyme-catalyzed formal [4 + 2] cycloaddition, *J. Am. Chem. Soc.*, 2015, **137**, 3494–3497.
47. Z. Zhang, G. A. Hudson, N. Mahanta, J. I. Tietz, W. A. van der Donk and D. A. Mitchell, Biosynthetic timing and substrate specificity for the thiopeptide thiomuracin, *J. Am. Chem. Soc.*, 2016, **138**, 15511–15514.
48. F. F. Zhang, C. X. Li and W. L. Kelly, Thiostrepton variants containing a contracted quinaldic acid macrocycle result from mutagenesis of the second residue, *ACS Chem. Biol.*, 2016, **11**, 415–424.
49. C. X. Li, F. F. Zhang and W. L. Kelly, Heterologous production of thiostrepton A and biosynthetic engineering of thiostrepton analogs, *Mol. BioSyst.*, 2011, **7**, 82–90.
50. X. Luo, C. Zambaldo, T. Liu, Y. Zhang, W. Xuan, C. Wang, S. A. Reed, P. Y. Yang, R. E. Wang, T. Javahishvili, P. G. Schultz and T. S. Young, Recombinant thiopeptides containing noncanonical amino acids, *Proc. Natl. Acad. Sci. U. S. A.*, 2016, **113**, 3615–3620.
51. Y. Kobayashi, M. Ichioka, T. Hirose, K. Nagai, A. Matsumoto, H. Matsui, H. Hanaki, R. Masuma, Y. Takahashi, S. Omura and T. Sunazuka, Bottromycin derivatives: efficient chemical modifications of the ester moiety and evaluation of anti-MRSA and anti-VRE activities, *Bioorg. Med. Chem. Lett.*, 2010, **20**, 6116–6120.
52. L. Huo, S. Rachid, M. Stadler, S. C. Wenzel and R. Muller, Synthetic biotechnology to study and engineer ribosomal bottromycin biosynthesis, *Chem. Biol.*, 2012, **19**, 1278–1287.
53. G. Mann, L. J. Huo, S. Adam, B. Nardone, J. Vendome, N. J. Westwood, R. Muller and J. Koehnke, Structure and substrate recognition of the bottromycin maturation enzyme BotP, *ChemBioChem*, 2016, **17**, 2286–2292.
54. Y. P. Hou, M. D. B. Tianero, J. C. Kwan, T. P. Wyche, C. R. Michel, G. A. Ellis, E. Vazquez-Rivera, D. R. Braun, W. E. Rose, E. W. Schmidt and T. S. Bugni, Structure and biosynthesis of the antibiotic bottromycin D, *Org. Lett.*, 2012, **14**, 5050–5053.

55. W. J. K. Crone, N. M. Vior, J. Santos-Aberturas, L. G. Schmitz, F. J. Leeper and A. W. Truman, Dissecting bottromycin biosynthesis using comparative untargeted metabolomics, *Angew. Chem., Int. Ed.*, 2016, **55**, 9638–9642.
56. W. J. K. Crone, F. J. Leeper and A. W. Truman, Identification and characterisation of the gene cluster for the anti-MRSA antibiotic bottromycin: expanding the biosynthetic diversity of ribosomal peptides, *Chem. Sci.*, 2012, **3**, 3516–3521.
57. N. H. Tan and J. Zhou, Plant cyclopeptides, *Chem. Rev.*, 2006, **106**, 840–895.
58. C. Jennings, J. West, C. Waine, D. Craik and M. Anderson, Biosynthesis and insecticidal properties of plant cyclotides: the cyclic knotted proteins from Oldenlandia affinis, *Proc. Natl. Acad. Sci. U. S. A.*, 2001, **98**, 10614–10619.
59. J. P. Tam, S. Wang, K. H. Wong and W. L. Tan, Antimicrobial peptides from plants, *Pharmaceuticals (Basel)*, 2015, **8**, 711–757.
60. A. Gould, Y. B. Ji, T. L. Aboye and J. A. Camarero, Cyclotides, a novel ultrastable polypeptide scaffold for drug discovery, *Curr. Pharm. Des.*, 2011, **17**, 4294–4307.
61. G. K. Nguyen, X. Hemu, J. P. Quek and J. P. Tam, Butelase-mediated macrocyclization of d-amino-acid-containing peptides, *Angew. Chem., Int., Ed. Engl.*, 2016, **55**, 12802–12806.
62. Y. Cao, G. K. Nguyen, S. Chuah, J. P. Tam and C. F. Liu, Butelase-mediated ligation as an efficient bioconjugation method for the synthesis of peptide dendrimers, *Bioconjugate Chem.*, 2016, **27**, 2592–2596.
63. G. K. Nguyen, A. Kam, S. Loo, A. E. Jansson, L. X. Pan and J. P. Tam, Butelase 1: a versatile ligase for peptide and protein macrocyclization, *J. Am. Chem. Soc.*, 2015, **137**, 15398–15401.
64. G. K. T. Nguyen, S. Wang, Y. Qiu, X. Hemu, Y. Lian and J. P. Tam, Butelase 1 is an Asx-specific ligase enabling peptide macrocyclization and synthesis, *Nat. Chem. Biol.*, 2014, **10**, 732–738.
65. Y. H. Huang, S. T. Henriques, C. K. Wang, L. Thorstholm, N. L. Daly, Q. Kaas and D. J. Craik, Design of substrate-based BCR-ABL kinase inhibitors using the cyclotide scaffold, *Sci. Rep.*, 2015, **5**, 12974.
66. T. Aboye, C. J. Meeks, S. Majumder, A. Shekhtman, K. Rodgers and J. A. Camarero, Design of a MCoTI-based cyclotide with angiotensin (1-7)-like activity, *Molecules*, 2016, **21**, 152.
67. A. Sangphukieo, W. Nawae, T. Laomettachit, U. Supasitthimethee and M. Ruengjitchatchawalya, Computational design of hypothetical new peptides based on a cyclotide scaffold as HIV gp120 inhibitor, *PLoS One*, 2015, **10**, e0139562.
68. C. K. Wang, G. J. King, S. E. Northfield, P. G. Ojeda and D. J. Craik, Racemic and quasi-racemic X-ray structures of cyclic disulfide-rich peptide drug scaffolds, *Angew. Chem., Int. Ed. Engl.*, 2014, **53**, 11236–11241.
69. C. K. Wang, S. H. Hu, J. L. Martin, T. Sjogren, J. Hajdu, L. Bohlin, P. Claeson, U. Goransson, K. J. Rosengren, J. Tang, N. H. Tan and D. J. Craik, Combined X-ray and NMR analysis of the stability of the cyclotide cystine

knot fold that underpins its insecticidal activity and potential use as a drug scaffold, *J. Biol. Chem.*, 2009, **284**, 10672–10683.

70. M. Trabi and D. J. Craik, Tissue-specific expression of head-to-tail cyclized miniproteins in Violaceae and structure determination of the root cyclotide Viola hederacea root cyclotide1, *Plant Cell*, 2004, **16**, 2204–2216.
71. D. C. Ireland, M. L. Colgrave and D. J. Craik, A novel suite of cyclotides from Viola odorata: sequence variation and the implications for structure, function and stability, *Biochem. J.*, 2006, **400**, 1–12.
72. H. Morita, T. Kayashita, A. Uchida, K. Takeya and H. Itokawa, Cyclic peptides from higher plants. 33. Delavayins A–C, three new cyclic peptides from Stellaria delavayi, *J. Nat. Prod.*, 1997, **60**, 212–215.
73. B. Picur, M. Lisowski and I. Z. Siemion, A new cyclolinopeptide containing nonproteinaceous amino acid *N*-methyl-4-aminoproline, *Lett. Pept. Sci.*, 1998, **5**, 183–187.
74. N. H. Tan and J. Zhou, Plant cyclopeptides, *Chem. Rev.*, 2006, **106**, 840–895.
75. Y. S. Yun, H. Morita, K. Takeya and H. Itokawa, Cyclic peptides from higher plants. 34. Segetalins G and H, structures and estrogen-like activity of cyclic pentapeptides from Vaccaria segetalis, *J. Nat. Prod.*, 1997, **60**, 216–218.
76. H. Morita, M. Eda, T. Iizuka, Y. Hirasawa, M. Sekiguchi, Y. S. Yun, H. Itokawa and K. Takeya, Structure of a new cyclic nonapeptide, segetalin F, and vasorelaxant activity of segetalins from Vaccaria segetalis, *Bioorg. Med. Chem. Lett.*, 2006, **16**, 4458–4461.
77. C. J. S. Barber, P. T. Pujara, D. W. Reed, S. Chiwocha, H. X. Zhang and P. S. Covello, The two-step biosynthesis of cyclic peptides from linear precursors in a member of the plant family caryophyllaceae involves cyclization by a serine protease-like enzyme, *J. Biol. Chem.*, 2013, **288**, 12500–12510.
78. J. D. Walton, H. E. Hallen-Adams and H. Luo, Ribosomal biosynthesis of the cyclic peptide toxins of Amanita mushrooms, *Biopolymers*, 2010, **94**, 659–664.
79. M. Umemura, N. Nagano, H. Koike, J. Kawano, T. Ishii, Y. Miyamura, M. Kikuchi, K. Tamano, J. J. Yu, K. Shin-ya and M. Machida, Characterization of the biosynthetic gene cluster for the ribosomally synthesized cyclic peptide ustiloxin B in Aspergillus flavus, *Fungal Genet. Biol.*, 2014, **68**, 23–30.
80. S. H. Eren, Y. Demirel, S. Ugurlu, I. Korkmaz, C. Aktas and F. M. Guven, Mushroom poisoning: retrospective analysis of 294 cases, *Clinics (Sao Paulo)*, 2010, **65**, 491–496.
81. J. A. Pulman, K. L. Childs, R. M. Sgambelluri and J. D. Walton, Expansion and diversification of the MSDIN family of cyclic peptide genes in the poisonous agarics *Amanita phalloides* and *A. bisporigera*, *BMC Genomics*, 2016, **17**, 1038.
82. H. Luo, H. E. H. Len-Adams, J. S. Scott-Craig and J. D. Walton, Ribosomal biosynthesis of alpha-amanitin in Galerina marginata, *Fungal Genet. Biol.*, 2012, **49**, 123–129.

83. H. Luo, S. Y. Hong, R. M. Sgambelluri, E. Angelos, X. Li and J. D. Walton, Peptide macrocyclization catalyzed by a prolyl oligopeptidase involved in alpha-amanitin biosynthesis, *Chem. Biol.*, 2014, **21**, 1610–1617.
84. C. M. Czekster and J. H. Naismith, Kinetic landscape of a peptide bond-forming prolyl oligopeptidase, *Biochemistry*, 2017, **56**, 2086–2095.
85. N. Nagano, M. Umemura, M. Izumikawa, J. Kawano, T. Ishii, M. Kikuchi, K. Tomii, T. Kumagai, A. Yoshimi, M. Machida, K. Abe, K. Shin-ya and K. Asai, Class of cyclic ribosomal peptide synthetic genes in filamentous fungi, *Fungal Genet. Biol.*, 2016, **86**, 58–70.
86. M. Umemura, N. Nagano, H. Koike, J. Kawano, T. Ishii, Y. Miyamura, M. Kikuchi, K. Tamano, J. Yu, K. Shin-ya and M. Machida, Characterization of the biosynthetic gene cluster for the ribosomally synthesized cyclic peptide ustiloxin B in Aspergillus flavus, *Fungal Genet. Biol.*, 2014, **68**, 23–30.
87. W. Ding, W. Q. Liu, Y. Jia, Y. Li, W. A. van der Donk and Q. Zhang, Biosynthetic investigation of phomopsins reveals a widespread pathway for ribosomal natural products in Ascomycetes, *Proc. Natl. Acad. Sci. U. S. A.*, 2016, **113**, 3521–3526.
88. A. Cormier, M. Marchand, R. B. Ravelli, M. Knossow and B. Gigant, Structural insight into the inhibition of tubulin by vinca domain peptide ligands, *EMBO Rep.*, 2008, **9**, 1101–1106.
89. Y. Ye, A. Minami, Y. Igarashi, M. Izumikawa, M. Umemura, N. Nagano, M. Machida, T. Kawahara, K. Shin-Ya, K. Gomi and H. Oikawa, Unveiling the biosynthetic pathway of the ribosomally synthesized and post-translationally modified peptide ustiloxin b in filamentous fungi, *Angew. Chem., Int. Ed. Engl.*, 2016, **55**, 8072–8075.
90. H. Luo, S. Y. Hong, R. M. Sgambelluri, E. Angelos, X. Li and J. D. Walton, Peptide macrocyclization catalyzed by a prolyl oligopeptidase involved in alpha-amanitin biosynthesis, *Chem. Biol.*, 2014, **21**, 1610–1617.
91. W. Ding, W. Q. Liu, Y. Jia, Y. Li, W. A. van der Donk and Q. Zhang, Biosynthetic investigation of phomopsins reveals a widespread pathway for ribosomal natural products in Ascomycetes, *Proc. Natl. Acad. Sci.*, 2016, **113**, 3521–3526.
92. Y. Ye, A. Minami, Y. Igarashi, M. Izumikawa, M. Umemura, N. Nagano, M. Machida, T. Kawahara, K. Shin-ya, K. Gomi and H. Oikawa, Unveiling the biosynthetic pathway of the ribosomally synthesized and post-translationally modified peptide Ustiloxin b in filamentous fungi, *Angew. Chem., Int. Ed.*, 2016, **55**, 8072–8075.
93. T. Tsukui, N. Nagano, M. Umemura, T. Kumagai, G. Terai, M. Machida and K. Asai, Ustiloxins, fungal cyclic peptides, are ribosomally synthesized in Ustilaginoidea virens, *Bioinformatics*, 2015, **31**, 981–985.

CHAPTER 3

Thioesterase Domain-mediated Macrocyclization of Non-ribosomal Peptides

SHO KONNO, LAËTITIA MISSON AND MICHAEL D. BURKART*

Department of Chemistry and Biochemistry, University of California, San Diego, 9500 Gilman Drive, La Jolla, CA 92093-0358, USA
*E-mail: mburkart@ucsd.edu

3.1 Introduction

Microorganism-produced non-ribosomal peptides (NRPs) display a notable range of biological activities due to the diverse and unique structures they demonstrate. Their activities range from antibiotic (tyrocidine A, daptomycin, vancomycin), to immunosuppressant (cyclosporin A), anticancer (bleomycin), and anthelmintic (avermectin).[1] Most of these peptide-based natural products have macrocyclic structures that are responsible for their unique biological activities. First, the macrocyclization of the peptide backbone strongly constrains the conformation of the otherwise inherently flexible peptides, and intramolecular hydrogen bonds further constrict potential conformations. The resulting semi-rigid structures are not only specific and selective toward the binding of target proteins, but also more membrane permeable and more resistant to proteolytic hydrolysis.[2,3] In addition, unlike small molecule inhibitors (≤500 Da), these larger

Chemical Biology No. 6
Cyclic Peptides: From Bioorganic Synthesis to Applications
Edited by Jesko Koehnke, James Naismith and Wilfred A. van der Donk

Published by the Royal Society of Chemistry, www.rsc.org

molecules (500–2000 Da) could potentially disrupt protein–protein interactions ($K_d = 10^{-7\pm3}$ M^{-1}) and involve interactions over large surface areas (700–1500 $Å^2$).[4,5]

Macrocyclic NRPs are biosynthesized by large, highly versatile, multifunctional mega-enzymes known as non-ribosomal peptide synthetases (NRPSs). The biosynthetic machinery of NRPSs involves a so-called thiotemplate mechanism analogous to that in fatty acid synthases (FASs) and polyketide synthases (PKSs) (Figure 3.1).[6–11] As they are thoroughly reviewed elsewhere, we will only briefly describe these synthases. NRPSs are made up of multiple modules, each responsible for incorporation of a monomer building block within the NRP. Modules are further subdivided into functional domains, each catalyzes a reaction in the NRPS assembly line. A single module typically contains at least three essential domains, adenylation (A), peptidyl carrier protein (PCP; also referred to as thiolation, T) and condensation (C) domains. A domains are responsible for selection and activation of a substrate amino acid by ATP to generate an aminoacyl-AMP. PCP domains play a central role in the NRP assembly line as chaperones of the growing metabolite. Each amino acid is loaded onto the PCP by attack of the thiol terminus of a 4′-phosphopantetheine (PPant) arm on the *holo*-PCP to the aminoacyl-AMP intermediate prepared by the adjacent A domain. Once loaded, each amino acid is condensed with a downstream amino acid to form a peptide bond. During the NRP assembly line, the amino acids of aminoacyl-*S*-PCP can be further modified by tailoring domains such as epimerization, methyltransferase, formylation, heterocyclization, oxidation/reduction, and halogenation domains.[10]

Termination of NRP biosynthesis is commonly performed by thioesterase (TE) domains located at the C-terminus of most NRPSs. TE domains play a role in releasing full-length precursor peptides from NRPSs either through hydrolysis with a water molecule or through macrocyclization with intramolecular nucleophilic attack by free amino or hydroxy groups.[11] Because of their key role in product release and substrate macrocyclization, engineering TEs for the production of non-natural macrocyclic NRPs is a promising approach toward preparing novel biologically active materials. Therefore, the mechanisms and structures of the TE domains, as well as their interaction with the PCP domains, have been extensively studied since the 2000s.[12–15] While the successful synthesis of truncated cyclopeptides through relocation of the TE domain has been reported, a general strategy to design macrocyclic peptide analogues has not yet been established.[16] As a result, efforts to re-engineer NRPSs have been performed on a case-by-case basis. In this chapter, we describe the general mechanism of macrocyclic NRP biosynthesis, emphasizing the mechanistic role of TE domains. We also review recent advances in the engineering of TE domains to understand their structures and interactions with PCP domains, as well as efforts to modify TEs to synthesize "non-natural" products.

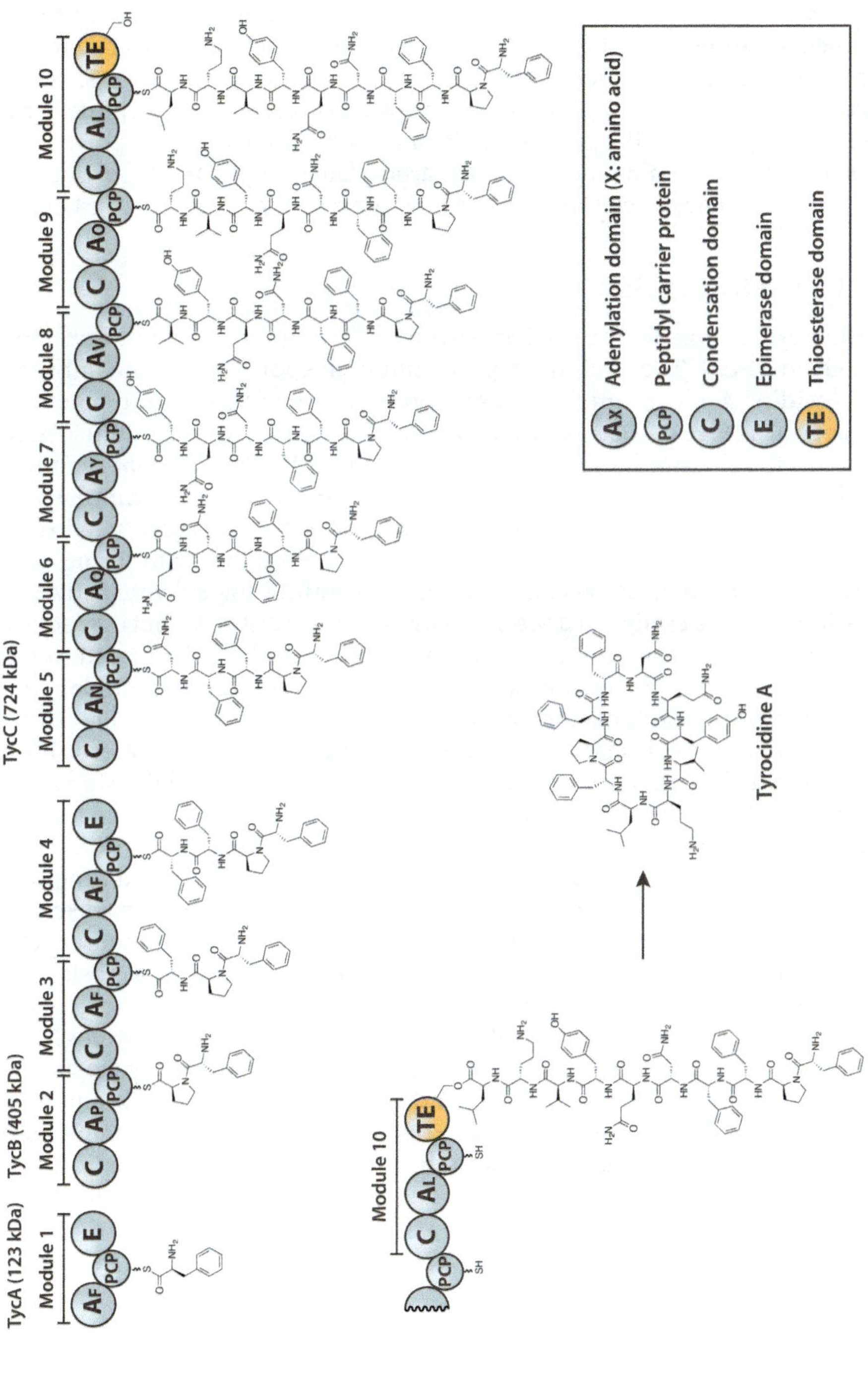

Figure 3.1 Biosynthesis of tyrocidine A by the tyrocidine synthetases TycA, B and C.

3.2 Types of Macrocyclic Non-ribosomal Peptide

Macrocyclic NRPs can be classified according to the nature of the bonds forming their main chain.[11] Typically, macrocyclization occurs in a head-to-tail fashion between the N- and C-termini of a linear substrate. In some cases, ester, thioester, and imine functionalities are also present in the peptide backbone. These electrophilic groups can be attacked by nucleophiles present on the substrate, like amino, hydroxy and carboxylate groups. In this section, we categorize the macrocyclic NRPs according to the nature of the main chain.

3.2.1 Cyclic Peptides

Peptide cyclization is achieved in two distinct manners, head-to-tail and side-chain-to-tail. The immunosuppressant cyclosporin A and the antibiotics tyrocidine A and gramicidin S are representative of head-to-tail macrocyclic peptides (Figure 3.2a). Cyclosporin A, a secondary metabolite isolated from the fungi *Cylindrocarpon lucidum* Booth and *Tolypocladium inflatum*, consists of eleven amino acids and contains ten non-natural amino acids, including one D-amino acid and six *N*-methylated amino acids.[17] Cyclosporin A specifically binds to the immunophilin cyclophilin, interfering with protein–protein interactions and consequently inhibiting enzyme activity.[18] Tyrocidine A is a cyclic decapeptide with an anti-parallel β-sheet structure isolated from the bacterium *Bacillus brevis*. Gramicidin S is a head-to-tail cyclic decapeptide, which consists of a dimer of iterative pentapeptides containing the non-proteinogenic amino acids D-Phe and ornithine. The structure of gramicidin S is also an anti-parallel β-sheet. Both antibiotics exhibit an amphipathic character, which derives from the anti-parallel β-sheet and the hydrophobic amino acids and ornithine residues. This structure specificity enables broad-spectrum antibiotic activity by disrupting membranes.[19,20]

In some cases, macrocyclization occurs between an amine-containing side chain of a Lys or of a non-proteinogenic amino acid in the polypeptide chain and the C-terminal carboxylic acid. Bacitracin is a branched macrocyclic NRP that is bridged by the ε-amino group of a lysine residue at position 6 and the C-terminal asparagine.[21] Bacitracin inhibits the enzymatic dephosphorylation of C_{55}-isoprenyl pyrophosphate and consequently peptidoglycan biosynthesis in the bacterial cell wall.[22] Polymyxin is another cyclic antibiotic composed of a hexapeptide ring and a tripeptide side chain bound to a fatty acid tail.[23] Polymyxin includes three amine-containing side chains from diaminobutylic acid (Dab) residues in the cyclopeptide, and cyclization specifically occurs between the L-Dab residue at position 4 and the C-terminal L-Thr, providing a lariat structure.

3.2.2 Cyclic Depsipeptides

Depsipeptides are defined as peptides in which an ester bond replaces one or more amide bonds. These ester bonds originate not only from the hydroxy side chain of the proteinogenic amino acids (*e.g.*, serine, threonine and

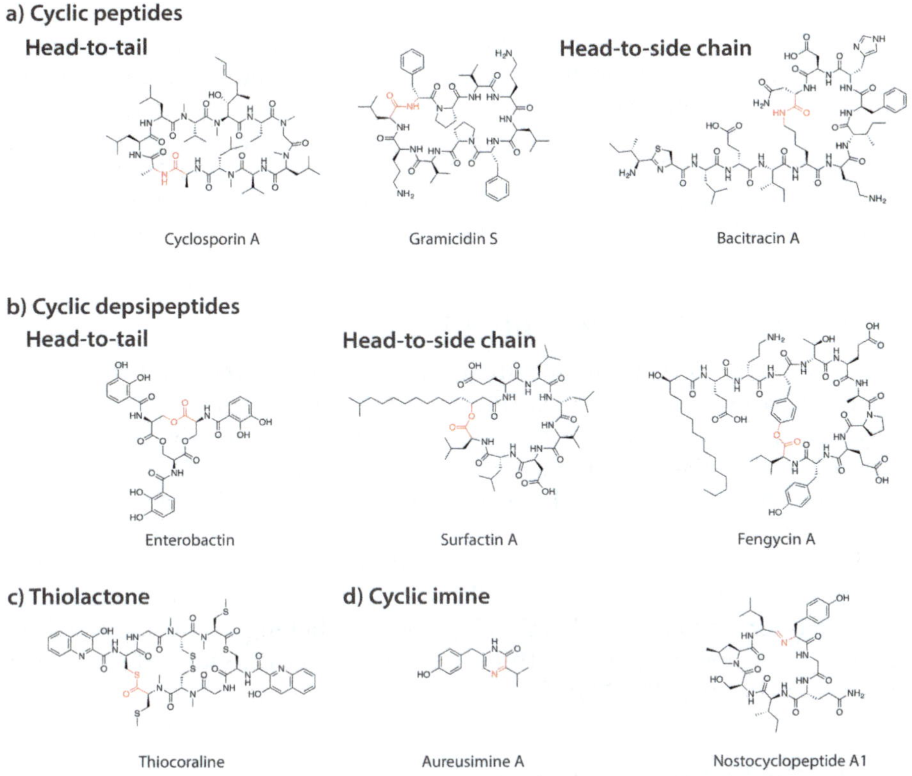

Figure 3.2 Structures of the representative microorganism-produced NRPs. Structural features and macrocyclization points are shown in red. (a) Structures of cyclic peptides. (b) Structures of cyclic depsipeptides. (c) Structure of a thiolactone peptide. (d) Structures of imine-containing cyclic peptides.

tyrosine) but also from hydroxy groups found in non-proteinogenic amino acids, fatty acids, and hydroxy acids. Many bioactive macrocyclic depsipeptides have been identified in microorganisms, such as fungi and bacteria.[24] As an example, surfactins produced by various strains of *Bacillus subtilis* are cyclic lipodepsipeptides, which consist of seven amino acids and branched fatty acid chains of different length (Figure 3.2b).[25,26] The ester bond of surfactin is found between the β-hydroxy group of the N-terminus of the fatty acid tail and the C-terminal of the carboxylic acid of L-Leu. Surfactin is a strong biosurfactant with amphiphilic properties derived from the fatty acid tail and two acidic amino acids, L-Asp and L-Glu. Fengycins also belong to a family of lipodepsipeptides. Fengycin consists of a decapeptide and a fatty acid tail, where eight out of the ten amino acids form the macrolactone ring between the phenoxy group of D-Tyr and the C-terminal carboxylic acid of L-Ile. The chain is formed with the dipeptide formed by D-Orn and L-Glu, the latter being ester bound to a hydrophobic tail.[26] Oligomeric cyclic depsipeptides can act as iron chelators. Enterobactin, produced by Gram-negative bacteria such as *Escherichia coli*, is a cyclic depsipeptide formed from a trimer of L-Ser linked to 2,3-dihydroxybenzoic acid. It binds Fe(III) with a high affinity and is the strongest known siderophore.[27] When the cells lack ferrous iron, they secrete enterobactin to the extracellular environment to chelate Fe(III), and then a system of membrane proteins and esterases transports the enterobactin–Fe(III) complex and releases the ferric ion in the bacteria cell.[28] Some fungi also produce cyclooligomer peptides such as beauvericin and PF1022.[29] These cyclic peptides show a wide variety of biological activities, including antibiotic, insecticidal, anthelminthic and herbicidal, as well as cytotoxic activities.

3.2.3 Cyclic Thiodepsipeptides

Thiocoraline is a cyclic thiodepsipeptide that was isolated from the actinomycete *Micromonospora marina* (Figure 3.2c). Thiocoraline consists of two 3-hydroxyquinaldic acid moieties, two rare *S*-methylated L-Cys moieties, an intramolecular disulfide bond, and a thiodepsipeptide backbone. The two thioester linkages are formed between the thiol side chains of Cys residues and the carboxylic acids of the two *S*-methylated L-Cys residues.[30] Thiocoraline is a potent anticancer agent exhibiting strong anti-proliferative activities against several cancer cell lines *in vitro*, as well as human carcinoma xenografts *in vivo*.[31] This thiodepsipeptide belongs to the family of bis-intercalators; however, the rare thioester linkage has, thus far, only been seen in thiocoraline and BE-22179.[32]

3.2.4 Cyclic Imino Peptides

Some cyclic NRPs possess a head-to-tail imine bond in the peptide backbone. The macrocyclic nostocyclopeptides A1 and A2 were isolated from the cyanobacterium *Nostoc* sp. ATCC53789 and consist of a cyclic heptapeptide that possess an imine linkage (Figure 3.2d).[33] The aureusimines are cyclic dipeptides found in the bacterium *Staphylococcus aureus*, and have a unique pyrazinone structure that most likely derives from the spontaneous cyclization of the dipeptidyl-aldehyde to form a dihydroxypyrazinone, which is subsequently oxidized to a pyrazinone.[34,35] The biosynthetic enzymes of these imine-containing natural products possess a reductase (R) domain in the termination module instead of a TE domain.[35,36] The R domains reductively cleave the thioester C–S, yielding *holo*-PCP and a peptidyl-aldehyde in a NAD(P)H-dependent manner. Spontaneous condensation of the resulting aldehyde with the N-terminal or the amine side chain of one of the residues is believed to occur to provide the cyclic imine peptide.

3.3 Biosynthesis of Macrocyclic NRPs

A wide variety of NRPs are synthesized by multi-modular biosynthetic assembly systems. The modules consist of the core A, PCP and C domains and non-essential domains such as methylation and cyclization domains.

The combination of each domain in the module leads to the structural diversity of NRPs outlined in the previous section. Peptide release is carried out by either TE, R or condensation-like (C_T) domains located at the C-terminus of the modular NRPSs.

3.3.1 NRP Biosynthesis

NRP biosynthesis starts with the selection of a building block by the A domain.[6–9] The A domain selects a cognate amino acid from a monomer pool and activates it as an aminoacyl-AMP at the expense of ATP. The substrates of A domains are not only the 20 proteinogenic amino acids, but also non-proteinogenic amino acids, α-hydroxy amino acids, aryl acids and other acids. The activated amino acid is transferred onto the thiol group of PPant by the action of the A domain. The latter is post-translationally loaded onto the hydroxy side chain of a conserved serine residue of the PCP domain by 4′-phosphopantetheinyl transferase (PPTase) (Figure 3.3a).[37] C domains mediate the formation of new peptide bonds by catalyzing the nucleophilic attack of the amine group of downstream aminoacyl-S-PCP toward upstream aminoacyl- or peptidyl-S-PCP. Finally, the linear polypeptide appended on the terminal PCP domain is loaded onto the TE domain, followed by nucleophilic attack of water or intramolecular cyclization (Figure 3.3b).

3.3.2 Thioesterase Domains

TE domains (~280 amino acids) are responsible for releasing the products from NRPS modules in the final stage of the NRP assembly line. TE domains belong to the α/β hydrolase superfamily, which includes lipases, esterases and proteases.[38] TE domains are classified in two types; type I TE domains (TE-I), which are integrated into modular NRPSs, and discrete type II TE (TE-II) domains.[10] While TE-I and TE-II domains structurally and functionally differ, both types have the same highly conserved catalytic triad of Ser–His–Asp as commonly found in serine hydrolases and proteases.[39]

3.3.2.1 Type I TE Domains

TE-Is are located in the terminal modules of multi-modular NRPSs. A number of crystal structures of TE-Is have been solved and their structural characteristics identified. Although TE-I domains are structurally similar to typical α/β hydrolases, they possess a distinct active site cavity and "lid" region distinctive from serine hydrolases. Among TE-Is, these lid regions are highly variable in sequence and may be responsible for the substrate selectivity of TE-Is. Bruner *et al.* reported the structure of the excised surfactin synthetase C (SrfAC) TE-I from *Bacillus subtilis* JH642 and showed the overall TE structure in two distinct conformations, open (O) and closed (C) (Figure 3.3c).[40] The SrfAC TE-I is a globular domain with a bowl-shaped active site

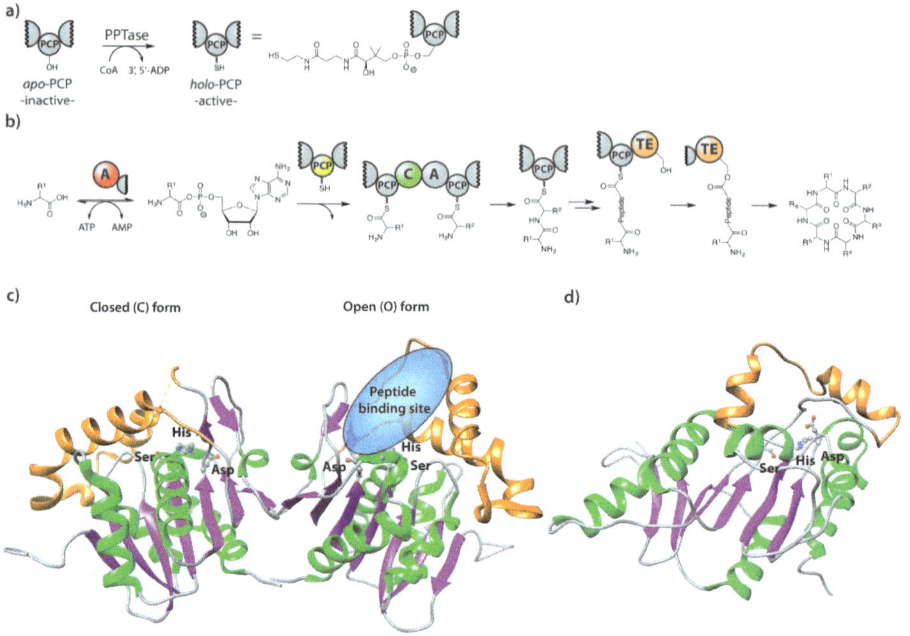

Figure 3.3 Catalytic domains in NRP biosynthesis. (a) Post-translational modification of *apo*-PCP to *holo*-PCP by PPTase. The PPant unit is attached to a conserved Ser residue of PCP. (b) Schematic illustration of the reactions catalyzed by each domain. (c) Crystal structures of the TE domain of the surfactin synthetase SrfAC from *Bacillus subtilis*. The two conformations are shown in open (O) and closed (C) forms. (d) Structure of the type II TE domain of the surfactin synthetase SrfAD from *Bacillus subtilis*. For (c) and (d), the α-helices and β-sheets are shown in green and the lid region is colored in orange. The catalytic triad is displayed as ball and stick models (magenta).

cavity and an α/β hydrolase-like fold missing strand β1, helix αD, and part of αE. The structure of SrfAC TE-I consists of five α-helices and seven β-strands, with a characteristic three-helix lid region between strands β6 and β7. The lid region undergoes significant structural rearrangement when the TE domain transitions from the (O) and (C) conformations. In the (O) conformation, the lid is folded back, which allows peptidyl-PCPs to access the active site of the TE domain. Conversely, in the (C) conformation the lid moves to cover the active site in order to prevent solvent from reaching it, ensuring that substrate release occurs *via* intramolecular cyclization of the substrate rather than hydrolysis. The dynamic of the lid region of the SrfAC TE domain clearly suggests that it has an important role in both substrate recognition and macrocyclization catalysis.

The structure of the excised fengycin synthetase B (FenB) TE-I has also been reported. FenB is structurally similar to SrfAC TE-I, only varying from SrfAC TE-I in its lid region.[41] The lid of FenB TE-I is shorter than that of

SrfAC TE-I and lacks the αL1. In addition, unlike SrfAC TE-I, the FenB TE domain was crystallized exclusively in its open state, such that substrates have access to the active site of the TE domain. The lid region of FenB TE-I may not play a role in substrate recognition, but rather prevent the binding of unwanted macromolecular substrates. Interestingly, the sequence alignments of a selection of cyclic NRPS TE domains show significant differences in the length of the lid regions among each TE domain. However, no correlation between the length of the lid region and the members and mode of macrocyclic products was found. Although the exact function of the lid region remains unclear, continued efforts to characterize the structures of TE-I domains could help determine its role more definitively.

3.3.2.2 *Type II TE Domains*

TE-IIs are discrete hydrolysis enzymes found in many NRPS gene clusters, which, unlike TE-Is, are not covalently bound to the synthases. Although TE-IIs are not essential in NRP biosynthesis, these enzymes often play an important housekeeping role in NRPS biosynthesis.[42] Several studies have demonstrated that TE-IIs hydrolyze unwanted substrates off of the PCP, as the loading of nonreactive substrates can inactivate NRPSs. TE-II proteins associated with SrfAD, BacT and TycF in surfactin, bacitracin and tyrocidine synthesis, respectively, have been shown to restore the NRPS activity by removing substrates from misprimed peptidyl-PCPs. Such mispriming can be caused by PPTases, as well as non-cognate amino acids loaded onto the PCP domains in NRPS modules.[43,44] In addition, the endogenous CoA pool exists mostly in the acyl form, such as acetyl-CoA (~80%), in bacteria. TE-II deacylates such PCPs, enabling their loading.[45] Deletion of the TE-II locus of SrfAD reduced surfactin production by 84% in the organism, demonstrating the importance of these TEs.[46] In addition, heterologous co-expression in *E. coli* of valinomycin synthetases along with a TE-II encoded in the valinomycin synthetase gene cluster led to an 8.8-fold increase in valinomycin production.[47] In rare cases, TE-IIs also play a role in intermediate and product release.[48]

The structures of several TE-II domains have been resolved. These enzymes also adopt a typical α/β hydrolase fold.[49] Comparison of the structure of SrfAC TE-I and SrfAD TE-II domains demonstrates that SrfAD TE-II has a shorter lid region with a helix-turn-helix-loop motif that partially occludes the active site (Figure 3.3d). In addition, the SrfAD TE-II domain features a shallower cavity than the SrfAC TE-I domain, suggesting that it evolved to specifically accommodate small substrates.

3.3.3 Other Termination Domains

TE domain-catalyzed peptide cyclization is the most common mode for termination of bacterial NRP synthesis. However, the TE domain is absent in the vast majority of fungal NRPSs, so another enzymatic route is used to catalyze cyclization and product release. The final module of the fungal modular enzymes consists instead of releasing domains such as reductase (R) or condensation-like (C_T) domains.[50–52]

3.3.3.1 Reductase Domains

R domains (~350 amino acids) are found in fungal NRPS terminal modules and are responsible for the release of the polypeptide chain bound to the final PCP domain and their successive macrocyclization. The reductive cleavage of the thioester bond on PCP by the R domain requires NAD(P)H as a cofactor.[53] The resulting linear C-terminal alcohol or aldehyde moiety of the freed peptide then non-enzymatically cyclizes *via* attack of either a side chain or an N-terminal amino group to form a stable macrocyclic imine. Wyatt and co-workers reported the structural and functional characterization of the peptidyl aldehyde-producing R domain of aureusimine synthetase (AusA).[54] The *in vitro* chemoenzymatic synthesis of aureusimine analogues in this study resulted in both pyrazinone products and diketopiperazine derivatives. The AusA R domain possesses substrate preference for some amino acid pairs, and it favors diketopiperazine formation through attack of the free amine onto the thioester of the first amino acid. The sequence of the AusA R domain is similar to the sequences of members of the short-chain dehydrogenase/reductase superfamily, which includes other terminal NRPS R domains. The crystal structure of AusA shows that the reductase adopts a Rossmann fold that contains a central β-sheet surrounded by α-helices and possesses a conserved Thr–Tyr–Lys active site catalytic triad. Further structural and biochemical studies are needed to fully assess the mechanism of R domain-catalyzed reductive thioester bond cleavage.

3.3.3.2 Condensation(-like) Domains

In addition to R domains, some fungal NRPSs like cyclosporine or tryptoquialanine synthases use condensation-like (C_T) terminal domains to release the macrocyclic products.[55,56] The *in vitro* reconstitution and biochemical characterization of the trimodular TqaA (tryptoquialanine synthetase) have demonstrated that the C_T domain catalyzes the release of a linear product from the PCP domain that subsequently undergoes spontaneous cyclization.[56] Efforts to probe the ability of C_T domains to process mimics of the natural acyl-S-enzyme substrate of TqaA, like peptidyl-SNAC and peptidyl-CoA, demonstrate that C_T-catalyzed cyclization requires interactions with the partner PCP domain. Substrate mimics are not recognized as substrates, and only the peptide tethered to the natural PCP partner was recognized as a substrate. This C_T domain dependence differs from TE domains, which accept both peptidyl-SNAC and peptidyl-PCP as substrates.[57] While preparing this manuscript, the crystal structures of the TqaA C_T and PCP–C_T domains were published.[58] The overall structure of the TqaA C_T domain adopts a pseudo-dimeric open sandwich (V shaped) form, which shows high similarity to canonical C domains. These structures show that C_T features a unique N-terminal α1 helix and a shorter α2 helix than is observed in canonical C domains. As a result, the acceptor site and the solvent channel are occluded, which prevents nucleophilic attack on the peptide substrate by the solvent or other adventitious nucleophiles. Since the C_T domain is optimized for intramolecular cyclic condensation reactions and not linear peptide release, the acceptor site of the C_T domain is likely vestigial.

3.4 Mechanistic Insights into TE Domain-catalyzed Peptide Cyclization and Release

TE domains catalyze two-step processes. In the first step, substrate peptides are loaded onto the TE domain by transesterification of the PCP thioester. Loading is then followed by product release.[59] The general reaction mechanism is similar to the one of serine hydrolases apart from the situation of product release by macrocyclization.

3.4.1 Loading Step

The nucleophilic active site Ser of TE domains attacks the carbonyl group of the thioester of the peptidyl-S-PCP to form peptidyl-*O*-TE (Figure 3.4). This serine residue is activated as a nucleophile by hydrogen bonding with nearby His and Asp residues. In this catalytic triad, His acts as a general base deprotonating Ser as it reacts with the thioester moiety, thereby increasing the nucleophilicity of its hydroxy group. Reaction of this His–Asp dyad-activated hydroxy group with the thioester yields an anionic peptide-*O*-TE tetrahedral adduct that is stabilized by an oxyanion hole, involving two backbone amide N–H bonds. This tetrahedral intermediate then collapses with release of the thiol-containing *holo*-PCP.

3.4.2 Releasing Step

The second step of the TE-catalyzed reaction is the release of the TE-bound peptide (Figure 3.4). The peptide-*O*-TE adduct is cleaved by attack of a nucleophile on the ester carbonyl, again forming a tetrahedral intermediate. Collapse of the tetrahedral intermediate releases the product. When the nucleophile is a water molecule, linear peptides are released, as observed in the biosynthesis of vancomycin and the β-lactam antibiotics.[60,61] TE domains also recruit intramolecular nucleophiles such as the side chain amine, hydroxy and thiol groups of amino acids and other monomers to release cyclic products. In the absence of a TE domain, the peptide thioester is recalcitrant toward decomposition in aqueous solutions. This stability of the peptide thioester bond demonstrates that the TE domain is essential to catalyze the hydrolysis of this bond.[62,63] Mechanistic insight into the TE-catalyzed macrocyclization, which competes with hydrolysis, remains poorly understood. Considering the

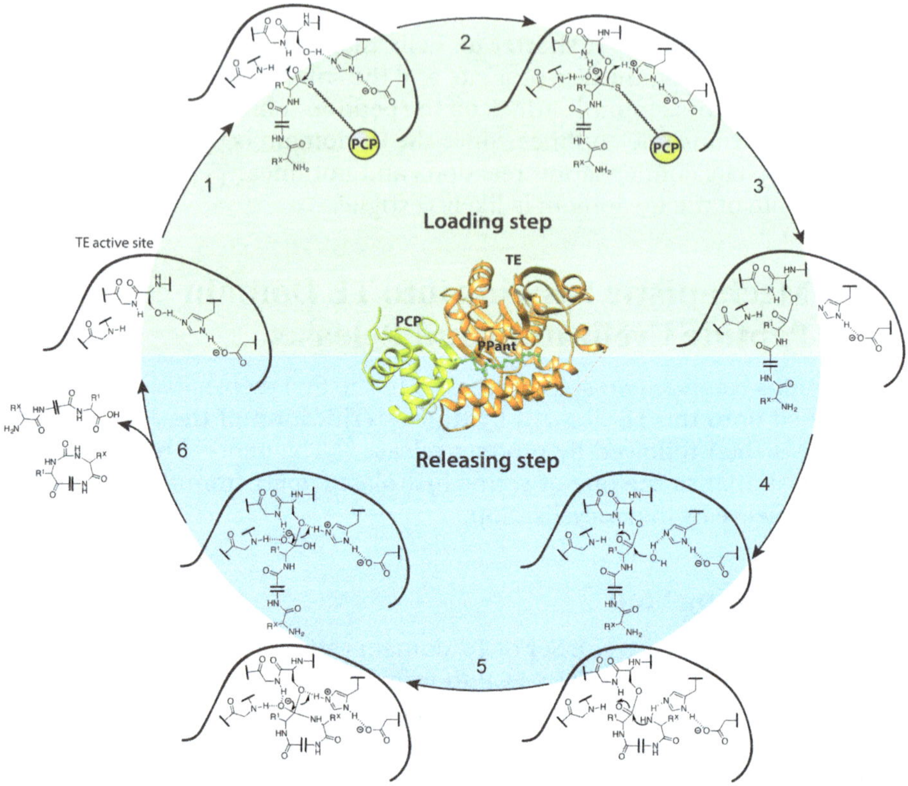

Figure 3.4 Mechanism of TE-mediated reaction for peptide loading and release. [1] A mature peptide appended on the PPant of PCP reaches the active site of the TE domain. [2] A Ser residue activated by the His–Asp dyad attacks the thioester bond of the PPant of the PCP. [3] The anionic tetrahedral intermediate is stabilized by an oxyanion hole (red). Then the thioether bond is broken and the PPant is removed from the TE domain. [4] The ester bond is either attacked by a water molecule or by an intramolecular nucleophilic residue. [5] An oxyanion hole (red) stabilizes the tetrahedral intermediate again. [6] The peptide is released from the TE domain and the active site of the TE domain is reactivated.

substrate preference, amino acid identity, and stereoselectivity of each cyclization, a general mechanism for this complex reaction may not exist, and instead distinct TEs likely operate through related, but distinct mechanisms.

3.5 The Application of TE-I Domains for Synthesis of Cyclic Peptide Analogues

Because truncated TE-I domains remain active *in vitro*, the use of these domains in combinatorial synthesis of non-natural macrocyclic peptides has emerged. Many studies have been carried out to characterize the function and structure of TE domains, as well as to develop chemoenzymatic approaches to produce NRP analogues.

3.5.1 Excised TE-I Domains

The TycC TE domain was the first excised TE-I domain characterized experimentally. The Walsh and Marahiel groups investigated the activities of the excised TycC TE domain towards the macrocyclization of precursor peptides.[57] They demonstrated that the excised TycC TE domain efficiently catalyzes the macrocyclization of linear tyrocidine-SNAC substrates and accelerates hydrolysis and cyclization by factors of 1×10^6 and 2×10^7, respectively. In addition, the TycC TE domain showed a high tolerance regarding the ring size of the final macrocycle, catalyzing the formation of 6- to 14-membered rings with a similar kinetic efficiency.[62] The TycC TE domain is also quite permissive in terms of the nature of substrate specificity, and a variety of synthetic peptidyl-SNAC substrates could be processed by the enzyme (Figure 3.5a). For each substrate, each amino acid in the tyrocidine decapeptide sequence was replaced with Ala, which led to the identification of the key 'recognition residues' of the peptide substrates. For the TycC TE domain, the D-Phe1 and Orn9 positions of the peptide substrate are essential for macrocyclization. If D-Phe1 is replaced with L-Phe or D-Ala, no cyclization occurs; instead the peptidyl-SNAC substrate is hydrolyzed by the enzyme, yielding linear oligopeptides as reaction products. Interestingly, hydrolysis occurs with kinetic parameters similar to the native tyrocidine substrate. Therefore, the replacement of D-Phe1 with its enantiomer or a smaller aliphatic residue affects not the loading of the substrate but the release of the product. Both Walsh *et al.* and Marahiel *et al.* also showed that the TycC TE domain could cyclize peptide substrates with different amino acid lengths, as well as the gramicidin S precursor. Further investigations on the TycC TE domain using tyrocidine peptidomimetics in which amino acids are substituted or amide bonds are replaced with PEG moieties or ester bonds confirmed that some backbone functionalities are important for macrocyclization.[63] Moreover, macrocyclization catalyzed by the TycC TE domain can be enhanced by addition of a non-ionic detergent, resulting in 150- to 300-fold increases in the yields of cyclic products.[64] Other excised TE-Is, from the gramicidin S synthetase (GrsB) and SrfAC, have also been expressed and their catalytic activities for macrocyclization of each substrate peptide probed.[62,65,66]

The key substrate residues for the macrocyclization catalyzed by the LicC TE domain of lichenysin lipopeptide synthetase were identified using peptidyl-SNAC analogues.[67] These synthetic analogues with different configurations of the β-hydroxy fatty acid and an amine group instead of the hydroxy group of lichenysin showed the importance of the stereochemistry of the β-hydroxy fatty acid and the presence of the hydroxy group for the enzymatic macrocyclization. Furthermore, an Ala scan of the substrate peptides led to the identification of Asp5 and Ile7 as the key 'substrate residues' recognized by the LicC TE domain.

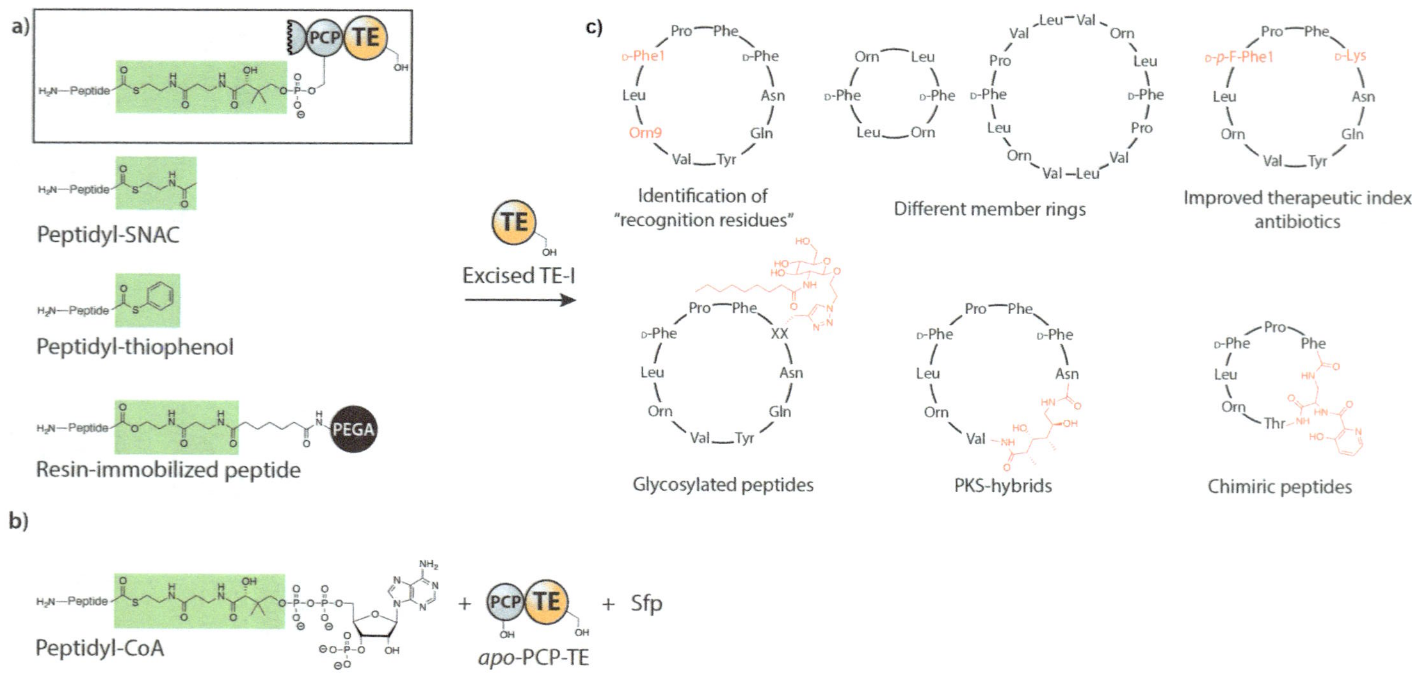

Figure 3.5 Chemoenzymatic strategies using excised TE-I for producing NRP analogues. (a) Peptide thio- or oxoesters inspired from PPant functionality as substrates of the excised TE-I. (b) An approach with the *apo*-PCP–TE didomain and peptidyl-CoA utilizing post-translational modification by PPTase. (c) The representative tyrocidine analogues synthesized by the chemoenzymatic approach.

Almost all excised TE-Is can react with the peptidyl-SNAC, apart from the TE-Is from the fengycin synthetase (Fen) and the mycosubtilin synthetase (Myc).[68] To study the reaction of these excised TE-Is, peptidyl-CoA substrates, which can be loaded on the *apo*-PCP by PPTase, were used instead of peptidyl-SNAC (Figure 3.5b). Subsequently, the peptidyl-thiophenol was developed to overcome the limitations of peptidyl-SNAC as a substrate mimic (Figure 3.5a).[69] The improved leaving group ability of the thiophenol enables the peptidyl-thiophenol substrate to react with the TE-I of both Fen and Myc. With these new substrate mimics in hand, the excised TE-I domains present an attractive means of generating cyclic NRP derivatives and macrocyclic peptide libraries by cyclizing oligopeptides prepared using solid-phase peptide synthesis (SPPS).

3.5.2 Chemoenzymatic Approaches to Generate Natural Product Analogues

The macrocyclization of linear peptides in organic synthesis is often difficult because intermolecular oligomerizations often occur in competition with intramolecular macrocyclization. Traditionally, macrocyclization of peptides is carried out at low substrate concentrations (10^{-3}–10^{-4} M) to avoid this intermolecular reaction. In these cases, large volumes of organic solvents can be required. Several chemical methods have been developed to facilitate the synthesis of cyclic peptides.[70,71] The first example of a chemoenzymatic strategy for the preparation of macrocyclic peptides was reported by Walsh.[72] Here a library of tyrocidine derivatives was prepared using a combination of SPPS and the excised TycC TE domain (Figure 3.5a). A collection of more than 300 linear decapeptide analogues of tyrocidine immobilized on polyethylene glycol amide (PEGA) resin through an ester linkage was prepared by automated SPPS in a 96-well format. These immobilized peptides were then cyclized using the TycC TE domain. Assays of cyclization showed that TycC only reacts with substrates that feature D-Phe – not its enantiomer – at position 4. This chemoenzymatic synthesis of a library of macropeptides identified two new bioactive compounds featuring D-*p*-F-Phe and basic amino acids D-Arg or D-Lys at the 1 and 4 positions of the peptide, respectively (Figure 3.5c). These compounds demonstrated broad-spectrum activities against Gram-positive and Gram-negative bacteria, as well as reducing the haemolytic activity against human erythrocytes.

These chemoenzymatic strategies using the TycC TE domain led to the synthesis of a number of glycosylated, (*E*)-alkene isosteric, PKS-hybrid and chimeric tyrocidine peptide analogues, and remarkably novel integrin binding peptides (Figure 3.5c).[73–78] To overcome streptogramin resistance, a series of chimeric streptogramin-tyrocidine peptides were prepared by SPPS and cyclized using the TycC TE domain.[77] The chimeric peptides showed broad-spectrum antibacterial activities including against streptogramin resistant species. Moreover, the modes of action of both the parent streptogramin and tyrocidine differ. Therefore, this chimeric macrocyclic peptide can be defined as a new class of antibiotics. Another chemoenzymatic approach to obtain streptogramin B analogues was conducted by Marahiel.[79] The problem of epimerization at the C-terminal L-Phg7 is recurrent in the total chemical synthesis of streptogramin B. To achieve the synthesis of enantiopure streptogramin B analogues, the TE-I domain of the pristinamycin synthetase SnbDE was employed. SnbDE TE-I also showed remarkable substrate promiscuity, tolerating modification at all seven positions of the substrate, permitting the generation of new streptogramin B analogues. Of particular note, SnbDE TE-I was shown to cyclize the only the L-isomer for Phg at position 7; but the linear substrate racemizes in aqueous reaction conditions. The solution was found in the cyclization of L-Phg7-PLP-SNAC, which shifts the equilibrium of epimerization to provide a dynamic kinetic resolution.

Daptomycin is an antibiotic used in the treatment of infections caused by Gram-positive bacteria and belongs to a family of acidic lipopeptides that includes calcium-dependent antibiotics (CDAs).[80] These branched depsipeptides possesses a 5-amino acid macrolactone ring. Based upon this structure similarity, the CDA TE-I domain showed surprising substrate tolerance to the substitution of multiple residues in the peptide backbone, and multiple daptomycin analogues have been prepared using the excised TE-I from CDA synthetase.[81] The same group carried out the synthesis of analogues of the acidic lipopeptide A54145.[82] The truncated PCP–TE didomains from daptomycin (Dap), CDA and A54145 synthetases were also expressed. These were evaluated for their substrate specificities and the ratios of the rates of cyclization to hydrolysis. The A54145 PCP–TE could generate 9 to 11-membered lactone rings, showing high cyclizing efficiency for the A54145 analogue with a Val at position 13, with less than 10% of the hydrolysis product formed. The activity of synthetic daptomycin analogues against *B. subtilis* revealed that the residues Asp7 and Asp9 are essential to maintain the antibiotic activity. The replacement of both Asp residues in A54145 also led to the complete loss of antibacterial activity.[81,82]

On-resin enzymatic peptide macrocyclization has proven to be a convenient method to generate macrocyclic peptide libraries. The enzymatic accessibility of on-resin peptides also has been investigated by evaluating the activities of hydrolases and proteases towards immobilized peptide substrates.[83,84] On-resin macrocyclization strategies have been successfully used to synthesize new macrocyclic compounds with TycC, GrsB and cryptophycin synthetase TE-I.[72,76,85,86] In all cases, the yields of 'non-natural' cyclic peptides were lower than the yields of the cyclized native product. However, sufficient materials for evaluation of biological activities were prepared. Nonetheless, improvement of the efficiency and substrate promiscuity of excised TE-I domains would facilitate the preparation of highly diverse cyclic peptide libraries.

3.6 Insight into the Interaction Between the TE-I and PCP Domains

PCP domains play a central role in modular synthetases by shuttling the growing peptidyl substrates/intermediates between catalytic domains in type I NRPSs. Understanding the interactions between PCP and TE domains, as well as other partner domains, is a prerequisite to engineering novel and efficient biosynthetic pathways for cyclic peptides.

3.6.1 Interaction with the *apo*-PCP Domain

Frueh *et al.* reported the structure of the PCP–TE didomain of the enterobactin synthetase EntF (C–A–PCP–TE) subunit and the interactions between the two partner proteins using protein NMR (Figure 3.6a).[87] To obtain the homologous *apo*-protein, the active site Ser48 of the PCP domain was mutated to Ala. This point mutation had no effect on the chemical shift compared to the native *apo*-PCP–TE. The overall structure of the didomain exhibited a well-defined relative orientation, including a hydrophobic domain interface. Here the triple-helix bundle of the PCP domain is wedged between the globular core and the lid (α4TE-α5TE) region of the TE domain. The two helices of the lid region in the TE domain appear to cover both the PCP and TE active sites. A distance of 17 Å between Ser48Ala and the active site Ser180 of the TE domain was estimated, which enables the 20 Å PPant prosthetic arm to access the TE active site. The surface of interaction of the PCP domain was identified as involving residues found in the

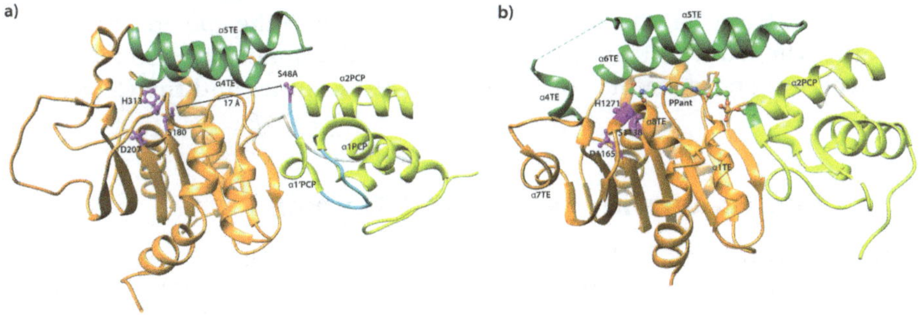

Figure 3.6 Structures of the PCP–TE didomain of the enterobactin synthetase EntF from *Escherichia coli* in the *apo* and *holo* forms where the PCP, TE and lid region of the TE domain are colored in yellow, orange and forest green, respectively. (a) Structure of the *apo*-PCP–TE didomain. S48A and the catalytic triad residues of the TE domain are displayed as ball and stick models (magenta). L3PCP is depicted in cyan. (b) Structure of the *holo*-PCP–TE didomain. The α-hydroxyamide PPant (green) and the catalytic triad residues of the TE domain (magenta) are displayed as ball and stick models.

beginning of α2PCP, including S48A toward the end of L3PCP and the single turn helix α1′PCP. This unique structure of α1′PCP has been observed in related FAS and PKS acyl carrier proteins, but is absent in other excised PCP domains.[88–90] This absence suggests that the motif may be responsible for recognition of TE-I domains. In fact, point mutations of either Phe41 or Phe42 in the α1′PCP resulted in the disruption of the interactions between PCP–TE and the unfolding of the PCP domain. The PCP of the PCP–TE didomain forms a conformation resembling the A/H state,[90] although the conformation was expected to be similar to the *apo* state. This indicates that the TE domain appears to stabilize the PCP in the A/H state, in which more hydrophobic amino acid residues protrude to form the PCP–TE interface. Internal mobility of the PCP domain was observed, which leads to the low intensity of the NMR signals for L3PCP and fast NH exchange. Reduced intensities of the NMR signals were also shown in many residues of the lid region of the TE domain, which indicates the slow modulation of this lid region.

In this study, the interactions and dynamics between PCP–TE and other partner enzymes were also reported. The titration of the promiscuous PPTase Sfp against a solution of PCP–TE led to a change in the conformation to the open form of the PCP–TE didomain, in which PCP:TE interactions are either partially or completely disrupted. This result indicates that Sfp is able to access the PCP's conserved Ser residue by inducing dissociation of the PCP–TE dimer. Also investigated was the interaction between the EntF C domain and the PCP–TE didomain. As the interaction sites of PCP with both the C and TE domains do not overlap, major chemical shift changes by addition of the C domain to PCP–TE were observed at the C–PCP binding interface, whereas minor changes in chemical shift were observed on the TE domain. However, this study was performed with the *apo*-enzymes. It is clear that the loaded PPant arm of the *holo*-PCP domain will also influence the structure and dynamics of both the PCP–TE dimer and the C–PCP–TE complex.

3.6.2 Interaction with the *holo*-PCP Domain

In the biosynthesis of NRPs, all PCP domains undergo PPTase-catalyzed post-translational modification to form the active form of PCP (*holo*-PCP), which has a PPant prosthetic arm on a conserved Ser residue.[37] Therefore, insights into the structural basis of the *holo*-PCP–TE domain provide a better understanding of the interactions, mechanism and conformation of these domains during NRP assembly. The carrier protein is often disordered in crystal structures due to its dynamic nature. To obtain the crystal structure of the *holo*-PCP–TE domain, Liu *et al.* applied a synthetic CoA analogue with an α-chloro-amide moiety, which would form a covalent bond with the active site Ser or His of the TE domain (Figure 3.6b).[91] Utilizing the α-chloro-amide-CoA cross-linker, they were able to crystallize the *crypto*-PCP–TE domain of the enterobactin synthase EntF and solve the structure with 1.9 Å resolution. The two proteins were observed in the asymmetric unit with a slight difference in the lid region, where α5TE of monomer B is disordered. Although the disorder in the lid region has been observed in all crystallized TE domains in NRPSs without the substrate, in monomer A only 4 residues (residues 1176–1180) could not be resolved. The phosphopantetheine arm attached to Ser1006 of PCP lies in a channel leading toward the catalytic triad of the TE domain. The cofactor forms a number of interactions with the TE domain. In addition to the interaction of the phosphate group of the PPant arm and a water molecule, Phe1077 and Trp1079 of the α1TE, as well as Ile1213 of the lid α6TE form the hydrophobic pocket around the gem-dimethyl moiety of the PPant arm. Mutation of Trp1079 abolished enterobactin production. The two amide groups of the cofactor form hydrogen bonds with Ala1074 and Ser1075 and are in van der Waals contact with Gln1080 of the TE domain. Although the α-chloro-amide warhead of the cross-linker was expected to form a covalent bond with the catalytic triad of the TE domain, the mass spectral analysis of the *crypto*-PCP–TE showed that the structure was 18 Da heavier than expected for a covalently modified *crypto*-PCP–TE complex. This mass increase suggests that the α-chloro-amide moiety is hydrolyzed either after forming the covalent linkage with the catalytic triad or during the time needed for crystallization.

Compared to the NMR structure of the *apo*-PCP–TE EntF described in the previous paragraph, the TE domain of the *holo*-PCP–TE reveals much more about the structure of the didomain. Indeed, the helix α4TE is formed in the lid region and the helices α7TE and α8TE in the core region. The structure of the *holo*-PCP–TE also shows that the α1 and α2PCP helices move toward the TE domain, while α5TE approaches the PCP domain. In addition, the interactions between the lid region of the TE and the PCP domains can be more easily distinguished in the presence of the cross-linker than in the *apo*-protein. To summarize, cross-linkers can facilitate the resolution of these domains in their catalytically active states.

3.7 Summary and Outlook

Natural macrocyclic NRPs and their derivatives display a broad range of biological activity and are promising medicinal resources. Yet they are underused drugs compared to conventional synthetic small molecules, in part due to the complexity of their preparation.[92] A better understanding of the mechanism, dynamics, conformation and interactions of NRPS modules and each domain will facilitate the application of NRPS enzymes to produce bioactive natural product analogues.[93] In the synthesis of new macrocyclic NRPs, the macrocyclization of linear peptides will remain a challenging step. Although it has been demonstrated that TE domains have great potential for producing natural product mimetic macrocyclic peptides, we cannot yet engineer TE domains for specialized cyclization events. The restricted substrate specificities and low catalytic efficiencies of these enzymes discourage organic chemists from utilizing a chemoenzymatic approach for cyclic peptide synthesis. Thus far, studies on TE-mediated macrocyclization have focused primarily on substrate recognition from the perspective of the substrate. However there remains a lack of information regarding the residues or regions of the TE domain that are essential for the macrocyclization to occur.[40,41] Therefore, the identification of the specific amino acid residues in the TE macrocyclization region, as well as the interactions between the substrate peptide and the TE domain, may enable future engineering and design of versatile TE domains for the development of novel macrocyclic NRP analogues.

Acknowledgements

This work was supported by the National Institute of Health R01 GM095970. S. K. is supported by an Uehara Research Fellowship from the Uehara Memorial Foundation, and L. M. is supported by an Early Postdoc. Mobility Fellowship from the Swiss National Science Foundation. We thank Dr Ashay Patel for proof reading.

References

1. D. Schwarzer, R. Finking and M. A. Marahiel, *Nat. Prod. Rep.*, 2003, **20**, 275.
2. J. Rizo and L. M. Gierasch, *Annu. Rev. Biochem.*, 1992, **61**, 387.
3. E. M. Driggers, S. P. Hale, J. Lee and N. K. Terrett, *Nat. Rev. Drug Discovery*, 2008, 7, 608.
4. J. A. Robinson, *ChemBioChem*, 2009, **10**, 971.
5. D. Reichmann, O. Rahat, M. Cohen, H. Neuvirth and G. Schreiber, *Curr. Opin. Struct. Biol.*, 2007, **17**, 67.
6. H. von Döhren, U. Keller, J. Vater and R. Zocher, *Chem. Rev.*, 1997, **97**, 2675.
7. M. A. Marahiel, T. Stachelhaus and H. D. Mootz, *Chem. Rev.*, 1997, **97**, 2651.
8. H. D. Mootz, D. Schwarzer and M. A. Marahiel, *ChemBioChem*, 2002, **3**, 490.
9. R. Finking and M. A. Marahiel, *Annu. Rev. Microbiol.*, 2004, **58**, 453.
10. G. H. Hur, C. R. Vickery and M. D. Burkart, *Nat. Prod. Rep.*, 2012, **29**, 1074.
11. S. A. Sieber and M. A. Marahiel, *J. Bacteriol.*, 2003, **185**, 7036.
12. S. C. Tsai, H. Lu, D. E. Cane, C. Khosla and R. M. Stroud, *Biochemistry*, 2002, **41**, 12598.
13. D. L. Akey, J. D. Kittendorf, J. W. Giraldes, R. A. Fecik, D. H. Sherman and J. L. Smith, *Nat. Chem. Biol.*, 2006, **2**, 537.
14. J. W. Giraldes, D. L. Akey, J. D. Kittendorf, D. H. Sherman, J. L. Smith and R. A. Fecik, *Nat. Chem. Biol.*, 2006, **2**, 531.

15. J. B. Scaglione, D. L. Akey, R. Sullivan, J. D. Kittendorf, C. M. Rath, E. S. Kim, J. L. Smith and D. H. Sherman, *Angew. Chem., Int. Ed.*, 2010, **49**, 5726.
16. L. Gao, H. Liu, Z. Ma, J. Han, Z. Lu, C. Dai, F. Lv and X. Bie, *Sci. Rep.*, 2016, **6**, 38467.
17. S. A. Survase, L. D. Kagliwal, U. S. Annapure and R. S. Singhal, *Biotechnol. Adv.*, 2011, **29**, 418.
18. S. Matsuda and S. Koyasu, *Immunopharmacology*, 2000, **47**, 119.
19. M. Kuo and W. A. Gibbons, *Biophys. J.*, 1980, **32**, 807.
20. E. J. Prenner, R. N. A. H. Lewis and R. N. McElhaney, *Biochim. Biophys. Acta*, 1999, **1462**, 201.
21. G. A. Brewer, in *Analytical Profiles of Drug Substances*, ed. K. Florey, Academic Press, New York, 1980, vol. 9, p. 1.
22. K. J. Stone and J. L. Strominger, *Proc. Natl. Acad. Sci. U. S. A.*, 1971, **68**, 3223.
23. A. P. Zavascki, L. Z. Goldani, J. Li and R. L. Nation, *J. Antimicrob. Chemother.*, 2007, **60**, 1206.
24. J. Kitagaki, G. Shi, S. Miyauchi, S. Murakami and Y. Yang, *Anticancer Drugs*, 2015, **26**, 259.
25. N. S. Shaligram and R. S. Singhal, *Food Technol. Biotechnol.*, 2010, **48**, 119.
26. M. Ongena and P. Jacques, *Trends Microbiol.*, 2008, **16**, 115.
27. C. J. Carrano and K. N. Raymond, *J. Am. Chem. Soc.*, 1979, **101**, 5401.
28. K. N. Raymond, E. A. Dertz and S. S. Kim, *Proc. Natl. Acad. Sci. U. S. A.*, 2003, **100**, 3584.
29. R. Süssmuth, J. Müller, H. von Döhren and I. Molnár, *Nat. Prod. Rep.*, 2011, **28**, 99.
30. F. Romero, F. Espliego, J. Pérez Baz, T. García de Quesada, D. Grávalos, F. de la Calle and J. L. Fernández-Puentes, *J. Antibiot. (Tokyo)*, 1997, **50**, 734.
31. D. Bergamaschi, M. Faretta, S. Ronzoni, S. Taverna, P. De Feudis, M. Bonfanti, G. Guidi, M. Faircloth, J. Jimeno, M. D'Incalci and E. Erba, *Eur. J. Histochem.*, 1997, **41**(suppl. 2), 63.
32. O. E. Zolova, A. S. A. Mady and S. Garneau-Tsodikova, *Biopolymers*, 2010, **93**, 777.
33. T. Golakoti, W. Y. Yoshida, S. Chaganty and R. E. Moore, *J. Nat. Prod.*, 2001, **64**, 54.
34. M. A. Wyatt, W. Wang, C. M. Roux, F. C. Beasley, D. E. Heinrichs, P. M. Dunman and N. A. Magarvey, *Science*, 2010, **329**, 294.
35. M. Zimmermann and M. A. Fischbach, *Chem. Biol.*, 2010, **17**, 925.
36. J. E. Becker, R. E. Moore and B. S. Moore, *Gene*, 2004, **325**, 35.
37. J. Beld, E. C. Sonnenschein, C. R. Vickery, J. P. Noel and M. D. Burkart, *Nat. Prod. Rep.*, 2014, **31**, 61.
38. M. Nardini and B. W. Dijkstra, *Curr. Opin. Struct. Biol.*, 1999, **9**, 732.
39. U. Linne, D. Schwarzer, G. N. Schroeder and M. A. Marahiel, *Eur. J. Biochem.*, 2004, **271**, 1536.
40. S. D. Bruner, T. Weber, R. M. Kohli, D. Schwarzer, M. A. Marahiel, C. T. Walsh and M. T. Stubbs, *Structure*, 2002, **10**, 301.

41. S. A. Samel, B. Wagner, M. A. Marahiel and L. O. Essen, *J. Mol. Biol.*, 2006, **359**, 876.
42. M. Kotowska and K. Pawlik, *Appl. Microbiol. Biotechnol.*, 2014, **98**, 7735.
43. D. Schwarzer, H. D. Mootz, U. Linne and M. A. Marahiel, *Proc. Natl. Acad. Sci. U. S. A.*, 2002, **99**, 14083.
44. E. Yeh, R. M. Kohli, S. D. Bruner and C. T. Walsh, *ChemBioChem*, 2004, **5**, 1290.
45. D. S. Vallari, S. Jackowski and C. O. Rock, *J. Biol. Chem.*, 1987, **262**, 2468.
46. A. Schneider and M. A. Marahiel, *Arch. Microbiol.*, 1998, **169**, 404.
47. J. Li, J. Jaitzig, L. Theuer, O. E. Legala, R. D. Süssmuth and P. Neubauer, *J. Biotechnol.*, 2015, **193**, 16.
48. A. O. Zabala, Y. H. Chooi, M. S. Choi, H. C. Lin and Y. Tang, *ACS Chem. Biol.*, 2014, **9**, 1576.
49. A. Koglin, F. Löhr, F. Bernhard, V. V. Rogov, D. P. Frueh, E. R. Strieter, M. R. Mofid, P. Güntert, G. Wagner, C. T. Walsh, M. A. Marahiel and V. Dötsch, *Nature*, 2008, **454**, 907.
50. F. Kopp and M. A. Marahiel, *Nat. Prod. Rep.*, 2007, **24**, 735.
51. L. Du and L. Lou, *Nat. Prod. Rep.*, 2010, **27**, 255.
52. H. von Döhren, *Fungal Genet. Biol.*, 2009, **46**, S45.
53. F. Kopp, C. Mahlert, J. Grünewald and M. A. Marahiel, *J. Am. Chem. Soc.*, 2006, **128**, 16478.
54. M. A. Wyatt, M. C. Y. Mok, M. Junop and N. A. Magarvey, *ChemBioChem*, 2012, **13**, 2408.
55. K. Eisfeld, in *The Mycota XV. Physiology and Genetics: Selected Basic and Applied Aspects*, ed. T. Anke and D. Weber, Springer-Verlag, Berlin, 2009, p. 305.
56. X. Gao, S. W. Haynes, B. D. Ames, P. Wang, L. P. Vien, C. T. Walsh and Y. Tang, *Nat. Chem. Biol.*, 2012, **8**, 823.
57. J. W. Trauger, R. M. Kohli, H. D. Mootz, M. A. Marahiel and C. T. Walsh, *Nature*, 2000, **407**, 215.
58. J. Zhang, N. Liu, R. A. Cacho, Z. Gong, Z. Liu, W. Qin, C. Tang, Y. Tang and J. Zhou, *Nat. Chem. Biol.*, 2016, **12**, 1001.
59. M. E. Horsman, T. P. A. Hari and C. N. Boddy, *Nat. Prod. Rep.*, 2016, **33**, 183.
60. B. K. Hubbard and C. T. Walsh, *Angew. Chem., Int. Ed.*, 2003, **42**, 730.
61. M. F. Byford, J. E. Baldwin, C. Y. Shiau and C. J. Schofield, *Chem. Rev.*, 1997, **97**, 2631.
62. R. M. Kohli, J. W. Trauger, D. Shwarzer, M. A. Marahiel and C. T. Walsh, *Biochemistry*, 2001, **40**, 7099.
63. J. W. Trauger, R. M. Kohli and C. T. Walsh, *Biochemistry*, 2001, **40**, 7092.
64. E. Yeh, H. Lin, S. L. Clugston, R. M. Kohli and C. T. Walsh, *Chem. Biol.*, 2004, **11**, 1573.
65. C. C. Tseng, S. D. Bruner, R. M. Kohli, M. A. Marahiel, C. T. Walsh and S. A. Sieber, *Biochemistry*, 2002, **41**, 13350.
66. K. M. Hoyer, C. Mahlert and M. A. Marahiel, *Chem. Biol.*, 2007, **14**, 13.

67. S. Cao, Y. Yang, N. L. Ng and Z. Guo, *Bioorg. Med. Chem. Lett.*, 2005, **15**, 2595.
68. S. A. Sieber, C. T. Walsh and M. A. Marahiel, *J. Am. Chem. Soc.*, 2003, **125**, 10862.
69. S. A. Sieber, J. Tao, C. T. Walsh and M. A. Marahiel, *Angew. Chem., Int. Ed.*, 2004, **43**, 493.
70. J. S. Davies, *J. Pept. Sci.*, 2003, **9**, 471.
71. C. J. White and A. K. Yudin, *Nat. Chem.*, 2011, **3**, 509.
72. R. M. Kohli, C. T. Walsh and M. D. Burkart, *Nature*, 2002, **418**, 658.
73. H. Lin and C. T. Walsh, *J. Am. Chem. Soc.*, 2004, **126**, 13998.
74. H. Lin, D. A. Thayer, C. H. Wong and C. T. Walsh, *Chem. Biol.*, 2004, **11**, 1635.
75. D. Garbe, S. A. Sieber, N. G. Bandur, U. Koert and M. A. Marahiel, *ChemBioChem*, 2004, **5**, 1000.
76. R. M. Kohli, M. D. Burke, J. Tao and C. T. Walsh, *J. Am. Chem. Soc.*, 2003, **125**, 7160.
77. T. A. Mukhtar, K. P. Koteva and G. D. Wright, *Chem. Biol.*, 2005, **12**, 229.
78. R. M. Kohli, J. Takagi and C. T. Walsh, *Proc. Natl. Acad. Sci. U. S. A.*, 2002, **99**, 1247.
79. C. Mahlert, S. A. Sieber, J. Grünewald and M. A. Marahiel, *J. Am. Chem. Soc.*, 2005, **127**, 9571.
80. M. Strieker and M. A. Marahiel, *ChemBioChem*, 2009, **10**, 607.
81. J. Grünewald, S. A. Sieber, C. Mahlert, U. Linne and M. A. Marahiel, *J. Am. Chem. Soc.*, 2004, **126**, 17025.
82. F. Kopp, J. Grünewald, C. Mahlert and M. A. Marahiel, *Biochemistry*, 2006, **45**, 10474.
83. J. Vágner, G. Barany, K. S. Lam, V. Krchňák, N. F. Sepetov, J. A. Ostrem, P. Strop and M. Lebl, *Proc. Natl. Acad. Sci. U. S. A.*, 1996, **93**, 8194.
84. J. Kress, R. Zanaletti, A. Amour, M. Ladlow, J. G. Frey and M. Bradley, *Chem.–Eur. J.*, 2002, **8**, 3769.
85. X. Wu, X. Bu, K. M. Wong, W. Yan and Z. Guo, *Org. Lett.*, 2003, **5**, 1749.
86. W. Seufert, Z. Q. Beck and D. H. Sherman, *Angew. Chem., Int. Ed.*, 2007, **46**, 9298.
87. D. P. Frueh, H. Arthanari, A. Koglin, D. A. Vosburg, A. E. Bennett, C. T. Walsh and G. Wagner, *Nature*, 2008, **454**, 903.
88. G. A. Zornetzer, B. G. Fox and J. L. Markley, *Biochemistry*, 2006, **45**, 5217.
89. S. C. Findlow, C. Winsor, T. J. Simpson, J. Crosby and M. P. Crump, *Biochemistry*, 2003, **42**, 8423.
90. A. Koglin, M. R. Mofid, F. Löhr, B. Schäfer, V. V. Rogov, M. M. Blum, T. Mittag, M. A. Marahiel, F. Bernhard and V. Dötsch, *Science*, 2006, **312**, 273.
91. Y. Liu, T. Zheng and S. D. Bruner, *Chem. Biol.*, 2011, **18**, 1482.
92. N. K. Terrett, *Drug Discovery Today: Technol.*, 2010, **7**, e97.
93. M. Winn, J. K. Fyans, Y. Zhuo and J. Micklefield, *Nat. Prod. Rep.*, 2016, **33**, 317.

CHAPTER 4

The Biosynthetic Machinery and Its Potential to Deliver Unnatural Cyclic Peptides

RASHED S. AL TOMA, NATALIA A. JUNGMANN AND RODERICH D. SÜSSMUTH*

Institut für Chemie, Technische Universität Berlin, 10623 Berlin, Germany
*E-mail: suessmuth@chem.tu-berlin.de

4.1 Non-natural Cyclic RiPPs – Expanding the Structural Space and Activities

From a structural viewpoint, many ribosomally synthesized and post-translationally modified peptide (RiPP) structures exhibit either a linear, mono- or even multicyclic architecture of the peptide backbone. Cyclization as a structural feature, *i.e.* the presence of macrolactone or macrolactam rings, restricts certain conformations and thus is crucial in conferring bioactivity and other molecular properties that secure the peptide's function, *e.g.* higher stability to proteolytic degradation or facilitated membrane transport.

Among the various members of the RiPPs, many comprise ring systems and in fact they represent the vast majority, including the complex polycyclic lanthipeptides, the lasso peptides, and the branched cyclic thiopeptides, as well as bottromycins, cyanobactins, microviridins, sactipeptides, amatoxins, phallotoxins, cyclotides, orbitides, and glycocins, in addition to some

Chemical Biology No. 6
Cyclic Peptides: From Bioorganic Synthesis to Applications
Edited by Jesko Koehnke, James Naismith and Wilfred A. van der Donk

Published by the Royal Society of Chemistry, www.rsc.org

members of the conopeptides and autoinducing peptides.[1] The core biosynthesis of all these ribosomally synthesized peptides first produces a linear precursor peptide with an approximate size limit of 10 kDa, which then undergoes post-translational modification (PTM) guided to a specific part by a leader peptide. Generally, the ribosomal synthesis of these peptides commonly involves a set of building blocks of 20 canonical amino acids (cAAs), upon which the structural diversity can be massively increased by means of PTMs.

Recently, it has been realized that the introduction of additional amino acids into the above basic set, *i.e.* expansion of the biosynthetic repertoire by so-called non-canonical amino acids (ncAAs), potentially could further increase the structural diversity and give rise to additional novel chemical functionalities (Figures 4.1 and 4.2). Hence, with this appealing approach,

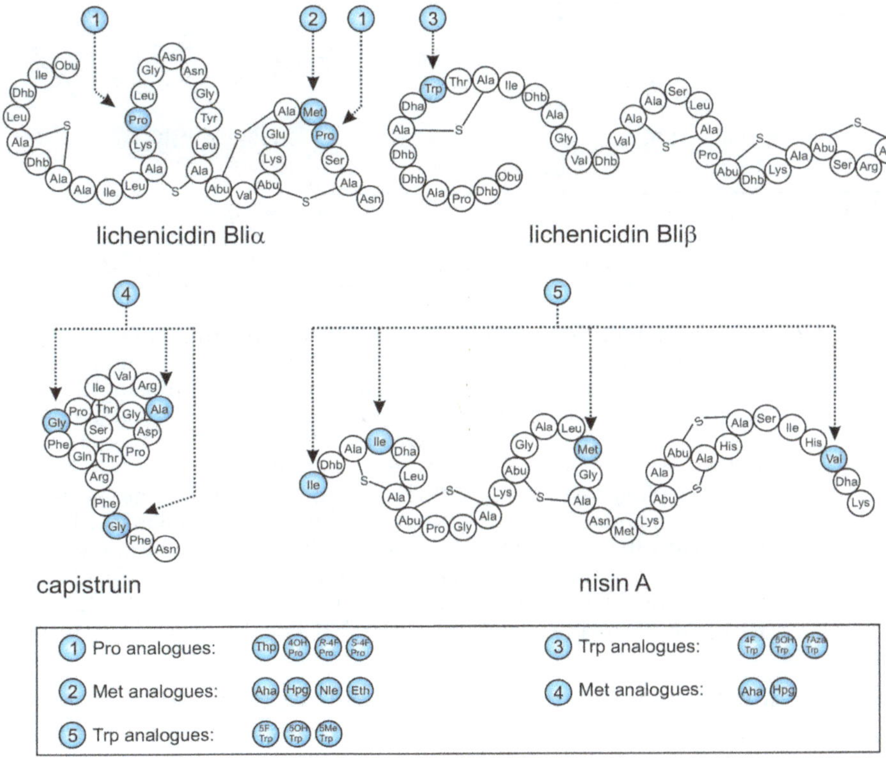

Figure 4.1 Schematic representation of cyclic RiPPs: lichenicidin, capistruin, and nisin A, of which the non-natural analogues have been generated by ncAA incorporation using the SPI approach. Thp: L-thioproline, S-4F-Pro: L-*cis*-4-fluoroproline, *R*-4F-Pro: L-*trans*-4-fluoroproline, 4OH-Pro: L-*trans*-4-hydroxyproline. Aha: L-azidohomoalanine, Hpg: L-homopropargylglycine, Nle: L-norleucine, Eth: L-ethionine. 5OH-Trp: L-5-hydroxytryptophan, 4F-Trp: L-4-fluorotryptophan, 5F-Trp: L-5-fluorotryptophan, 5Me-Trp: L-5-methyltryptophan, 7Aza-Trp: L-7-azatryptophan.

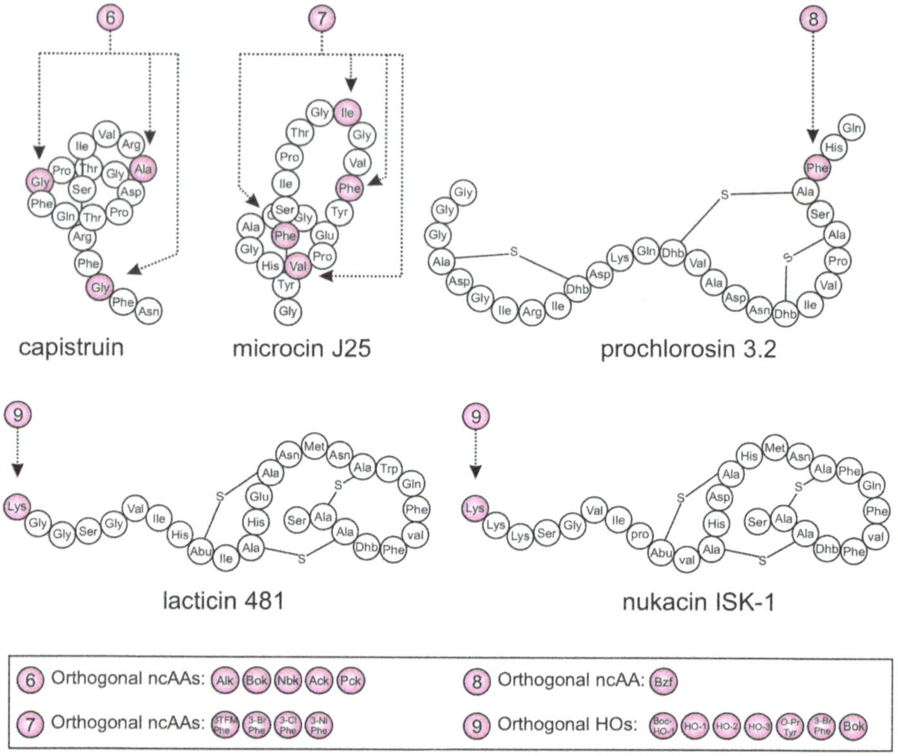

Figure 4.2 Schematic representation of cyclic RiPPs: capistruin, microcin J25, prochlorosin, lacticin 481, and nukacin ISK-1, of which the non-natural analogues have been generated by ncAA incorporation using the SCS approach. Nbk: *N*ε-5-norbornene-2-yloxycarbonyl-L-lysine, Alk: *N*ε-Alloc-L-lysine, Ack: *N*ε-2-azidoethyloxycarbonyl-L-lysine, Pck: *N*ε-2-propyn-1-yloxycarbonyl-L-lysine, Bok *N*ε-Boc-L-lysine, 3TFM-Phe: 3-trifluoromethyl-L-phenylalanine, 3-Br-Phe: 3-bromo-L-phenylalanine, 3-Cl-Phe: 3-chloro-L-phenylalanine, 3-Ni-Phe: 3-nitro-L-phenylalanine, Bzf: 4-benzoyl-L-phenylalanine, HO-1: (S)-6-amino-2-hydroxyhexanoic acid, Boc-HO-1: (S)-6-(Boc-amino)-2-hydroxyhexanoic acid, HO-2: (S)-3-(3-bromophenyl)-2-hydroxypropanoic acid, HO-3: (S)-2-hydroxy-3-(4-(prop-2-yn-1-yloxy)phenyl)propanoic acid, and *O*-Pr-Tyr (*O*-propargyl-L-tyrosine).

novel functional groups could allow for subsequent conjugation reactions of labels or other ligands conferring new properties and functions to the molecule.

As an alternative to the biosynthesis of RiPPs, Nature has evolved another peptide biosynthesis pathway, performed by non-ribosomal peptide synthetases (NRPSs), which deliver an even greater structural diversity. Remarkably, genes or operons in the respective gene clusters dedicated to the biosynthesis of non-proteinogenic amino acids provide additional building blocks, which could be of interest for incorporation in RiPP biosynthesis. Processing domains on NRPSs can introduce additional structural modifications, *e.g.* epimerization, and methylation. The recruitment of *in trans* acting enzymes and a post-NRPS tailoring event can lead to an enormous structural diversity. Attempts to exploit and manipulate the modularity of RiPP systems to alter and to design novel peptide structures have revealed these systems to be even more sophisticated than expected (further details below, Section 4.2).

While previous attempts to increase the structural diversity of *in vivo* produced RiPPs were mainly directed toward site-directed mutagenesis (SDM) of the peptide sequence, more recently,[2–6] more advanced technical methods are at one's disposal to perform the *in vivo* generation of non-natural RiPPs by the incorporation of non-canonical amino acids.[5–8] This *in vivo* incorporation of ncAAs into RiPPs is mainly achieved by two approaches: engineering the genetic code by supplementation-based incorporation, and expanding the genetic code by stop codon suppression.

4.1.1 The Supplementation-based Incorporation Approach

One main methodology used for genetic code engineering has been termed "selective pressure incorporation (SPI)", also known as "supplementation -based incorporation" or "*in vivo* residue-specific substitution".[7,9,10] Nature has evolved aminoacyl-tRNA synthetases (aaRSs) to distinguish between cAAs in order to assure accurate translation into proteins and peptides. However, aaRSs show a significant promiscuity towards isostructural ncAAs. Hence, genetic code engineering is mainly based on the property of substrate tolerance displayed by several aaRSs towards ncAAs (Figure 4.3). This principle has been examined worldwide by many scientists for various different ncAAs.[11–13] Interestingly, what we now call genetic code engineering was started before the genetic code itself was decrypted (1961–1964),[14–17] and even before Crick proposed his "central dogma of molecular biology" (1958).[18] In 1957, Cowie and Cohen[19] used a Met-auxotrophic *E. coli* strain to globally substitute Met with selenomethionine (SeMet). Ten years later, Schlesinger performed the first study on an isolated protein congener, where he substituted His with triazolealanine in alkaline phosphatase.[20] Rediscovery and amelioration of genetic code engineering by the SPI approach performed by Tirrell[21] and Budisa[12] in the mid-90s paved the way for extended applications in the field of protein engineering, X-ray crystallography and nuclear magnetic resonance (NMR) studies, *e.g.* the structural investigation of proteins by incorporation of SeMet and ^{19}F-labeled amino acids.[9]

For genetic code engineering, the use of auxotrophic cells is mandatory and this affords the inactivation of biosynthesis pathways of the corresponding cAA. Briefly, the auxotrophs are grown until the second half of the exponential phase. Then the dedicated cAA is depleted in the growth medium by starvation. Subsequently, the new replacing ncAA is added and the expression of the gene of interest is induced (in the case of an inducible system). The cAA depletion can be achieved by a media shift technique,[5,21] or by calibrating the cAA concentration in the used minimal medium to be totally

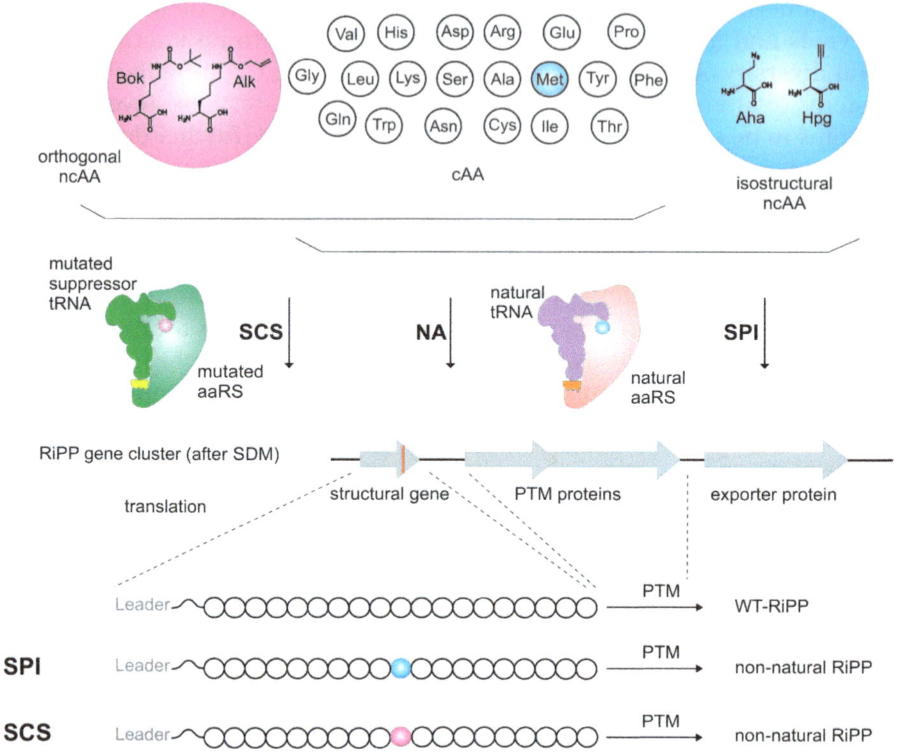

Figure 4.3 General scheme for *in vivo* non-natural RiPP production using the SPI and SCS approaches. NA: normal aminoacylation, SDM: site-directed mutagenesis, SPI: selective pressure incorporation, and SCS: stop codon suppression.

consumed by the mid-point of the exponential phase.[6,12] The media shift procedure is usually performed by growing the cells until the required growth point is reached, upon which the cells are harvested by centrifugation and resuspended in a minimal medium that lacks the cAA to be exchanged.[9] The cells are grown further in the minimal medium for a short starvation period in order to deplete the target cAA from the cells, and then the ncAA to be incorporated can be added.

A high incorporation efficiency achieved by this methodology is strongly dependent on the control of the (inducible) expression system, such that the expression of the target gene should only start when the cAA is depleted and only the ncAA is available.[22,23]

Unlike proteins, to-date there are only a few examples of small peptides from RiPP families containing non-natural amino acids. One early example is the class II two-component lantibiotic lichenicidin from the lanthipeptide family (Figure 4.1), containing lanthionine (Lan) as a characteristic structural motif. Lichenicidin is naturally produced by the Gram-positive bacterial strain *Bacillus licheniformis* I89 and consists of the two peptides Bliα and Bliβ. As a lantibiotic, lichenicidin has antimicrobial activities against *Listeria monocytogenes*, methicillin-resistant *Staphylococcus aureus*, and vancomycin-resistant *Enterococcus* strains. Bliα and Bliβ synergistically interact at nanomolar concentration against the targeted bacteria. Due to structural homologies with other two-component lantibiotics, lichenicidin is proposed to have a similar mode of action,[24] where Bliα binds to the cell wall precursor lipid II, thus inhibiting cell wall biosynthesis, followed by pore formation in the cytoplasmic membrane *via* aggregation of the β-subunit to the Bliα-lipid II complex. The gene cluster of lichenicidin is composed of 14 genes that regulate its biosynthesis. Upon gene expression of structural genes *licA1* and *licA2*, the resulting precursor peptides, Bliα and Bliβ, undergo post-translational modifications by lantibiotic synthetases LicM1 and LicM2. These modifying enzymes recognize the N-terminal leader peptide of LiA1 and LicA2 and install Ser/Thr dehydrations and Lan/MeLan bridges in the C-terminal core peptides. These PTMs are followed by export of the modified precursor peptides outside of the cell by the transporter enzyme LicT. LicT concomitantly has proteolytic activity targeting the double glycine motif, removing the leader peptide and releasing the mature Bliα and the premature Bliβ.[25] Premature Bliβ still requires additional processing by the protease LicP in the periplasmic space for removing an N-terminal hexapeptide and ultimately releasing the mature Bliβ. Other genes in the lichenicidin gene cluster encode self-immunity proteins (*licFGEHI*) or transcriptional regulators (*licXRY*). While previously it proved difficult to heterologously express lanthipeptide pathways in *E. coli* as a versatile host, nearly the complete *lic* gene cluster was successfully cloned into the fosmid pLic5, and heterologously expressed in the Gram-negative bacterium *E. coli*.[25]

The successful establishment of the heterologous expression system of the lichenicidin gene cluster in *E. coli* paved the way for the incorporation of ncAAs into the two component subunits by SPI, since from *E. coli* several auxotrophic strains are available. Thus, isostructural analogues of Pro (L-thioproline (Thp), L-*cis*-4-fluoroproline ((4S-F)-Pro), L-*trans*-4-fluoroproline ((4*R*-F)-Pro), and L-*trans*-4-hydroxyproline ((4*R*-OH)-Pro)), and Met (L-homopropargylglycine (Hpg), L-azidohomoalanine (Aha), L-norleucine (Nle) and L-ethionine (Eth)) (Figure 4.4) were incorporated into Bliα by using the pro-auxotrophic *E. coli* JM83 (Pro^-) strain and the Met auxotrophic *E. coli* B834 (Met^-) strain, respectively.[5,7] Trp surrogates (L-5-hydroxytryptophan ((5-OH)-Trp), L-4-fluorotryptophan ((4-F)-Trp), and L-7-azatryptophan ((7-Aza)-Trp) (Figure 4.4)) were incorporated into Bliβ by using the Trp auxotrophic *E. coli* WP2 (ATCC 49980) (Trp^-) strain.[5] The purified congeners were analyzed by high resolution HPLC-ESI-MS, and incorporation of the ncAAs was detected based on the characteristic mass shift of the molecular masses of Bliα and Bliβ. The incorporation of the Pro, Met and Trp surrogates into the dedicated positions within the B-C-D rings of Bliα and Bliβ was further confirmed by analysis of fragment spectra from HR-ESI-MS/MS. The mass spectrometric analysis revealed high levels of incorporation for all the tested amino acid

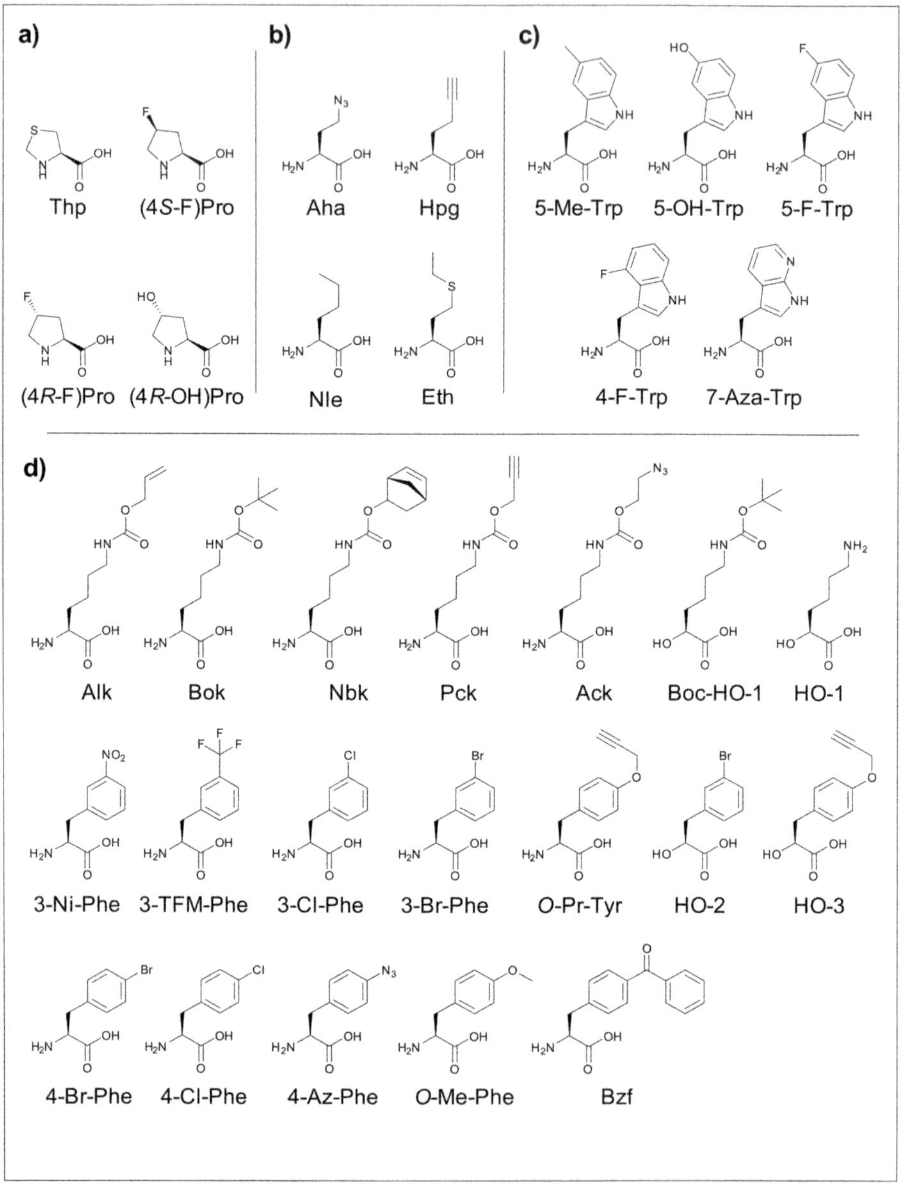

Figure 4.4 Chemical structure of ncAAs incorporated into cyclic RiPPs *in vivo*. (a–c) Pro, Met, and Trp isostructural analogues (respectively) incorporated using the SPI approach, and (d) orthogonal ncAAs incorporated using the SCS approach.

analogues. Subsequently the congeneric peptides were tested for their antimicrobial activities, together with the complementary wild-type peptide, in an agar-well diffusion assay against *Micrococcus luteus*. The majority of the congeners retained their antimicrobial activities. However, an enhancement in the antimicrobial activity was noticed for L-thioproline (Thp)-containing Bliα combined with the wild-type Bliβ, while a complete loss of the bioactivity occurred upon the incorporation of L-*cis*-4-fluoroproline ((4S-F)-Pro) into Bliα.[7] As a proof of principle, an *in vitro* bio-orthogonal post-biosynthetic structural diversification was applied on one of the Bliα congeners containing Hpg instead of Met28. Hence a Cu^{I}-catalyzed Huisgen 1,3-dipolar cycloaddition (click reaction)[26] was performed according to protocols described previously.[27] Synthetic coupling to the Bliα peptide was achieved with 1-azido-1-deoxy-β-D-glucopyranoside and azido-fluorescein, thus obtaining a glycol-diversified and a dye-labeled lanthipeptide.[5] Recently, another example was published involving the incorporation of 3 different analogues of Trp (5-fluorotryptophan (5-F-Trp), 5-hydroxytryptophan (5-OH-Trp), and 5-methyltryptophan (5-Me-Trp)) at 4 different positions of the lanthipeptide nisin A (Ile1, Ile4, Met17, and Val32) (Figure 4.1).[28] Interestingly, this study used *Lactococcus lactis* as a heterologous expression host, while to date all other studies that used the SPI approach to generate unnatural cyclic peptides have used *E. coli* as a heterologous expression host. However, all produced nisin A congeners exhibited lower antimicrobial activities in comparison with the wild-type peptide.[28]

4.1.2 Genetic Code Expansion

Although a wide variety of ncAAs have been introduced into proteins by the above-mentioned genetic code engineering,[11,23] this approach is confined to the incorporation of isostructural analogues of respective cAAs, and is not suited for more sterically demanding amino acids. Furthermore, the number of amino acids simultaneously used in the same pool for the same culture is still 20. In contrast, genetic code expansion aims to increase this number by adding an orthogonal ncAA under *in vivo* conditions to the standard 20 amino acid pool. The assignment of this ncAA in response to nonsense (stop) codons requires its recognition by an orthogonal aminoacyl RNA synthetase:transfer RNA_{sup} (aaRS:$tRNA_{sup}$) system. The orthogonality of the aaRS:$tRNA_{sup}$ pair is an essential requirement for expanding the genetic code, particularly in order to avoid cross-reactivity with the endogenous host components. In other words, the endogenous host tRNAs must not interact with the aaRS that aminoacylates the cognate $tRNA_{sup}$ with the ncAA, and the endogenous aaRSs must not interact with the $tRNA_{sup}$ that recognizes the nonsense codon. A further requirement is a high efficiency of $tRNA_{sup}$ in reading the nonsense codon of choice for proper integration of a non-canonical amino acid into the peptide of interest.[22]

The first attempts to develop a stop codon suppression (SCS) approach (Figure 4.3) were started in 1997 by Schultz and co-workers using approaches of rational design and directed evolution to generate a GlnRS:$tRNA^{Gln}{}_{CUA}$ in *E. coli*. However, initially, this enterprise was not entirely successful since the employed GlnRS was not orthogonal and was still able to recognize the endogenous $tRNA^{Gln}$.[29,30] One year later, Furter was able to evolve the first successful amber stop codon suppression system in *E. coli* (PheRS:$tRNA^{Phe}{}_{CUA}$) using the yeast *Saccharomyces cerevisiae*, and he was able to assign 4-fluoro-Phe in response to an amber stop codon.[31] The Schultz group then followed this approach and established a successful orthogonal scGlnRS:sc$tRNA^{Gln}{}_{CUA}$ system.[32] Since then, a number of orthogonal aaRS:tRNA pairs have been developed, of which a few have gained greater attention. The most important pairs are the PylRS:$tRNA^{Pyl}$-based ones, which are naturally orthogonal. These are from bacteria or archaea (*Methanosarcina mazei*,[33] *Methanosarcina barkeri*,[34,35] and *Desulfitobacterium hafniense*[36]) and are highly efficient in nonsense codon suppression and incorporation of the amino acid pyrrolysine (Pyl). Furthermore, the high substrate promiscuity exhibited by $tRNA^{Pyl}$ towards ncAAs allowed this suppression system to charge structurally different non-canonical Pyl derivatives and even Lys analogues, whilst at the same time keeping its orthogonality against other cAAs.[11,37]

While the majority of PylRS:$tRNA^{Pyl}$ pairs have been developed against the amber stop codon (TAG),[33,35,36] there are other PylRS:$tRNA^{Pyl}$ pairs that have been developed to address the other stop codons (opal (TGA) and ochre (TAA)),[35] and even quadruplet codons are known.[34] Various screening strategies have been employed for the development of successful aaRS:$tRNA_{sup}$ pairs including the recruitment of positive (based on a β-lactamase gene or a chloramphenicol acetyl transferase gene that contains a stop codon) and negative selection markers (*e.g.* the toxic barnase gene).[38–42] Fluorescence activated cell sorting (FACS) was also used for improving the incorporation efficiency of ncAAs.[43] More recently, attempts have been devoted to engineering the translation machinery for the development of stop codon suppression systems,[44–46] even for the incorporation of ncAAs into mammalian cells by a PylRS:$tRNA_{CUA}$ pair (engineering the eRF1).[47] A comprehensive overview and discussion of the most recent advances in genetic code expansion has been highlighted by Davis and co-workers.[11]

The RiPP capistruin is a prominent member of the lasso peptide family (Figures 4.1 and 4.2), and belongs to the class II lasso peptides. It is produced by the Gram-negative bacterium *Burkholderia thailandensis* E264, and its biosynthesis is genetically encoded by the *capABCD* gene cluster.[48] Capistruin is composed of 19 amino acid residues. The N-terminal nine amino acids form a macrolactam ring, through which the C-terminal tail is threaded. The overall course of capistruin biosynthesis follows that of other RiPPs: the expression of the structural gene *capA* yields a precursor peptide (47 residues) composed of a leader peptide (28 residues) and a core peptide region (19 residues), that undergoes post-translational modifications to yield the mature bioactive lasso-folded capistruin.[48] The PTM machinery consists of the cysteine protease (CapB), responsible for cleaving the core peptide from the leader peptide, and the macrolactam synthase (CapC), responsible for isopeptide bond formation. The latter occurs *via* peptide bond formation between the carboxylic side chain of Asp9 and the N-terminal amino group of Gly1, which has been liberated upon cleavage of the leader peptide.[48,49] The knot-like peptide structure (lasso) contains six residues of the C-terminal tail trapped under the ring, while the remaining four residues form the loop. This structure is firmly locked by the bulky side chains of Arg11 and Arg15. The export and immunity protein (CapD) then exports the mature lasso peptide outside of the cell.[48,49] Capistruin shows antimicrobial activities against closely related *Burkholderia* and *Pseudomonas* strains (MIC 12–150 μM), as well as against the *E. coli* 363 wild-type strain (MIC 25 μM).[48] The main target of capistruin (similar to microcin J25) known to date is the inhibition of bacterial RNA polymerase (RNAP).[50–52] In order to make capistruin amenable to the incorporation of ncAAs, the *capABCD* gene cluster of capistruin was cloned from the chromosomal DNA of *B. thailandensis* E264 into expression plasmids and heterologously expressed in *E. coli*.[6,7]

Engineering and expanding the genetic code by SPI and SCS approaches was used for co-translational incorporation of ncAAs into the lasso peptide capistruin. Two bioisosteric analogues of Met, L-homopropargylglycine (Hpg) and L-azidohomoalanine (Aha), were used in the SPI approach (Figure 4.1), and five orthogonal ncAAs (Nbk (*N*ε-5-norbornene-2-yloxycarbonyl-L-lysine), Alk (*N*ε-Alloc-L-lysine), Ack (*N*ε-2-azidoethyloxycarbonyl-L-lysine), Pck (*N*ε-2-propyn-1-yloxycarbonyl-L-lysine), or Bok (*N*ε-Boc-L-lysine)) were employed for site-directed incorporation using the SCS method (Figures 4.2 and 4.4). Site-directed mutagenesis (SDM) was used to modify the structural gene *capA* and substitute the Met codon or amber stop codon corresponding to the chosen positions within the core peptide of capistruin. All tested ncAAs were successfully co-translationally incorporated in three strategic positions of the peptide. Interestingly, based on HR-ESI-MS analytics, the SCS approach in this particular example was proven to be more successful compared to SPI. Interestingly, a significant increase in the antimicrobial activity was obtained for some newly generated capistruin congeners in comparison with the wild-type peptide. The highest increase in antimicrobial activity was obtained with the capistruin congener Cap_Hpg17 containing the ncAA "Hpg" (L-homopropargylglycine) instead of Gly, where an increase of 100-fold was observed.[7] Finally, bio-orthogonal chemistry was applied for post-biosynthetic modifications of capistruin congener Cap_Alk10, containing the ncAA "Alk" (*N*ε-Alloc-L-lysine) instead of Ala. An *in vitro* metathesis reaction between Cap-Alk10 and allyl alcohol was achieved by using a 2nd generation Hoveyda–Grubbs catalyst as a mediator, and this was confirmed by means of HR-ESI-MS analysis.[6] Such applications of orthogonal chemistry for ncAA-containing RiPPs in both previous examples of lichenicidin and capistruin show the high potential of these approaches in increasing further their structural diversity.

Microcin J25 (Figure 4.2) is another lasso peptide where the biosynthesis pathway is similar to that of capistruin. The SCS approach was applied to incorporate 4 orthogonal analogues of Phe (3-trifluoromethyl-L-Phe (3-TFM-Phe), 3-bromo-L-Phe (3-Br-Phe), 3-chloro-L-Phe (3-Cl-Phe), and 3-nitro-L-Phe (3-Ni-Phe)) into the core peptide of microcin J25 (Figure 4.4).[8] The antimicrobial activity of the generated congeners was equivalent to that of the wild-type peptide. However, no bioorthogonal chemistry was applied in this example.

The incorporation of a ncAA into RiPPs was reported by van der Donk and co-workers, who used the SCS approach to replace Phe26 with 4-benzoyl-Phe in the class II lanthipeptide prochlorosin precursor peptide (ProcA 3.2) using *E. coli* as a heterologous expression host (Figure 4.2).[53] However, this system was partially *in vitro* since the congeneric prochlorosin precursor peptides were designed to have a TEV protease recognition site between the leader and the core peptides, which was used later to cut off the leader peptide *in vitro* after His-tag-based immobilized metal affinity chromatography (IMAC). Another example was published by Schultz and co-workers regarding the incorporation of the same ncAA (4-benzoyl-L-Phe) into cyclic peptides, where a method called "split-intein circular ligation of proteins and peptides" (SICLOPPS) was used combined with an amber stop codon suppression system in *E. coli*. Although the cyclic peptides produced by SICLOPPS are totally different in their origin from all other RiPPs, this technique enables the *in vivo* generation of genetically encoded cyclic peptide libraries of around 1×10^8 members. The SICLOPPS technique is based on cyclizing an extein (peptide of interest) upon splicing it from two flanking inteins (an N-terminal and a C-terminal intein) connecting the two inteins in a rearrangement process of the PCC6803 DnaE trans inteins (from *Synechocystis* sp. (Ssp)).[54–56] The aim of SICLOPPS in that experiment was to evolve cyclic peptides that inhibit HIV proteases.[57] Upon two rounds of cellular viability-based selection, two of the three isolated cyclic peptides were found to bear the keto amino acid "4-benzoyl-Phe" as a structural moiety. The most potent isolated cyclic peptide as an HIV protease inhibitor was found to have the sequence GIXVSL (*X* = 4-benzoyl-Phe) (Figure 4.5). The inhibition was believed to occur through the formation of a covalent Schiff base adduct of the 4-benzoyl-Phe residue on the side chain amino group of Lys12 of the HIV protease.[57]

The first application of the SCS approach for cyanobactins was published by Schmidt and co-workers in 2012,[58] where the group incorporated four orthogonal Phe analogues (4-Cl-Phe (4-chloro-L-phenylalanine), 4-Br-Phe (4-bromo-L-phenylalanine), *O*-Me-Phe (*O*-methyl-L-phenylalanine), and 4-Az-Phe (4-azido-L-phenylalanine)) replacing Leu5 in the marine natural product patellin 2, and replacing Phe6 with 4-Br-Phe in the anticancer peptide trunkamide. Similar to other cyanobactins, patellin 2 and trunkamide are head-to-tail macrocyclic peptides with a thiazole moiety, in addition to two Thr or Ser prenylations (Figure 4.5).[1,57] The heterologous expression system used for this study has shown an optimal incorporation of 4-chloro-Phe without induction. Thus, the peptide expression could not be fully controlled, a shortcoming that future research will try to avoid upon expanding the genetic code.

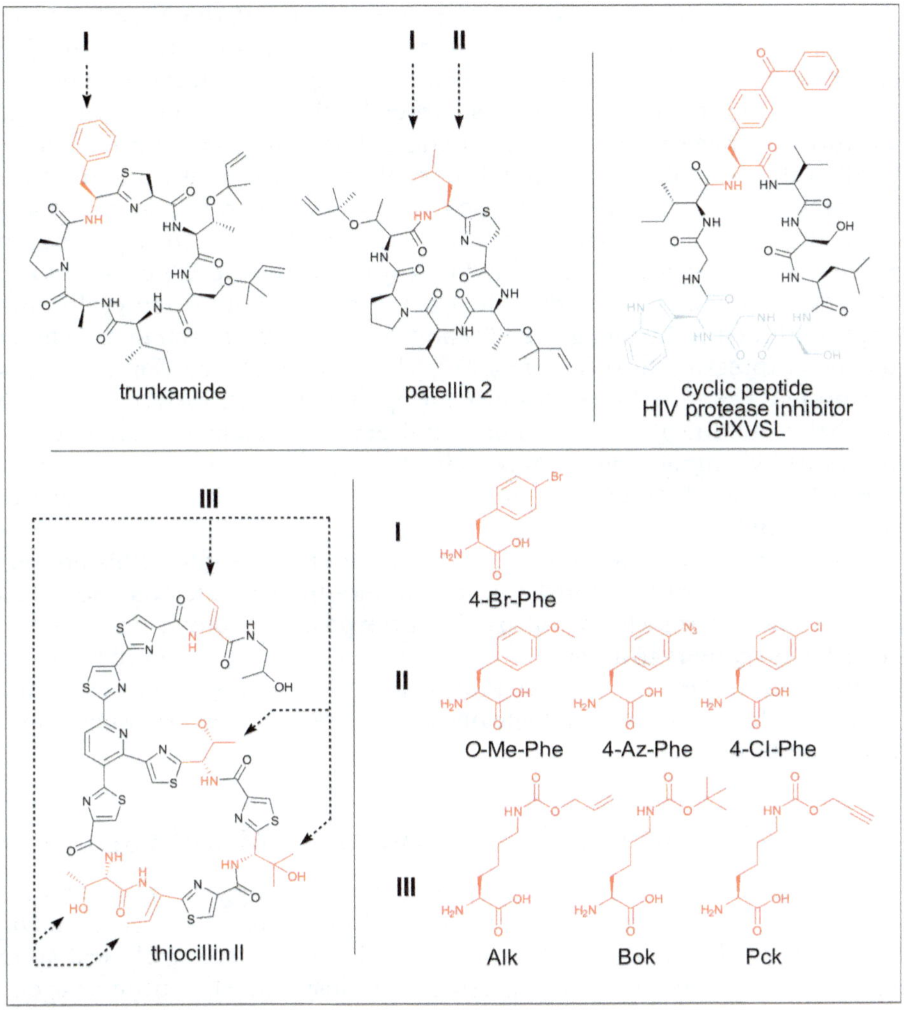

Figure 4.5 Structures of the cyclic peptide GIXVSL that inhibit HIV proteases, thiocillin II, patellin 2, and trunkamide, showing positions where non-canonical amino acids were incorporated using the SCS approach.

Interestingly, the SCS approach has not only been used to incorporate ncAAs into RiPPs, but also hydroxy acids.[59] This has been applied for two lanthipeptides, lacticin 481 and nukacin ISK-1 (Figure 4.2), where the hydroxy acids were incorporated into the precursor peptides at position 1 of the core peptide. The hydroxy acids that have been used are derivatives of Lys, Phe or Tyr; HO-1 ((S)-6-amino-2-hydroxyhexanoic acid), Boc-HO-1 ((S)-6-(Boc-amino)-2-hydroxyhexanoic acid), HO-2 ((S)-3-(3-bromophenyl)-2-hydroxypropanoic acid), and HO-3 ((S)-2-hydroxy-3-(4-(prop-2-yn-1-yloxy)phenyl)propanoic acid) (Figure 4.4). The researchers also incorporated the ncAA analogues of the tested hydroxy acids (Bok (*N*ε-Boc-L-lysine) as an analogue of Boc-HO-1,

3Br-Phe (3-bromo-L-phenylalanine) as an analogue of HO-2, and *O*-Pr-Tyr (*O*-propargyl-L-tyrosine) as an analogue of HO-3) (Figure 4.4). The incorporation of hydroxy acids results in an ester linkage between the core peptide and the leader peptide, which was subsequently cleaved later by a simple hydrolysis to release mature lanthipeptides, but with two C-terminal ends.[59]

All of the examples of lichenicidin, trunkamide, capistruin and microcin J25 previously mentioned have used *E. coli* as a heterologous expression host. Recently, Schultz and co-workers reported the first incorporation of three orthogonal ncAAs (Bok (*N*ε-Boc-L-lysine), Alk (*N*ε-Alloc-L-lysine), or Pck (*N*ε-2-propyn-1-yloxycarbonyl-L-lysine)) into the branched cyclic thiopeptide thiocillin by using the SCS approach in *Bacillus cereus* as a heterologous expression host (Figure 4.5).[60] The successful incorporation of the amino acid PcK into thiocillin substituting Thr3 was followed by a post-biosynthetic modification (a copper-catalyzed azide–alkyne cycloaddition reaction) by conjugating a fluorescein-PEG2_azide to the alkyne group of the Pck residue,[60] which is along the lines of previous contributions on lichenicidin.[5,6]

The previous examples of the incorporation of ncAAs into RiPPs not only show the ability of the translation machinery to accept ncAAs and to use them as building blocks, but also exhibit the positive promiscuity shown by the PTM machinery to accept these non-proteinogenic moieties in their substrate peptides. Hence, Synthetic Biology recruitment reveals the potential of *in vivo* production of new biologically active RiPPs from an expanded amino acid repertoire.

4.2 Cyclic NRPs with New-to-nature Modifications

Non-ribosomal peptides (NRPs) represent one of the biggest and structurally diverse groups of biologically active natural products. Their broad activity spectrum has been extensively studied and is widely used in clinical applications, such as in immunosuppressants (cyclosporin A), antitumor agents (bleomycin A2) or antibiotics of last resort (*e.g.* daptomycin).[61–63] Consequently, more recently, increasing efforts have been directed toward the generation of peptide analogues with altered bioactivity or improved physicochemical properties. While many drugs in therapeutic use are produced by chemical synthesis or semi-synthesis, the complexity of some structures motivates the search for alternative and sustainable methods of production, alongside the testing of novel structures that may be discovered for their bioactivities.[64] Hence, there is growing interest in using biochemical tools to generate novel peptide structures by making use of bioengineering approaches.[64–67]

Non-ribosomal peptides are commonly synthesized by large multimodular enzymes known as non-ribosomal peptide synthetases (NRPSs). In addition, they are often further modified by tailoring enzymes. The NRPS modules are divided into at least three essential catalytic domains (Figure 4.6). Selection and activation of the monomers takes place in the adenylation (A) domains and the monomers are further transferred onto phosphopantetheinyl carrier proteins (PCPs), with amide bond formation taking place in the condensation (C) domains.[68–70] In general, during the biosynthesis of the NRPs, each module is responsible for incorporation of one building block into the growing peptide chain, which explains the enormous size of the NRPS proteins

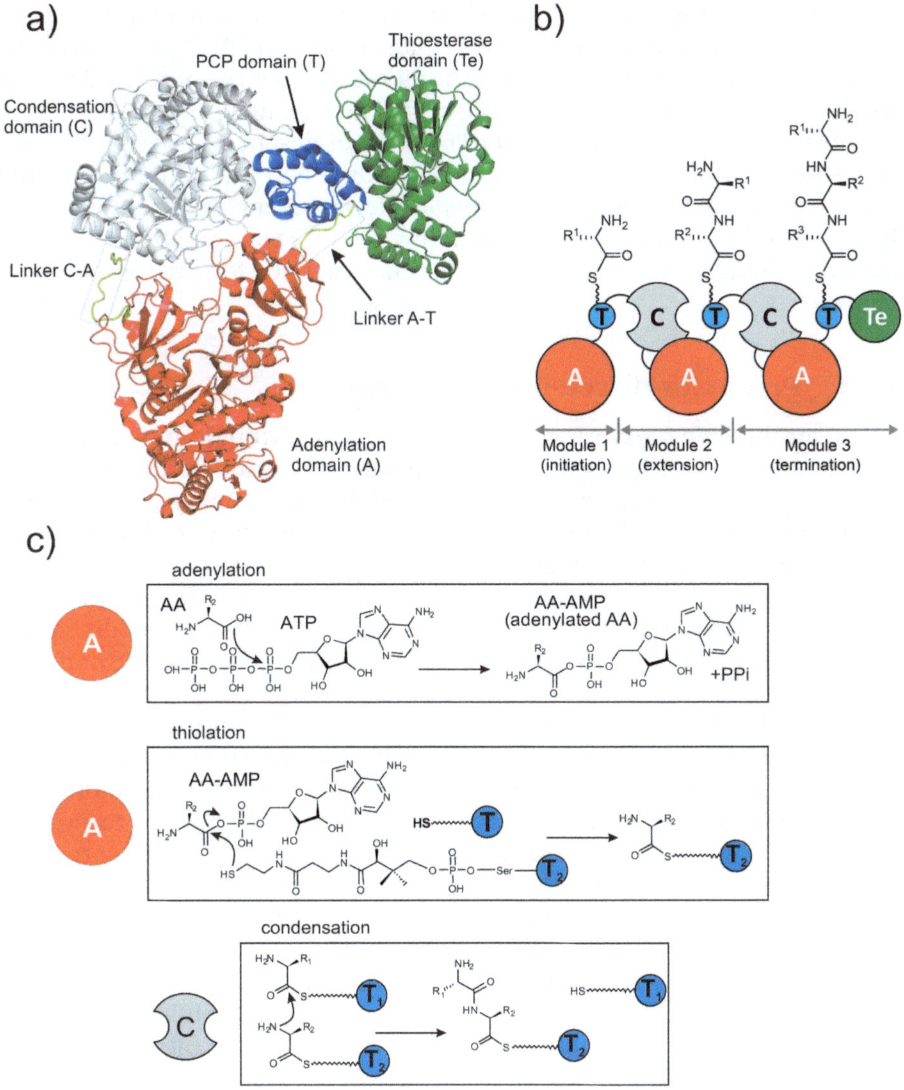

Figure 4.6 NRPS biosynthesis: (a) the crystal structure of the NRPS synthetase AB3403 (PDB: 4ZXH), clearly showing the condensation, adenylation, thiolation and thioesterification domains with the linkers between each of them, (b) a general schematic representation of NRP biosynthesis, and (c) catalytic events of adenylation, thiolation and condensation.

or complexes thereof, *e.g.* the cyclosporine synthetase is 1.6 MDa in size.[71] It is important to note that the substrates of NRPS are not limited to the 20 canonical amino acids. It is rather estimated that >500 monomers, including non-proteinogenic amino acids, fatty acids and α-hydroxy acids, are used, accounting for the impressive structural diversity synthesized by NRPSs.[72,73] A considerable number of attempts to expand the structural diversity of these peptides have taken advantage of the modularity of the enzyme complexes. Early research approaches focused on the incorporation of non-natural building blocks *via* precursor-directed biosynthesis (PDB)[74] or mutasynthesis/mutational biosynthesis (MBS).[75] Additionally, over the past few decades, combinatorial approaches have played an increasingly important role in the field, facilitating the application of rational design approaches to engineer NRPSs. This includes the swapping of entire NRPSs,[76] or modules[77] and domains,[78] module extension,[79] site-specific mutagenesis to modulate the specificity of the A-domains[80,81] and pathway level recombination.[82,83] In the following paragraphs, we will discuss the characteristic features of individual approaches.

4.2.1 Precursor-directed Biosynthesis

Precursor-directed biosynthesis (PDB) is a relatively straightforward method to generate new analogues of NRPs. In this derivatization approach, precursor analogues that can be competitively used during the biosynthesis assembly of the target peptide are added to the growth medium of the wild-type peptide producing strain (Figure 4.7).[74,75] As a result, a mixture of natural peptides and non-natural analogues is obtained. However, there are possible issues that have to be considered while using this method. A modified precursor will compete with the endogenous building block being the preferred substrate, thus often resulting in poor yields of the desired compound.[84] Consequently, the separation and purification of the unnatural analogue becomes often a considerable obstacle for subsequent structure–activity relationship studies. This might be compensated by increasing the amount of the added precursor,[74,85,86] while maintaining its concentration below toxic doses for the producing organism. Furthermore, some unnatural precursors could show low membrane permeability,[74] resulting in no or scant incorporation, which might be falsely interpreted as low incorporation by the biosynthetic machinery into the artificial precursor. Cell-free *in vitro* studies including experiments with purified synthetases (chemoenzymatics) might give valuable information on the principal accessibility of the new building block.[87–91] However, they are applicable only in a limited number of cases. A clear advantage of the PDB concept is its simplicity, as it does not require either sophisticated genetic manipulation methods to create mutants or chimeric strains. One of the first uses of the PDB approach to obtain novel analogues of NRPs was presented on ergot peptide alkaloids from *Claviceps purpurea*. Two research groups reported the successful incorporation of either D,L-*p*-chlorophenylalanine in place of D,L-phenylalanine,[92] or L-thiazolidine-4-carboxylic

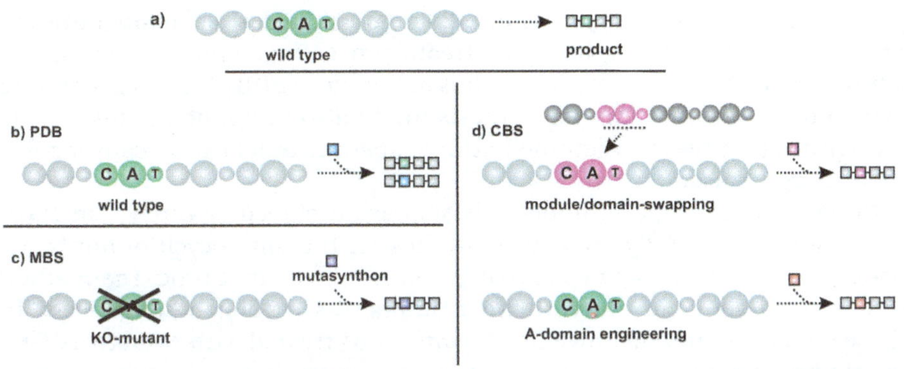

Figure 4.7 Comparison of wild-type biosynthesis (a) with approaches leading to the generation of unnatural NRP structures (b–d). Precursor-directed biosynthesis (PDB; b) is based on feeding of the wild-type bacteria resulting in generation of a product mixture. In contrast, in mutational biosynthesis (MBS; c) segments of NRPSs are inactivated and missing building blocks are supplemented to yield a single product. Combinatorial biosynthesis (CB; d) is based on the recent Synthetic Biology approaches used for module/domain swapping or A-domain engineering.

acid in place of proline residues in the L-amino acid derived side chain of ergopeptins.[93] A well-studied example of PDB is the generation of new derivatives of the immunosuppressant NRP cyclosporine.[94–98] First, the production of minor natural cyclosporines differing mainly in the amino acid in position 2 could be upregulated, when the amino acids of interest were fed into the culture broth of the wild-type strain *Tolypocladium inflatum*.[97] Later, the same group succeeded in generating six new cyclosporines with amino acid replacements in positions 1, 2 and 8.[94] Although efforts to exchange amino acids on the remaining positions *via* classical PDB failed, *in vitro* experiments with the purified cyclosporine synthetase were found to produce formerly undetectable analogues.[87,96,99,100] More recently, the PDB approach was used to elucidate the biosynthesis and generate new analogues of other NRPs like hormaomycin,[66,101,102] halogenated balhimycins[103] or enniatins,[89] emphasizing the utility of this method.

4.2.2 Mutasynthesis

An advanced variant of the PDB method is mutasynthesis, also called mutational biosynthesis (MBS). Unfortunately, despite many reviews[75,104–106] on this topic, the PDB term is often used to describe results obtained by applying MBS. The general idea of mutasynthesis is to create a mutant strain that is blocked at a particular step in the biosynthetic pathway; with subsequent incorporation of the missing precursor by the feeding of a synthetic analogue, the so-called "mutasynthon", the new product will be formed. In contrast, PDB makes use of a wild-type strain with a wild-type biosynthetic pathway. In this respect, successful MBS renders the desired analogue as the only product, facilitating its downstream processing, which is the biggest advantage over mixtures of metabolites obtained by PDB. Besides generating new-to-nature peptides, mutasynthesis might also be applied to investigate biosynthesis pathways when the natural substrate or its isotopically labeled derivatives are used.

Initially the concept of mutasynthesis was developed by Birch[107] in 1963, and it was successfully applied in studies on the aminocyclitol antibiotic neomycin produced by *Streptomyces fradiae*.[108,109] By that time, the method of choice to generate biosynthetic mutants was random mutagenesis. In these studies on new neomycins, *S. fradiae* was treated with *N*-methyl-*N*′-nitro-*N*-nitrosoguanidine, a mutagen that causes random point mutations in the chromosomal DNA. This was followed by laborious screening for suitable mutants and subsequent feeding of mutasynthons. Recent significant advancements in the genetic manipulation techniques in a number of cases have overcome this problem, allowing the generation of directed mutants in a comparably short time. However, one of the drawbacks of this approach is the need for fundamental knowledge about the biosynthetic gene cluster and, even more importantly, an appropriate genetic tractability of the producing organism. The latter might be overcome by an alternative approach of heterologous expression. In that case, the biosynthetic pathway lacking unwanted genes is cloned and expressed in a suitable host organism. However, as shown recently in a study on the generation of non-natural derivatives of the NRP beauvericin, not all synthetic building blocks incorporated into the mutant producer strain *Beauveria bassiana* were accepted by the heterologous host *Escherichia coli*.[90] To overcome this problem, and thus low production yields, a proper screening of possible hosts should be conducted.[110,111]

Over the past decades, mutasynthesis experiments have been conducted, mainly to generate new derivatives of polyketide antibiotics.[108,109,112–115] There are only a few examples on MBS applications to generate new derivatives of NRPs; among them, the most prominent examples are on the vancomycin-type glycopeptide antibiotic balhimycin and calcium-dependent antibiotics (CDAs).[116–118] Vancomycin is one of the antibiotics of last resort for the treatment of severe bacterial infections, including those caused by methicillin-resistant *Staphylococcus aureus* (MRSA).[119,120] The number of vancomycin-resistant bacterial strains detected over the past few years has forced scientists to seek new analogues.[119,121,122] The semi-synthesis approach is an option as shown for the glycopeptide derivative oritavancin approved in 2014 by the FDA for treatment of skin infections.[123] However, oritavancin shows only variation of the peripheral features of vancomycin, leaving the aglycon intact. Our group realized that the core part of glycopeptide antibiotics can be modified using mutasynthesis, which has been deeply investigated for the vancomycin-type glycopeptide balhimycin (Figure 4.8). This technique is mainly focused on two essential building blocks of the glycopeptide aglycon, formed by the non-proteinogenic amino acids β-hydroxytyrosine (Hty) and

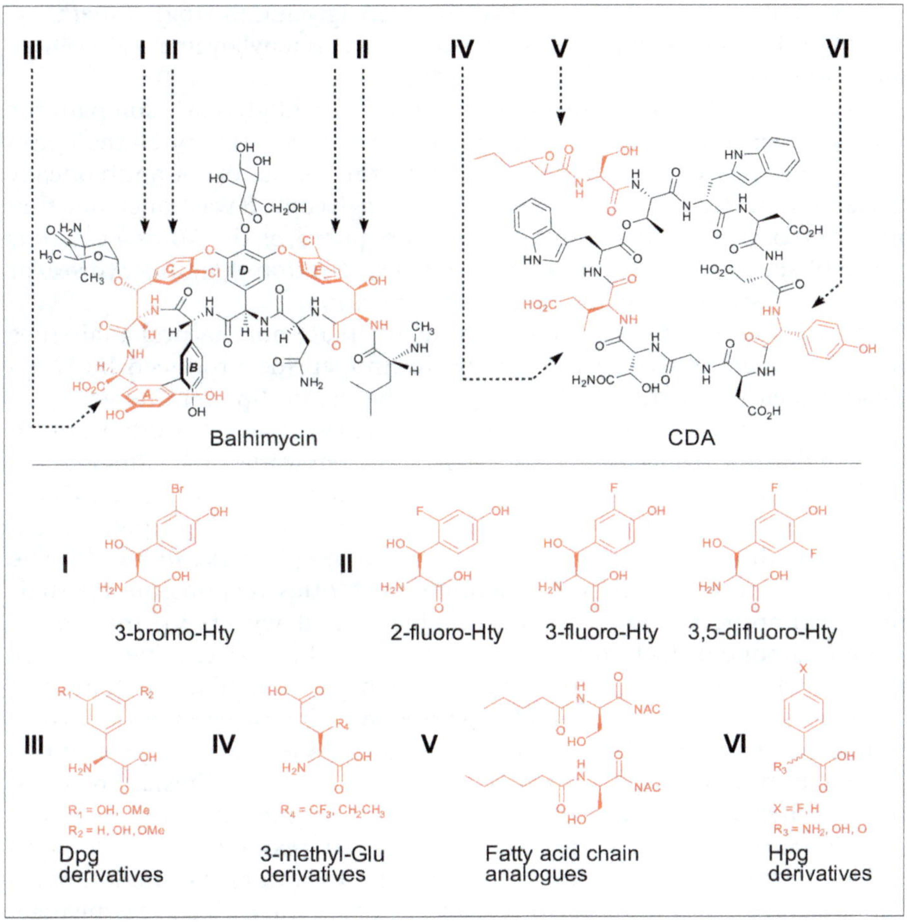

Figure 4.8 Chemical structures of balhimycin and CDA showing moieties that have been changed by PDB (Nr. I) or by mutasynthesis (Nr. II – VI).

the C-terminal 3,5-dihydroxyphenylglycine (Dpg). In the first mutasynthesis study, an in-frame deletion mutant of the *bhp* perhydrolase gene, deficient in Hty biosynthesis, was prepared.[124] After successful restoration of balhimycin production, upon feeding with Hty, supplementation with non-natural fluorinated Hty was conducted. Three mutasynthons (2-fluoro-, 3-fluoro- and 3,5-difluoro-β-hydroxytyrosine) resulted in the production of new fluorobalhimycins with antibiotic activity against *Bacillus subtilis*. D/L-Tyrosine and (fluoro-)phenylserines lacking the *para*-hydroxy group were not accepted as substrates, showing the importance of the Hty-moiety as a building block for balhimycin biosynthesis.[116] In subsequent studies, Dpg was exchanged by feeding analogues to the Δ*dpgA* mutant of *A. balhimycina*, in which the first step of Dpg biosynthesis is blocked.[125] Formation of balhimycin analogues was restored by feeding the following mono- and di-substituted hydroxy and methoxy substituents of Dpg: 3-hydroxyphenylglycine (3-Hpg), 3-methoxyphenylglycine (3-MeOPg), 3-hydroxy-5-methoxyphenylglycine (HMeOPg), and 3,5-dimethoxyphenylglycine (DMeOPg) (Figure 4.8).

Moreover, more than 20 metabolites with different glycosylation patterns were observed. Interestingly, employment of the mono-substituted analogues (3-Hpg and 3-MeOPg) resulted not only in the formation of antibiotically active tricyclic derivatives, but also in bicyclic aglycons. It was concluded that proper orientation of the aromatic rings is required, as the AB-ring can only be established by the oxygenase OxyC when an electron-donating hydroxy or methoxy substituent is placed in the *ortho*-position.[117]

The second example of mutasynthesis to create non-natural analogues of NRPs concerns CDA. CDA is a cyclic lipopeptide produced by *Streptomyces coelicolor* A3(2).[126] It belongs to the acidic lipopeptide group of non-ribosomally synthesized antimicrobials, together with other antibiotics including friulimycin, amphocin, and daptomycin.[127] Since 2003, the latter has been in clinical use for the treatment of skin and soft tissue infections caused by Gram-positive bacteria, including life-threatening methicillin- and vancomycin-resistant *Staphylococcus aureus*.[127] The common features of acidic lipopeptide antibiotics responsible for their biological properties are the N-terminal fatty acid acyl chain, and a decapeptide lactone or lactam ring with a number of acidic residues that are likely involved in calcium binding.[118] The high therapeutic potential of lipopeptide antibiotics to combat pathogens with resistance to current antibiotic treatments explains the trials to engineer future generations of their non-natural derivatives by use of *e.g.* mutasynthesis. For CDA, mutasynthesis was used to rationally perform engineering of three building blocks: the arylglycine residue,[118] glutamate residues[128] and the lipid moiety (Figure 4.8).[129] To test the incorporation of arylglycine-mutasynthons, a Δ*hmaS* mutant of *S. coelicolor* was generated, which is a gene inactivation mutant in the biosynthesis operon of 4-hydroxyphenylglycine (Hpg).[127] By feeding analogues bearing hydrogen or fluorine in the 4-position of the aromatic side chain, it was possible to restore CDA production yielding new derivatives (CDA2d and CDA2fb, respectively). More bulky para-substituents, such as chlorine and methoxy, failed to yield detectable amounts of CDA analogues. In the mutasynthesis approach, feeding of 3-MeGlu, a residue modulating toxicity and antimicrobial activity, has yielded two new variants of CDA, Et-CDA3b and CF_3-CDA3a/3b.[130,131] This was achieved by feeding 3-ethyl and 3-trifluoromethyl glutamic acid derivatives to a Δ*glmT* mutant.[128] Following this, mutasynthesis has also been used to modify the lipid moiety of CDA. In this approach, the peptidyl carrier protein (PCP) domain of module 1 of the CDA synthetase was blocked by mutation of the active site Ser to Ala to generate a mutant strain (Δ*PCP1*). A selection of synthetic *N*-acyl-L-serinyl *N*-acetylcysteamine (NAC) thioester analogues was fed to growing cultures of this mutant. Incorporation of short acyl chains (C5 and C6) could be detected, in contrast to the longer chains (C7 and C10), suggesting a tight specificity of the CDA assembly line for C6 or shorter acyl chains.[129] Unfortunately, low production yields in all described mutasynthesis approaches prevented the testing of the biological activity of the newly generated peptides.

In summary, the mutasynthesis approach proved to be a suitable method to generate new non-natural NRPs, which cannot be obtained easily by chemical methods. The biggest advantage of this method is the *in vivo* production of the metabolite of interest with the only limitation that the preferred mutasynthon has to be accepted by the biosynthetic machinery of the mutant strain. Hence, mutasynthesis has biotechnological potential for scale-up and extension of the potential drug candidate pool. However, mutasynthesis in many cases results in two to three orders of magnitude lower yields of the novel analogues.

4.2.3 Combinatorial Biosynthesis and Domain Engineering

Recent developments that can be summarized under the term Synthetic Biology have provided powerful tools to perform combinatorial biosynthesis (CBS) aiming at introducing structural diversity within non-ribosomal peptides. CBS refers to any engineering method with the aim to change the promiscuity of microbial anabolic pathways resulting in new chemical entities.[132] It takes advantage of the modular architecture characteristic of NRPS assembly lines. The CBS approaches investigated to date can be divided into three major strategies: site-directed mutagenesis of active sites or recognition domains of proteins (predominantly of adenylation domains), NRPS module and domain exchange, and engineering of tailoring enzymes.[67,83]

The least invasive approach is to incorporate point mutations in the active site of the adenylation domains. The specificity towards certain building blocks was shown to be encoded in a 10-amino acid code of highly conserved residues.[80,133] Changes of individual amino acids alter or relax the specificity of the A-domain binding pocket allowing the incorporation of non-native substrates. The first example of this approach to incorporate a non-natural amino acid into the structure of NRPs involved mutating the A-domain from module 10 of CDA synthetase.[81] Introduction of a single mutation (K278Q) changed the specificity of the A domain from Glu/3mGlu (3-methylglutamate) to Gln/3mGln (3-methylglutamine), yielding a CDA molecule with a Gln residue in position 10. To date, there are some more examples where residues of the A domain binding pocket have been engineered to yield NRPs with non-natural amino acids.[134,135] In the case of engineered gramicidin S[134] and anabaenopeptin[135] synthetases, researchers have focused on the incorporation of "clickable" amino acids to create NRPs for biorthogonal labelling.

The strategy of swapping entire subunits, modules and domains appears as a more classical method of combinatorial biosynthesis. However, it is even more challenging since this is severe interference with a complex multi-domain and multi-modular system. This approach is based on the modular character of the NRPS enzymes and the fact that the sequence of the non-ribosomal peptides is determined by the module and domain type. Hence, shuffling of parts of NRPSs should result in new, non-native peptides. In order to produce functional NRPS hybrid enzymes, it is of high importance to make an appropriate choice of the linker regions in the module and domains. This supposedly has a strong influence on protein folding and protein–protein interactions, and may also affect the proofreading activity towards the peptide product encoded within the selective C-domain acceptor site.[136,137] For NRPs, next to the Marahiel group (Germany), some of the pioneers in this field were researchers at Cubist Pharmaceuticals (USA) working on the lipocyclopeptide daptomycin, a drug with the trade name Cubicin® approved in 2003 by the FDA to treat skin infections caused by Gram-positive bacteria. Domain and module swapping with parts of related NRPSs of the lipopeptides A54145 and calcium-dependent antibiotics (CDAs) led to the generation of >70 compounds with different degrees of structural diversity and antibacterial activity.[138–141] Noteworthily, not only exchange or deletion of the modules, but also insertion or extension of additional modules was successfully adapted to expand product diversity. This was shown on the examples of balhimycin (*in vivo*)[79] and tyrocidine synthetases (*in vitro*),[142] respectively. One of the more recent studies focused on the generation of novel hybrid structures of the medically relevant molecules from the cyclodepsipeptide (CDP) class.[143] In this study, the hydroxy-acid-activating modules of three highly homologous cyclodepsipeptide synthetases (CDPSs) were combined, including enniatin (ESYN), PF1022 (PSYN) and beauvericin (BSYN) synthetases. Their products show broad activities as insecticidal, phytotoxic, anthelmintic, antibiotic, antifungal and cytostatic compounds for enniatin, anthelmintic compounds for PF1022 and antitumor, antibiotic, antifungal, insecticidal and anthelmintic compounds for beauvericin.[143] The N-terminal module of PSYN showed relatively high substrate tolerance accepting both aliphatic (lactate, D-Lac) and aromatic (phenyl lactate, D-Phe-Lac) amino acids.[91] Therefore, to create new CDP structures, the C-terminal module of PSYN was exchanged with modules of ESYN and BSYN, incorporating the amino acids L-Val or L-Phe, respectively. Two heterologous expression hosts were tested, a bacterial *Escherichia coli* strain and a fungal Tet-on expression system of *Aspergillus niger*,[111,144] with the latter showing much higher expression yields. Characterization of the peptides was performed by mass spectrometric peptide fingerprinting. Upon feeding a Lac/PheLac mixture to the *A. niger* strains carrying genes of chimeric CDPSs hPB-SYN and hPE-SYN (*h*ybrids *P*SYN-*B*SYN and *P*SYN-*E*SYN, respectively) all four expected Lac/PheLac enniatins and two beauvericin-type compounds were detected. This indicated that the proof-reading activity of the C_2 domain of enniatin synthetases is less restrictive than that of beauvericin.[143,145] Furthermore, only hexadepsipeptides were detected, which correspond to the size of wild-type enniatin and beauvericin, but not that of the octadepsipeptide type PF1022, suggesting that the ring-size is determined by the C-terminal module.[146] The three chimeric CDPs produced in the highest yields were tested for their nematicidal activity and showed either similar or slightly superior activity to the wild-type enniatin.[143] In future work, the authors plan to expand the structural diversity of CDPs by incorporation of non-natural α-hydroxy acids that could serve as substrates for post-biosynthetic modifications, such as cross-coupling reactions or click chemistry.

Module or domain swapping often yields insoluble proteins incapable of producing the non-natural peptides, which is mainly due to disruption of inter-domain linker regions or overall protein structure.

Engineering of tailoring enzymes is an important method to create structural diversity; however, it causes only minor changes, not affecting the constituents of the core peptide itself. The genes encoding enzymes responsible for halogenation,[147] glycosylation,[148,149] acylation and sulfation[150] are shuffled between the genomes of the strains producing similar NRPs to create new variants. Using this method, new chlorination patterns were shown for the lipopeptide enduracidin upon the exchange or insertion of the similar flavin-dependent halogenase from the related ramoplanin biosynthesis pathway.[147] The halogenase from the enduracidin biosynthesis pathway produces a dichlorinated compound, whereas the halogenase from the ramoplanin pathway acts only once. Complementation of the Δ*orf30* mutant of *Streptomyces fungicidicus*,[151] producing didechloroenduracidin, with the ramoplanin halogenase resulted in singly chlorinated enduracidin. Interestingly, complementation with the ramoplanin halogenase of the wild-type *S. fungicidicus* gave a new trichlorinated enduracidin, pointing to a possible cooperation of both halogenases. Although no improvement of antibacterial activity was observed, new non-natural entities could be obtained by this CB method.

The last method of combinatorial biosynthesis to generate non-natural cyclic peptides described here is directed evolution. It has already been suggested that the great diversity within the NRPs is due to evolutionary changes within the modular synthetases *via* point mutations, gene duplications, deletions or insertions.[65] Therefore, there are significant advantages of applying directed evolution approaches to combinatorial biosynthesis. This is mainly due to increase of the throughput by generation of a vast number of mutant variants in comparison to a single enzyme variant obtained by classical domain swapping or point mutation.[152–154] Moreover, directed evolution can be also applied when the mechanism and participating enzymes are not entirely elucidated. The first directed evolution approach was accomplished during studies on the production of enterobactin analogues. An A-domain incorporating Ser1 was exchanged for another NRP – syringomycin – A-domain of the same specificity. The 30-fold loss in activity of this chimeric domain could be restored with the use of directed evolution methods, in this specific case after rounds of random mutagenesis during mutagenic PCR.[152] Another example showed the utility of this method to generate new derivatives after manipulation of A-domains in the biosynthetic pathway of a NRPS/PKS hybrid peptide andrimid.[153] In this experiment, researchers were able to substitute Val at position 2 of the peptide with Ala, Phe, Leu, or Ile, and the correct substitution was confirmed by means of electrospray ionization (ESI) mass spectrometry.

In summary, combinatorial biosynthesis and domain engineering is a powerful approach to generate novel non-natural cyclic peptides with the potential to deliver compounds with enhanced bioactivities and pharmacological properties.[83,155] In contrast to polyketide synthases, the era of frequent manipulation of the NRPS biosynthetic pathways using the methods described above has started rather recently.[75,108,112,116–118,156,157] However, one should not forget about the common issues of CBS methods; it is often hard to predict what impact the modifications will have on the performance of the newly formed enzymes. Therefore, to fully exploit the potential of combinatorial biosynthesis, a better understanding of the mechanism and interaction of the key enzymes is required.[158]

Acknowledgements

We would like to thank the EU-project STREPSYNTH for the financial support. We would like also to thank Dr Patrick Durkin for his helpful proof-reading this chapter, and we would like to thank Dr Andi Mainz and Charlotte Steiniger for their helpful ideas with some figures.

References

1. P. G. Arnison, M. J. Bibb, G. Bierbaum, A. A. Bowers, T. S. Bugni, G. Bulaj, J. A. Camarero, D. J. Campopiano, G. L. Challis, J. Clardy, P. D. Cotter, D. J. Craik, M. Dawson, E. Dittmann, S. Donadio, P. C. Dorrestein, K. D. Entian, M. A. Fischbach, J. S. Garavelli, U. Goransson, C. W. Gruber, D. H. Haft, T. K. Hemscheidt, C. Hertweck, C. Hill, A. R. Horswill, M. Jaspars, W. L. Kelly, J. P. Klinman, O. P. Kuipers, A. J. Link, W. Liu, M. A. Marahiel, D. A. Mitchell, G. N. Moll, B. S. Moore, R. Muller, S. K. Nair, I. F. Nes, G. E. Norris, B. M. Olivera, H. Onaka, M. L. Patchett, J. Piel, M. J. Reaney, S. Rebuffat, R. P. Ross, H. G. Sahl, E. W. Schmidt, M. E. Selsted, K. Severinov, B. Shen, K. Sivonen, L. Smith, T. Stein, R. D. Süssmuth, J. R. Tagg, G. L. Tang, A. W. Truman, J. C. Vederas, C. T. Walsh, J. D. Walton, S. C. Wenzel, J. M. Willey and W. A. van der Donk, *Nat. Prod. Rep.*, 2013, **30**, 108–160.
2. M. Paul and W. A. v. d. Donk, *Mini-Rev. Org. Chem.*, 2005, **2**, 23–37.
3. N. Ghalit, J. F. Reichwein, H. W. Hilbers, E. Breukink, D. T. Rijkers and R. M. Liskamp, *ChemBioChem*, 2007, **8**, 1540–1554.
4. O. P. Kuipers, G. Bierbaum, B. Ottenwalder, H. M. Dodd, N. Horn, J. Metzger, T. Kupke, V. Gnau, R. Bongers, P. van den Bogaard, H. Kosters, H. S. Rollema, W. M. de Vos, R. J. Siezen, G. Jung, F. Gotz, H. G. Sahl and M. J. Gasson, *Antonie van Leeuwenhoek*, 1996, **69**, 161–169.
5. F. Oldach, R. Al Toma, A. Kuthning, T. Caetano, S. Mendo, N. Budisa and R. D. Süssmuth, *Angew. Chem., Int. Ed. Engl.*, 2012, **51**, 415–418.

6. R. S. Al Toma, A. Kuthning, M. P. Exner, A. Denisiuk, J. Ziegler, N. Budisa and R. D. Suessmuth, *ChemBioChem*, 2015, **16**, 503–509.
7. R. S. Al Toma, Engineering the Genetic Code and Expanding the Genetic Code for *In Vivo* Incorporation of Noncanonical Amino Acids into Biologically Active Ribosomal Peptides, PhD Thesis, Technische Universität Berlin, Berlin, Germany, 2015.
8. F. J. Piscotta, J. M. Tharp, W. R. Liu and A. J. Link, *Chem. Commun.*, 2015, **51**, 409–412.
9. N. Budisa, *Engineering the Genetic Code: Expanding the Amino Acid Repertoire for the Design of Novel Proteins*, Wiley, 2005.
10. N. Budisa, *Angew. Chem., Int. Ed. Engl.*, 2004, **43**, 6426–6463.
11. A. Dumas, L. Lercher, C. D. Spicer and B. G. Davis, *Chem. Sci.*, 2015, **6**, 50–69.
12. N. Budisa, B. Steipe, P. Demange, C. Eckerskorn, J. Kellermann and R. Huber, *Eur. J. Biochem.*, 1995, **230**, 788–796.
13. M. C. Hartman, K. Josephson, C. W. Lin and J. W. Szostak, *PLoS One*, 2007, **2**(10), e972.
14. F. H. Crick, L. Barnett, S. Brenner and R. J. Watts-Tobin, *Nature*, 1961, **192**, 1227–1232.
15. M. W. Nirenberg and J. H. Matthaei, *Proc. Natl. Acad. Sci. U. S. A.*, 1961, **47**, 1588–1602.
16. M. Nirenberg and P. Leder, *Science*, 1964, **145**, 1399–1407.
17. J. G. Moffatt and H. G. Khorana, *J. Am. Chem. Soc.*, 1961, **83**, 649–658.
18. F. H. Crick, *Symp. Soc. Exp. Biol.*, 1958, **12**, 138–163.
19. D. B. Cowie and G. N. Cohen, *Biochim. Biophys. Acta*, 1957, **26**, 252–261.
20. S. Schlesinger and M. J. Schlesinger, *J. Biol. Chem.*, 1967, **242**, 3369–3372.
21. E. Yoshikawa, M. J. Fournier, T. L. Mason and D. A. Tirrell, *Macromolecules*, 1994, **27**, 5471–5475.
22. M. G. Hösl, Expanding the Toolkit of Protein Engineering: Towards Multiple Simultaneous *In Vivo* Incorporation of Noncanonical Amino Acids, PhD Thesis, Technische Universität München, München, Germany, 2011.
23. M. G. Hösl and N. Budisa, *Curr. Opin. Biotechnol.*, 2012, **23**, 751–757.
24. A. Kuthning, E. Mosker and R. D. Süssmuth, *Appl. Microbiol. Biotechnol.*, 2015, **7**, 7.
25. T. Caetano, J. M. Krawczyk, E. Mosker, R. D. Süssmuth and S. Mendo, *Chem. Biol.*, 2011, **18**, 90–100.
26. E. Saxon and C. R. Bertozzi, *Science*, 2000, **287**, 2007–2010.
27. V. Hong, S. I. Presolski, C. Ma and M. G. Finn, *Angew. Chem., Int. Ed. Engl.*, 2009, **48**, 9879–9883.
28. L. Zhou, J. Shao, Q. Li, A. J. van Heel, M. P. de Vries, J. Broos and O. P. Kuipers, *Amino Acids*, 2016, **48**, 1309–1318.
29. D. R. Liu, T. J. Magliery and P. G. Schultz, *Chem. Biol.*, 1997, **4**, 685–691.
30. D. R. Liu, T. J. Magliery, M. Pastrnak and P. G. Schultz, *Proc. Natl. Acad. Sci. U. S. A.*, 1997, **94**, 10092–10097.
31. R. Furter, *Protein Sci.*, 1998, **7**, 419–426.
32. D. R. Liu and P. G. Schultz, *Proc. Natl. Acad. Sci. U. S. A.*, 1999, **96**, 4780–4785.

33. S. K. Blight, R. C. Larue, A. Mahapatra, D. G. Longstaff, E. Chang, G. Zhao, P. T. Kang, K. B. Green-Church, M. K. Chan and J. A. Krzycki, *Nature*, 2004, **431**, 333–335.
34. W. Wan, Y. Huang, Z. Wang, W. K. Russell, P. J. Pai, D. H. Russell and W. R. Liu, *Angew. Chem., Int. Ed. Engl.*, 2010, **49**, 3211–3214.
35. T. Yanagisawa, R. Ishii, R. Fukunaga, T. Kobayashi, K. Sakamoto and S. Yokoyama, *Chem. Biol.*, 2008, **15**, 1187–1197.
36. K. Nozawa, P. O'Donoghue, S. Gundllapalli, Y. Araiso, R. Ishitani, T. Umehara, D. Soll and O. Nureki, *Nature*, 2009, **457**, 1163–1167.
37. W. T. Li, A. Mahapatra, D. G. Longstaff, J. Bechtel, G. Zhao, P. T. Kang, M. K. Chan and J. A. Krzycki, *J. Mol. Biol.*, 2009, **385**, 1156–1164.
38. J. Xie, W. Liu and P. G. Schultz, *Angew. Chem., Int. Ed. Engl.*, 2007, **46**, 9239–9242.
39. H. Neumann, S. Y. Peak-Chew and J. W. Chin, *Nat. Chem. Biol.*, 2008, **4**, 232–234.
40. W. R. Liu, Y. S. Wang and W. Wan, *Mol. BioSyst.*, 2011, **7**, 38–47.
41. C. E. Melancon 3rd and P. G. Schultz, *Bioorg. Med. Chem. Lett.*, 2009, **19**, 3845–3847.
42. T. S. Young, I. Ahmad, J. A. Yin and P. G. Schultz, *J. Mol. Biol.*, 2010, **395**, 361–374.
43. S. M. Kuhn, M. Rubini, M. Fuhrmann, I. Theobald and A. Skerra, *J. Mol. Biol.*, 2010, **404**, 70–87.
44. O. Rackham and J. W. Chin, *Nat. Chem. Biol.*, 2005, **1**, 159–166.
45. K. Wang, H. Neumann, S. Y. Peak-Chew and J. W. Chin, *Nat. Biotechnol.*, 2007, **25**, 770–777.
46. Y. Huang, W. K. Russell, W. Wan, P. J. Pai, D. H. Russell and W. Liu, *Mol. BioSyst.*, 2010, **6**, 683–686.
47. W. H. Schmied, S. J. Elsasser, C. Uttamapinant and J. W. Chin, *J. Am. Chem. Soc.*, 2014, **136**, 15577–15583.
48. T. A. Knappe, U. Linne, S. Zirah, S. Rebuffat, X. Xie and M. A. Marahiel, *J. Am. Chem. Soc.*, 2008, **130**, 11446–11454.
49. T. A. Knappe, U. Linne, L. Robbel and M. A. Marahiel, *Chem. Biol.*, 2009, **16**, 1290–1298.
50. M. Tsunakawa, S.-L. Hu, Y. Hoshino, D. J. Detlefson, S. E. Hill, T. Furumai, R. J. White, M. Nishio and K. Kawano, *et al.*, *J. Antibiot.*, 1995, **48**, 433–434.
51. M. A. Delgado, M. R. Rintoul, R. N. Farias and R. A. Salomon, *J. Bacteriol.*, 2001, **183**, 4543–4550.
52. J. Mukhopadhyay, E. Sineva, J. Knight, R. M. Levy and R. H. Ebright, *Mol. Cell*, 2004, **14**, 739–751.
53. Y. Shi, X. Yang, N. Garg and W. A. van der Donk, *J. Am. Chem. Soc.*, 2011, **133**, 2338–2341.
54. E. L. Osher and A. Tavassoli, *Methods Mol. Biol.*, 2017, **1495**, 27–39.
55. C. P. Scott, E. Abel-Santos, M. Wall, D. C. Wahnon and S. J. Benkovic, *Proc. Natl. Acad. Sci. U. S. A.*, 1999, **96**, 13638–13643.
56. H. D. Mootz, *Split Inteins: Methods and Protocols*, Springer, 2017.

57. T. S. Young, D. D. Young, I. Ahmad, J. M. Louis, S. J. Benkovic and P. G. Schultz, *Proc. Natl. Acad. Sci. U. S. A.*, 2011, **108**, 11052–11056.
58. M. D. Tianero, M. S. Donia, T. S. Young, P. G. Schultz and E. W. Schmidt, *J. Am. Chem. Soc.*, 2012, **134**, 418–425.
59. N. A. Bindman, S. C. Bobeica, W. R. Liu and W. A. van der Donk, *J. Am. Chem. Soc.*, 2015, **137**, 6975–6978.
60. X. Luo, C. Zambaldo, T. Liu, Y. Zhang, W. Xuan, C. Wang, S. A. Reed, P. Y. Yang, R. E. Wang, T. Javahishvili, P. G. Schultz and T. S. Young, *Proc. Natl. Acad. Sci. U. S. A.*, 2016, **113**, 3615–3620.
61. D. J. Newman, G. M. Cragg and K. M. Snader, *J. Nat. Prod.*, 2003, **66**, 1022–1037.
62. E. A. Felnagle, E. E. Jackson, Y. A. Chan, A. M. Podevels, A. D. Berti, M. D. McMahon and M. G. Thomas, *Mol. Pharm.*, 2008, **5**, 191–211.
63. F. E. Koehn and G. T. Carter, *Nat. Rev. Drug Discovery*, 2005, **4**, 206–220.
64. F. R. Schmidt, *Appl. Microbiol. Biotechnol.*, 2003, **63**, 335–343.
65. D. E. Cane, C. T. Walsh and C. Khosla, *Science*, 1998, **282**, 63–68.
66. I. Höfer, M. Crüsemann, M. Radzom, B. Geers, D. Flachshaar, X. Cai, A. Zeeck and J. Piel, *Chem. Biol.*, 2011, **18**, 381–391.
67. M. Winn, J. K. Fyans, Y. Zhuo and J. Micklefield, *Nat. Prod. Rep.*, 2016, **33**, 317–347.
68. T. Stachelhaus, A. Hüser and M. A. Marahiel, *Chem. Biol.*, 1996, **3**, 913–921.
69. T. Stachelhaus, H. D. Mootz, V. Bergendahl and M. A. Marahiel, *J. Biol. Chem.*, 1998, **273**, 22773–22781.
70. R. Finking and M. A. Marahiel, *Annu. Rev. Microbiol.*, 2004, **58**, 453–488.
71. G. Weber, K. Schörgendorfer, E. Schneider-Scherzer and E. Leitner, *Curr. Genet.*, 1994, **26**, 120–125.
72. S. Caboche, M. Pupin, V. Leclère, A. Fontaine, P. Jacques and G. Kucherov, *Nucleic Acids Res.*, 2008, **36**, D326–D331.
73. C. T. Walsh, R. V. O'Brien and C. Khosla, *Angew. Chem., Int. Ed. Engl.*, 2013, **52**, 7098–7124.
74. R. Thiericke and J. Rohr, *Nat. Prod. Rep.*, 1993, **10**, 265–289.
75. S. Weist and R. D. Süssmuth, *Appl. Microbiol. Biotechnol.*, 2005, **68**, 141–150.
76. V. Miao, M. F. Coeffet-Le Gal, K. Nguyen, P. Brian, J. Penn, A. Whiting, J. Steele, D. Kau, S. Martin, R. Ford, T. Gibson, M. Bouchard, S. K. Wrigley and R. H. Baltz, *Chem. Biol.*, 2006, **13**, 269–276.
77. K. T. Nguyen, D. Ritz, J.-Q. Gu, D. Alexander, M. Chu, V. Miao, P. Brian and R. H. Baltz, *Proc. Natl. Acad. Sci. U. S. A.*, 2006, **103**, 17462–17467.
78. T. Stachelhaus, A. Schneider and M. A. Marahiel, *Science*, 1995, **269**, 69–72.
79. D. Butz, T. Schmiederer, B. Hadatsch, W. Wohlleben, T. Weber and R. D. Süssmuth, *ChemBioChem*, 2008, **9**, 1195–1200.
80. T. Stachelhaus, H. D. Mootz and M. A. Marahiel, *Chem. Biol.*, 1999, **6**, 493–505.

81. J. Thirlway, R. Lewis, L. Nunns, M. Al Nakeeb, M. Styles, A.-W. Struck, C. P. Smith and J. Micklefield, *Angew. Chem., Int. Ed. Engl.*, 2012, **51**, 7181–7184.
82. A. S. Eustáquio, D. O'Hagan and B. S. Moore, *J. Nat. Prod.*, 2010, **73**, 378–382.
83. H. Sun, Z. Liu, H. Zhao and E. L. Ang, *Drug Des., Dev. Ther.*, 2015, **9**, 823–833.
84. A. Kirschning, F. Taft and T. Knobloch, *Org. Biomol. Chem.*, 2007, **5**, 3245–3259.
85. S. N. Agathos and J. Lee, *Biotechnol. Prog.*, 1993, **9**, 54–63.
86. U. Keller, R. Zocher and H. Kleinkauf, *J. Gen. Microbiol.*, 1980, **118**, 485–494.
87. A. Lawen, R. Traber, R. Reuille and M. Ponelle, *Biochem. J.*, 1994, **300**, 395–399.
88. S. C. Feifel, T. Schmiederer, T. Hornbogen, H. Berg, R. D. Süssmuth and R. Zocher, *ChemBioChem*, 2007, **8**, 1767–1770.
89. M. Krause, A. Lindemann, M. Glinski, T. Hornbogen, G. Bonse, P. Jeschke, G. Thielking, W. Gau, H. Kleinkauf and R. Zocher, *J. Antibiot.*, 2001, **54**, 797–804.
90. D. Matthes, L. Richter, J. Müller, A. Denisiuk, S. C. Feifel, Y. Xu, P. Espinosa-Artiles, R. D. Süssmuth and I. Molnár, *Chem. Commun.*, 2012, **48**, 5674–5676.
91. J. Müller, S. C. Feifel, T. Schmiederer, R. Zocher and R. D. Süssmuth, *ChemBioChem*, 2009, **10**, 323–328.
92. E. Beacco, M. L. Bianchi, A. Minghetti and C. Spalla, *Experientia*, 1978, **34**, 1291–1293.
93. A. Baumert, D. Erge and D. Groger, *Planta Med.*, 1982, **44**, 122–123.
94. R. Traber, H. Hofmann and H. Kobel, *J. Antibiot. (Tokyo)*, 1989, **42**, 591–597.
95. A. A. Patchett, D. Taub, O. D. Hensens, R. T. Goegelman, L. Yang, F. Dumont, L. Peterson and N. H. Sigal, *J. Antibiot. (Tokyo)*, 1992, **45**, 94–102.
96. A. Lawen, R. Trader, D. Geyl, R. Zocher and H. Kleinkauf, *J. Antibiot. (Tokyo)*, 1989, **42**, 1283–1289.
97. H. Kobel and R. Traber, *Eur. J. Appl. Microbiol. Biotechnol.*, 1982, **14**, 237–240.
98. O. D. Hensens, R. F. White, R. T. Goegelman, E. S. Inamine and A. A. Patchett, *J. Antibiot. (Tokyo)*, 1992, **45**, 133–135.
99. A. Lawen, R. Traber and D. Geyl, *J. Biol. Chem.*, 1991, **266**, 15567–15570.
100. A. Lawen and R. Traber, *J. Biol. Chem.*, 1993, **268**, 20452–20465.
101. S. I. Kozhushkov, B. D. Zlatopolskiy, M. Brandl, P. Alvermann, M. Radzom, B. Geers, A. de Meijere and A. Zeeck, *Eur. J. Org. Chem.*, 2005, **2005**, 854–863.
102. B. D. Zlatopolskiy, M. Radzom, A. Zeeck and A. de Meijere, *Eur. J. Org. Chem.*, 2006, **2006**, 1525–1534.

103. B. Bister, D. Bischoff, G. J. Nicholson, S. Stockert, J. Wink, C. Brunati, S. Donadio, S. Pelzer, W. Wohlleben and R. D. Süssmuth, *ChemBioChem*, 2003, **4**, 658–662.
104. S. J. Daum and R. J. Lemke, *Annu. Rev. Microbiol.*, 1979, **33**, 241–265.
105. A. Kirschning and F. Hahn, *Angew. Chem., Int. Ed. Engl.*, 2012, **51**, 4012–4022.
106. S. Boecker, S. Zobel, V. Meyer and R. D. Süssmuth, *Fungal Genet. Biol.*, 2016, **89**, 89–101.
107. A. J. Birch, *Pure Appl. Chem.*, 1963, **7**, 527–537.
108. W. T. Shier, K. L. Rinehart Jr and D. Gottlieb, *Proc. Natl. Acad. Sci. U. S. A.*, 1969, **63**, 198–204.
109. K. L. Rinehart Jr, *Pure Appl. Chem.*, 1977, **49**, 1361–1384.
110. S. Zobel, J. Kumpfmüller, R. D. Süssmuth and T. Schweder, *Appl. Microbiol. Biotechnol.*, 2014, **99**, 681–691.
111. L. Richter, F. Wanka, S. Boecker, D. Storm, T. Kurt, Ö. Vural, R. Süßmuth and V. Meyer, *Fungal Biol. Biotechnol.*, 2014, **1**, 4.
112. C. J. Dutton, S. P. Gibson, A. C. Goudie, K. S. Holdom, M. S. Pacey, J. C. Ruddock, J. D. Bu'Lock and M. K. Richards, *J. Antibiot.*, 1991, **44**, 357–365.
113. U. Galm, M. A. Dessoy, J. Schmidt, L. A. Wessjohann and L. Heide, *Chem. Biol.*, 2014, **11**, 173–183.
114. S. Kitamura, H. Kase, Y. Odakura, T. Iida, K. Shirahata and K. Nakayama, *J. Antibiot.*, 1982, **35**, 94–96.
115. K. Takeda, A. Kinumaki, T. Furumai, T. Yamaguchi, S. Ohshima and Y. Ito, *J. Antibiot.*, 1978, **31**, 247–249.
116. S. Weist, B. Bister, O. Puk, D. Bischoff, S. Pelzer, G. J. Nicholson, W. Wohlleben, G. Jung and R. D. Süssmuth, *Angew. Chem., Int. Ed. Engl.*, 2002, **41**, 3383–3385.
117. S. Weist, C. Kittel, D. Bischoff, B. Bister, V. Pfeifer, G. J. Nicholson, W. Wohlleben and R. D. Süssmuth, *J. Am. Chem. Soc.*, 2004, **126**, 5942–5943.
118. Z. Hojati, C. Milne, B. Harvey, L. Gordon, M. Borg, F. Flett, B. Wilkinson, P. J. Sidebottom, B. A. Rudd, M. A. Hayes, C. P. Smith and J. Micklefield, *Chem. Biol.*, 2002, **9**, 1175–1187.
119. K. C. Nicolaou, C. N. C. Boddy, S. Bräse and N. Winssinger, *Angew. Chem., Int. Ed. Engl.*, 1999, **38**, 2096–2152.
120. B. K. Hubbard and C. T. Walsh, *Angew. Chem., Int. Ed. Engl.*, 2003, **42**, 730–765.
121. D. H. Williams and B. Bardsley, *Angew. Chem., Int. Ed. Engl.*, 1999, **38**, 1172–1193.
122. R. D. Süssmuth, *ChemBioChem*, 2002, **3**, 295–298.
123. A. Markham, *Drugs*, 2014, **74**, 1823–1828.
124. O. Puk, P. Huber, D. Bischoff, J. Recktenwald, G. Jung, R. D. Süssmuth, K. H. van Pée, W. Wohlleben and S. Pelzer, *Chem. Biol.*, 2002, **9**, 225–235.

125. V. Pfeifer, G. J. Nicholson, J. Ries, J. Recktenwald, A. B. Schefer, R. M. Shawky, J. Schröder, W. Wohlleben and S. Pelzer, *J. Biol. Chem.*, 2001, **276**, 38370–38377.
126. B. A. M. Rudd, Genetics of pigmented secondary metabolites in Streptomyces coelicolor A3(2), PhD Thesis, University of East Anglia, Norwich, U.K., 1978.
127. R. S. Al Toma, C. Brieke, M. J. Cryle and R. D. Süssmuth, *Nat. Prod. Rep.*, 2015, **32**, 1207–1235.
128. A. Powell, M. A. Nakeeb, B. Wilkinson and J. Micklefield, *Chem. Commun.*, 2007, 2683–2685.
129. A. Powell, M. Borg, B. Amir-Heidari, J. M. Neary, J. Thirlway, B. Wilkinson, C. P. Smith and J. Micklefield, *J. Am. Chem. Soc.*, 2007, **129**, 15182–15191.
130. C. Milne, A. Powell, J. Jim, M. Al Nakeeb, C. P. Smith and J. Micklefield, *J. Am. Chem. Soc.*, 2006, **128**, 11250–11259.
131. F. T. Counter, N. E. Allen, D. S. Fukuda, J. N. Hobbs, J. Ott, P. W. Ensminger, J. S. Mynderse, D. A. Preston and C. Y. E. Wu, *J. Antibiot.*, 1990, **43**, 616–622.
132. Y. Xu, T. Zhou, S. Zhang, P. Espinosa-Artiles, L. Wang, W. Zhang, M. Lin, A. A. L. Gunatilaka, J. Zhan and I. Molnár, *Proc. Natl. Acad. Sci. U. S. A.*, 2014, **111**, 12354–12359.
133. G. L. Challis, J. Ravel and C. A. Townsend, *Chem. Biol.*, 2000, **7**, 211–224.
134. H. Kries, R. Wachtel, A. Pabst, B. Wanner, D. Niquille and D. Hilvert, *Angew. Chem., Int. Ed. Engl.*, 2014, **53**, 10105–10108.
135. H. Kaljunen, S. H. Schiefelbein, D. Stummer, S. Kozak, R. Meijers, G. Christiansen and A. Rentmeister, *Angew. Chem., Int. Ed. Engl.*, 2015, **54**, 8833–8836.
136. M. J. Calcott and D. F. Ackerley, *Biotechnol. Lett.*, 2014, **36**, 2407–2416.
137. P. J. Belshaw, C. T. Walsh and T. Stachelhaus, *Science*, 1999, **284**, 486–489.
138. R. H. Baltz, V. Miao and S. K. Wrigley, *Nat. Prod. Rep.*, 2005, **22**, 717–741.
139. S. Doekel, M.-F. Coëffet-Le Gal, J.-Q. Gu, M. Chu, R. H. Baltz and P. Brian, *Microbiology*, 2008, **154**, 2872–2880.
140. K. T. Nguyen, X. He, D. C. Alexander, C. Li, J.-Q. Gu, C. Mascio, A. V. Praagh, L. Mortin, M. Chu, J. A. Silverman, P. Brian and R. H. Baltz, *Antimicrob. Agents Chemother.*, 2010, **54**, 1404–1413.
141. R. H. Baltz, P. Brian, V. Miao and S. K. Wrigley, *J. Ind. Microbiol. Biotechnol.*, 2006, **33**, 66–74.
142. H. D. Mootz, D. Schwarzer and M. A. Marahiel, *Proc. Natl. Acad. Sci. U. S. A.*, 2000, **97**, 5848–5853.
143. S. Zobel, S. Boecker, D. Kulke, D. Heimbach, V. Meyer and R. D. Süssmuth, *ChemBioChem*, 2016, **17**, 283–287.
144. V. Meyer, F. Wanka, J. van Gent, M. Arentshorst, C. A. van den Hondel and A. F. Ram, *Appl. Environ. Microbiol.*, 2011, **77**, 2975–2983.
145. S. Meyer, J. C. Kehr, A. Mainz, D. Dehm, D. Petras, R. D. Süssmuth and E. Dittmann, *Cell Chem. Biol.*, 2016, **23**, 462–471.
146. D. Yu, F. Xu, D. Gage and J. Zhan, *Chem. Commun.*, 2013, **49**, 6176–6178.

147. X. Yin, Y. Chen, L. Zhang, Y. Wang and T. M. Zabriskie, *J. Nat. Prod.*, 2010, **73**, 583–589.
148. S. Di Palo, R. Gandolfi, S. Jovetic, F. Marinelli, D. Romano and F. Molinari, *Enzyme Microb. Technol.*, 2007, **41**, 806–811.
149. S.-Y. Lyu, Y.-C. Liu, C.-Y. Chang, C.-J. Huang, Y.-H. Chiu, C.-M. Huang, N.-S. Hsu, K.-H. Lin, C.-J. Wu, M.-D. Tsai and T.-L. Li, *J. Am. Chem. Soc.*, 2014, **136**, 10989–10995.
150. L. Kalan, J. Perry, K. Koteva, M. Thaker and G. Wright, *J. Bacteriol.*, 2013, **195**, 167–171.
151. X. Yin and T. M. Zabriskie, *Microbiology*, 2006, **152**, 2969–2983.
152. M. A. Fischbach, J. R. Lai, E. D. Roche, C. T. Walsh and D. R. Liu, *Proc. Natl. Acad. Sci. U. S. A.*, 2007, **104**, 11951–11956.
153. B. S. Evans, Y. Chen, W. W. Metcalf, H. Zhao and N. L. Kelleher, *Chem. Biol.*, 2011, **18**, 601–607.
154. B. Villiers and F. Hollfelder, *Chem. Biol.*, 2011, **18**, 1290–1299.
155. R. J. Goss, S. Shankar and A. A. Fayad, *Nat. Prod. Rep.*, 2012, **29**, 870–889.
156. J. Staunton and B. Wilkinson, *Curr. Opin. Chem. Biol.*, 2001, **5**, 159–164.
157. H. G. Menzella, R. Reid, J. R. Carney, S. S. Chandran, S. J. Reisinger, K. G. Patel, D. A. Hopwood and D. V. Santi, *Nat. Biotechnol.*, 2005, **23**, 1171–1176.
158. U. T. Bornscheuer, G. W. Huisman, R. J. Kazlauskas, S. Lutz, J. C. Moore and K. Robins, *Nature*, 2012, **485**, 185–194.

CHAPTER 5

Modulation of Protein–Protein Interactions Using Cyclic Peptides

SALVADOR GUARDIOLA[a], SONIA CIUDAD[a], LAURA NEVOLA[b] AND ERNEST GIRALT*[a,c]

[a]Institute for Research in Biomedicine (IRB Barcelona), The Barcelona Institute of Science and Technology, Baldiri Reixac, 10, 08028 Barcelona, Spain; [b]IDP Pharma, Barcelona Science Park, Baldiri Reixac 10, 08028, Barcelona, Spain; [c]Department of Inorganic and Organic Chemistry, University of Barcelona, Spain
*E-mail: ernest.giralt@irbbarcelona.org

5.1 Introduction

The human protein–protein interactome is estimated to contain between 130 000 and 650 000 protein–protein interactions (PPIs), forming highly complex networks that constitute the basis of genotype–phenotype relationships in organisms.[1,2] However, despite enormous scientific effort, only around 10–15% of the human interactome is currently known,[3] and only a few PPIs have been the target of drug discovery efforts.

In contrast to conventional drug targets (*i.e.* enzymes/receptors), PPIs pose a considerable challenge due to their large contact areas[4] (1000–3000 Å^2) and the lack of well-defined cavities that characterize protein interfaces, their chemical space differing considerably from traditional drug-like compounds.[5]

Chemical Biology No. 6
Cyclic Peptides: From Bioorganic Synthesis to Applications
Edited by Jesko Koehnke, James Naismith and Wilfred A. van der Donk

Published by the Royal Society of Chemistry, www.rsc.org

In addition, there are no endogenous small PPI ligands that can be used as a starting point to design new inhibitors. Protein interactions are mediated by regions in which a small subset of residues, the most relevant for the interaction, are often delocalized over multiple epitopes.[6] Despite accounting for less than half of the interaction surface, these hotspots contribute most of the free energy of binding.[7] Although time-consuming and labor-intensive, alanine-scanning mutagenesis, consisting of systematic rounds of point mutations and binding experiments, has been the most robust method to map hotspot residues on protein interfaces. However, when PPI structural data (such as that achieved using X-ray crystallography or NMR) are available, the determination of hotspots is facilitated, and a number of computational methods (discussed later) have been developed for this purpose.

PPIs can be peptide-mediated (a continuous binding epitope, often adopting a precise secondary structure, accounts for the binding to the partner protein) or domain-mediated (Figure 5.1). In the latter, which are often found in globular proteins, the hotspots tend to be scattered along the protein sequence, being brought into spatial proximity only by protein folding.[8] Given the added difficulty of targeting such discontinuous protein–protein interfaces,[7] it comes as no surprise that most studies in the literature have been devoted to the modulation of peptide-mediated PPIs.

In this context, the use of linear peptide segments as PPI modulators has a major disadvantage, namely that their high degree of flexibility results in a negligible or random structure, thereby severely impairing recognition by a well-structured protein target. In addition, linear peptides are prone to protease degradation, low membrane permeability, and metabolic instability.[9] Peptide cyclization has been used extensively to overcome these problems, resulting in better mimetics of protein interfaces.[10] The development of

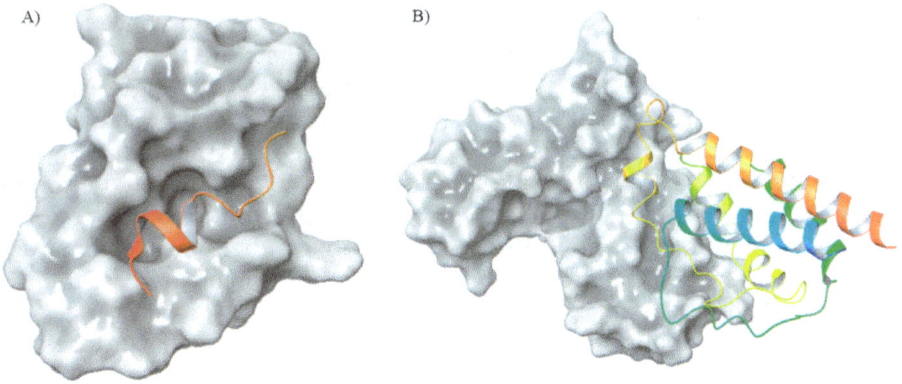

Figure 5.1 (A) A peptide-mediated interaction: a short helical motif of p53 (in orange cartoon) binds to the MDM2 oncoprotein (on the gray surface); PDB code: 1YCR. (B) The domain-mediated interaction of IL-2 (in cartoon) bound to its α-receptor (on the gray surface); note how several regions of IL-2 contribute to the binding interface; PDB code: 1Z92.

specific cyclic scaffolds, sometimes with additional molecular constraints, has allowed the structural and functional mimicry of protein surfaces with relatively small structures by means of robust synthetic methods. In the following sections, we will address some of the most successful strategies used to design and develop bioactive cyclic peptides, going from structure-driven approaches (mimicry of interface α-helices, β-strands, loops and turns) to *in vitro* surface display techniques, all the way through to computational and fragment-based tools.

5.2 Structure-based Design

5.2.1 "Classic" Cyclic Peptides

In Nature, examples of peptides with cyclic structures abound. These range from hormones like oxytocin and insulin to microorganism metabolites such as cyclosporine A or vancomycin, all of which are now available in the pharmaceutical market. In most cases, ring structures are formed by disulfide bridges between cysteine residues. Other types of cyclization are head-to-tail (the ring is formed *via* an amide bond between the N- and C-termini), side-chain-to-side-chain (between two side chain functional groups), and side-chain-to-head or side-chain-to-tail. In PPIs, one of the partners sometimes exposes a loop which makes the greatest contribution to the molecular recognition event. In these cases, loops with a ring-like structure cannot be targeted by small molecules or by linear peptides, which have to pay a high entropic price because of their flexibility. Indeed, a recent study demonstrated that some PPIs can be targeted only by cyclic molecules, disclosing a number of useful guidelines for macrocyclic drug design.[11]

In addition to backbone cyclization, the flexibility of peptides can be further constrained by other methods, such as the introduction of a proline ring or an *N*-methyl group into the structure. N. Fujii's group have recently developed a cyclic pentapeptide that selectively targets the CXC chemokine receptor 7 (CXCR7), a protein that closely resembles CXCR4 (Figure 5.2A).[12] Starting from a number of potent CXCR4 inhibitors, they systematically explored individual modifications on the pentapeptide scaffold to switch the activity from CXCR4 to CXCR7. By introducing an *N*-methylated Arg residue and mutating the Gly residue to Pro, they restricted the conformational space considerably and showed how the activity can be tuned for peptides with an identical combination of functional groups (phenol, two guanidine groups, and naphthalene).

Experimental data on the binding of a peptide ligand, such as those provided by X-ray crystallography and NMR, are crucial for understanding the molecular details of the interaction. The SPSB2 protein mediates the proteosomal degradation of inducible nitric oxide synthase (iNOS). Starting from the N-terminal region of iNOS, the work of Yap *et al.*[13] is a nice example of how a combination of biophysical tools can guide the design of potent cyclic PPI inhibitors. Eventually, the authors came up with a cyclic octapeptide that

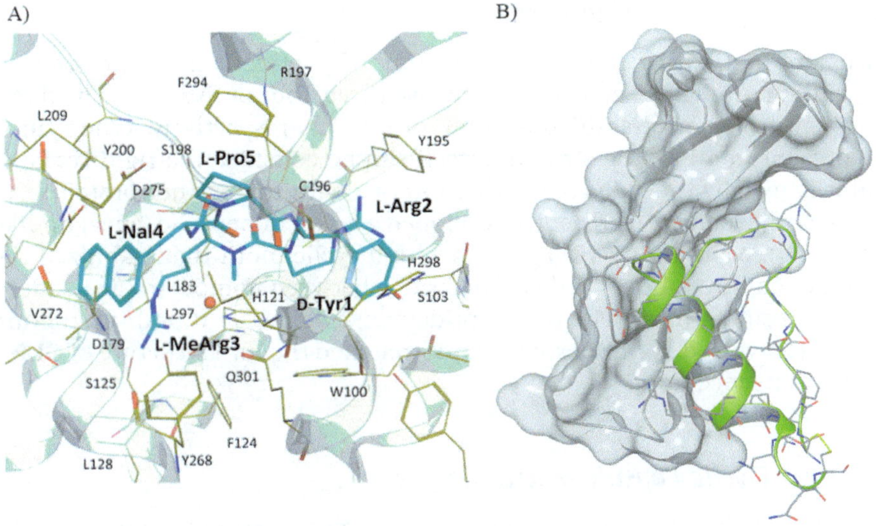

Figure 5.2 (A) Binding mode of a highly constrained cyclic peptide to CXCR7 (IC_{50} = 0.80 μM). The additional rigidity provided by Pro and N-MeArg make it selective for CXCR7 over CXCR4. Reproduced with permission from Oishi *et al.*, *J. Med. Chem.*, **2015**, 58, 5218–5225, http://pubs.acs.org/doi/abs/10.1021%2Facs.jmedchem.5b00216. Copyright (2015) American Chemical Society. (B) A cyclic miniprotein (in green cartoon), derived from the extracellular domain of EGFR, is able to bind EGF (on the gray surface) and disrupt the EGF–EGFR interaction.

reproduces the conformation of N-terminal iNOS and binds to SPSB2 with strong affinity (K_d = 4.4 nM).

Larger cyclic peptides with a well-defined secondary structure are able to mimic the hotspots of a particular receptor and maintain its original function. Using this approach, in our group we identified a cyclic subdomain in EGFR that was critical for the recognition of EGF.[14] This cyclic miniprotein had a very similar structure in solution to that of the receptor and was able to bind to EGF with a μM K_d (Figure 5.2B). Despite its lower affinity compared to the natural receptor, the peptide was found to effectively disrupt the EGF–EGFR interaction in cancer cells, thereby leading to a novel mechanism for anticancer therapy.

5.2.2 Secondary Structure Mimetics

As previously discussed, in almost 40% of all known PPIs, the interacting peptide in one of the partners adopts a specific secondary structure, normally an α-helix.[8] The stability of the secondary structure is closely linked to its ability to interact with the target protein. In this regard, rigidification of the peptide structure by cyclization is one of the main strategies to constrain the folding of peptides into well-defined secondary structures.

5.2.2.1 Cyclic Peptide Turns

Turns are motifs that reverse the direction of peptide helices and strands. Depending on their size, turns can involve three, four or five amino acid residues, and they are classified as γ-, β- or α-turns, respectively. When turns are present in a PPI, the structure itself can be considered a rigid scaffold, and it is usually the side chains that interact with the partner protein.[15] In this regard, Hoang *et al.* recently developed a series of peptidic glucagon-like peptide-1 receptor (GLP-1R) agonists with a cyclic motif that stabilized the bioactive conformation.[16] A disulfide bridge between homocysteine residues 2 and 5 resulted in a cyclic tetrapeptide adopting a type II β-turn structure which, as revealed in NMR experiments, matched the turn structure of GLP-1 bound to its receptor.

5.2.2.2 Cyclic Peptide β-strands

β-sheets consist of extended polypeptide chains (β-strands) connected by a precise hydrogen bonding network, with contiguous side chains facing opposite sides of the backbone. β-sheets are generally not flat, but rather the strands are twisted in a right-handed fashion, forming a 15–20° angle between them.[17] The β-sheet accounts for ≈30% of all protein secondary structure. Although the β-sheet was long considered a mere scaffold in protein architecture, there are now many examples of its involvement in protein–protein and protein–DNA interactions.[18] Many proteases (*e.g.* HIV-1 protease) work as dimers and are held together by the formation of inter-molecular β-sheets.[19,20] Likewise, β-sheet formation is involved in some oncogenic PPIs, like the Raf–Rap1 interaction that leads to tumor cell proliferation.[21,22] Finally, in many neurodegenerative diseases, such as Alzheimer's and Huntington's disease, soluble proteins form β-sheet structures that aggregate into oligomers and insoluble polymers.[23,24] Thus, peptides mimicking β-strands have vast potential as therapeutic agents, thus explaining the immense research effort dedicated to this field, ranging from early peptidomimetic scaffolds[25,26] to more complex structures that efficiently orientate the peptide side chains in a β-strand pattern.

Protease recognition of a β-strand motif has been successfully modulated using macrocyclic templates that stabilize short peptides in their bioactive conformation. This is exemplified by the work of Tyndall *et al.*,[27] whose design features a tripeptide chain that is successfully pre-organized by a cyclic scaffold into a high-affinity protease substrate conformation, leading to low nM antagonists of HIV-1 protease (Figure 5.3A). In larger β-sheet structures, artificial turns and template units are introduced to nucleate and stabilize the β-strands (see reviews[28,29]). Achiral α-aminoisobutyric acid and D-Pro are typically used in the $i+1$ position of the turn, in combination with Gly, Ala or other units in the $i+2$ position.[30–32] Two δ-linked ornithine β-turn mimics have also been widely used to promote β-sheet folding in water.[33–35]

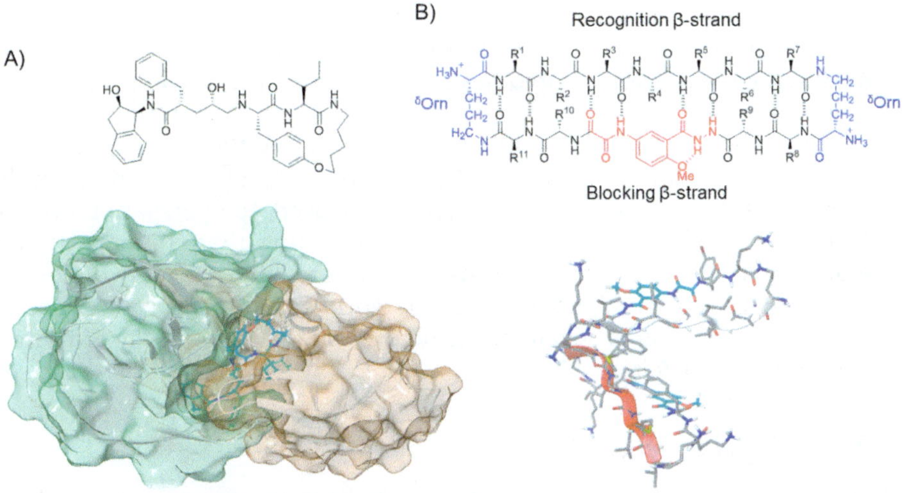

Figure 5.3 (A) A small cyclic scaffold is sufficient to pre-organize a tripeptide chain into its bioactive conformation (top). The crystal structure of the inhibitor bound to the HIV-1 protease confirms its strand conformation, which is key for protease recognition (bottom); PDB code: 1D4K. (B) Structural features of an amyloid β (Aβ)-sheet mimic that reduces amyloid protein toxicity. In the lower strand, the Hao unit acts as the aggregation blocker (top). The X-ray structure of the β-sheet peptide shows a dimer in which the two recognition β-strands come together in an antiparallel fashion.

Parallel β-sheet macrocycles are synthetically more demanding and have received less attention.[29,36] In all cases, peptide solubility, strand length, and amino acid pairing play an important role in the stability of the designed β-sheet and should be carefully inspected.[37–39]

The amyloid protein aggregation that characterizes many devastating diseases, such as Alzheimer's disease, Parkinson's disease and type II diabetes, involves a variety of β-sheet structures in which native proteins aggregate to form oligomers and fibrils.[40–42] To interfere with aggregation, a number of β-sheet mimetic cyclic peptides have been developed as competitive inhibitors of amyloid fibrillation[43–47] (see also reviews[48–50]). Such inhibitors typically feature a parallel β-sheet structure, in which one of the strands is complementary to the amyloid sequence, thus promoting recognition of the aggregated protein, and the other one carries a modification that impedes further growth of the fibril. Along these lines, Nowick's group has achieved remarkable results, ranging from the synthesis of simple β-sheet templates[51,52] to the development of a refined cyclic β-sheet scaffold that dimerizes in aqueous solution.[53] A Hao unit, an unnatural amino acid that mimics a tripeptide β-strand, is introduced into one of the strands in order to reduce its hydrogen bond functionality, thus preventing the aggregation of the peptide. Both strands, oriented in an antiparallel fashion, are cyclized by two δ-turns that restrain the conformation of the entire structure (Figure 5.3B). In a seminal work,[54] this scaffold was used to display a range of 18 amyloidogenic sequences derived from amyloid proteins, including the amyloid β (Aβ) peptide involved in Alzheimer's disease. β-sheet folding in water was in most cases confirmed by NMR experiments, which revealed the robustness of this scaffold to a variety of sequences. Remarkably, one of these compounds, featuring residues 16–22 of the Aβ peptide, delayed Aβ aggregation even at sub-stoichiometric concentrations, thus confirming that this structurally pre-organized artificial β-sheet binds more strongly to the early oligomers than the Aβ unstructured monomers. This work is a promising precedent for the future application of secondary structure mimetic peptides to efficiently reduce amyloid toxicity.

5.2.2.3 *Cyclic Helical Peptides*

First discovered by L. Pauling in 1951,[55] α-helices comprise the most abundant class of protein secondary structure, mediating >60% of all PPIs gathered in the PDB.[56] The α-helix, featuring 3.6 residues per turn ($i + 4 \rightarrow i$ backbone amide hydrogen bonds), accounts for *ca.* 90% of all helical structures. The 3_{10}-helix ($i + 3 \rightarrow i$ hydrogen bonding) and the π-helix ($i + 5 \rightarrow i$) are rarely found in protein interfaces and are not addressed here. While helices located deep in the core of proteins are crucial for the stability of their tertiary structure, exposed α-helices on the protein surface are vital for the molecular recognition of other proteins and nucleic acids. In such a context, a ligand that successfully reproduces the key features of the helix is likely to act as a competitive inhibitor of the interaction. In a detailed study of helical interfaces,[56] Jochim and Arora distinguished the following three main types of interaction: (a) proteins with a well-defined cleft for helix binding, such as p53/HDM2[57] and Bcl-xL/Bak[58] (Figure 5.1A); (b) extended interfaces mediated by a number of helical motifs, which often involve more than one side of the helix, such as the NOTCH transcription factor[59] and the hexameric gp41;[60] and (c) low-affinity transient interactions lacking well-defined hotspots. Of the three classes, a rational design approach, based on the interface structure, would be most feasible for the first class. This chapter will focus exclusively on cyclic peptides, although there are excellent reviews covering other types of α-helical scaffolds.[61–63]

Decreasing the structural freedom of linear peptides, by introducing a specific cyclic constraint to promote helicity, enhances not only the binding properties (due to the reduced entropic cost of adopting the bioactive conformation), but also the resistance to protease degradation, serum stability, and cell membrane permeability.[64,65] Numerous cyclization strategies, usually through the side chains of two residues, have been developed for helix stabilization (Figure 5.4).

Lactam Bridges. Since their first application in the early 90s,[66,67] lactam bridges have been the approach most widely used to stabilize α-helical peptides. A covalent side chain-to-side chain amide bond between lysine and

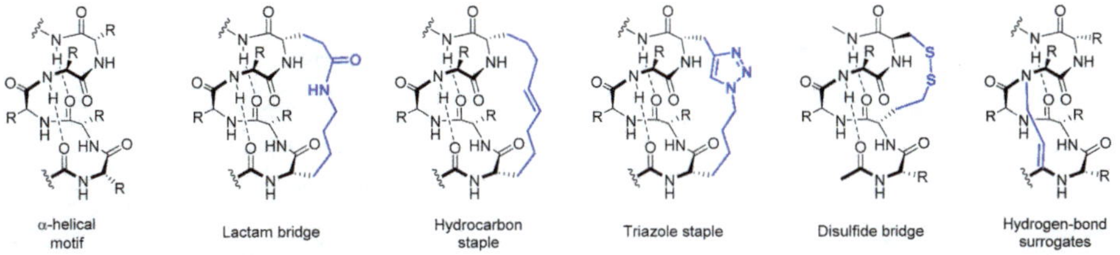

Figure 5.4 Schematic representation of the main approaches used for covalent helix stabilization. The artificial element is shown in blue.

glutamic/aspartic residues located on the same side of the helix (positions $i+4 \rightarrow i$ - one turn, $i+7 \rightarrow i$ - two turns, $i+11 \rightarrow i$ - three turns) can promote the helical folding of the peptide. For shorter peptides, which often lack a secondary structure, Fairlie and co-workers performed a comparative study of several cyclization linkers (Figure 5.4). They found that a Lys1 → Asp5 lactam linker conferred the greatest helicity to a model pentapeptide in water.[68] Using this approach, this group successfully developed constrained α-helices that effectively mimic helical epitopes of viral, bacterial and human proteins, thus acting as competitive inhibitors.[69] For instance, a 17-residue helical peptide of HIV-1 Rev, which binds to RNA, is responsible for the transport of viral mRNA from the nucleus to the cytoplasm of HIV-infected host cells. By introducing two lactam-bridged cycles in the Rev peptide sequence, they obtained a highly helical bicyclic peptide that showed nM affinity for its target RNA. In these examples, helix pre-organization resulted in up to a 10 000-fold enhancement of affinity, compared to the linear analogues, as well as increased stability to serum proteases.

Longer lactam linkers, $i+7 \rightarrow i$ and $i+11 \rightarrow i$, often containing unsaturated bonds and aromatic rings, have also been explored, but their application for the modulation of PPIs has remained limited.[70,71]

Ring-closing Metathesis. Hydrocarbon-stapled peptides were pioneered by Verdine and co-workers in 2000.[72] Olefinic staples have since proved to be a valuable scaffold for targeting challenging PPIs and have become one of the most thoroughly studied and successful peptide stapling techniques. The biggest advantage of this technique is that it endows peptides with enhanced ability to penetrate cellular membranes,[73] with staple-type and formal charge being the two most relevant parameters for cell penetration.[65] The staple is usually introduced on the non-interacting face of the helix, although in some cases the hydrophobic character of this linker has been used to achieve contacts with the binding cleft of the partner protein.[74,75] However, prediction of the most stabilizing architecture and optimal site for staple introduction is generally complex, so a detailed understanding of the binding process is crucial for success.[76] Usually, α-methylated amino acids are used in the bridging amino acids, as they are known to better stabilize α-helical conformations,[77,78] although a recent example showed that they are not essential for helicity.[79] Also, the length of the linker and absolute configuration of the

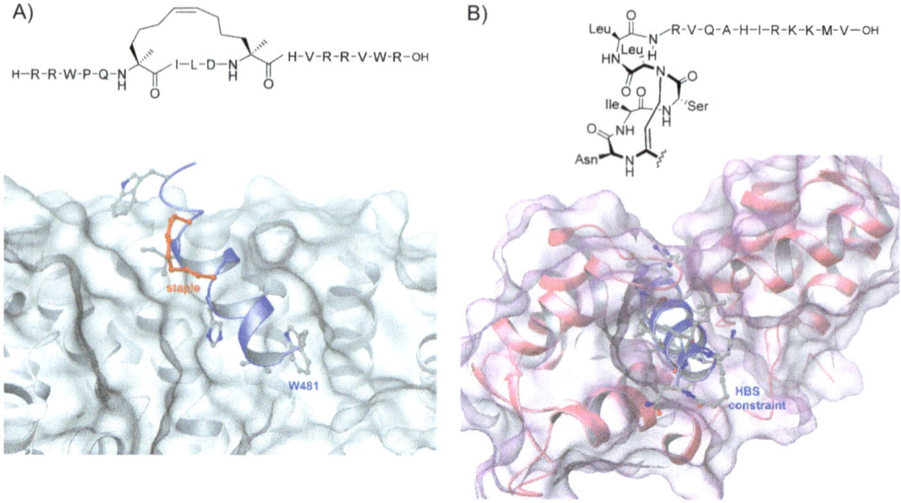

Figure 5.5 (A) X-ray structure of β-catenin (on the gray surface) in complex with the stapled peptide aStAx-35, showing that it binds at the same site as axin; PDB code: 4DJS. (B) A myoA-derived HBS peptide binds to a very narrow groove in the MTIP protein, with all three faces of the helix contributing to the PPI; PDB code: 4MZL.

α-methyl α-alkenyl residues introduced have been examined in depth. In the $i + 4 \rightarrow i$ staple, two S-configured building blocks are required, together with an 8-atom staple. For the longer $i + 7 \rightarrow i$ system, a cross-link containing 11 carbon atoms is preferred, and R and S amino acids should be used at positions i and $i + 7$, respectively.[72,78]

Despite its relatively recent introduction, this stapling tool has been successfully applied by chemical biologists to target disease-relevant PPIs for which few or no ligands exist.[80–85] The work of Grossmann *et al.* on the inhibition of the oncogenic Wnt signaling pathway is an elegant example of structure-based rational design.[86] β-Catenin is the central hub in the Wnt pathway, binding to several peptide ligands through extended PPIs that have remained elusive to small molecule drugs. A series of stapled peptides were designed based on the helical region of axin bound to β-catenin, and their binding potency and cell permeability were improved through systematic screening of stapling positions and phage-display optimization. The best candidate (Figure 5.5A) was able to antagonize β-catenin with a K_d in the low nM range and to selectively inhibit the growth of Wnt-dependent cancer cells.

Selective modulation of transcription factors is also a major challenge in the field of drug discovery, due to the intracellular location of these molecules and the general absence of surface cavities suitable for binding.[87] An important advance in this field was the rational design of stapled α-helical peptides acting as direct antagonists of the oncogenic transcription factor NOTCH1.[88] The best candidate was an $i \rightarrow i + 4$ stapled peptide, which showed 94% helicity in water. This peptide inhibited NOTCH1 by competing with the MAML1 coactivator. Also, it proved to be cell-permeable, and it suppressed NOTCH1 signaling in leukemia cells and halted the progression of this malignancy in animals. This case exemplifies the benefits of peptide stapling in terms of enhancing cell permeability, target affinity, proteolytic resistance, and plasma half-life. Not surprisingly, two stapled peptides, ALRN-6924 and ALRN-5281, have reached clinical trials for the treatment of solid tumors and growth hormone deficiency, respectively.[76]

More recently, highly constrained bicyclic stapled peptides, developed from previously described stapled peptides,[89] were designed as Rab8 GTPase inhibitors.[90] The additional cycle was shown to further stabilize the bioactive conformation, resulting in a 3-fold increase in binding affinity to Rab8 GTPase, compared to the monocyclic analogue. Obtained through a novel orthogonal ring-closing alkyne and olefin metathesis approach, this kind of bicyclic architecture results in highly rigid conformations and has the potential to target especially challenging PPIs.

Hydrogen-bonding Surrogates (HBSs). Canonical α-helices are defined by 13-membered hydrogen-bonded α-turns. The HBS approach aims to constrain this feature by introducing a covalent N-terminal $i \rightarrow i + 4$ cross-link. In 1991, the pioneering work of Kemp showed how a constrained Pro–Pro derivative at the N-terminus nucleated the formation of an α-helix[91]—a discovery that inspired the development of the first HBS approach, which used a covalent $i \rightarrow i + 4$ hydrazone cross-link.[92] Since then, the properties of HBS peptides have been improved by using more stable and readily synthesizable alkenyl and thioether linkers. As with the other helical scaffolds, these peptides have enhanced proteolytic resistance and cell permeability,[93] with the advantage that the topological features of the helix are not perturbed by artificial modification, as all faces are untethered. This strategy has been successfully applied by Arora's group to a number of targets, such as HIV gp41,[94] Bcl-xL,[95] p53,[96] HIF-1α[97] and Ras.[98] In the latter, HBS-pre-organized cell-permeable α-helices were designed to mimic the wild-type Sos helical domain that interacts with Ras. The candidates that showed the greatest binding to Ras were characterized by FP and ^{1}H–^{15}N HSQC NMR, and cellular assays proved their ability to block Ras signaling in response to receptor tyrosine kinase activation.

In a recent study, Douse and co-workers designed a set of stapled and HBS peptides mimicking the helical tail of the malaria parasite invasion motor myosin (myoA) protein, a particularly challenging interaction, as revealed by the X-ray structure of the complex.[99] In this regard, myoA is buried in a very narrow groove inside its partner protein MTIP, with all three faces of the helix contributing to the PPI.[100] Strikingly, the more encumbered stapled peptide binds to MTIP with a similar affinity to the HBS analogue (IC_{50}s in the low μM range for both). High resolution X-ray structures reveal that the pentenyl glycine staple nicely replaces the interactions of Val807 and Ile811 in myoA and that the mode of binding shown by the HBS peptide matches that observed for the native protein (Figure 5.5B).

Cycloadditions. Although less exploited than the above mentioned strategies, the azide-alkyne 1,3-dipolar cycloaddition reaction, generally between an azide-functionalized norleucine and a propargylalanine residue, at the *i* and *i* + 4 positions, has been applied to develop helix-constrained peptides.[101] According to Kawamoto *et al.*, D-propargylalanine yields more helical peptides than its L isomer.[102] In their work, they targeted the interaction between β-catenin and B-cell CLL/lymphoma 9 (BCL9) using triazole-stapled BCL9-derived α-helical peptides. After an extensive optimization of the positions, length, and chirality of the linker, they generated a double-stapled BCL9, which showed >90% helicity in aqueous solution, was stable to proteolytic degradation, and demonstrated potent binding to β-catenin (K_i = 0.41 μM). In a recent study, also by Wang's group,[103] the crystal structure of the RAP1/TRF2 complex was the basis for the rational design of triazole-stapled peptides targeting the RAP1/TRF2 interaction. Their most potent analogue showed a K_i value of 7 nM, which is >100 times more potent than the corresponding linear TRF2 peptide.

Using the same strategy, D. R. Spring and co-workers developed an alternative stapling method involving two separate components, the peptide and a linker, which are combined to form the final stapled peptide. Starting from a single linear precursor, this efficient double-click method was applied to generate a range of stapled peptides that target the widely studied p53/MDM2 interaction.[104] Cell permeability and inhibitory potency were significantly improved by introducing cationic groups to the staple linkage, without modifying the peptide sequence. Additional modifications also led to a marked effect on cellular activity.[105] This two-component double-click approach was recently described in detail.[106]

Redox Reactions. Disulfide-bridged cyclic peptides, generally between cysteine residues spaced three amino acids apart, have been used to stabilize helical conformations. A D- and L-cysteine at positions *i* and *i* + 3, respectively, is the optimal combination for helicity.[107] This strategy was applied to constrain short peptides containing the LXXLL nuclear receptor box motif, which is known to adopt an α-helical conformation upon binding to estrogen receptor α. The resulting analogues showed a low helical character in aqueous solution, but an X-ray structure confirmed that the peptide does bind in the expected α-helical conformation, in a clear example of receptor-induced folding of the peptide ligand. Interestingly, the $i \rightarrow i + 3$ disulfide staple was more potent (K_i = 25 nM) than the $i \rightarrow i + 4$ bridged lactam.

Although disulfide bridging was one of the first stapling techniques reported, it has encountered limited application, its major drawback being its intrinsic instability in reducing environments, which hampers its applicability in *in vivo* models. A straightforward solution is the replacement of the disulfide bond by a more stable thioether analogue[108]—an approach that is used in an example, shown below, by Gerona-Navarro *et al.*[109] Another reversible reaction, recently developed by Horne and co-workers,[110] involves the formation of an oxime bond between the side chains of modified residues placed at the *i* and *i* + 4 positions. However, its applicability to PPI modulation remains to be seen.

Photoswitchable Peptides. The introduction of a photo-sensitive cross-linker between the side chains of residues placed at one, two or three turn distance in the α-helix allows for reversible light-mediated control of the peptide conformation. In peptide helices designed to modulate PPIs, this property opens up the possibility to regulate their activity, kinetics, and site of action with temporal and spatial precision using light as an external stimulus.[111,112] The photoinduced folding and unfolding mechanisms have recently been elucidated in great detail using molecular dynamics (MD) modeling, which has shown that the interactions between the peptide and the cross-linker play a key role in regulating the change in the secondary structure of the peptide.[113] The first example of a light-induced conformational transition of a peptide reported the attachment of an azobenzene moiety to the carboxyl side chain of a poly-Glu peptide—a strategy that, years later, inspired Allemann's group to develop several photocontrollable DNA-binding helical peptides.[114,115] In the field of PPIs, the same group introduced an azobenzene cross-linker *via* cysteine residues on various positions of the BH3 helical domains of Bak and BID to disrupt their interaction with the anti-apoptotic Bcl-xL.[116]

This approach was applied by Nevola *et al.* to photoregulate clathrin-mediated endocytosis in living cells by engineering peptides based on the C-terminal helical domain of β-arrestin.[117] In the design strategy, residues separated by one, two or three helix turns on the non-interacting face of the helix were replaced by cysteine pairs that were then cross-linked with an azobenzene moiety, in such a way that the stability of the helix could be reversibly altered using 380 nm or 500 nm light. The affinities of these cyclic peptide inhibitors for their target protein were dependent on the *trans*/*cis* isomerization of the azobenzene unit, thus allowing the selective and temporal modulation of membrane receptor internalization in living cells using light (Figure 5.6).

To date, peptides with conformations that can be reversibly controlled by light are modified by introducing non-peptidic photoswitches, such as azobenzenes, into the side chains. Recently, however, a diarylethene scaffold was introduced into the backbone of an antimicrobial cyclic peptide, and the biological activity of this peptide was shown to be effectively regulated by light.[118] Indeed, substantial advances can be expected in the coming years with respect to the specific spatiotemporal modulation of protein complexes in response to external stimuli such as light.

As many of the above-mentioned examples show, some degree of optimization is usually needed to find a helix-constrained peptide with the desired properties. Nevertheless, it has recently been reported that the most helical peptide does not always provide the highest target affinity.[119,120] This observation is not limited to any one stapling technique, but addresses mechanistic questions regarding the binding mechanism and the effects of structural pre-organization on protein–ligand interactions. Generally, constraining a peptide leads to entropically favorable binding, which is counterbalanced by an often opposing enthalpy change attributed to the newly formed unfavorable interactions involving the backbone and side chains. This effect can be dominant when the ligand binding process is governed

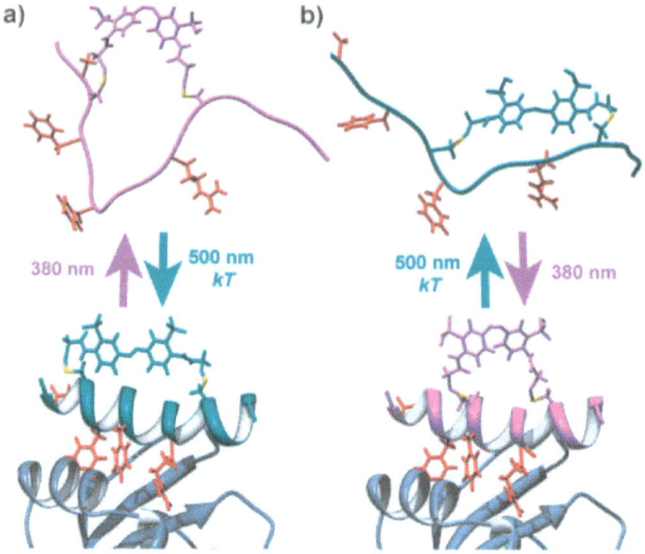

Figure 5.6 Photoswitchable peptides targeting the AP2 protein. (a) The azobenzene cross-linker, conjugated at cysteine residues *i* and *i* + 11, favors binding in the *trans* configuration. (b) The opposite effect is observed by conjugating the cross-linker at positions *i* and *i* + 7. Reproduced with permission from Nevola *et al.*, Light-Regulated Stapled Peptides to Inhibit Protein–Protein Interactions Involved in Clathrin-Mediated Endocytosis, *Angew. Chem., Int. Ed.*, Copyright © 2013 WILEY-VCH Verlag GmbH & Co. KGaA, Weinheim.

by an induced fit mechanism, *i.e.* when the unstructured peptide is recognized by the target protein and subsequently folds on the binding surface. In this context, fixing the ligand conformation will limit the number of ways through which the protein can bind, and this may impair binding potency. In conclusion, in order to select the best strategy to stabilize the peptide bioactive conformation, a detailed investigation of the binding process is crucial to ensure that an optimal approach and degree of pre-organization are achieved.

5.3 *In silico* Approaches

High-resolution protein structures have become the cornerstone for the design of peptides mimicking protein interface domains; however, there are still far more predicted PPIs in the human interactome than complexes available in the PDB. In the absence of experimental structures, homology models and mutagenesis data can provide a solid foundation for the start of a drug discovery process. Even when access to high-quality structures of protein complexes is available, large protein interfaces—with a few hotspots spread over the contact surface—are extremely challenging. Computational strategies can be invaluable in guiding the rational design of peptide inhibitors—a process that covers a broad spectrum of applications, ranging from the identification of relevant PPIs to the design and docking of peptide binders, through to the use of virtual libraries and the conformational prediction of structures.

As previously mentioned, most PPIs take place on large interfaces in which a few residues, often present in disconnected segments, provide most of the binding energy.[121] Therefore the identification of such hotspots is the first step towards narrowing the search space of relevant interface peptides. In this regard, experimental methods based on systematic mutagenesis and affinity determination provide reliable data; however, they are slow, laborious and costly. A number of computational tools, often relying on ΔΔG or ΔSASA metrics, have been developed for measuring the relative importance of interface residues. ΔΔG measures the change in binding energy upon mutation of a residue to alanine, and it can be computationally predicted from MD trajectories through MM-PBSA or MM-GBSA calculations.[122,123] High positive ΔΔG values reveal the critical residues for binding. More importantly, ΔΔG values can be directly correlated to affinity constants (K_d or IC_{50}). In this regard, MM-PBSA outperforms MM-GBSA, even for relatively short and inexpensive MD simulations,[124] although the latter may not be accurate enough to be used alone for predictive drug design.[125] On the other hand, ΔSASA measures the change in solvent-accessible surface area of two proteins upon binding, and it can be readily decomposed on a per-residue contribution. Hotspots tend to be deeply buried inside the interface, thus ΔSASA is widely used in hot-spot prediction because it is more straightforward to compute than ΔΔG, in spite of its relation with affinity constants being more distant. A selection of online servers and databases that provide useful information about hotspots in protein interfaces are shown in Table 5.1.

For protein–protein interfaces that lack high-affinity hotspots and/or well-defined secondary structures, virtual screening can provide starting hits that are unlikely to be derived from a rational design-based approach (Figure 5.7). In addition to the target structure, a library of candidate compounds is needed for virtual screening. Although many libraries contain small molecules and fragments, virtual libraries of cyclic peptides are rare, and the most direct way to create them is by combinatorial generation of the desired amino acid repertoire. Alternatively, the software CycloPs is a useful tool to generate cycle-constrained peptide libraries containing natural and non-natural amino acids.[142] Once the libraries have been generated, three-dimensional conformations have to be calculated for all molecules and those with the lowest energy should be identified. For this task, a number of high-quality software packages are available, both commercial, such as Schrödinger, MOE and OEChem, and open-source, such as RDKit, OpenBabel and CDK. Online servers have also been developed for this purpose, some of which can handle cyclic structures, such as PepLook,[143] Pep-Fold[144] and PEPstrMOD.[145] Nonetheless, molecules with a large number of rotatable bonds are difficult

Table 5.1 Selection of online computational tools that provide useful data regarding protein–protein interactions.

Name	Description
FTMap[126]	Hotspot mapping server that calculates the binding energy score for a series of small organic probes, identifying regions on the protein surface that contribute the most to ligand ΔG.
PredHS[127]	Webserver that uses optimally structural neighborhood properties to predict hotspots on PPI interfaces
LoopFinder[128]	Identifies interface loops in protein–protein complexes from the PDB that might be amenable to mimicry by cyclic peptides
iPred[129]	JAVA-based software tool that predicts potential protein–protein interface areas and hotspot residues
PCRPi[130]	Webserver that predicts hotspot residues relying on the integration of 7 variables that account for energy, structural and evolutionary information, using Bayesian networks
HSPred[131]	Webserver that uses a support vector machine-based method to predict hotspot residues
Robetta[132]	Webserver that provides both *ab initio* and comparative models of protein domains
KFC2 [133]	A web-based machine-learning approach for predicting binding hotspots
PocketQuery[134]	Web service that clusters residues at the interface of PPIs and yields $\Delta\Delta$G and ΔSASA metrics
HippDB,[135] SippDB,[136] DippDB[137]	Secondary structure databases, catalogued by the Arora lab, providing computational Ala scanning and ΔSASA values for interfacial residues
TIMBAL[138]	Database containing small molecules that modulate a number of PPIs
PIFACE[139]	Database that clusters PPIs on the basis of their interface structures
PINT,[140] SKEMPI[141]	Databases that contain thermodynamic experimental data for a number of PPIs

to model computationally, so special attention must be paid in order to generate accurate structures.

In the next step, the predicted topologies for the ligands are docked on the binding region of the target. Docking consists of two iterative steps: pose generation, which positions the ligand within the predefined binding site, and scoring, which estimates the binding affinity of each pose and ranks them. Generally, the protein is assumed to be rigid during the docking process, and only the ligand is allowed a certain degree of flexibility. The wide range of

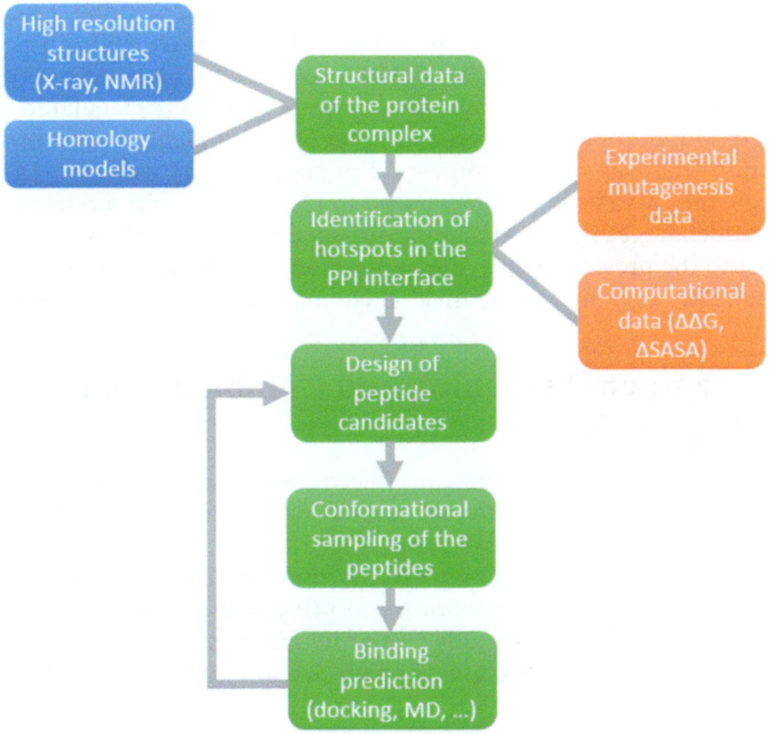

Figure 5.7 Schematic workflow of a computational approach to design structure-based peptide inhibitors of protein–protein interactions.

docking methods available, together with their limitations and future challenges, have been reviewed in detail.[146,147]

The bioinformatics approach has made a strong entry into the field of PPIs in recent years. A number of success stories have provided chemical tools in general, and cyclic peptides in particular, for targets previously considered undruggable. The transcriptional activation of p53 target genes is mediated by the interaction between the CREB binding protein (CBP) and p53. Starting from the NMR structure of this complex,[148] Gerona-Navarro *et al.* identified a key β-turn motif in p53 that interacts with CBD.[109] A MD simulation of the β-turn binding to p53 suggested that introducing a cyclic constraint in the linear peptide would stabilize the bioactive conformation. Indeed, the replacement of two non-interacting residues by cysteines yielded a disulfide-bridged active peptide (IC_{50} = 22.6 μM). The expected mode of binding was confirmed by MD simulation and NMR studies. With this privileged information in hand, the peptide could be optimized to yield a more potent and stable thioether analogue that effectively disrupted the p53–CBP interaction in carcinoma cells.

Yin and co-workers used the Rosetta modeling software (recently upgraded to support diverse cyclic constraints and D-amino acids)[149] to design a series of cyclic peptides able to modulate the TLR4–MD2 interaction—a key event in the activation of the innate immune response. Starting from peptides derived from the binding region of MD2 (a protruding loop constrained by a disulfide bridge), theoretical affinities towards TLR4 were calculated *in silico* and were found to be even lower than that of the full-length MD2 protein, a prediction that was further confirmed in experimental assays.[150] In a later article,[151] they also used the Rosetta program to derive and model macrocyclic peptides targeting the MD2-binding region of TLR. Interestingly, and in contrast to the previous study, the peptides obtained had an agonistic effect on TLR4, although the precise mechanism of activation was not elucidated.

5.4 Fragment Screening and Combinatorial Approaches

Fragment-based strategies are an effective way to target PPIs, especially in cases where there is limited structural information and/or the relevant hotspots are scattered across a disorganized region. Fragment-based methods have been widely exploited by the pharmaceutical industry, usually identifying low-affinity fragments (<300 MW) that can be then optimized or merged to obtain potent leads.[152] Cyclic peptides are large and structurally complex, thus explaining why there are few cases of fragment-based discovery of such compounds; however, the structure–activity relationship information gathered in fragment screens can be extremely useful for the efficient design of peptides. For instance, using this approach, Fesik's group identified two series of fragments with a moderate affinity for the DNA-binding protein RPA.[153] One fragment included a 3,4-dichlorophenyl moiety that made key interactions with the protein. In the continuation of this work, this valuable input was applied to refine a stapled peptide derived from ATRIP, an endogenous α-helical protein that mediates the DNA damage response.[154] In addition to traditional optimization methods (Ala scan, removal of irrelevant residues, use of distinct stapling patterns), the introduction of a 3,4-dichlorophenyl moiety, previously determined to occupy the same binding hotspot as the Phe side chain of the ATRIP peptide, produced a 175-fold increase in binding affinity. This approach reflects how fragment screening in a large peptide can identify specific mutations that enhance its binding affinity.

Another related approach is combinatorial target-guided synthesis, which comprises the discovery of peptide ligands, selected from peptide libraries, by *in situ* click chemistry in the presence of the target protein. Developed by Heath's group, this strategy has been successful in identifying linear or macrocyclic peptides able to bind specific epitopes in proteins with antibody-like affinities.[155–157] In this approach, a synthetic epitope of the target (9–30 amino acid fragment) is prepared with a terminally appended biotin, as well as an azidolysine click handle. Next, one-bead-one-compound peptide libraries (comprehensive in 18 natural or non-natural amino acids, ≈2×10^6 sequences) featuring a propargylglycine click handle, are screened, first against a scrambled version of the epitope (to discard non-specific binders) and then against the synthetic epitope itself. Library elements that bind the synthetic epitope in the correct orientation undergo the cycloaddition reaction and are covalently linked to the epitope. Finally, the library is washed to remove non-covalently bound peptides and hit peptides are synthesized in bulk and tested against the full-length protein to determine affinity values (Figure 5.8). In a recent study, macrocyclic peptide libraries (one of them containing stapled peptides) were used to target 12 therapeutically relevant protein targets, achieving affinities in the pM–nM range in most cases.[158] Interestingly, macrocycles yield average $\log EC_{50}$ values of >7, while for the linear libraries that value is <6, proving their superior performance in targeting PPIs.

In a similar way, the covalently bound ligand can be used as an anchor to perform the screening of a second peptide library; *in situ* triazole formation leads to a bi-ligand and the process can be repeated to generate a high affinity tri-ligand. This strategy was used to develop a highly potent inhibitor of botulinum neurotoxin.[159] Two macrocyclic peptides, which bind different subdomains of the neurotoxin, were identified from respective peptide libraries. An *in situ* click screen was then performed with a library of linkers that would connect the two macrocycles in the optimal orientation in order to enhance the binding avidity. The resulting divalent inhibitor exhibited an

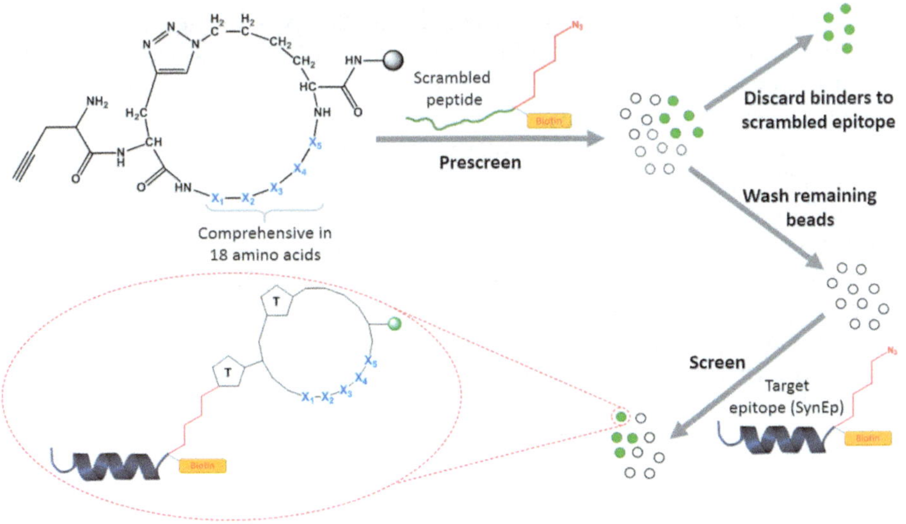

Figure 5.8 General strategy to develop high-affinity cyclic peptides against specific epitopes using target-guided combinatorial chemistry. Adapted from Das *et al.*, A General Synthetic Approach for Designing Epitope Targeted Macrocyclic Peptide Ligands, *Angew. Chem., Int. Ed.*, John Wiley and Sons. © 2015 WILEY-VCH Verlag GmbH & Co. KGaA, Weinheim.

IC_{50} value of 165 pM and acted like a Trojan horse in cells, being internalized by the neurotoxin endocytic machinery and, once in the cytosol, blocking the catalytic activity of the toxin.

5.5 *In vitro* Methods

Although chemical synthetic methods have improved significantly both in scale and coupling efficiency, they still have some important drawbacks. Conventional peptide libraries generated using Fmoc or Boc chemistry are prohibitively expensive for the library sizes required for drug discovery purposes. In contrast, recombinant display systems are able to screen libraries that are orders of magnitude larger than those obtained by chemical synthesis. These display technologies rely on the principle of directed evolution, which is based on iterative cycles of gene mutagenesis, expression and screening or selection in order to perform an optimal sparse sampling of a vast multidimensional sequence space. In these systems, a genotype–phenotype link is exhibited by displaying the peptide with its genetic material encapsulated in a virus particle (*e.g.* phage display), or on the surface of a cell that contains genetic information (*e.g.* yeast or bacterial display), or directly linked to the nucleic acid through non-covalent or covalent interactions (*e.g.* ribosome display, mRNA display and CIS display) (Figure 5.9).

Encoded linear peptide libraries are several orders of magnitude larger than classical chemical libraries, are more readily screened, and give rise to higher affinity ligands. However, their use as drugs is limited by rotational flexibility and by the poor number of natural amino acids. However, these limitations can be overcome by means of chemical modification.[160] In this regard, several strategies have been developed to obtain cyclic and bicyclic encoded peptide libraries. One of the most common is disulfide-bridged peptides.[161] However, it is known that such bonds are susceptible to reduction and to disulfide exchange reactions.

Roberts *et al.* cyclized encoded peptides with an amide bond-forming chemical linker by using disuccinimidyl glutarate (DSG) to connect the α-amino group at the peptide N-terminus and the ε-amino group of a lysine residue (Figure 5.10a). Recently, this group has applied this strategy for the *in vitro* evolution of serum-stable peptide ligand antagonists of Gai1.[162]

The use of unnatural amino acids containing chemical handles for cyclization is another way to obtain cyclic peptides. Suga *et al.* developed modified amino acids with a chloroacetamide functional group that spontaneously reacts with the sulfhydryl group of a cysteine in the same peptide, forming a thioether linkage (Figure 5.10b).[163] Using this methodology, cyclic peptides were isolated for a range of targets such as VEGF receptor 2.[164] They developed other cyclization strategies with unnatural amino acids

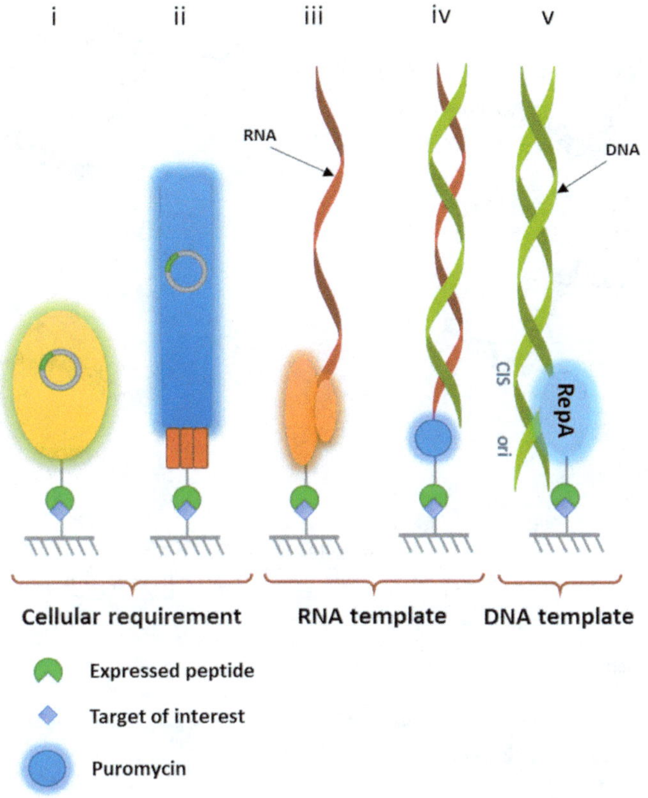

Figure 5.9 Diagram showing different display technologies where the peptide is expressed: (i) on the surface of a yeast or bacterium (yeast/bacterial display); (ii) in a bacteriophage (phage display); (iii) in complex with a ribosome (ribosome display); (iv) linked to RNA through puromycin (mRNA display); or (v) bound to DNA through the cis activity of a DNA binding protein (CIS display). (Adapted from C. G. Ullman and L. Frigotto, *In vitro* methods for peptide display and their applications, *Briefings in Functional Genomics*, 2011, **10**(3), 125–134, by permission of Oxford University Press.)

also for application to mRNA-encoded peptide libraries, such as oxidative coupling of 5-hydroxytryptophan and benzylamine linked to the α-amino group of phenylalanine (Figure 5.10c)[165] or the Michael addition of cysteine to dehydrobutyrine.[166]

Another approach for cyclizing mRNA-encoded peptide libraries was developed by Timmerman *et al.* This method uses the reagent α,α′-dibromo-*m*-xylene to efficiently cyclize peptides with cysteines in aqueous solution (Figure 5.10d). Szostak *et al.* showed first that the reagent was suitable to cyclize translated peptides *in vitro*. They applied this methodology to cyclize peptide

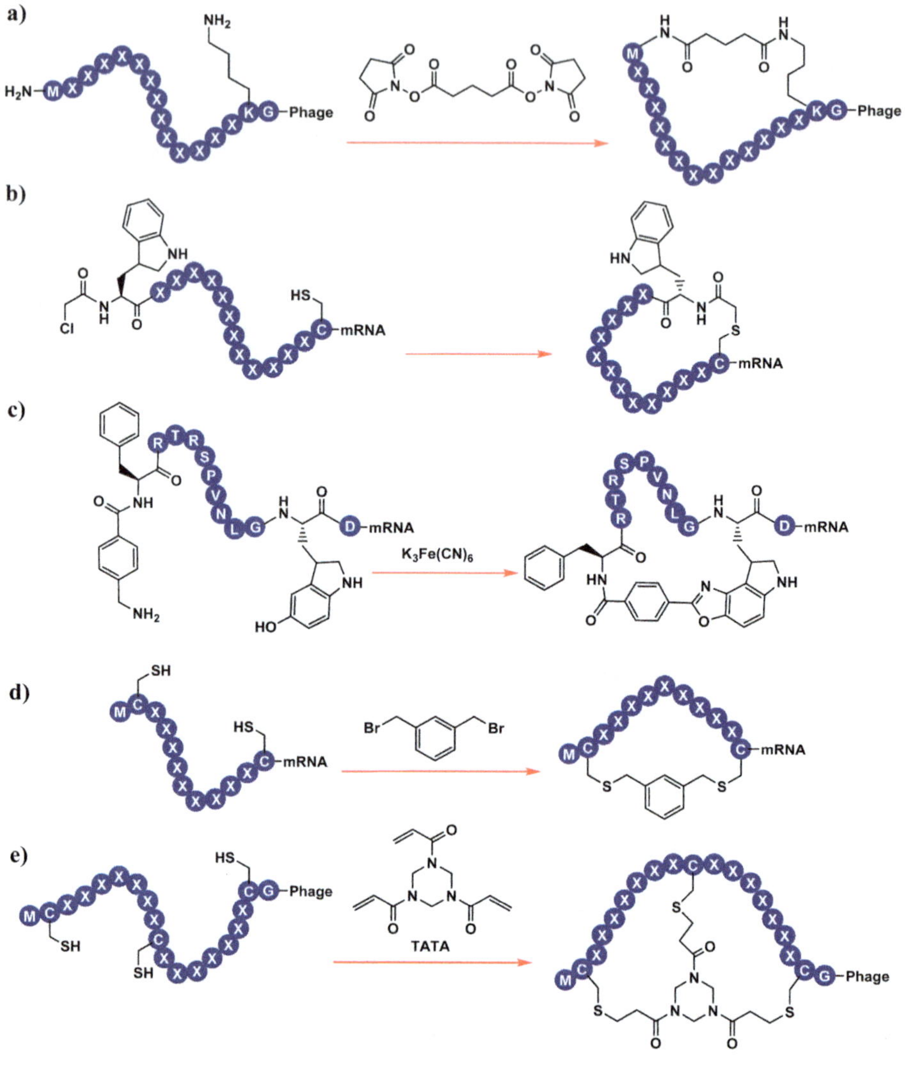

Figure 5.10 Chemical modifications performed to obtain cyclic and bicyclic encoded peptide libraries. (a) Cyclization through an amide bond-forming chemical linker; (b) and (c) use of unnatural amino acids containing chemical handles for cyclization; (d) and (e) cyclization through cysteine residues using different reagents to obtain cyclic and bicyclic peptides.

mRNA display libraries containing unnatural amino acids, in order to isolate thrombin inhibitors based on peptides with natural (K_d = 1.5 nM) and unnatural amino acids (K_d = 20 nM).[167] Winter and Heinis developed a phage display-based methodology for encoding and screening combinatorial libraries of bicyclic peptides by cyclizing peptides with the sequence ACX_6CX_6CG displayed on a phage with the chemical linker 1,3,5-tris(bromomethyl)-benzene

(TBMB).[168] These kinds of libraries have been described with all combinations of ring sizes between three and six amino acids, as well as with novel thiol-reactive trivalent chemical linkers 1,3,5-triacryloyl-1,3,5-triazinane (TATA) and *N*,*N*′,*N*″-(benzene-1,3,5-triyl)-tris(2-bromoacetamide) (TBAB). These strategies were successfully applied to obtain entirely different families of ligands against the target uPA (Figure 5.10e).[169]

More recently, light-responsive peptide ligands were developed by screening large combinatorial libraries using display techniques. Ito *et al.* introduced an azobenzene moiety into linear mRNA-encoded peptides using the unnatural amino acid ε-(lysine)-azobenzene.[170] The groups of Derda and Heinis independently developed phage display screening methods for peptide libraries cyclized with an azobenzene linker.[171–173] Split intein circular ligation of peptides and proteins (SICLOPPs) has also been used to generate cyclic peptide libraries and cyclic proteins.[174,175]

5.5.1 Cellular Approaches

5.5.1.1 Phage Display

Phage display was created by G. Smith in 1985[176] as a method for presenting polypeptides on the surface of lysogenic filamentous bacteriophages. This approach is now one of the most effective ways to produce large peptide libraries containing up to 10×10^{10} variants.

Phage display has been widely used to build diverse peptide libraries for high-throughput screening, and it is a common tool in the field of PPIs. This methodology has recently been used to discover bicyclic and α-helical cyclic peptides against β-catenin. For the design of bicyclic peptides, phages displaying $>4 \times 10^9$ peptides of the format ACX_6X_6CG were prepared and then cyclized with three chemical linkers, namely TBMB, TATA and TBAB, to obtain three libraries with a total diversity exceeding 12 billion peptide macrocycles.[177] After biopanning against β-catenin streptavidin, around 170 000, 69 000, and 84 000 clones isolated from the three peptide libraries (TBMB, TATA and TBAB, respectively) were sequenced, revealing many diverse consensus sequences. For peptides cyclized with TBMB, TATA, and TBAB, 12, 5, and 4 distinct consensus sequences were found. Representative peptides of the consensus groups were chemically synthesized and labeled with fluorescein in the N termini to allow measurement of binding to β-catenin by fluorescence polarization. Of the 22 peptides tested, four were observed to bind with single-digit μM K_d values, and two showed K_d values between 20 and 30 μM. They also performed competition studies and found that the peptide macrocycles interact with β-catenin *via* diverse peptide binding motifs and that they bind to various regions of the target protein. The same group also developed α-helical peptide ligands for the same target using phage display.[178] They designed phage-display libraries based on the axin α-helix, which was mutated to different extents based on structural data using the axin/β-catenin co-crystal structure[179] and sequence–activity relationship data reported by Grossmann *et al.*[180] In order to obtain the cyclic peptides, the cysteines of the purified phages were reduced and modified by incubation with 100 μM DBMB or CDCB. They synthesized the peptides obtained from phage display with an N-terminal fluorescein and measured their binding to β-catenin as (i) peptides stabilized with DBMB, (ii) peptides stabilized with CDCB, (iii) linear peptides, and (iv) peptides cyclized by a disulfide bridge. Fluorescence polarization assays revealed the linker-stabilized peptides to show good binding, with dissociation constants in the low nanomolar range. The best ligands bound β-catenin with a K_d of 5.4 and 5.2 nM, which was more than 200-fold greater than that of the starting point.

This technique has experienced considerable improvements, such as proteomic-peptide phage display (ProP-PD). This approach couples bioinformatics, oligonucleotide arrays, and peptide phage display to explore the interactome of the human PDZ domain, and it has been applied in particular to study cellular and pathogen–host PPIs.[181]

One of the most recent advances in this methodology is the use of "silent barcoding", developed by Derda's group to introduce post-translational modifications in phage-display libraries. This strategy consists of encoding the same amino acid sequence with a distinct set of genetic codons. It allows simple encoding of a relatively small number of modifications (2–100), thus permitting rapid diversification of readily available phage-display peptide libraries. It is envisaged that silent barcoding will find future applications in selections that use phage-displayed or mRNA-displayed peptide-derived macrocycles (Figure 5.11).[182]

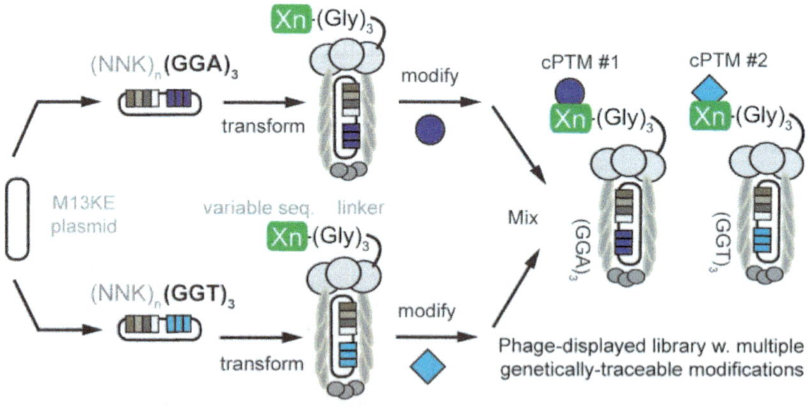

Figure 5.11 Redundant codons are used to encode chemical post-translational modification with silent barcodes — *e.g.*, $(GGA)_3$ or $(GGT)_3$ — to trace multiple modifications within a mixed modified library. Libraries bearing different barcodes are chemically modified and then combined to yield a mixed library of chemically modified peptides. Reproduced with permission from ref. 182. Copyright (2016) American Chemical Society.

5.5.1.2 Yeast and Bacterial Display

Yeast cell display is an alternative method to phage display.[183] Foreign peptides and proteins (including growth factors, antibody fragments, and cell surface receptors) are usually fused to the Aga2p agglutinin subunit on the surface of the yeast cell.[184] Bacterial display is a technically related method whereby peptides or proteins can be expressed on the surface of Gram-positive or Gram-negative cells in fusion with outer membrane proteins or on the bacterial flagella.[185]

Recently, Daugherty's group developed a method for the *de novo* discovery of bioactive cyclic peptides using the bacterial display technique.[186] As a proof of concept, they used libraries of the X_3CX_{12}, $X_4CX_7CX_4$, and XCX_7CX form containing either one or two cysteine residues to identify cyclic peptides with desired specificity and affinity towards arbitrary target proteins.

Camarero's group was the first to describe the recombinant expression of fully folded bioactive cyclotides inside live yeast cells using intracellular protein *trans*-splicing in combination with a highly efficient split-intein.[187] They successfully used yeast display to produce the naturally occurring cyclotide MCoTI-I and the engineered bioactive cyclotide MCoCP4, which was shown to reduce the toxicity of human α-synuclein in live yeast cells. These achievements demonstrate the potential of using yeast to perform phenotypic screening of genetically encoded cyclotide-based libraries in eukaryotic cells (Figure 5.12).

5.5.2 Non-cellular Approaches

5.5.2.1 Ribosome Display and mRNA Display

Ribosome display is an *in vitro* selection and evolution technology for proteins and peptides from large libraries.[188] mRNA display, like ribosome display, allows for the identification of polypeptide sequences with desired properties from both a natural protein library and a combinatorial peptide library.[189] Both methodologies are useful for the production of peptide libraries. However, for the synthesis of macrocyclic peptides using the translation machinery, post-translational modifications have to be developed. A reliable method for the construction of macrocyclic peptides is based on the concept of manipulating the genetic code, an approach known as genetic code reprogramming. In this strategy, designated codons are made vacant and then reassigned to non-proteinogenic amino acids.

One of the most used methodologies for genetic code reprograming involves 'flexible' tRNA acylation ribozymes, known as flexizymes.[190] Developed by Suga *et al.*, flexizymes facilitate the preparation of a wide array of non-proteinogenic aminoacyl tRNAs with almost unlimited choice. The combination of a custom-made *in vitro* translation system with flexizymes, referred to as the FIT (Flexible *In vitro* Translation) system, allows the

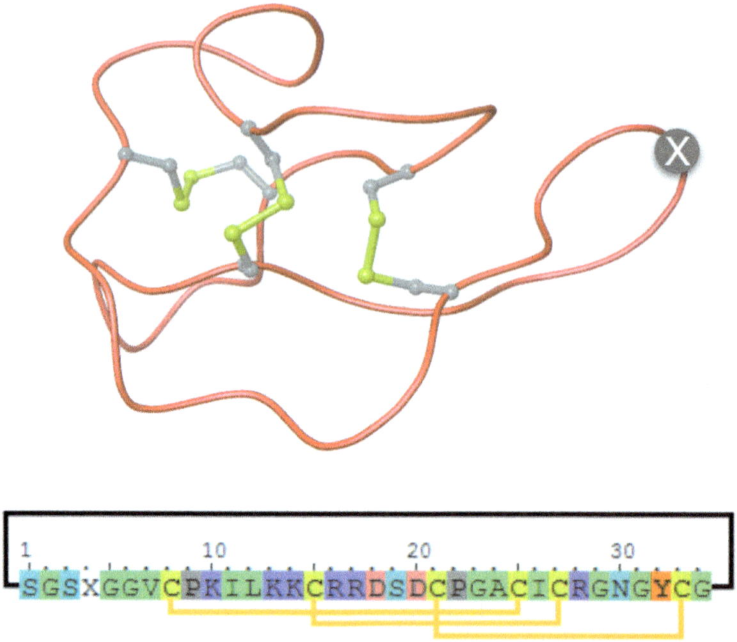

Figure 5.12 Tertiary structure of the cyclotide MCoTI-II (PDB code: 1IB9) and primary structures of the cyclotides used in Camarero's group, MCoTI-I (X = D) and MCoCP4 (X = SLATWAVG). The CP4-derived peptide was grafted onto loop 6, marked with an X. The backbone-cyclized peptide is stabilized by the three-disulfide bonds (shown in yellow).

ribosomal synthesis of macrocyclic peptides using non-proteinogenic amino acids capable of cross-linking with other proteinogenic or non-proteinogenic residues.

Many successful selections of macrocyclic peptides using both ribosome display and mRNA display and their variants have been reported. Szostak's group has reported an example of the capacity of selection approaches to identify macrocyclic peptides against sortase A (K_d = 3 mM), in which they used the lanthionine-like intramolecular cyclization method.[191] The FIT system has also been coupled to mRNA display, referred to as the RaPID (Random non-standard Peptides Integrated Discovery) system, and applied for the selection of highly isoform-selective inhibitors against the protein kinase Akt2 (IC_{50} = 110 nM and >250-fold and 40-fold higher IC_{50} for Akt1 and Akt2, respectively).[192] Macrocyclic peptide inhibitors were also selected against a eukaryotic ABC-drug transporter (IC_{50} = 65 nM) using this strategy.[193]

A major strength of the RaPID system is the ability to express *N*-methyl-macrocyclic peptides and the selection of active species. One example of such inhibitors has been published against VEGFR2.[164] In that study, an updated method of RaPID, referred to as TRAP display,[194] was used

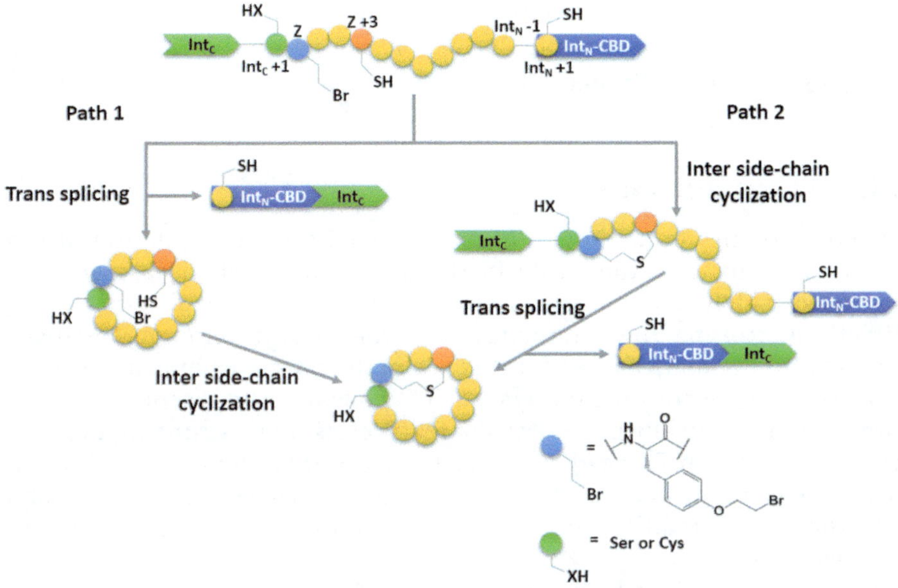

Figure 5.13 Strategy for the ribosomal synthesis of thioether-bridged bicyclic peptides in *E. coli*. From N to C, the linear precursor polypeptide comprises the C-terminal domain of split intein DnaE (IntC), a Ser or Cys residue at the IntC+1 site, the unnatural amino acid O-(2-bromoethyl)-tyrosine (O2beY or 'Z'), a variable target sequence containing the reactive cysteine (purple), and the DnaE N-terminal domain fused to a chitin binding domain (IntN-CBD). The two envisioned pathways leading to the bicyclic product are indicated. Adapted from N. Bionda and R. Fasan, Ribosomal Synthesis of Natural Product Like Bicyclic Peptides in *Escherichia coli*, *ChemBioChem*, John Wiley and Sons, © 2015 WILEY-VCH Verlag GmbH & Co. KGaA, Weinheim.

to yield a potent inhibitor (K_d = 33 nM). In recent work, Fasan's group reported the use of ribosome display, together with split intein-catalyzed head-to-tail cyclization, to synthesize natural-like macrocyclic peptides (Figure 5.13).[195]

5.5.2.2 CIS Display

CIS display technology relies on DNA-based systems. Such systems have the advantage of speed and stability over RNA templates, as the DNA template is less sensitive to degradation. Libraries can be quickly generated by standard PCR procedures and outputs are rapidly analyzed by next-generation sequencing.[196]

Ullman's group has recently reported the use of this CIS display system for the selection of a high-affinity WW domain against the extracellular region of VEGFR-2. The isolated inhibitor has low nM affinity to VEGFR-2 and inhibits the binding of human VEGF to its receptor. The structure is amenable to cyclization to improve its proteolytic stability and has advantages over larger protein scaffolds since it can be synthesized chemically in high yields, thus offering potential for therapeutic and non-therapeutic applications.[197]

5.6 Final Remarks

Protein–protein interactions (PPIs) are crucial for the regulation of biological systems, and their role in the development of disease states is nowadays widely documented. Despite this, small molecules that act by directly disrupting the interaction between two proteins are relatively rare. One of the main causes for this lies in the physicochemical nature of PPIs. Unlike conventional targets, such as enzymes and receptors, protein interfaces are typically large, flat and featureless. In this context, peptides have emerged as privileged scaffolds to target PPIs; their size and modularity allows them to accurately mimic protein surfaces, and at the same time they can be readily obtained in the laboratory, both by synthetic and by *in vitro* methods.

However, the use of linear peptides as PPI modulators has a chief disadvantage, namely that their high degree of flexibility results in negligible or random structures, thereby impairing recognition by a well-structured protein target. To overcome this issue, researchers have exploited the use of cyclic constraints and non-natural amino acids, which promote the folding of peptides into well-defined topologies. In addition to having enhanced binding affinity, cyclic peptides successfully tackle other drawbacks of linear peptides, such as their susceptibility to proteolytic degradation and their low cell-membrane permeability.

As we have shown, a great number of cyclization strategies have been developed to promote the folding of peptides into precise secondary structures. The resulting designed peptides are able to adopt a bioactive conformation at a low enthropic cost, and their effectivity has been validated in several therapeutically relevant PPIs. In most reported examples, the availability of high-resolution structural data, together with a detailed study of the binding process, provides a solid foundation for the development of cyclic PPI modulators. In addition, the development of modern computational tools and fragment-based screening have enabled the characterization of PPIs and the optimization of the overall design process. Finally, genetic code reprogramming and directed evolution techniques have granted access to a vast diversity of macrocyclic peptides generated by *in vitro* methods.

On the whole, cyclic peptides have evolved into a promising class of PPI modulators, both as chemical biology tools and therapeutics. In fact, several constrained peptides have already entered clinical trials. Thus, although considerable effort is required to achieve selective molecular recognition, cyclic peptides are expected to achieve great success in the modulation of challenging PPIs.

Acknowledgements

This study was funded by MINECO-FEDER (BIO 2013-40716-R) and Generalitat de Catalunya (XRB and 2014SGR-521). S. Guardiola and S. Ciudad hold a La Caixa/IRB Barcelona PhD fellowship.

References

1. K. Venkatesan, J. F. Rual, A. Vazquez, U. Stelzl, I. Lemmens, T. Hirozane-Kishikawa, T. Hao, M. Zenkner, X. Xin, K. I. Goh, M. A. Yildirim, N. Simonis, K. Heinzmann, F. Gebreab, J. M. Sahalie, S. Cevik, C. Simon, A. S. de Smet, E. Dann, A. Smolyar, A. Vinayagam, H. Yu, D. Szeto, H. Borick, A. Dricot, N. Klitgord, R. R. Murray, C. Lin, M. Lalowski, J. Timm, K. Rau, C. Boone, P. Braun, M. E. Cusick, F. P. Roth, D. E. Hill, J. Tavernier, E. E. Wanker, A. L. Barabasi and M. Vidal, *Nat. Methods*, 2009, **6**, 83.
2. M. P. Stumpf, T. Thorne, E. de Silva, R. Stewart, H. J. An, M. Lappe and C. Wiuf, *Proc. Natl. Acad. Sci. U. S. A.*, 2008, **105**, 6959.
3. M. Vidal, *Sci. Signaling*, 2016, **9**, eg7.
4. S. Jones and J. M. Thornton, *Proc. Natl. Acad. Sci. U. S. A.*, 1996, **93**, 13.
5. C. M. Labbe, G. Laconde, M. A. Kuenemann, B. O. Villoutreix and O. Sperandio, *Drug. Discovery Today*, 2013, **18**, 958.
6. M. R. Arkin, M. Randal, W. L. DeLano, J. Hyde, T. N. Luong, J. D. Oslob, D. R. Raphael, L. Taylor, J. Wang, R. S. McDowell, J. A. Wells and A. C. Braisted, *Proc. Natl. Acad. Sci. U. S. A.*, 2003, **100**, 1603.
7. J. A. Wells and C. L. McClendon, *Nature*, 2007, **450**, 1001.
8. N. London, B. Raveh and O. Schueler-Furman, *Curr. Opin. Chem. Biol.*, 2013, **17**, 952.
9. H. J. Lee, *Arch. Pharmacal Res.*, 2002, **25**, 572.
10. D. J. Craik, D. P. Fairlie, S. Liras and D. Price, *Chem. Biol. Drug Des.*, 2013, **81**, 136.
11. E. A. Villar, D. Beglov, S. Chennamadhavuni, J. A. Porco Jr, D. Kozakov, S. Vajda and A. Whitty, *Nat. Chem. Biol.*, 2014, **10**, 723.
12. S. Oishi, T. Kuroyanagi, T. Kubo, N. Montpas, Y. Yoshikawa, R. Misu, Y. Kobayashi, H. Ohno, N. Heveker, T. Furuya and N. Fujii, *J. Med. Chem.*, 2015, **58**, 5218.
13. B. K. Yap, E. W. Leung, H. Yagi, C. A. Galea, S. Chhabra, D. K. Chalmers, S. E. Nicholson, P. E. Thompson and R. S. Norton, *J. Med. Chem.*, 2014, **57**, 7006.
14. S. Guardiola, M. Díaz-Lobo, J. Seco, J. García, L. Nevola and E. Giralt, *ChemBioChem*, 2016, **17**, 702.
15. G. Ruiz-Gomez, J. D. Tyndall, B. Pfeiffer, G. Abbenante and D. P. Fairlie, *Chem. Rev.*, 2010, **110**, PR1.
16. H. N. Hoang, K. Song, T. A. Hill, D. R. Derksen, D. J. Edmonds, W. M. Kok, C. Limberakis, S. Liras, P. M. Loria, V. Mascitti, A. M. Mathiowetz, J. M. Mitchell, D. W. Piotrowski, D. A. Price, R. V. Stanton, J. Y. Suen, J. M. Withka, D. A. Griffith and D. P. Fairlie, *J. Med. Chem.*, 2015, **58**, 4080.

17. B. K. Ho and P. M. Curmi, *J. Mol. Biol.*, 2002, **317**, 291.
18. W. A. Loughlin, J. D. Tyndall, M. P. Glenn and D. P. Fairlie, *Chem. Rev.*, 2004, **104**, 6085.
19. A. Wlodawer, M. Miller, M. Jaskolski, B. K. Sathyanarayana, E. Baldwin, I. T. Weber, L. M. Selk, L. Clawson, J. Schneider and S. B. Kent, *Science*, 1989, **245**, 616.
20. P. K. Madala, J. D. Tyndall, T. Nall and D. P. Fairlie, *Chem. Rev.*, 2010, **110**, PR1.
21. J. Avruch, X. F. Zhang and J. M. Kyriakis, *Trends Biochem. Sci.*, 1994, **19**, 279.
22. S. R. Sprang, *Structure*, 1995, **3**, 641.
23. J. C. Rochet and P. T. Lansbury Jr, *Curr. Opin. Struct. Biol.*, 2000, **10**, 60.
24. A. Ciechanover and Y. T. Kwon, *Exp. Mol. Med.*, 2015, **47**, e147.
25. S. F. Martin, R. E. Austin, C. J. Oalmann, W. R. Baker, S. L. Condon, E. deLara, S. H. Rosenberg, K. P. Spina, H. H. Stein and J. Cohen, *et al.*, *J. Med. Chem.*, 1992, **35**, 1710.
26. A. B. Smith, M. C. Guzman, P. A. Sprengeler, T. P. Keenan, R. C. Holcomb, J. L. Wood, P. J. Carroll and R. Hirschmann, *J. Am. Chem. Soc.*, 1994, **116**, 9947.
27. J. D. Tyndall, R. C. Reid, D. P. Tyssen, D. K. Jardine, B. Todd, M. Passmore, D. R. March, L. K. Pattenden, D. A. Bergman, D. Alewood, S. H. Hu, P. F. Alewood, C. J. Birch, J. L. Martin and D. P. Fairlie, *J. Med. Chem.*, 2000, **43**, 3495.
28. O. Khakshoor and J. S. Nowick, *Curr. Opin. Chem. Biol.*, 2008, **12**, 722.
29. P. N. Cheng, J. D. Pham and J. S. Nowick, *J. Am. Chem. Soc.*, 2013, **135**, 5477.
30. S. Aravinda, N. Shamala, R. Rajkishore, H. N. Gopi and P. Balaram, *Angew. Chem., Int. Ed.*, 2002, **41**, 3863.
31. L. R. Masterson, M. A. Etienne, F. Porcelli, G. Barany, R. P. Hammer and G. Veglia, *Biopolymers*, 2007, **88**, 746.
32. R. Rai, P. G. Vasudev, K. Ananda, S. Raghothama, N. Shamala, I. L. Karle and P. Balaram, *Chemistry*, 2007, **13**, 5917.
33. O. Khakshoor, B. Demeler and J. S. Nowick, *J. Am. Chem. Soc.*, 2007, **129**, 5558.
34. S. Levin and J. S. Nowick, *J. Am. Chem. Soc.*, 2007, **129**, 13043.
35. R. Spencer, K. H. Chen, G. Manuel and J. S. Nowick, *Eur. J. Org. Chem.*, 2013, **2013**, 3523.
36. F. Freire and S. H. Gellman, *J. Am. Chem. Soc.*, 2009, **131**, 7970.
37. J. S. Nowick, *Acc. Chem. Res.*, 2008, **41**, 1319.
38. F. Freire, A. M. Almeida, J. D. Fisk, J. D. Steinkruger and S. H. Gellman, *Angew. Chem., Int. Ed.*, 2011, **50**, 8735.
39. H. M. Fooks, A. C. Martin, D. N. Woolfson, R. B. Sessions and E. G. Hutchinson, *J. Mol. Biol.*, 2006, **356**, 32.
40. S. A. Sievers, J. Karanicolas, H. W. Chang, A. Zhao, L. Jiang, O. Zirafi, J. T. Stevens, J. Munch, D. Baker and D. Eisenberg, *Nature*, 2011, **475**, 96.

41. L. E. Buchanan, E. B. Dunkelberger, H. Q. Tran, P. N. Cheng, C. C. Chiu, P. Cao, D. P. Raleigh, J. J. de Pablo, J. S. Nowick and M. T. Zanni, *Proc. Natl. Acad. Sci. U. S. A.*, 2013, **110**, 19285.
42. J. D. Pham, R. K. Spencer, K. H. Chen and J. S. Nowick, *J. Am. Chem. Soc.*, 2014, **136**, 12682.
43. J. Zheng, C. Liu, M. R. Sawaya, B. Vadla, S. Khan, R. J. Woods, D. Eisenberg, W. J. Goux and J. S. Nowick, *J. Am. Chem. Soc.*, 2011, **133**, 3144.
44. J. Zheng, A. M. Baghkhanian and J. S. Nowick, *J. Am. Chem. Soc.*, 2013, **135**, 6846.
45. E. A. Mirecka, H. Shaykhalishahi, A. Gauhar, S. Akgul, J. Lecher, D. Willbold, M. Stoldt and W. Hoyer, *Angew. Chem., Int. Ed.*, 2014, **53**, 4227.
46. T. Arai, D. Sasaki, T. Araya, T. Sato, Y. Sohma and M. Kanai, *ChemBioChem*, 2014, **15**, 2577.
47. P. Y. Cho, G. Joshi, M. D. Boersma, J. A. Johnson and R. M. Murphy, *ACS Chem. Neurosci.*, 2015, **6**, 778.
48. T. Hard and C. Lendel, *J. Mol. Biol.*, 2012, **421**, 441.
49. B. Bulic, M. Pickhardt and E. Mandelkow, *J. Med. Chem.*, 2013, **56**, 4135.
50. J. Luo and J. P. Abrahams, *Chemistry*, 2014, **20**, 2410.
51. J. S. Nowick, E. M. Smith and G. Noronha, *J. Org. Chem.*, 1995, **60**, 7386.
52. D. L. Holmes, E. M. Smith and J. S. Nowick, *J. Am. Chem. Soc.*, 1997, **119**, 7665.
53. R. J. Woods, J. O. Brower, E. Castellanos, M. Hashemzadeh, O. Khakshoor, W. A. Russu and J. S. Nowick, *J. Am. Chem. Soc.*, 2007, **129**, 2548.
54. P. N. Cheng, C. Liu, M. Zhao, D. Eisenberg and J. S. Nowick, *Nat. Chem.*, 2012, **4**, 927.
55. L. Pauling, R. B. Corey and H. R. Branson, *Proc. Natl. Acad. Sci. U. S. A.*, 1951, **37**, 205.
56. A. L. Jochim and P. S. Arora, *ACS Chem. Biol.*, 2010, **5**, 919.
57. P. H. Kussie, S. Gorina, V. Marechal, B. Elenbaas, J. Moreau, A. J. Levine and N. P. Pavletich, *Science*, 1996, **274**, 948.
58. M. Sattler, H. Liang, D. Nettesheim, R. P. Meadows, J. E. Harlan, M. Eberstadt, H. S. Yoon, S. B. Shuker, B. S. Chang, A. J. Minn, C. B. Thompson and S. W. Fesik, *Science*, 1997, **275**, 983.
59. Y. Nam, P. Sliz, L. Song, J. C. Aster and S. C. Blacklow, *Cell*, 2006, **124**, 973.
60. D. C. Chan, D. Fass, J. M. Berger and P. S. Kim, *Cell*, 1997, **89**, 263.
61. V. Azzarito, K. Long, N. S. Murphy and A. J. Wilson, *Nat. Chem.*, 2013, **5**, 161.
62. M. A. Klein, *ACS Med. Chem. Lett.*, 2014, **5**, 838.
63. R. Dharanipragada, *Future Med. Chem.*, 2013, **5**, 831.
64. G. H. Bird, E. Gavathiotis, J. L. LaBelle, S. G. Katz and L. D. Walensky, *ACS Chem. Biol.*, 2014, **9**, 831.
65. Q. Chu, R. E. Moellering, G. J. Hilinski, Y.-W. Kim, T. N. Grossmann, J. T. H. Yeh and G. L. Verdine, *MedChemComm*, 2015, **6**, 111.

66. A. M. Felix, E. P. Heimer, C. T. Wang, T. J. Lambros, A. Fournier, T. F. Mowles, S. Maines, R. M. Campbell, B. B. Wegrzynski and V. Toome, *et al.*, *Int. J. Pept. Protein Res.*, 1988, **32**, 441.
67. G. Osapay and J. W. Taylor, *J. Am. Chem. Soc.*, 1990, **112**, 6046.
68. A. D. de Araujo, H. N. Hoang, W. M. Kok, F. Diness, P. Gupta, T. A. Hill, R. W. Driver, D. A. Price, S. Liras and D. P. Fairlie, *Angew. Chem., Int. Ed.*, 2014, **53**, 6965.
69. R. S. Harrison, N. E. Shepherd, H. N. Hoang, G. Ruiz-Gomez, T. A. Hill, R. W. Driver, V. S. Desai, P. R. Young, G. Abbenante and D. P. Fairlie, *Proc. Natl. Acad. Sci. U. S. A.*, 2010, **107**, 11686.
70. K. Fujimoto, N. Oimoto, K. Katsuno and M. Inouye, *Chem. Commun. (Camb.)*, 2004, 1280.
71. K. Fujimoto, M. Kajino and M. Inouye, *Chemistry*, 2008, **14**, 857.
72. C. E. Schafmeister, J. Po and G. L. Verdine, *J. Am. Chem. Soc.*, 2000, **122**, 5891.
73. C. Bechara and S. Sagan, *FEBS Lett.*, 2013, **587**, 1693.
74. S. Baek, P. S. Kutchukian, G. L. Verdine, R. Huber, T. A. Holak, K. W. Lee and G. M. Popowicz, *J. Am. Chem, Soc.*, 2012, **134**, 103.
75. C. Phillips, L. R. Roberts, M. Schade, R. Bazin, A. Bent, N. L. Davies, R. Moore, A. D. Pannifer, A. R. Pickford, S. H. Prior, C. M. Read, A. Scott, D. G. Brown, B. Xu and S. L. Irving, *J. Am. Chem. Soc.*, 2011, **133**, 9696.
76. P. M. Cromm, J. Spiegel and T. N. Grossmann, *ACS Chem. Biol.*, 2015, **10**, 1362.
77. C. Toniolo, M. Crisma, F. Formaggio, G. Valle, G. Cavicchioni, G. Precigoux, A. Aubry and J. Kamphuis, *Biopolymers*, 1993, **33**, 1061.
78. Y. W. Kim, P. S. Kutchukian and G. L. Verdine, *Org. Lett.*, 2010, **12**, 3046.
79. D. J. Yeo, S. L. Warriner and A. J. Wilson, *Chem. Commun. (Camb.)*, 2013, **49**, 9131.
80. L. D. Walensky, A. L. Kung, I. Escher, T. J. Malia, S. Barbuto, R. D. Wright, G. Wagner, G. L. Verdine and S. J. Korsmeyer, *Science*, 2004, **305**, 1466.
81. M. L. Stewart, E. Fire, A. E. Keating and L. D. Walensky, *Nat. Chem. Biol.*, 2010, **6**, 595.
82. F. Bernal, M. Wade, M. Godes, T. N. Davis, D. G. Whitehead, A. L. Kung, G. M. Wahl and L. D. Walensky, *Cancer Cell*, 2010, **18**, 411.
83. Y. S. Chang, B. Graves, V. Guerlavais, C. Tovar, K. Packman, K. H. To, K. A. Olson, K. Kesavan, P. Gangurde, A. Mukherjee, T. Baker, K. Darlak, C. Elkin, Z. Filipovic, F. Z. Qureshi, H. Cai, P. Berry, E. Feyfant, X. E. Shi, J. Horstick, D. A. Annis, A. M. Manning, N. Fotouhi, H. Nash, L. T. Vassilev and T. K. Sawyer, *Proc. Natl. Acad. Sci. U. S. A.*, 2013, **110**, E3445.
84. C. J. Brown, S. T. Quah, J. Jong, A. M. Goh, P. C. Chiam, K. H. Khoo, M. L. Choong, M. A. Lee, L. Yurlova, K. Zolghadr, T. L. Joseph, C. S. Verma and D. P. Lane, *ACS Chem. Biol.*, 2013, **8**, 506.
85. L. D. Walensky and G. H. Bird, *J. Med. Chem.*, 2014, **57**, 6275.
86. T. N. Grossmann, J. T. Yeh, B. R. Bowman, Q. Chu, R. E. Moellering and G. L. Verdine, *Proc. Natl. Acad. Sci. U. S. A.*, 2012, **109**, 17942.

87. J. E. Darnell Jr, *Nat. Rev. Cancer*, 2002, **2**, 740.
88. R. E. Moellering, M. Cornejo, T. N. Davis, C. Del Bianco, J. C. Aster, S. C. Blacklow, A. L. Kung, D. G. Gilliland, G. L. Verdine and J. E. Bradner, *Nature*, 2009, **462**, 182.
89. J. Spiegel, P. M. Cromm, A. Itzen, R. S. Goody, T. N. Grossmann and H. Waldmann, *Angew. Chem., Int. Ed.*, 2014, **53**, 2498.
90. P. M. Cromm, S. Schaubach, J. Spiegel, A. Furstner, T. N. Grossmann and H. Waldmann, *Nat. Commun.*, 2016, **7**, 11300.
91. D. S. Kemp, J. G. Boyd and C. C. Muendel, *Nature*, 1991, **352**, 451.
92. E. Cabezas and A. C. Satterthwait, *J. Am. Chem. Soc.*, 1999, **121**, 3862.
93. L. K. Henchey, A. L. Jochim and P. S. Arora, *Curr. Opin. Chem. Biol.*, 2008, **12**, 692.
94. D. Wang, M. Lu and P. S. Arora, *Angew. Chem., Int. Ed.*, 2008, **47**, 1879.
95. D. Wang, W. Liao and P. S. Arora, *Angew. Chem., Int. Ed.*, 2005, **44**, 6525.
96. L. K. Henchey, J. R. Porter, I. Ghosh and P. S. Arora, *ChemBioChem*, 2010, **11**, 2104.
97. L. K. Henchey, S. Kushal, R. Dubey, R. N. Chapman, B. Z. Olenyuk and P. S. Arora, *J. Am. Chem. Soc.*, 2010, **132**, 941.
98. A. Patgiri, K. K. Yadav, P. S. Arora and D. Bar-Sagi, *Nat. Chem. Biol.*, 2011, **7**, 585.
99. J. Bosch, S. Turley, C. M. Roach, T. M. Daly, L. W. Bergman and W. G. Hol, *J. Mol. Biol.*, 2007, **372**, 77.
100. C. H. Douse, S. J. Maas, J. C. Thomas, J. A. Garnett, Y. Sun, E. Cota and E. W. Tate, *ACS Chem. Biol.*, 2014, **9**, 2204.
101. S. Cantel, C. Isaad Ale, M. Scrima, J. J. Levy, R. D. DiMarchi, P. Rovero, J. A. Halperin, A. M. D'Ursi, A. M. Papini and M. Chorev, *J. Org. Chem.*, 2008, **73**, 5663.
102. S. A. Kawamoto, A. Coleska, X. Ran, H. Yi, C. Y. Yang and S. Wang, *J. Med. Chem.*, 2012, **55**, 1137.
103. X. Ran, L. Liu, C. Y. Yang, J. Lu, Y. Chen, M. Lei and S. Wang, *J. Med. Chem.*, 2016, **59**, 328.
104. Y. H. Lau, P. de Andrade, S.-T. Quah, M. Rossmann, L. Laraia, N. Skold, T. J. Sum, P. J. E. Rowling, T. L. Joseph, C. Verma, M. Hyvonen, L. S. Itzhaki, A. R. Venkitaraman, C. J. Brown, D. P. Lane and D. R. Spring, *Chem. Sci.*, 2014, **5**, 1804.
105. Y. H. Lau, P. de Andrade, N. Skold, G. J. McKenzie, A. R. Venkitaraman, C. Verma, D. P. Lane and D. R. Spring, *Org. Biomol. Chem.*, 2014, **12**, 4074.
106. Y. H. Lau, Y. Wu, P. de Andrade, W. R. Galloway and D. R. Spring, *Nat. Protoc.*, 2015, **10**, 585.
107. M. Pellegrini, M. Royo, M. Chorev and D. F. Mierke, *J. Pept. Res.*, 1997, **49**, 404.
108. F. M. Brunel and P. E. Dawson, *Chem. Commun. (Camb.)*, 2005, 2552.
109. G. Gerona-Navarro, R. Yoel, S. Mujtaba, A. Frasca, J. Patel, L. Zeng, A. N. Plotnikov, R. Osman and M. M. Zhou, *J. Am. Chem. Soc.*, 2011, **133**, 2040.

110. C. M. Haney, M. T. Loch and W. S. Horne, *Chem. Commun. (Camb.)*, 2011, **47**, 10915.
111. P. Gorostiza and E. Y. Isacoff, *Science*, 2008, **322**, 395.
112. R. Ferreira, J. R. Nilsson, C. Solano, J. Andreasson and M. Grotli, *Sci. Rep.*, 2015, **5**, 9769.
113. S. H. Xia, G. Cui, W. H. Fang and W. Thiel, *Angew. Chem., Int. Ed.*, 2016, **55**, 2067.
114. L. Guerrero, O. S. Smart, G. A. Woolley and R. K. Allemann, *J. Am. Chem. Soc.*, 2005, **127**, 15624.
115. L. Guerrero, O. S. Smart, C. J. Weston, D. C. Burns, G. A. Woolley and R. K. Allemann, *Angew. Chem., Int. Ed.*, 2005, **44**, 7778.
116. S. Kneissl, E. J. Loveridge, C. Williams, M. P. Crump and R. K. Allemann, *ChemBioChem*, 2008, **9**, 3046.
117. L. Nevola, A. Martin-Quiros, K. Eckelt, N. Camarero, S. Tosi, A. Llobet, E. Giralt and P. Gorostiza, *Angew. Chem., Int. Ed.*, 2013, **52**, 7704.
118. O. Babii, S. Afonin, M. Berditsch, S. Reibetaer, P. K. Mykhailiuk, V. S. Kubyshkin, T. Steinbrecher, A. S. Ulrich and I. V. Komarov, *Angew. Chem., Int. Ed.*, 2014, **53**, 3392.
119. B. R. Green, B. D. Klein, H. K. Lee, M. D. Smith, H. Steve White and G. Bulaj, *Bioorg. Med. Chem.*, 2013, **21**, 303.
120. F. Giordanetto, J. D. Revell, L. Knerr, M. Hostettler, A. Paunovic, C. Priest, A. Janefeldt and A. Gill, *ACS Med. Chem. Lett.*, 2013, **4**, 1163.
121. J. Janin, R. P. Bahadur and P. Chakrabarti, *Q. Rev. Biophys.*, 2008, **41**, 133.
122. F. Fogolari, A. Brigo and H. Molinari, *Biophys. J.*, 2003, **85**, 159.
123. R. T. Bradshaw, B. H. Patel, E. W. Tate, R. J. Leatherbarrow and I. R. Gould, *Protein Eng., Des. Sel.*, 2011, **24**, 197.
124. T. Hou, J. Wang, Y. Li and W. Wang, *J. Chem. Inf. Model.*, 2011, **51**, 69.
125. S. Genheden and U. Ryde, *Expert Opin. Drug Discovery*, 2015, **10**, 449.
126. D. Kozakov, L. E. Grove, D. R. Hall, T. Bohnuud, S. E. Mottarella, L. Luo, B. Xia, D. Beglov and S. Vajda, *Nat. Protoc.*, 2015, **10**, 733.
127. L. Deng, Q. C. Zhang, Z. Chen, Y. Meng, J. Guan and S. Zhou, *Nucleic Acids Res.*, 2014, **42**, W290.
128. J. Gavenonis, B. A. Sheneman, T. R. Siegert, M. R. Eshelman and J. A. Kritzer, *Nat. Chem. Biol.*, 2014, **10**, 716.
129. T. Geppert, B. Hoy, S. Wessler and G. Schneider, *Chem. Biol.*, 2011, **18**, 344.
130. J. Segura Mora, S. A. Assi and N. Fernandez-Fuentes, *PLoS One*, 2010, **5**, e12352.
131. S. Lise, D. Buchan, M. Pontil and D. T. Jones, *PLoS One*, 2011, **6**, e16774.
132. D. E. Kim, D. Chivian and D. Baker, *Nucleic Acids Res.*, 2004, **32**, W526.
133. X. Zhu and J. C. Mitchell, *Proteins*, 2011, **79**, 2671.
134. D. R. Koes and C. J. Camacho, *Nucleic Acids Res.*, 2012, **40**, W387.
135. C. M. Bergey, A. M. Watkins and P. S. Arora, *Bioinformatics*, 2013, **29**, 2806.
136. A. M. Watkins and P. S. Arora, *ACS Chem. Biol.*, 2014, **9**, 1747.

137. A. M. Watkins, M. G. Wuo and P. S. Arora, *J. Am. Chem. Soc.*, 2015, **137**, 11622.
138. A. P. Higueruelo, H. Jubb and T. L. Blundell, *Database (Oxford)*, 2013, **2013**, bat039.
139. E. Cukuroglu, A. Gursoy, R. Nussinov and O. Keskin, *PLoS One*, 2014, **9**, e86738.
140. M. D. Kumar and M. M. Gromiha, *Nucleic Acids Res.*, 2006, **34**, D195.
141. I. H. Moal and J. Fernandez-Recio, *Bioinformatics*, 2012, **28**, 2600.
142. F. J. Duffy, M. Verniere, M. Devocelle, E. Bernard, D. C. Shields and A. J. Chubb, *J. Chem. Inf. Model.*, 2011, **51**, 829.
143. J. Beaufays, L. Lins, A. Thomas and R. Brasseur, *J. Pept. Sci.*, 2012, **18**, 17.
144. Y. Shen, J. Maupetit, P. Derreumaux and P. Tuffery, *J. Chem. Theory Comput.*, 2014, **10**, 4745.
145. S. Singh, H. Singh, A. Tuknait, K. Chaudhary, B. Singh, S. Kumaran and G. P. Raghava, *Biol. Direct*, 2015, **10**, 73.
146. N. London, B. Raveh and O. Schueler-Furman, *Curr. Opin. Struct. Biol.*, 2013, **23**, 894.
147. R. Sable and S. Jois, *Molecules*, 2015, **20**, 11569.
148. S. Mujtaba, Y. He, L. Zeng, S. Yan, O. Plotnikova, Sachchidanand, R. Sanchez, N. J. Zeleznik-Le, Z. Ronai and M. M. Zhou, *Mol. Cell*, 2004, **13**, 251.
149. G. Bhardwaj, V. K. Mulligan, C. D. Bahl, J. M. Gilmore, P. J. Harvey, O. Cheneval, G. W. Buchko, S. V. Pulavarti, Q. Kaas, A. Eletsky, P. S. Huang, W. A. Johnsen, P. J. Greisen, G. J. Rocklin, Y. Song, T. W. Linsky, A. Watkins, S. A. Rettie, X. Xu, L. P. Carter, R. Bonneau, J. M. Olson, E. Coutsias, C. E. Correnti, T. Szyperski, D. J. Craik and D. Baker, *Nature*, 2016, **538**, 329.
150. P. F. Slivka, M. Shridhar, G. I. Lee, D. W. Sammond, M. R. Hutchinson, A. J. Martinko, M. M. Buchanan, P. W. Sholar, J. J. Kearney, J. A. Harrison, L. R. Watkins and H. Yin, *ChemBioChem*, 2009, **10**, 645.
151. M. Gao, N. London, K. Cheng, R. Tamura, J. Jin, O. Schueler-Furman and H. Yin, *Tetrahedron*, 2014, **70**, 7664.
152. C. W. Murray and D. C. Rees, *Nat. Chem.*, 2009, **1**, 187.
153. J. D. Patrone, J. P. Kennedy, A. O. Frank, M. D. Feldkamp, B. Vangamudi, N. F. Pelz, O. W. Rossanese, A. G. Waterson, W. J. Chazin and S. W. Fesik, *ACS Med. Chem. Lett.*, 2013, **4**, 601.
154. A. O. Frank, B. Vangamudi, M. D. Feldkamp, E. M. Souza-Fagundes, J. W. Luzwick, D. Cortez, E. T. Olejniczak, A. G. Waterson, O. W. Rossanese, W. J. Chazin and S. W. Fesik, *J. Med. Chem.*, 2014, **57**, 2455.
155. H. D. Agnew, R. D. Rohde, S. W. Millward, A. Nag, W. S. Yeo, J. E. Hein, S. M. Pitram, A. A. Tariq, V. M. Burns, R. J. Krom, V. V. Fokin, K. B. Sharpless and J. R. Heath, *Angew. Chem., Int. Ed.*, 2009, **48**, 4944.
156. S. W. Millward, R. K. Henning, G. A. Kwong, S. Pitram, H. D. Agnew, K. M. Deyle, A. Nag, J. Hein, S. S. Lee, J. Lim, J. A. Pfeilsticker, K. B. Sharpless and J. R. Heath, *J. Am. Chem. Soc.*, 2011, **133**, 18280.

157. S. W. Millward, H. D. Agnew, B. Lai, S. S. Lee, J. Lim, A. Nag, S. Pitram, R. Rohde and J. R. Heath, *Integr. Biol.*, 2013, **5**, 87.
158. S. Das, A. Nag, J. Liang, D. N. Bunck, A. Umeda, B. Farrow, M. B. Coppock, D. A. Sarkes, A. S. Finch, H. D. Agnew, S. Pitram, B. Lai, M. B. Yu, A. K. Museth, K. M. Deyle, B. Lepe, F. P. Rodriguez-Rivera, A. McCarthy, B. Alvarez-Villalonga, A. Chen, J. Heath, D. N. Stratis-Cullum and J. R. Heath, *Angew. Chem., Int. Ed.*, 2015, **54**, 13219.
159. B. Farrow, M. Wong, J. Malette, B. Lai, K. M. Deyle, S. Das, A. Nag, H. D. Agnew and J. R. Heath, *Angew. Chem., Int. Ed.*, 2015, **54**, 7114.
160. C. Heinis and G. Winter, *Curr. Opin. Chem. Biol.*, 2015, **26**, 89.
161. S. Chen and C. Heinis, *Biotherapeutics: Recent Developments Using Chemical and Molecular Biology*, The Royal Society of Chemistry, 2013, p. 241.
162. S. M. Howell, S. V. Fiacco, T. T. Takahashi, F. Jalali-Yazdi, S. W. Millward, B. Hu, P. Wang and R. W. Roberts, *Sci. Rep.*, 2014, **4**, 6008.
163. Y. Goto, A. Ohta, Y. Sako, Y. Yamagishi, H. Murakami and H. Suga, *ACS Chem. Biol.*, 2008, **3**, 120.
164. T. Kawakami, T. Ishizawa, T. Fujino, P. C. Reid, H. Suga and H. Murakami, *ACS Chem. Biol.*, 2013, **8**, 1205.
165. Y. Yamagishi, H. Ashigai, Y. Goto, H. Murakami and H. Suga, *ChemBioChem*, 2009, **10**, 1469.
166. Y. Goto, K. Iwasaki, K. Torikai, H. Murakami and H. Suga, *Chem. Commun. (Camb.)*, 2009, 3419.
167. Y. V. Guillen Schlippe, M. C. T. Hartman, K. Josephson and J. W. Szostak, *J. Am. Chem. Soc.*, 2012, **134**, 10469.
168. C. Heinis, T. Rutherford, S. Freund and G. Winter, *Nat. Chem. Biol.*, 2009, **5**, 502.
169. S. Chen, D. Bertoldo, A. Angelini, F. Pojer and C. Heinis, *Angew. Chem., Int. Ed.*, 2014, **53**, 1602.
170. M. Liu, S. Tada, M. Ito, H. Abe and Y. Ito, *Chem. Commun. (Camb.)*, 2012, **48**, 11871.
171. M. R. Jafari, L. Deng, P. I. Kitov, S. Ng, W. L. Matochko, K. F. Tjhung, A. Zeberoff, A. Elias, J. S. Klassen and R. Derda, *ACS Chem. Biol.*, 2014, **9**, 443.
172. S. Bellotto, S. Chen, I. Rentero Rebollo, H. A. Wegner and C. Heinis, *J. Am. Chem. Soc.*, 2014, **136**, 5880.
173. S. Kalhor-Monfared, M. R. Jafari, J. T. Patterson, P. I. Kitov, J. J. Dwyer, J. M. Nuss and R. Derda, *Chem. Sci.*, 2016, **7**, 3785.
174. J. E. Townend and A. Tavassoli, *ACS Chem. Biol.*, 2016, **11**, 1624.
175. K. R. Lennard and A. Tavassoli, *Chemistry*, 2014, **20**, 10608.
176. G. Smith, *Science*, 1985, **228**, 1315.
177. D. Bertoldo, M. M. G. Khan, P. Dessen, W. Held, J. Huelsken and C. Heinis, *ChemMedChem*, 2016, **11**, 834.
178. P. Diderich, D. Bertoldo, P. Dessen, M. M. Khan, I. Pizzitola, W. Held, J. Huelsken and C. Heinis, *ACS Chem. Biol.*, 2016, **11**, 1422.
179. Y. Xing, W. K. Clements, D. Kimelman and W. Xu, *Genes Dev.*, 2003, **17**, 2753.

180. T. N. Grossmann, J. T.-H. Yeh, B. R. Bowman, Q. Chu, R. E. Moellering and G. L. Verdine, *Proc. Natl. Acad. Sci. U. S. A.*, 2012, **109**, 17942.
181. Y. Ivarsson, R. Arnold, M. McLaughlin, S. Nim, R. Joshi, D. Ray, B. Liu, J. Teyra, T. Pawson, J. Moffat, S. S.-C. Li, S. S. Sidhu and P. M. Kim, *Proc. Natl. Acad. Sci. U. S. A.*, 2014, **111**, 2542.
182. K. F. Tjhung, P. I. Kitov, S. Ng, E. N. Kitova, L. Deng, J. S. Klassen and R. Derda, *J. Am. Chem. Soc.*, 2016, **138**, 32.
183. E. T. Boder and K. D. Wittrup, *Nat. Biotechnol.*, 1997, **15**, 553.
184. D. R. Bowley, A. F. Labrijn, M. B. Zwick and D. R. Burton, *Protein Eng., Des. Sel.*, 2007, **20**, 81.
185. J. A. Getz, T. D. Schoep and P. S. Daugherty, in *Methods in Enzymology*, ed. K. D. Wittrup and L. V. Gregory, Academic Press, 2012, vol. 503, p. 75.
186. A. V. Shivange and P. S. Daugherty, in *Peptide Libraries: Methods and Protocols*, ed. R. Derda, Springer, New York, p. 139.
187. K. Jagadish, A. Gould, R. Borra, S. Majumder, Z. Mushtaq, A. Shekhtman and J. A. Camarero, *Angew. Chem., Int. Ed.*, 2015, **54**, 8390.
188. C. Zahnd, P. Amstutz and A. Pluckthun, *Nat. Methods*, 2007, **4**, 269.
189. R. Wang, S. W. Cotten and R. Liu, *Methods Mol. Biol.*, 2012, **805**, 87.
190. Y. Goto, T. Katoh and H. Suga, *Nat. Protoc.*, 2011, **6**, 779.
191. F. T. Hofmann, J. W. Szostak and F. P. Seebeck, *J. Am. Chem. Soc.*, 2012, **134**, 8038.
192. Y. Hayashi, J. Morimoto and H. Suga, *ACS Chem. Biol.*, 2012, **7**, 607.
193. A. Kodan, T. Yamaguchi, T. Nakatsu, K. Sakiyama, C. J. Hipolito, A. Fujioka, R. Hirokane, K. Ikeguchi, B. Watanabe, J. Hiratake, Y. Kimura, H. Suga, K. Ueda and H. Kato, *Proc. Natl. Acad. Sci. U. S. A.*, 2014, **111**, 4049.
194. T. Ishizawa, T. Kawakami, P. C. Reid and H. Murakami, *J. Am. Chem. Soc.*, 2013, **135**, 5433.
195. N. Bionda and R. Fasan, *ChemBioChem*, 2015, **16**, 2011.
196. C. G. Ullman, L. Frigotto and R. N. Cooley, *Briefings Funct. Genomics*, 2011, **10**, 125.
197. S. Patel, P. Mathonet, A. M. Jaulent and C. G. Ullman, *Protein Eng., Des. Sel.*, 2013, **26**, 307.

CHAPTER 6

Biology and Synthesis of the Argyrins

L. MILLBRODT[a,b] AND M. KALESSE*[a,b,c]

[a]Leibniz Universität Hannover, Institute for Organic Chemistry, Schneiderberg 1B, 30167 Hannover, Germany; [b]Centre of Biomolecular Drug Research (BMWZ), Schneiderberg 38, 30167 Hannover, Germany; [c]Helmholtz Zentrum für Infektionsforschung (HZI), Inhoffenstr 7, 38124 Braunschweig, Germany
*E-mail: markus.kalesse@oci.uni-hannover.de

6.1 Introduction

Argyrins A and B (6.1 and 6.2, Figure 6.1) were originally isolated by Selva *et al.* from *Actinoplanes* sp. as A21459 A and B; however, at that time, the structural assignment wrongly positioned both tryptophan moieties and the position of the methoxy group.[1] The myxobacterial *Archangium* genus, which has been a rich source for bioactive natural products, such as gephyronic acid, the melithiazols, tubulysin and the archazolids, is also the origin of argyrins C–H. These argyrins were originally isolated from *Archangium gephyra*, strain Ar 8082, and then later also found in other strains thereof, as well as in strains of the genus *Cystobacter*. So far, eight different argyrins (**3**–**10**) are known, which consist of dehydroalanine, sarcosine, 2-(1′-aminoethyl)-thiazole-4-carboxylic acid, tryptophan, 4-methoxy tryptophan, glycin and either alanine or homoalanine.[2,3]

Chemical Biology No. 6
Cyclic Peptides: From Bioorganic Synthesis to Applications
Edited by Jesko Koehnke, James Naismith and Wilfred A. van der Donk

Published by the Royal Society of Chemistry, www.rsc.org

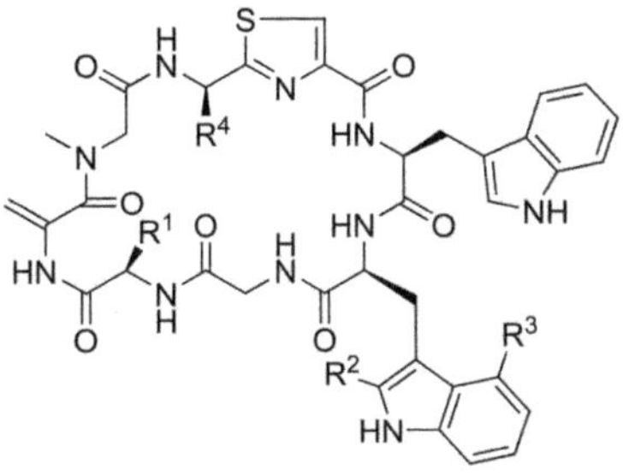

Argyrin A (**6.1**): R^1=Me; R^2=H; R^3=OMe; R^4=Me
Argyrin B (**6.2**): R^1=Et; R^2=H; R^3=OMe; R^4=Me
Argyrin C (**6.3**): R^1=Me; R^2=Me; R^3=OMe; R^4=Me
Argyrin D (**6.4**): R^1=Et; R^2=Me; R^3=OMe; R^4=Me
Argyrin E (**6.5**): R^1=Me; R^2=H; R^3=H; R^4=Me
Argyrin F (**6.6**): R^1=Me; R^2=H; R^3=OMe; R^4=CH_2OH
Argyrin G (**6.7**): R^1=Et; R^2=H; R^3=OMe; R^4=CH_2OH
Argyrin H (**6.8**): R^1=Me; R^2=H; R^3=OMe; R^4=H

Figure 6.1 Structure of all natural argyrins.

Table 6.1 Structure activity relationship of all natural argyrins.

	IC_{50} [μg/mL]		
Argyrins	*P. aeruginosa*	*P. acidovorans*	KB-3.1
A	0.12	0.07	0.1
B	0.08	0.05	0.3
C	0.1	0.05	0.1
D	0.14	0.07	0.3
E	1.4	1.5	10
F	5	5	1.5
G	4	4	>20
H	0.8	0.5	>20

6.2 Biological Activity

So far, the argyrins have exhibited three distinctly different biological activities. After discovering the first argyrins, their high antimicrobial activity, especially against *Clostridium difficile*, *Mycoplasma gallisepticum* and *Neisseria caviae*, had scientists hoping that they had found a highly active antibiotic for Gram-positive and Gram-negative bacteria. All argyrins are remarkably active against intrinsically resistant *Pseudomonas* sp. at the nano- to micromolar scale (Table 6.1), with argyrins A to D exhibiting extraordinary IC_{50} values against *P. aeruginosa* and *P. acidovorans*. The drop in activity in an SAR-dataset indicates the importance of the 4-methoxy-group of the tryptophan unit, as well as the methyl group next to the thiazole moiety. The argyrins, therefore, are one of the few natural product classes that exhibit antibiotic activity against Gram-negative bacteria.[3]

Even multidrug-resistant *P. aeruginosa* strains are susceptible to treatment with the argyrins, which combined with the lack of cross-resistance implied a new mode of inhibitory activity. By analyzing the genome of argyrin-resistant *Pseudomonas* strains, it has been shown that each strain has a singular SNP within the *fusA1*-gene of elongation factor G. Three possible gain-of-function mutations on domain III have been found within the β-sheet opposite to domain IV and therefore are in close proximity to the binding sites of another *Pseudomonas* antibiotic, fusidic acid (**11**, Figure 6.2). Opposite to

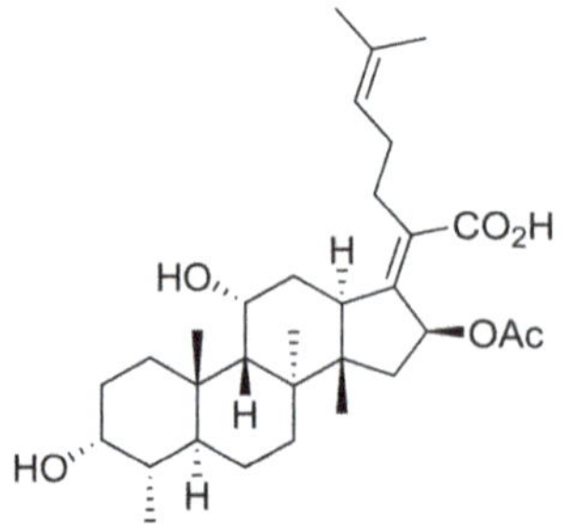

Figure 6.2 Fusidic acid (**11**).

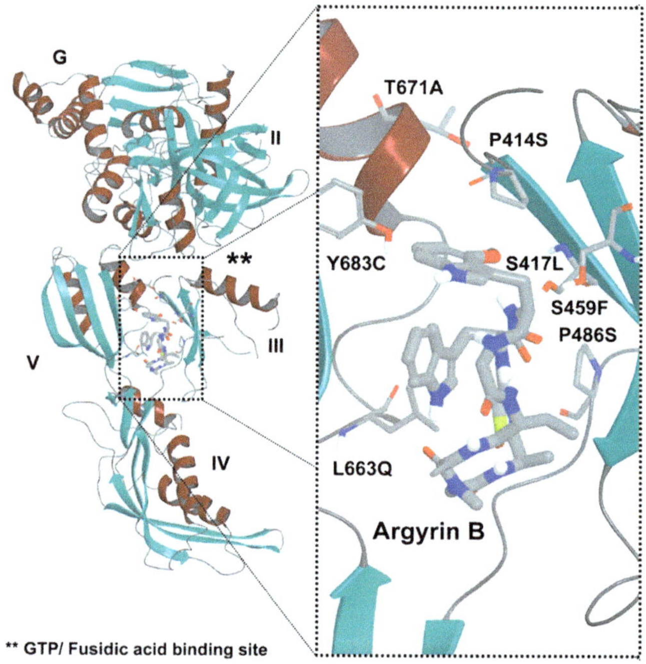

Figure 6.3 The binding sites of argyrin B (**4**) and fusidic acid (**11**) in EF-G.[5]

fusidic acid, which seems to trap EF-G after GTP hydrolysis and translocation, the argyrins are expected to lock EF-G in a distinct conformation or stop the enzyme from binding to the ribosome (Figure 6.3). Two other activating mutations are located at the contact interface between domain III and V and seem to stabilize the interdomain contact between domains III and V. This is supported by a corresponding SNP in *Salmonella typhimurium*, which induces fusidic acid-resistance.[4,5]

Another property of these cyclic peptides is their activity as an immunosuppressant, providing an opportunity to treat autoimmune diseases and

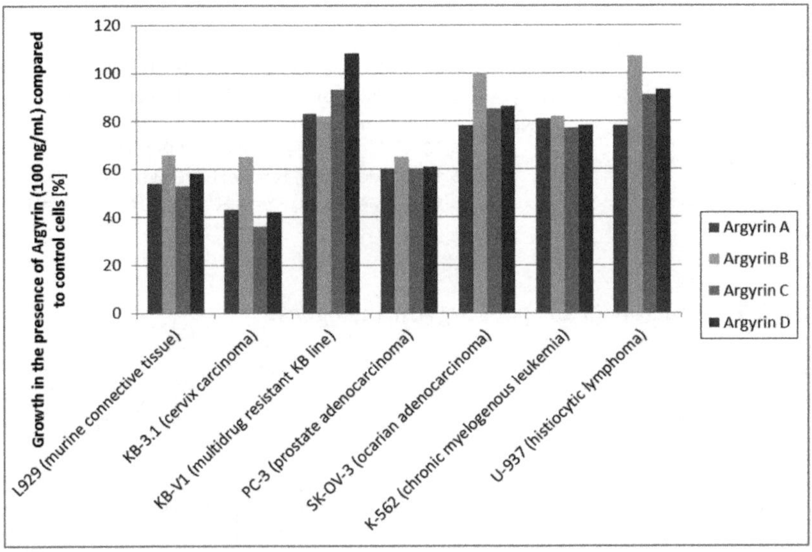

Figure 6.4 Argyrin-induced growth inhibition of mammalian cell lines.[3] Reproduced from B. Nyfeler *et al.*, *PLoS One*, 2012, 7, e42657.

ease solid organ xenotransplantation. Argyrin B very effectively inhibits murine T-cell-independent antibody formation, as well as the IgG production of human B-cells.[3]

Next to their antimicrobial properties and their role as immunosuppressants, the argyrins incompletely inhibit the growth of several mammalian cell lines; they seem to be especially effective against a cell line of cervical carcinoma (Figure 6.4). Evaluation of the proteasome inhibition and antiproliferative activities against SW-480 colon carcinoma cells of all argyrins, as well as six analogues, showed the necessity of the *exo*-methylene group, as well as of the 4-methoxy group of tryptophan Trp^2 for the biological activity. Argyrin A and F herein have been found to be the most potent derivatives among all natural argyrins and additional synthetic derivatives (Figure 6.5). Remarkably, xenotransplanted tumors do not restart growing after a 21 day treatment with argyrin F, while argyrin A-treated tumor growth resumed.[6]

The antitumor activities of the argyrins are in direct relation to their inhibition of the proteasome. Since $p27^{kip1}$ is one of the most frequently dysregulated cyclin kinase inhibitors in tumor cells without genetically encoded defunctionalization, the argyrins' ability to stop $p27^{kip1}$-degradation can be seen as a leap towards a new possible treatment of cancer in patients with normal expression. It has been shown that argyrin A can induce tumor regression and even apoptosis at a similar level to

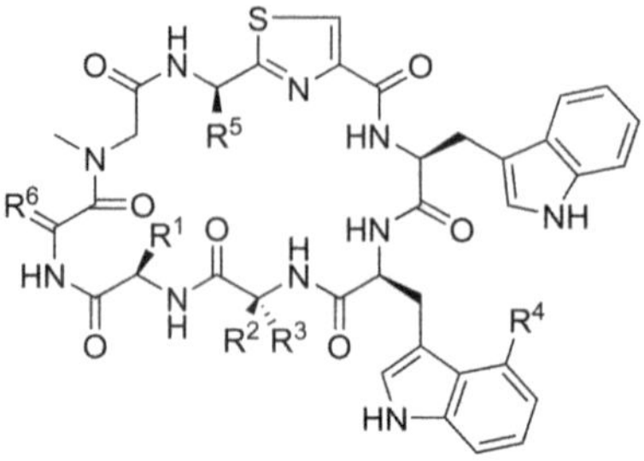

6.10 R^1=Et, R^2=CH_2OH, R^3=H, R^4=H, R^5=CH_2OH, R^6=CH_2
6.11 R^1=Et, R^2=H, R^3=H, R^4=H, R^5=CH_2OH, R^6=CH_2
6.12 R^1=Et, R^2=H, R^3=H, R^4=H, R^5=Me, R^6=CH_2
6.13 R^1=Me, R^2=Me, R^3=H, R^4=OMe, R^5=CH_2OH, R^6=CH_2
6.14 R^1=Me, R^2=H, R^3=Me, R^4=OMe, R^5=CH_2OH, R^6=CH_2
6.15 R^1=Me, R^2=H, R^3=H, R^4=OMe, R^5=Me, R^6=H_2

Figure 6.5 Structure of all tested argyrin analogues.[6]

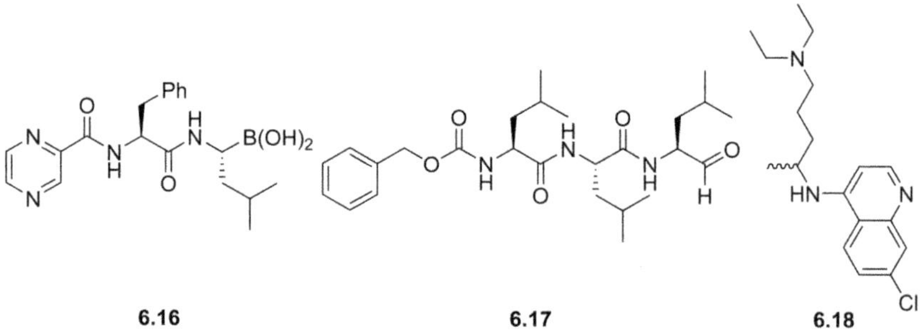

6.16 **6.17** **6.18**

Figure 6.6 Known proteasome inhibitors.

bortezomib (**18**, Figure 6.6), a clinically used p27^{kip1}-independent proteasome inhibitor, without having the same serious side effects, such as malaise and weight-loss. Astonishingly, even tumor cells that have been brought to cell-cycle arrest *in vitro* could be treated afterwards *in vivo* by the same compound. It has also been observed that even a single injection of argyrin A inhibits angiogenesis and interferes with the focal adhesion of tumor endothelial cells. This seems to be due to the fact that the small GTPase RhoA is directly controlled by p27^{kip1} and therefore is prevented from facilitating the growth of new vessels for the tumor cells.[7]

Structural analysis of the argyrins has shown that the *exo*-methylene group adds conformational fixation, because argyrin analogues lacking this moiety show broadened signals in their spectra. To check the structure of the argyrin/S20 proteasome complex, neither crystallographic nor direct NMR data could be obtained, so a competition experiment between argyrin and the β-subunit ligand MG-132 (**19**), as well as the α-unit ligand chloroquine (**20**), was set up. Due to the changed cross-peak intensities of the transferred Nuclear Overhauser Effect (NOE) peaks, it was concluded that argyrin A binds to the proteasome β-subunit of the human proteasome. A molecular modelling approach assuming that the bioactive conformation of argyrin A is the same as the free peptide and using the structure of a humanized yeast proteasome *holo* structure showed that the relevance of the methoxy-group of Trp^2 can be explained by hydrogen-bonding with

S118 in the caspase- and chymotrypsin-like binding pockets.[8] Further insights gained by modelling include the hydrogen bonding of T1 and G47 anchoring the argyrins at Gly[3] and D-Ala[4] at the bottom of the S1 pocket, as well as an illustration of the improved activity of argyrin F, which does not take the solubility into account, explained by a hydrogen bond between the OH next to Ala-Thiaz[7] and G23, therefore stabilizing the complex. Further docking simulations showed a high selectivity of argyrin A towards the β1 site, while argyrin F is more selective towards β2 and highly specific for β5 due to the specific interactions at the active site of the proteasome. Argyrin A forms a π–π-interaction with Y168 of β1, and argyrin F interacts with Y129 at β2, as well as forming an ionic hydrogen bond with D51 and G116, therefore having a stronger impact than usual on hydrogen bonding.[9]

Having determined the details of argyrin binding to the proteasome, it became possible to design analogues *in silico* and check their theoretical binding affinities to the three binding sites of the proteasome. By modifying D-Ala[4] with hydrogen-bond donors in the α-position, a bonding interaction to E22 of β1 could be induced. This calculated interaction would be specific for β1 and not be relevant in binding to the β2 and β5 sites. The addition of a hydroxyl and an amine moiety, respectively, supported this hypothesis by displaying a 30- to 67-fold increase in binding specificity compared to argyrin A (Table 6.2).[9] Müller *et al.* also postulated the possibility of binding

Table 6.2 Calculated inhibition constants and specificity parameters for argyrin A analogues.[9]

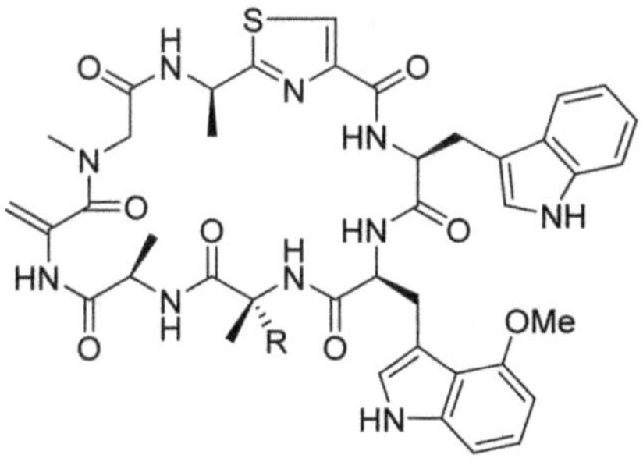

6.3

	K [μM]			Specificity parameter		
Compound	β1	β2	β3	β1	β2	β3
Argyrin A (R = H)	2.5	133	0.7	1.0	1.0	1.0
Argyrin F (**8**)	9.8	296	0.5	0.4	1.3	12.1
R = OH	0.4	310	1.2	30.2	0.1	0.2
R = NH_2	1.6	572	6.3	66.7	1.4	0.3
R = NHMe	3.2	546	2.4	11.8	1.1	1.4

to the mitochondrial EF-G, which shares 40% of its sequence with the bacterial EF-G.[4]

6.3 Synthesis

6.3.1 Biosynthesis

The biosynthesis of the argyrins (Figure 6.7) was described by Müller *et al.*,[10] who proposed the start of the biosynthesis by loading of alanine followed by serine. Condensation with glycine and subsequent action of a methyl transferase introduces the sarcosine unit. After condensation of another alanine, the heterocyclization domain forms a thiazoline ring by condensation with cysteine, which is afterwards oxidized to the thiazole. Addition of two tryptophan moieties and another glycine provides the linear, PCP-bound precursor. Upon activation of the thioesterase, cyclization takes place to generate pre-argyrin (**21**), which is post-translationally modified by elimination of the serine hydroxyl group, as well as by modification of the tryptophan.

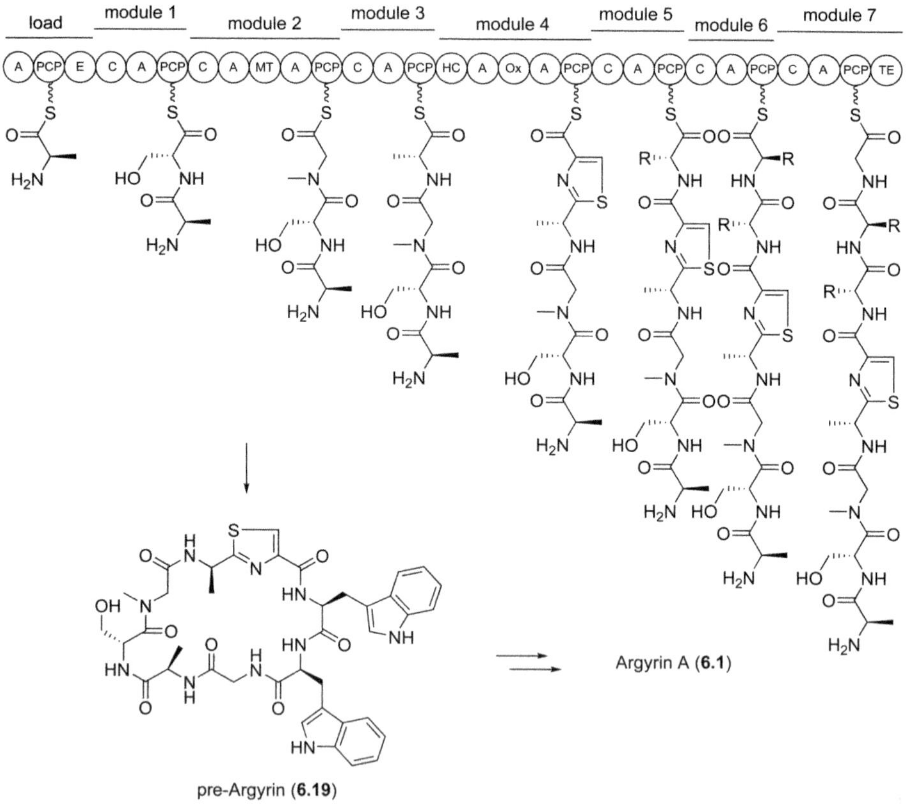

Figure 6.7 Proposed biosynthesis of the argyrins (R = 3-methylindole moiety).[10]

6.3.2 Ley's Total Synthesis

In 2001, Ley *et al.*[11] completed their total synthesis of argyrin B, which they retrosynthetically divided into three fragments (Figure 6.8): the thiazole unit (**22**), as well as the tripeptides **23** and **24**. The 4-methoxytryptophan moiety was generated by enzymatic kinetic resolution, while the installation of the dehydroalanine moiety was achieved by elimination of a selenophenyl-substituted amino acid.

For the synthesis of thiazole fragment **22** (Scheme 6.1), Boc-protected alanine (**25**) was transformed into the corresponding thioamide (**26**), which was then treated with bromoethyl pyruvate and TFAA. For the bistryptophan fragment **24** (Scheme 6.2), a strategy for the synthesis of the unusual 4-methoxy tryptophan was needed, which was found by treating racemic **27** with penicillin G acylase. Cbz-protection and subsequent coupling with glycine methyl ester led to dipeptide **29**, which could be further transformed by hydrogenolytic deprotection and coupling with Cbz-protected tryptophan to the tripeptide **24**. The last tripeptide fragment (Scheme 6.3) was synthesized by starting from Boc-protected serine (**30**), which was first converted to the β-lactone **31** using Mitsunobu conditions, and then reopened to the selenide for coupling

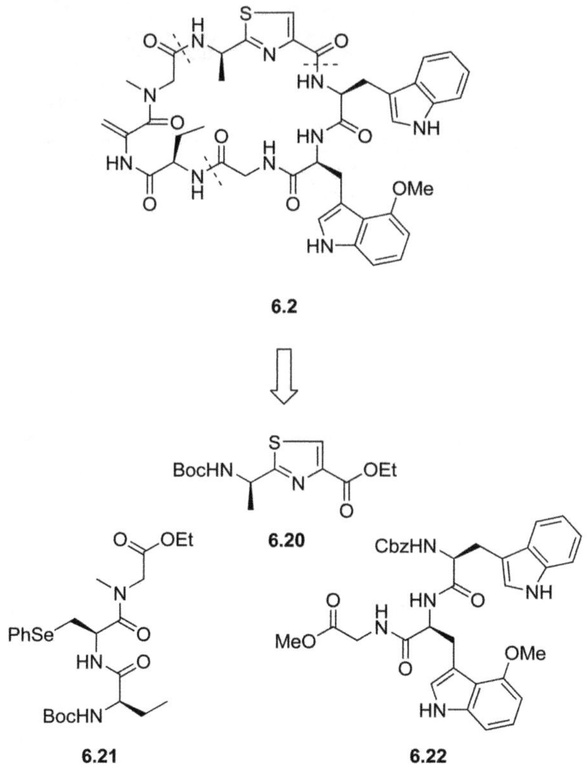

Figure 6.8 Retrosynthetic analysis of argyrin B.

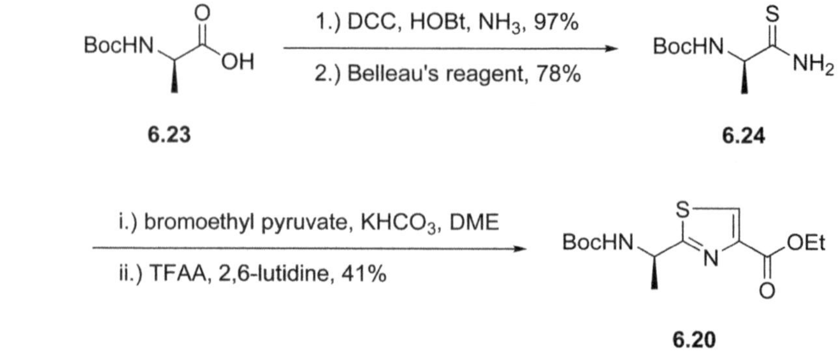

Scheme 6.1 Synthesis of the thiazole fragment **22** of argyrin B (**4**).

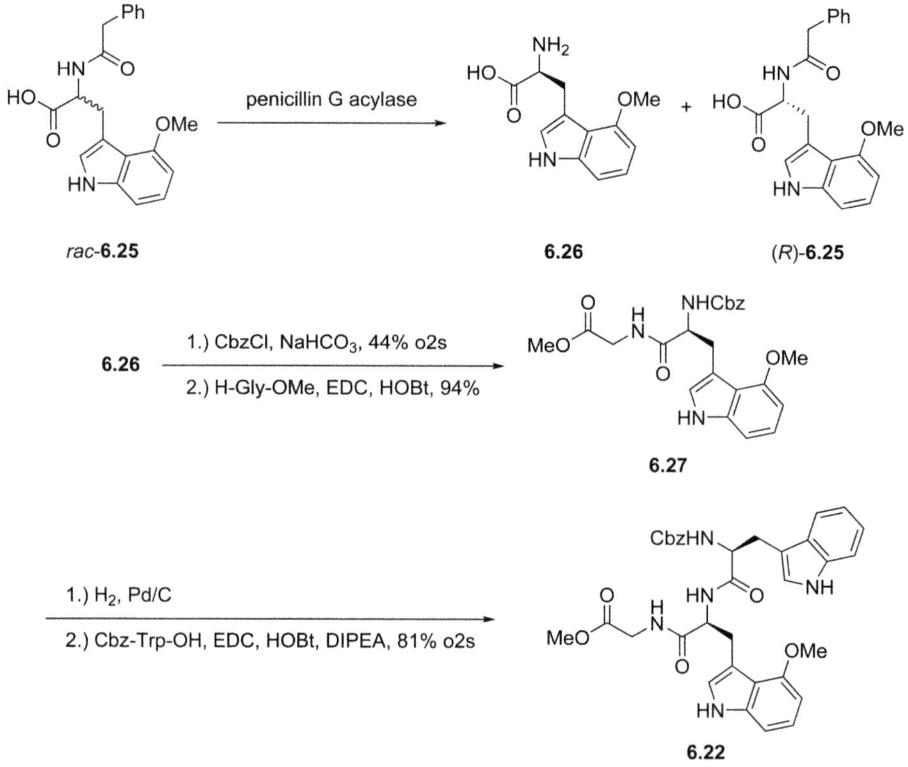

Scheme 6.2 Synthesis of the southeastern fragment **24** of argyrin B (**5**).

with ethyl sarcosine, deprotected and coupled with Boc-protected aminobutyric acid to form the selenide fragment **22**.

With the fragments in hand, **22** could then be coupled with **24** by deprotection and subsequent coupling (Scheme 6.4). The thiazole-tripeptide fragment **34** was saponified and the tripeptidic selenide was deprotected in order

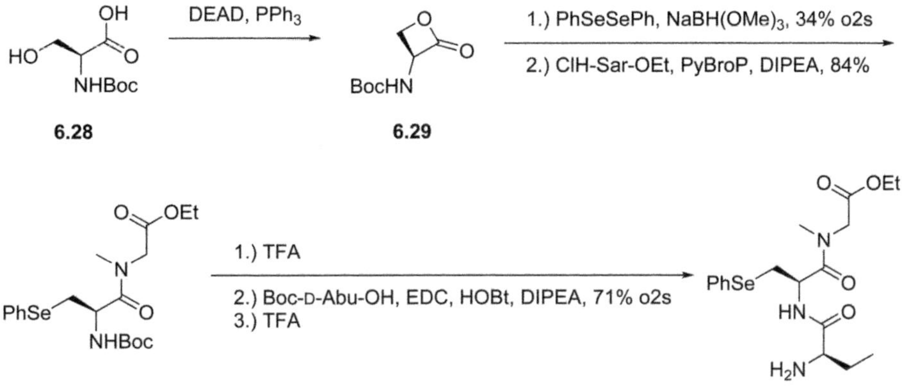

Scheme 6.3 Synthesis of the deprotected western fragment **33** of argyrin B (**4**).

to be coupled to the linear precursor **35**. Deprotection of the acid and amino termini and subsequent macrolactamization gave macropeptide **36**, which could be eliminated to argyrin B (**4**) using periodate oxidation and bicarbonate treatment.

6.3.3 Kalesse's Total Synthesis

The Kalesse total synthesis of argyrin F[6] (**8**) uses an analogous retrosynthetic approach compared to Ley *et al.*, but differs in the construction of the building blocks. The thiazole fragment **39** was synthesized from *N*-Boc-*O*-*t*Bu-serine (**37**) in an analogous manner *via* transformation to the thioamide (Scheme 6.5). Due to the unavailability of the previously used acylase, a catalytic asymmetric hydrogenation (Scheme 6.6) of the olefin **40** led to the fully protected 4-methoxy tryptophan **41** with 99% *ee*. The installation of the dehydroalanine moiety (Scheme 6.7) was already performed from dipeptide **43** by copper(I)-catalyzed elimination and was followed by coupling with sarcosine. Interestingly, the dehydroalanine moiety is chemically stabilized by substitution with an electron-withdrawing group at the amino moiety.

With the fragments in hand, the thiazole **39** and the tripeptide **24** were deprotected and coupled to form fragment **46** (Scheme 6.8). After its saponification, coupling with the dehydroalanine fragment **45** led to the linear precursor **47**. Deprotection and macrolactamization gave **48**, which was transformed to argyrin F (**8**) by acid-catalyzed *t*Bu-ether cleavage.

6.3.4 Jiang's Total Syntheses

For the argyrins A and E, a different retrosynthetic approach was pursued by Jiang's group[12] (Figure 6.9). Their three fragments were the tripeptide **49**, the thiazole-peptide hybrid **48** and the dipeptides **51** and **52**, respectively. For the

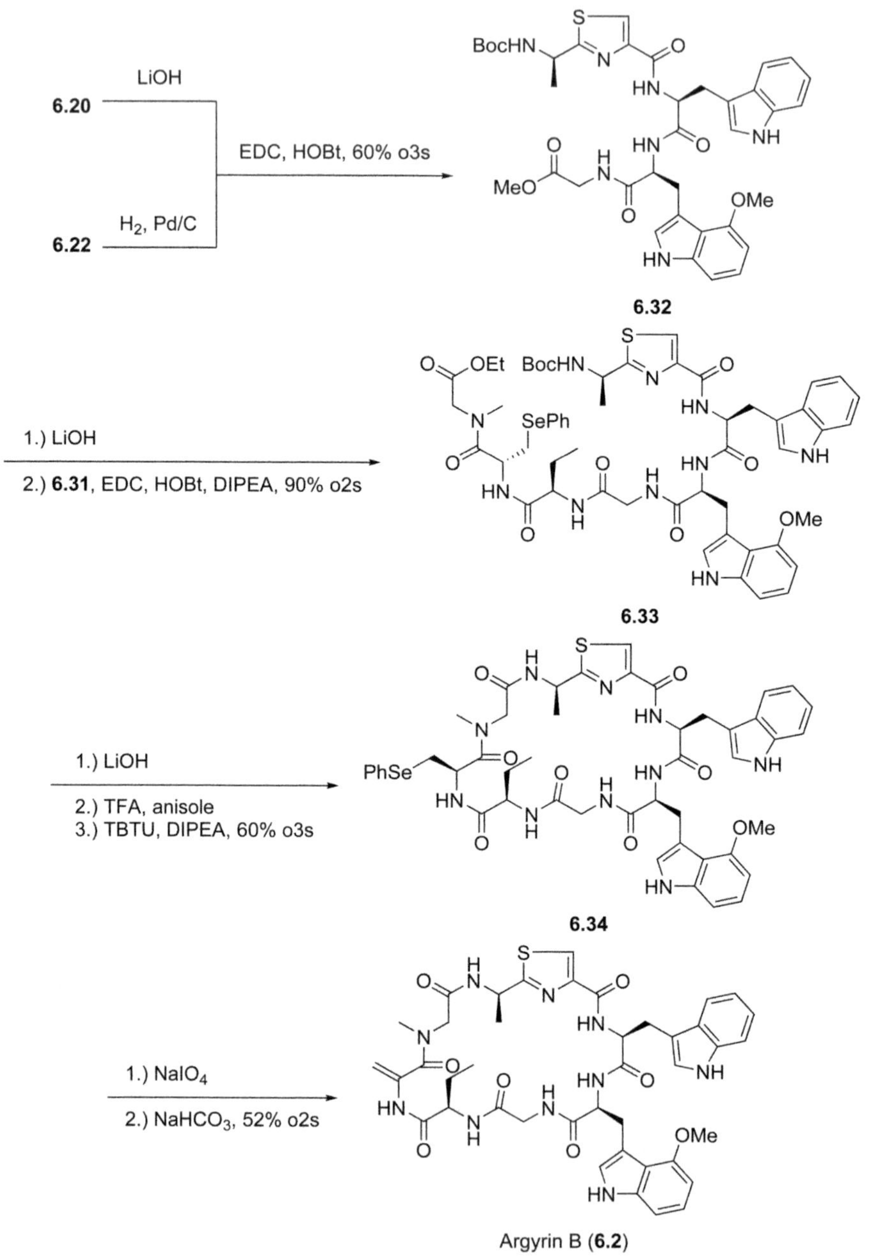

Scheme 6.4 Synthesis of argyrin B (**4**).

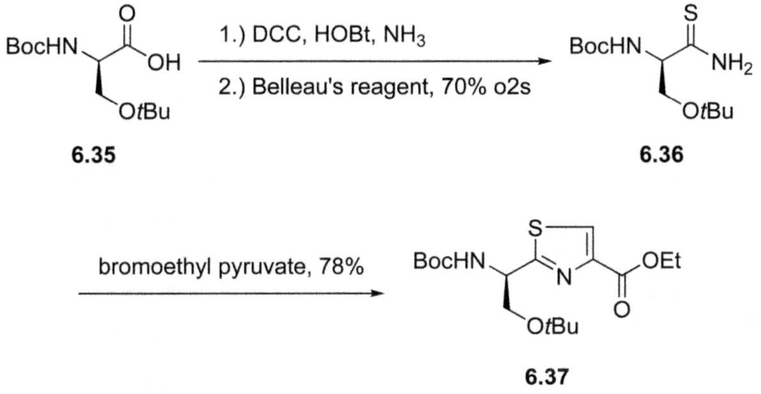

Scheme 6.5 Synthesis of the thiazole **39**.

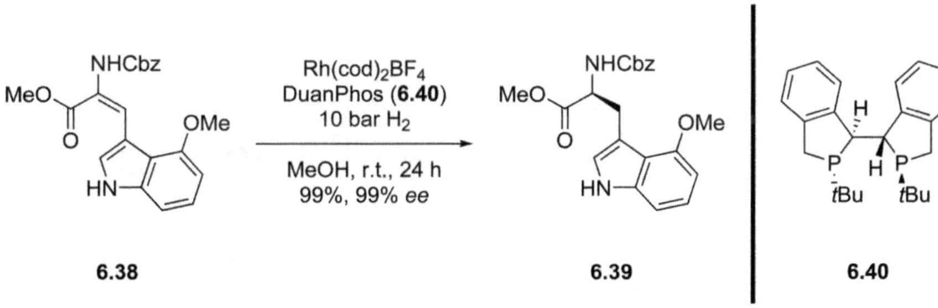

Scheme 6.6 Catalytic, asymmetric hydrogenation of dehydrotryptophan **40**.

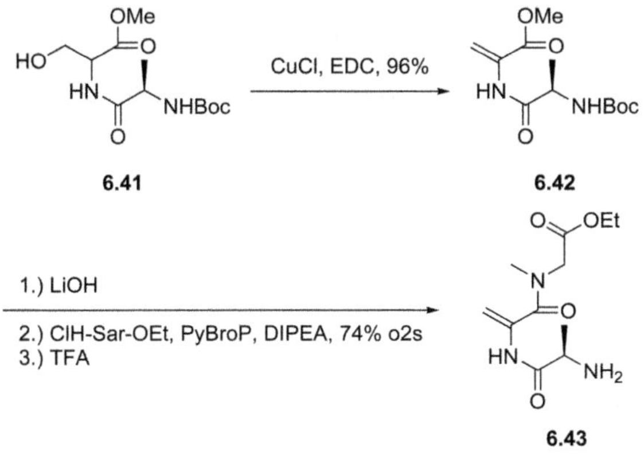

Scheme 6.7 Synthesis of the tripeptide **45**.

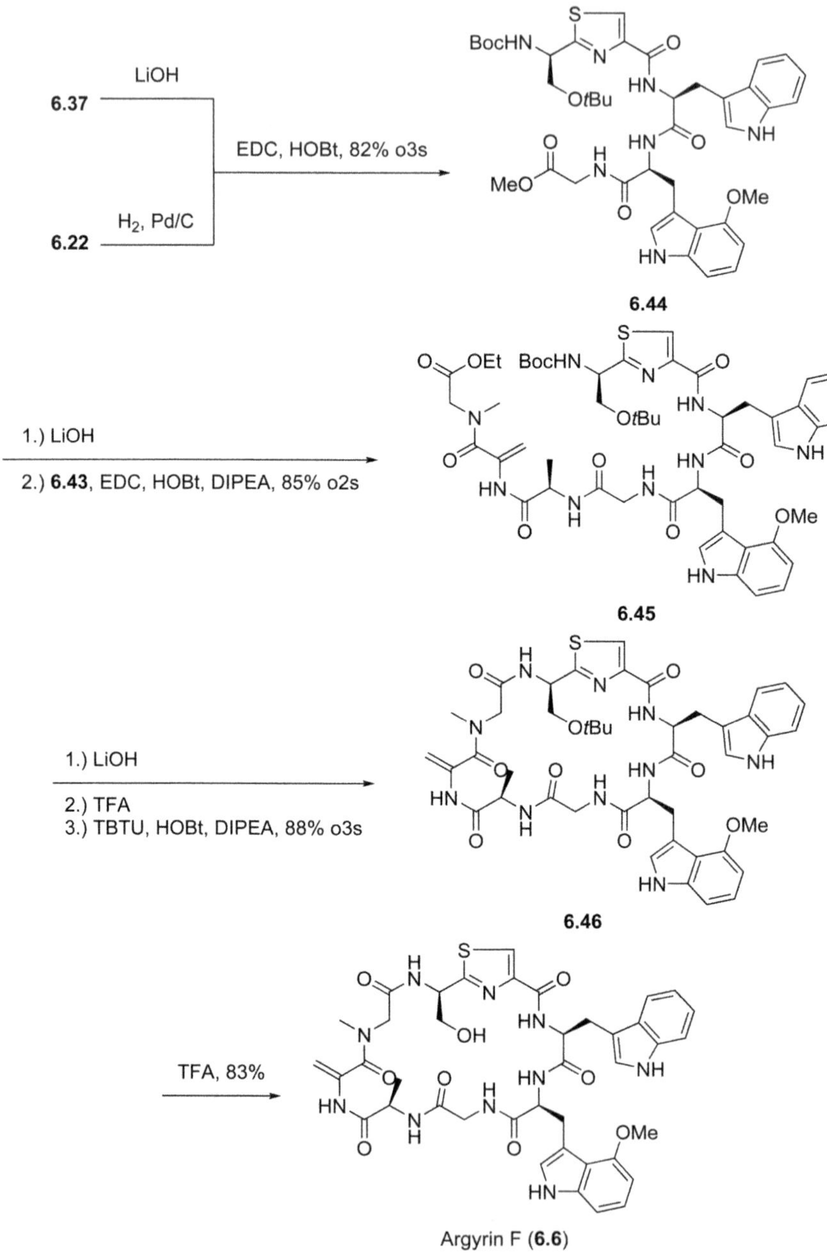

Scheme 6.8 Synthesis of argyrin F (8).

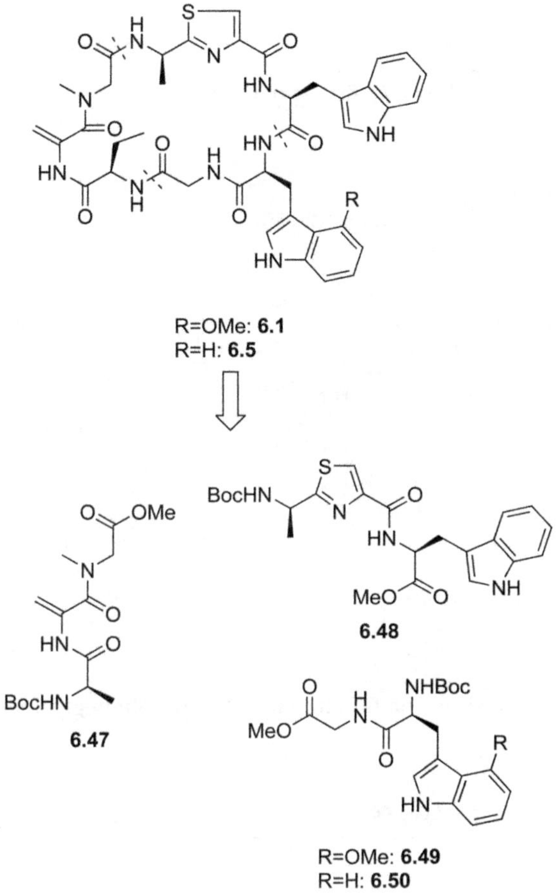

Figure 6.9 Jiang *et al.*'s retrosynthetic approach to argyrins A and E.

first fragment (Scheme 6.9), the thiazole **22** was saponified and coupled with protected tryptophan to give the dipeptide fragment **50**.

Protected 4-methoxy tryptophan **56** was prepared starting from 3-methoxy aniline (**53**, Scheme 6.10). Boc-protection, *ortho*-lithiation, iodination and finally deprotection provided the 1,2,3-functionalized aniline **54**, which could be converted to indole **56** by palladium(0)-catalyzed heteroannulation with aldehyde **55**. A single Boc-deprotection, saponification and coupling with glycine methyl ester gave dipeptide **57**. The simple tryptophan–glycine dipeptide was prepared by coupling of Boc-protected tryptophan (**58**, Scheme 6.11) with glycine methyl ester. Deprotection of both tryptophans gave the amines **57** and **59**, respectively.

The synthesis of the final fragment (**49**) started from Boc-protected alanine (**25**, Scheme 6.12), which was coupled with serine methyl ester. The dipeptide **43** was subjected to mesylate-mediated elimination, saponification and coupling with sarcosine ethyl ester.

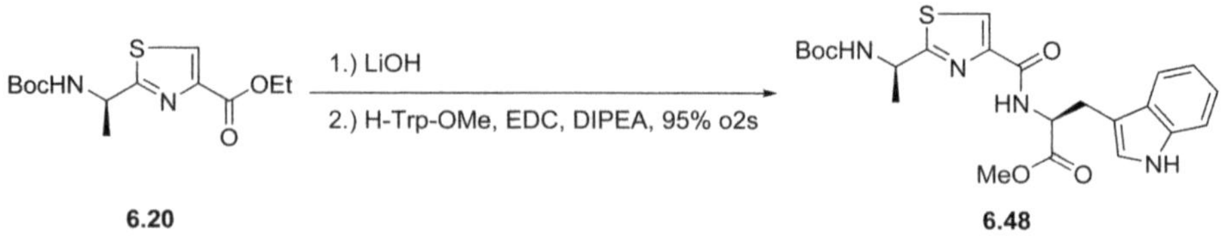

Scheme 6.9 Assembly of the dipeptide **50**.

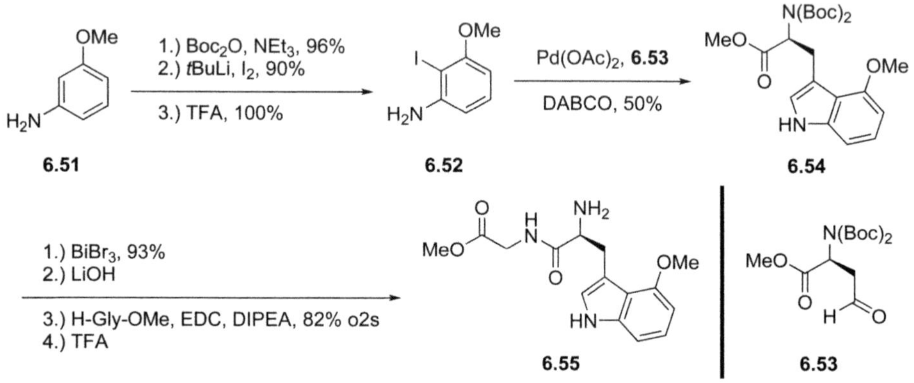

Scheme 6.10 Synthesis of the functionalized tryptophan–glycine dipeptide **57**.

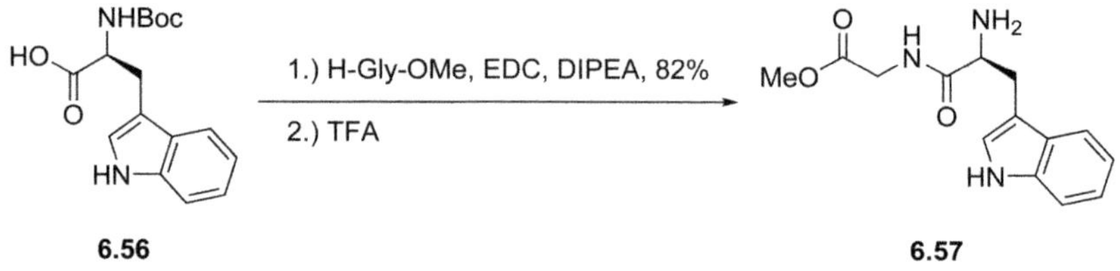

Scheme 6.11 Synthesis of methyl tryptophylglycinate (**59**).

Coupling of the deprotected analogues of **49** and **50** gave the thiazole-containing tetrapeptide. After deprotection of the acid **60**, coupling gave the linear precursors **61** and **62**, which were converted to the desired argyrins by global deprotection and macrolactamization (Scheme 6.13).

6.3.5 Chan's Approach to Argyrin Analogues

Other than the previous approaches, Chan *et al.* synthesized argyrin A, as well as several derivatives, by solid-phase peptide synthesis (SPPS).[13] Therefore they started from differently substituted indoles **63** (Scheme 6.14), which were subjected to Vilsmeier–Haack formylation and homologation by an enol-Wittig reaction. An asymmetric Strecker synthesis gave the desired α-aminonitriles **67** in favor of the desired diastereomer after testing three

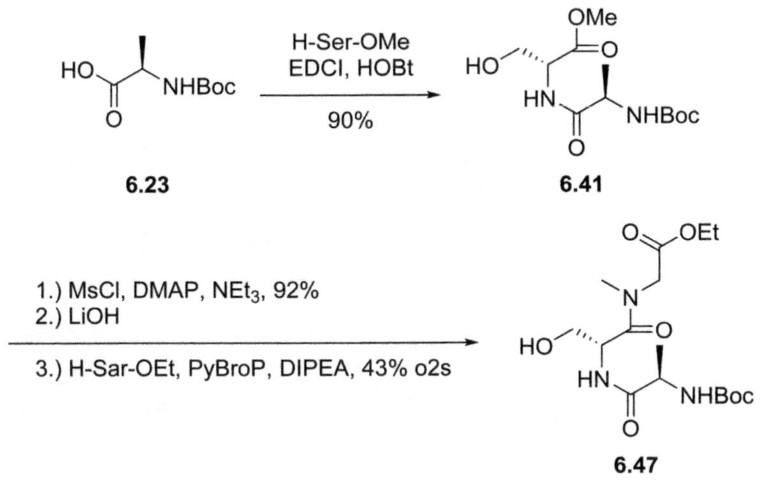

Scheme 6.12 Synthesis of the tripeptide segment **49**.

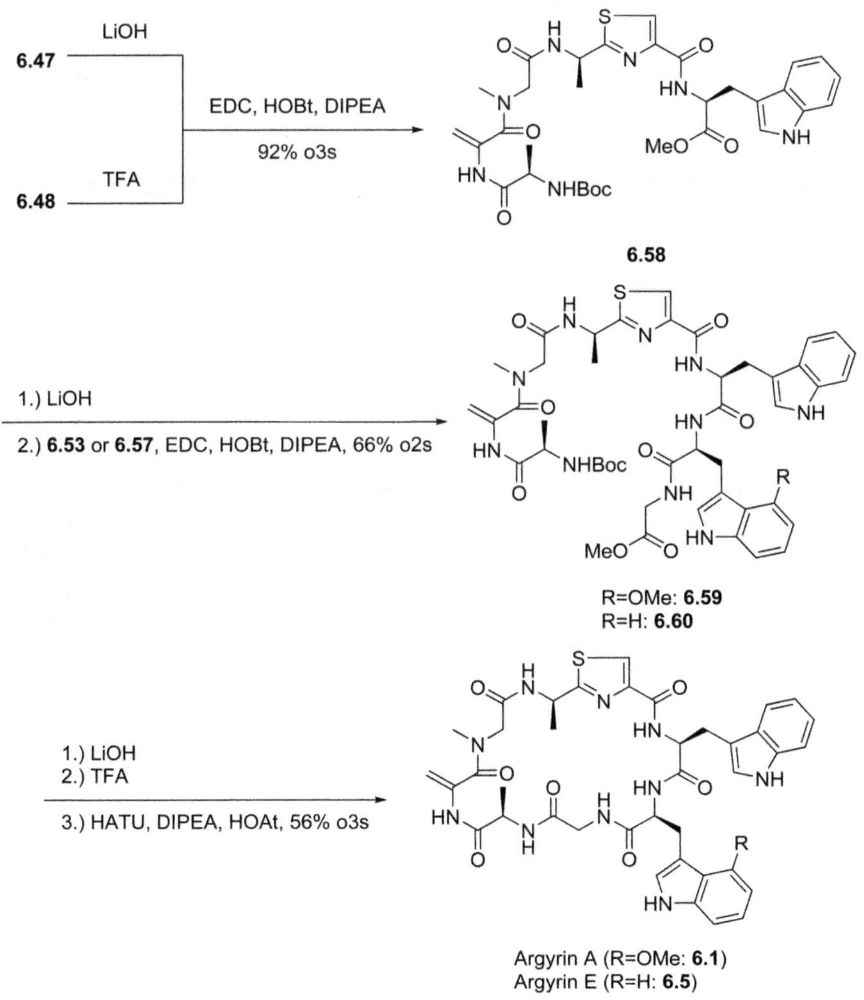

Scheme 6.13 Synthesis of the argyrins A (**3**) and F (**7**).

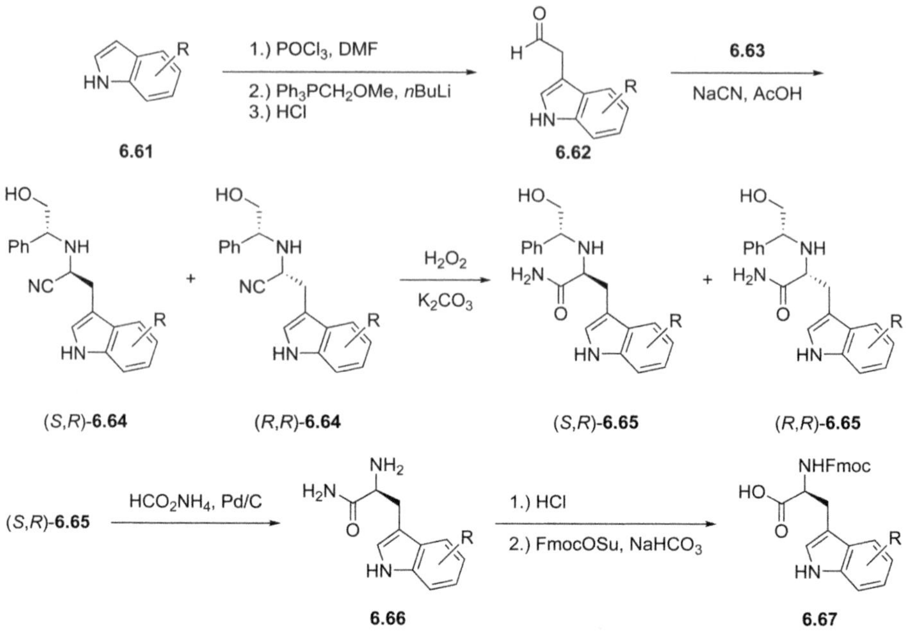

Scheme 6.14 Asymmetric synthesis of substituted *N*-protected tryptophans **69**.

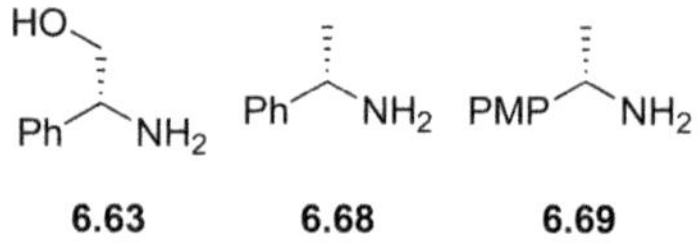

Figure 6.10 Chiral auxiliaries for asymmetric Strecker syntheses.

auxiliaries, **65**, **70** and **71** (Figure 6.10). Oxidation of the nitrile and hydrogenolytic cleavage of the auxiliary gave the aminoamides **68**, which could be transformed to the corresponding Fmoc-protected amino acids **69**. Hydrogenolytic cleavage of the halogenated tryptophans leads to dehalogenated indole cores, so the use of the acidolytically cleavable *para*-methoxy-α-benzylamine (**71**) as a chiral auxiliary was inevitable.

The SPPS (Scheme 6.15) started from Fmoc-protected sarcosine loaded onto a 2-chlorotrityl polystyrene resin and led to the linear precursor still attached to the resin. Acid treatment cleaved both the protecting group and the resin from the oligopeptide, which was then lactamized and eliminated to form argyrins A and E, and differently substituted derivatives.

In conclusion, the argyrins are promising bioactive peptides with antibiotic, antitumor and immunosuppressant properties. These properties are linked to their interaction with the bacterial elongation factor G, the human proteasome and eventually unidentified cellular targets. Altogether, four total syntheses have been achieved by the groups of Ley, Kalesse, Jiang and Chan.

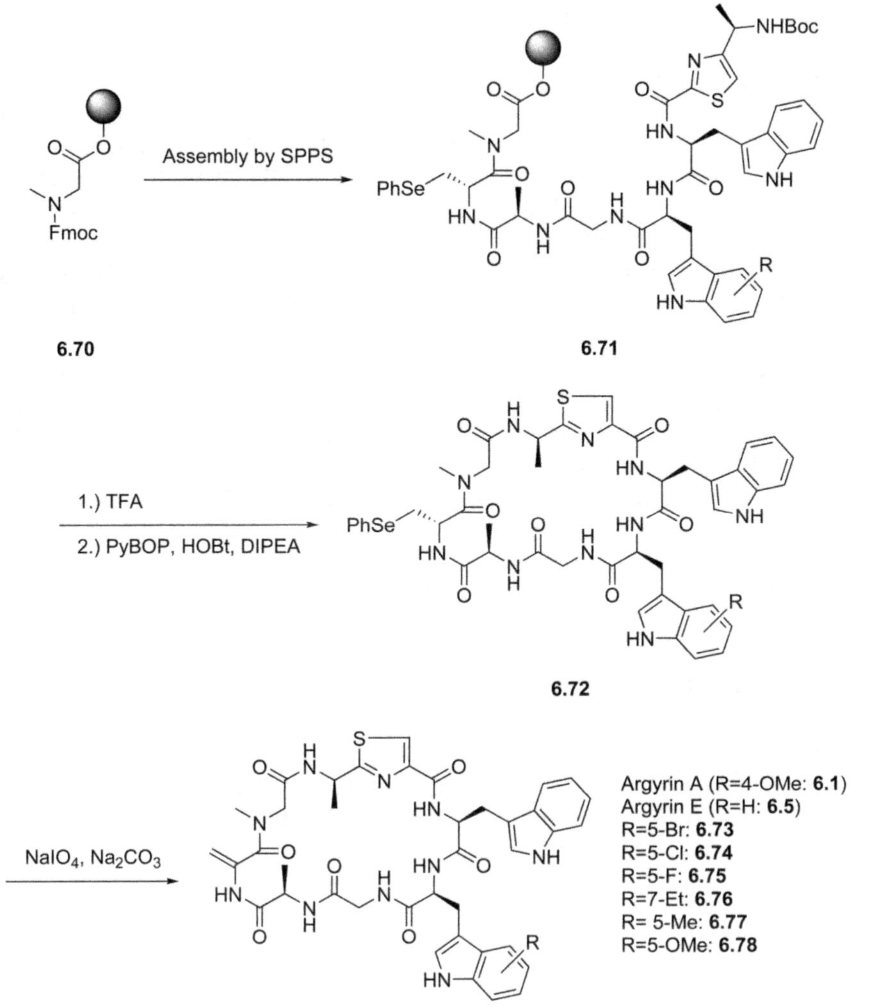

Scheme 6.15 SPPS-assisted total syntheses of argyrins A (3) and E (7), and derivatives thereof (**75**–**80**).

References

1. (a) P. Ferrari, K. Vékey, M. Galimberti, G. G. Gallo, E. Selva and L. F. Zerilli, *J. Antibiot.*, 1996, **49**, 150–154; (b) E. Selva, L. Gastaldo, G. S. Saddler, G. Toppo, P. Ferrari, G. Carniti and B. P. Goldstein, *J. Antibiot.*, 1996, **49**, 145–149.
2. L. Vollbrecht, H. Steinmetz, G. Höfle, L. Oberer, G. Rihs, G. Bovermann and P. von Matt, *J. Antibiot.*, 2002, **55**, 715–721.
3. F. Sasse, H. Steinmetz, T. Schupp, F. Petersen, K. Memmert, H. Hofmann, C. Heusser, V. Brinkmann, P. von Matt, G. Höfle and H. Reichenbach, *J. Antibiot.*, 2002, **55**, 543–551.

4. P. Bielecki, P. Lukat, K. Husecken, A. Dotsch, H. Steinmetz, R. W. Hartmann, R. Müller and S. Haussler, *ChemBioChem*, 2012, **13**, 2339–2345.
5. B. Nyfeler, D. Hoepfner, D. Palestrant, C. A. Kirby, L. Whitehead, R. Yu, G. Deng, R. E. Caughlan, A. L. Woods, A. K. Jones, S. W. Barnes, J. R. Walker, S. Gaulis, E. Hauy, S. M. Brachmann, P. Krastel, C. Studer, R. Riedl, D. Estoppey, T. Aust, N. R. Movva, Z. Wang, M. Salcius, G. A. Michaud, G. McAllister, L. O. Murphy, J. A. Tallarico, C. J. Wilson and C. R. Dean, *PLoS One*, 2012, **7**, e42657.
6. L. Bülow, I. Nickeleit, A.-K. Girbig, T. Brodmann, A. Rentsch, U. Eggert, F. Sasse, H. Steinmetz, R. Frank, T. Carlomagno, N. P. Malek and M. Kalesse, *ChemMedChem*, 2010, **5**, 832–836.
7. I. Nickeleit, S. Zender, F. Sasse, R. Geffers, G. Brandes, I. Sorensen, H. Steinmetz, S. Kubicka, T. Carlomagno, D. Menche, I. Gutgemann, J. Buer, A. Gossler, M. P. Manns, M. Kalesse, R. Frank and N. P. Malek, *Cancer Cell*, 2008, **14**, 23–35.
8. (a) B. Stauch, B. Simon, T. Basile, G. Schneider, N. P. Malek, M. Kalesse and T. Carlomagno, *Angew. Chem.*, 2010, **122**, 4026–4030; (b) B. Stauch, B. Simon, T. Basile, G. Schneider, N. P. Malek, M. Kalesse and T. Carlomagno, *Angew. Chem., Int. Ed.*, 2010, **49**, 3934–3938.
9. E. Z. Loizidou and C. D. Zeinalipour-Yazdi, *Chem. Biol. Drug Des.*, 2014, **84**, 99–107.
10. R. Müller, S. Wenzel and R. Garcia, EP2141242A1, 2008.
11. (a) S. V. Ley and A. Priour, *Eur. J. Org. Chem.*, 2002, **2002**, 3995–4004; (b) S. V. Ley, A. Priour and C. Heusser, *Org. Lett.*, 2002, **4**, 711–714.
12. W. Wu, Z. Li, G. Zhou and S. Jiang, *Tetrahedron Lett.*, 2011, **52**, 2488–2491.
13. C.-H. Chen, S. Genapathy, P. M. Fischer and W. C. Chan, *Org. Biomol. Chem.*, 2014, **12**, 9764–9768.

CHAPTER 7

Peptide Cross-links Catalyzed by Metalloenzymes in Natural Product Biosynthesis

MARCUS I. GIBSON[a], CLARISSA C. FORNERIS[a] AND MOHAMMAD R. SEYEDSAYAMDOST*[a,b]

[a]Princeton University, Department of Chemistry, Princeton, NJ, 08544, USA; [b]Princeton University, Department of Molecular Biology, Princeton, NJ, 08544, USA
*E-mail: mrseyed@princeton.edu

7.1 Introduction

Understanding the structure of a given molecule is essential in deducing how it functions in the biological world. A biochemist knows that structure imparts function, and that without any particular structure, there is nothing special about a given assortment of atoms. Small molecules are generally rigid, lacking many degrees of freedom due to stereochemical, multiple bonding, and steric constraints. This rigidity imparts function by locking the conformation of the molecule, thus facilitating reproducible interactions with macromolecules, in a typical lock-and-key-type association. Proteins, on the other hand, are large polymers with thousands of degrees of freedom. At this scale, individual stereochemical and steric constraints matter less in maintaining a particular global conformation. Rather, hundreds of weak, non-bonding interactions within proteins cumulatively impart secondary,

Chemical Biology No. 6
Cyclic Peptides: From Bioorganic Synthesis to Applications
Edited by Jesko Koehnke, James Naismith and Wilfred A. van der Donk

Published by the Royal Society of Chemistry, www.rsc.org

tertiary, and quaternary structures. When a protein becomes denatured and loses its structure due to changes in environmental conditions, it also loses its ability to function. Between these two extremes, however, lies a valley of conformational flexibility where most small peptides are found. Small peptides (<50 amino acids, approximately) are large enough that rotational flexibility gives rise to hundreds of degrees of freedom, but small enough that secondary and tertiary structures cannot play a stabilizing role in the peptide conformation. This lack of conformational stability can be a significant problem if the peptide must carry out a specific function that is important for the fitness of the producing organism. For this reason, many organisms install covalent cross-links into small peptides to reduce the degrees of freedom and impart to these molecules a rigid structure, and hence function.

A number of different types of cross-links have been observed in peptidic natural products throughout the years, often involving amide, ester, thioether, or thioester bonds.[1] This includes N–C macrocyclization of peptides and the related construction of lasso peptides, formation of thiazole and oxazole rings, creation of thioester and oxo-ester cyclized products, and synthesis of the thioether-containing lanthipeptides, where a cysteinyl sulfur atom is covalently bonded to the β-carbon of another residue. These, and most peptide cross-links studied to date, are formed by enzymes that utilize various kinds of heterolytic chemistries. There are some important peptide cross-links, however, whose biogenesis necessitates radical chemistry. Included in this class are some of our most potent antibiotics, the biosyntheses of which make use of transition metals to carry out difficult transformations. Here, we describe three main classes of transition-metalloenzymes and the radical mechanisms that they use to install complex C–N, C–S, C–O, and C–C covalent bonds, which in turn impart fascinating architectural complexity and exquisite bioactivity to their molecular products. These three classes of enzymes are mononuclear non-heme iron enzymes, cytochrome P450s, and radical *S*-adenosylmethionine (SAM) enzymes. These enzyme families are responsible for some of the most important chemical transformations of the last century, notably the biosynthesis of penicillins and glycopeptides (*e.g.* vancomycin). Less important from a medical perspective, but nonetheless intriguing biochemically, are the sactipeptide antibiotics, pyrroloquinoline quinone, and streptide (Figure 7.1). We will discuss the molecules that contain these novel cross-links, the structural features of the metalloenzymes involved in their biosyntheses, and the chemical mechanisms by which the radical transformations take place.

7.2 Penicillin Antibiotics

Penicillins are archetypal antibiotics, and the first microbially-derived molecules approved to treat bacterial infections.[2] The discovery of penicillin G by Sir Alexander Fleming has acquired legendary status as one of the key events of the 20th century, and its story of serendipity has been a source of motivation and encouragement to scientists of all ranks.[3] Since its initial discovery,

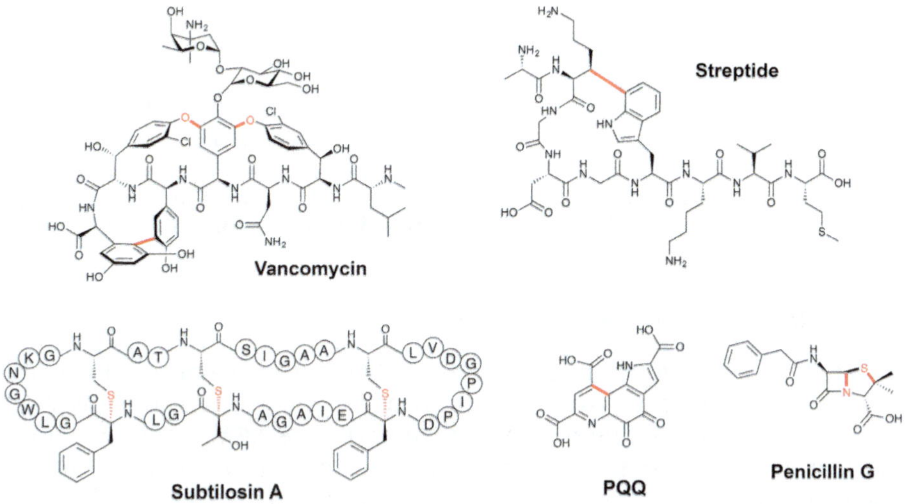

Figure 7.1 Structures of selected peptidic natural products biosynthesized by metalloenzyme-mediated radical reactions. All cross-links discussed are highlighted in red.

the class of penicillin antibiotics has expanded, as variations on the original motif have been found elsewhere in Nature, and second, third, and fourth generation penicillins have been semi-synthesized in the laboratory to combat the spread of penicillin resistance.[4]

7.2.1 Penicillin Biosynthesis

The fundamental motif common to all penicillins and related antibiotics is the β-lactam ring, a strained four-membered nitrogen-containing heterocycle, which is often fused with a five- or six-membered ring that may contain a chalcogen atom.[5,6] Different mechanisms have been implicated in β-lactam ring formation,[4] but the most common, and arguably the most elegant, is one that generates the fused thiazolidine and β-lactam rings of isopenicillin N by a single multi-step reaction.[7] Despite the relatively long history of penicillin's use in clinical settings, only in the last few decades have the mysteries of this transformation begun to be unraveled, with new results still being published even within the last year.

Penicillins are synthesized from δ-(L-α-aminoadipoyl)-L-cysteinyl-D-valine (ACV), a non-ribosomally-synthesized tripeptide that includes two non-canonical amino acids (Scheme 7.1).[8,9] The core penam structure, which comprises the fused β-lactam and thiazolidine rings, is formed directly from this starting peptide by making covalent bonds between the amide nitrogen of D-valine and the β-carbon of L-cysteine, as well as the β-carbon of D-valine and the γ-sulfur of L-cysteine.[8,10,11] The enzyme responsible of the creation of these bonds is isopenicillin N synthase (IPNS), and it uses non-heme iron to catalyze the reaction shown in Scheme 7.1.[12]

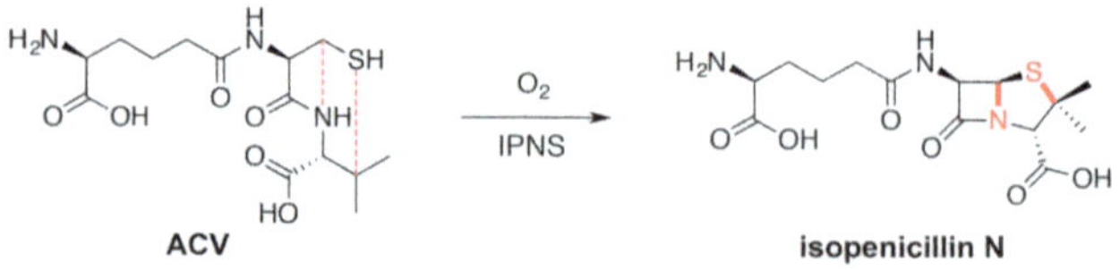

Scheme 7.1 IPNS catalyzes the oxidation of ACV to isopenicillin N by installing intramolecular cross-links, denoted in red, to form a bicyclic penam structure.

7.2.2 Isopenicillin N Synthase

IPNS falls into the canonical family of non-heme iron enzymes, which consist of a cupin fold that protects the active site and provides the His–Asp–His facial triad for binding the iron co-factor (Figure 7.2A).[13–15] Unlike other non-heme iron enzymes, IPNS does not need a sacrificial co-factor, such as α-ketoglutarate, to generate the high-valent Fe(IV)-oxo intermediate.[13,16] This is because IPNS, by catalyzing a four-electron oxidation of ACV, is able to use ACV as both a substrate and co-factor. This enzymatic reaction has been extensively studied using isotope effects, substrate analogues, X-ray crystallography, and various spectroscopic techniques to support a mechanism proposed two decades ago.[7,17]

7.2.3 IPNS Mechanism

IPNS was first purified and characterized by the research group of Jack Baldwin in 1984.[12] Over the next decade, researchers led by Baldwin would carry out countless enzymatic assays with IPNS, using a variety of techniques to assess the details regarding its reaction mechanism. A number of thorough reviews offer an in-depth treatment of the early mechanistic insights.[18,19] The consensus mechanism that has emerged from these experiments begins with ACV and dioxygen binding to the iron co-factor in the active site of IPNS (Scheme 7.2, step *a*).[7,20] ACV binds to Fe(II) *via* the cysteinyl sulfur *trans* from a histidine ligand, with a water molecule occupying the fourth equatorial coordination site.[21–24] O_2 binds to the axial site of iron and generates the oxidized Fe(III)-superoxo intermediate. Anoxic crystal structures of IPNS revealed this starting ligand arrangement, using NO as a catalytically inert substrate mimic for O_2 (Figure 7.2B), and were consistent with solution spectroscopic data.[7] Recently, stopped-flow spectroscopic experiments using ACV with deuterated cysteine have offered the first direct experimental observations of this initial reaction complex with ACV bound to the Fe(III)-superoxo species.[17]

The superoxo radical initiates catalysis by abstracting a hydrogen atom from the β-carbon of cysteine (Scheme 7.2, step *b*), which rapidly re-hybridizes the sulfur–iron bond, generating an Fe(II)-hydroperoxo intermediate and a thialdehyde (Scheme 7.2, step *c*).[7,17,25] Though this intermediate has not been directly observed, it has been supported by computational studies using

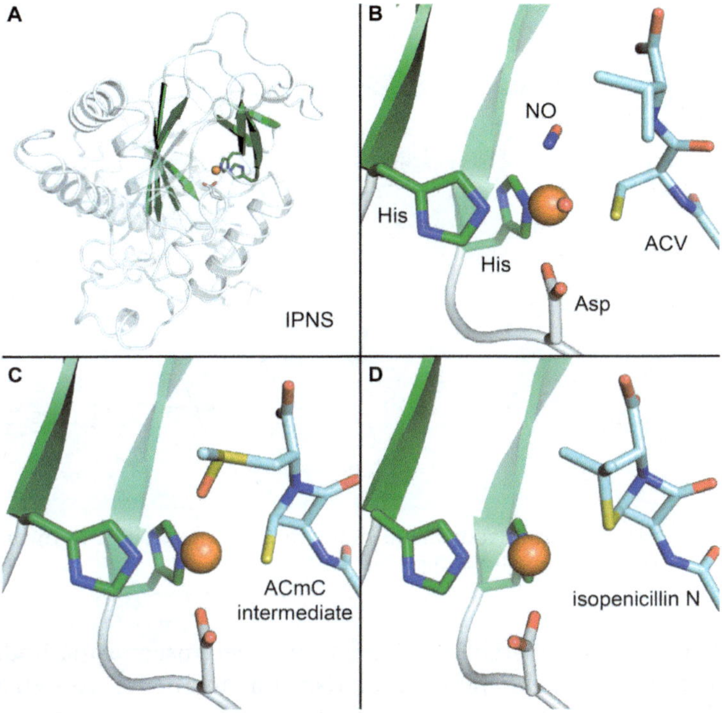

Figure 7.2 X-ray crystal structures of IPNS reveal protein structures and intermediates during the catalytic cycle. IPNS is rendered in a ribbon diagram with important residues and substrates shown as sticks. Oxygen is colored red; nitrogen, blue; sulfur, yellow. Iron is shown as an orange sphere, and the water ligand is shown as a small red sphere. Protein carbons are colored white and dark green, whereas ACV carbons are colored light blue. (**A**) The non-heme iron active site is located within a cupin barrel (colored dark green), with Fe ligated by two equatorial histidine residues and one axial aspartate residue. Exogenous ligands are omitted for clarity (PDB ID: 1BLZ).[7] (**B**) The structure of IPNS with ACV and NO bound in the active site revealed a six-coordinate Fe atom with NO oriented toward the β-carbon of the Cys residue in ACV (PDB ID: 1BLZ).[7] (**C**) Incubation of IPNS with O_2 and ACmC allowed formation of the β-lactam ring, followed by nucleophilic attack by the methyl-Cys sulfur onto the reactive Fe(IV)-oxo intermediate, thus forming a monocyclic sulfoxide. This sulfoxide product was captured in the structure (PDB ID: 1QJF).[29] (**D**) Crystal structure of IPNS with isopenicillin N ligated to the Fe center *via* the thiazolidine sulfur (PDB ID: 1QJE).[29]

combined QM/MM methods.[26–28] The precise order of events in the following step is unclear, but can be best described as a concerted circular flow of electrons, in which formation of the reactive Fe(IV)-oxo intermediate is coupled to heterolytic O–O bond cleavage, whereby the resulting hydroxide ion facilitates β-lactam ring construction (Scheme 7.2, step *d*). The Fe(IV)-oxo

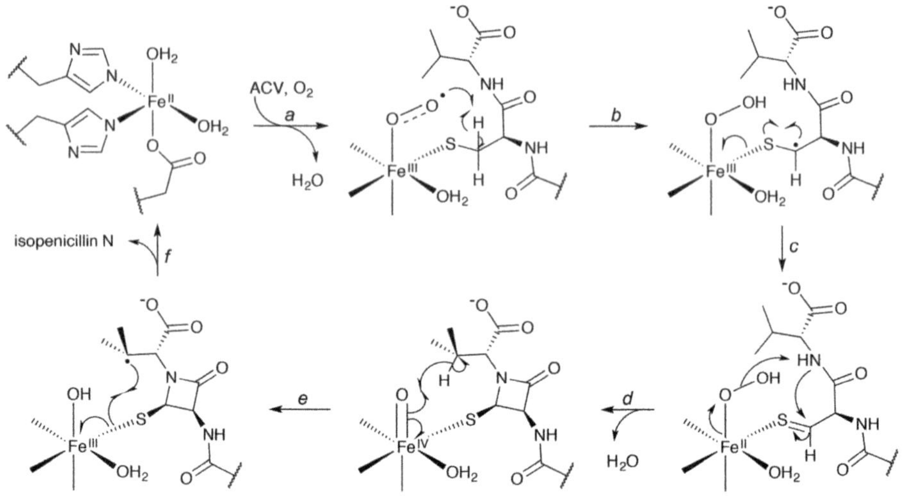

Scheme 7.2 The proposed radical mechanism for isopenicillin N formation by IPNS. See text for a description.

intermediate has been characterized by various spectroscopic methods,[17] and fits well within the established mechanism for non-heme iron enzymes.[16] Initial evidence for this specific intermediate came from a combination of biochemical and crystallographic studies. Using the ACV analogue δ-(L-α-aminoadipoyl)-L-cysteinyl-L-*S*-methylcysteine (ACmC), Baldwin and co-workers were able to trap the Fe(IV)-oxo intermediate *via* nucleophilic attack by the methylcysteine sulfur, which gave the monocyclic sulfoxide product, ligated to the now reduced Fe(II) co-factor (Figure 7.2C).[29] It was not unreasonable to assume that only an Fe(IV)-oxo intermediate would be electrophilic enough to attract the weak thioether nucleophile, thereby providing indirect evidence for this intermediate.

At this point in the catalytic cycle (Scheme 7.2, steps *a–d*), the ACV substrate has undergone the loss of two electrons and primed the Fe catalyst for creating the second ring in the penam structure. The potent Fe(IV)-oxo intermediate next abstracts the hydrogen from the β-carbon of valine (Scheme 7.2, step *e*), forming a Fe(III)-hydroxo species and a substrate-based tertiary radical. Finally, this tertiary radical recombines with an electron from the Fe-bound sulfur atom, forming the thiazolidine ring of the isopenicillin N product and regenerating the Fe(II) catalyst (Scheme 7.2, step *f*). This IPNS–product complex has also been captured crystallographically (Figure 7.2D). The last two steps of the catalytic cycle account for the final two electrons that are removed from the substrate for the complete reduction of O_2 to two molecules of water. Together, the combined biochemical and structural analyses of IPNS have provided a detailed picture of how this enzyme installs a C–C and C–S bond in a regio- and stereospecific fashion.

7.2.4 Impact of Penicillin and Its Biosynthesis

The importance of penicillin cannot be overstated. Its biggest impact, without a doubt, has been the countless lives saved through its clinical use, as well as the motivation it provided to discover and develop subsequent natural product antibiotics. In addition, the impetus to understand the biosynthesis of penicillin nurtured the field of natural product biosynthesis, especially the involvement of metalloenzymes that create bonds at unactivated carbon centers. Metalloenzymes come in all shapes and sizes, and among these, mononuclear non-heme iron enzymes are seemingly simple, requiring no complex co-factors or rare metals. But even within this class of enzymes, IPNS is uniquely elegant, requiring only a single iron atom to affect the transfer of four electrons from its substrate, ACV, to dioxygen. By using the full oxidative power of O_2, IPNS is able to form two peptide cross-links – one of these generating the strained and potently bioactive β-lactam ring. While penicillin and IPNS have been in the spotlight now for some time, the track record suggests that there is still more to learn and discover from this important molecule and the enzymes that make it.

7.3 Glycopeptide Antibiotics

By the early 1950s, not long after the introduction of penicillin, erythromycin, and tetracycline, resistant strains of *Staphylococcus* began to appear, against which the available therapies were ineffective.[30] In response to this growing crisis, Eli Lilly and Co. began a worldwide search for new antibiotic-producing organisms. After receiving soil samples from the jungles of Borneo, Edmund Kornfeld, a chemist at Eli Lilly, isolated *Amycolatopsis orientalis* (initially classified as *Streptomyces orientalis*), which produced a molecule that was active against antibiotic-resistant staphylococci. This molecule, once purified, became known as vancomycin, and it is still in use today as a drug-of-last-resort against methicillin-resistant *Staphylococcus aureus* (MRSA), though reduced susceptibility and even resistance have led to a decline in its efficacy.[31–34]

7.3.1 Oxy Enzymes in Glycopeptide Biosynthesis

Vancomycin is the founding member of the class of molecules known as glycopeptide antibiotics. Like ACV in penicillin biosynthesis, glycopeptides are the products of non-ribosomal peptide synthetases (NRPSs). By utilizing NRPS pathways rather than ribosomal synthesis, bacteria are able to incorporate more than the 20 canonical amino acids into the growing peptide chain of natural products.[35,36] Vancomycin, for example, is a 7mer that contains six unnatural amino acids. Its sequence, prior to modifications, consists of D-Leu–D-Tyr–L-Asn–D-Hpg–D-Hpg–L-Tyr–L-Dpg, where Hpg and Dpg are 4-hydroxyphenylglycine and 3,5-dihydroxyphenylglycine, respectively. The central Hpg in vancomycin becomes adorned with an *O*-linked vancosamine–glucose disaccharide.[37,38] Before glycosylation, the vancomycin aglycone undergoes a number of site-specific modifications, the order of which is still being investigated. In addition to *N*-methylation of D-Leu, each Tyr residue is hydroxylated at the pro-*R* position of the β-carbon and chlorinated at the 3-position on the ring. Most impressive, however, are a series of oxidative cross-coupling reactions involving the aromatic side chains: two C–O–C bonds, known as bisaryl ether cross-links, connect the central Hpg at the 3- and 5-positions to the phenol-hydroxyl groups in both Tyr side chains (Figure 7.1). Equally impressive is an aromatic carbon–carbon linkage, a biaryl cross-link, formed between the fifth and seventh residues, Hpg and Dpg, at the 3- and 2-positions, respectively. This characteristic cross-stitching provides vancomycin with its rigid, cup-shaped topology, which is essential for its potent antibiotic activity. While other glycopeptides may differ in amino acid composition, sugars, and modifications, the aromatic side chain cross-links are a structural hallmark and essential feature of this particular class of natural products.

The biaryl and bisaryl ether bonds found in glycopeptides are installed by heme-utilizing cytochrome P450 enzymes, generally referred to as CYP, and in GPA biosynthesis, as Oxy enzymes.[39,40] In vancomycin biosynthesis, the two bisaryl ether linkages to the central Hpg residue are installed by $OxyB_{van}$ and $OxyA_{van}$, in that order.[41–46] The last cross-link, which connects Hpg5 to Dpg7, is installed by $OxyC_{van}$ (Scheme 7.3). Glycopeptides may have as few as two (complestatin) or as many as four (teicoplanin) biaryl connections, and the biosynthetic gene cluster for each glycopeptide generally includes one Oxy enzyme per cross-link. The one exception is the recently sequenced biosynthetic gene cluster of kistamicin A, which only encodes two Oxy enzymes for the installation of three cross-links, thus implicating a bifunctional Oxy enzyme that may carry out two disparate aromatic cross-coupling reactions.[47]

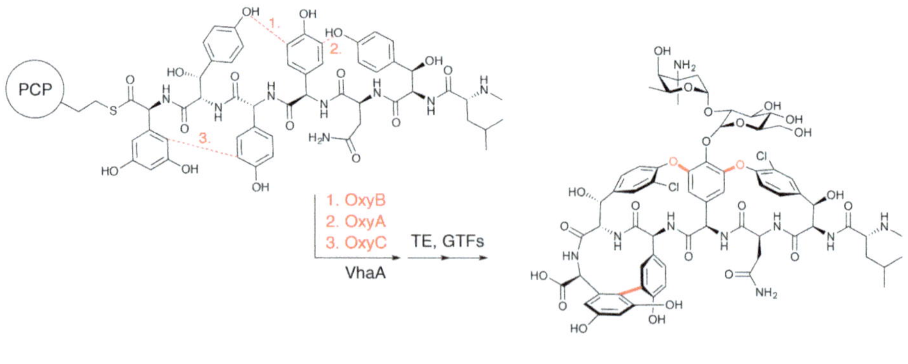

Scheme 7.3 Bisaryl ether and biaryl cross-links are inserted by OxyB, OxyA, and OxyC enzymes during vancomycin biosynthesis. These cross-linking reactions occur on the PCP-tethered peptide substrate. Though the order needs to be determined, it is thought that the cross-linked Tyr residues are then chlorinated by VhaA, and the peptide is cleaved from the PCP domain by a thioesterase (TE), and glycosylated *via* glycosyltransferases (GTFs), resulting in the final vancomycin product.

7.3.2 Structural Characterization of Oxy Enzymes

$OxyB_{van}$ from *A. orientalis* was the first enzyme within this family to be structurally characterized (Figure 7.3A). It was found to be homologous to other CYP enzymes, albeit with a larger active site cavity for the accommodation of the large peptide substrate.[48] Like other CYPs, OxyB is a predominantly α-helical protein, with two conserved helices that play an important role in catalytic turnover. The L-helix is located below the heme and provides the proximal Cys ligand that is a conserved feature of all CYPs. The I-helix lies above the heme co-factor, spanning the length of the enzyme, and provides a series of important catalytic residues, including Ala and Asn residues that, *via* backbone (Ala) and side chain (Asn) amides, provide hydrogen-bonding interactions to the distal water ligand of the heme (Figure 7.3B). In other CYPs, substrate binding induces a rotation of the I-helix, destabilizing the distal water ligand, and thus opening up the sixth coordination site for dioxygen binding.[49] The Asn residue of the I-helix is also thought to play an important role in shuttling protons to the active site to facilitate the reduction of dioxygen and the formation of the potent Compound I reaction intermediate, described below.[50,51]

Since the structural characterization of $OxyB_{van}$, the structures of $OxyC_{van}$,[52] $OxyE_{tei}$,[53] and $OxyA_{tei}$,[54] the latter two corresponding to CYPs responsible for the second and third cyclizations in teicoplanin biosynthesis,[55] have been

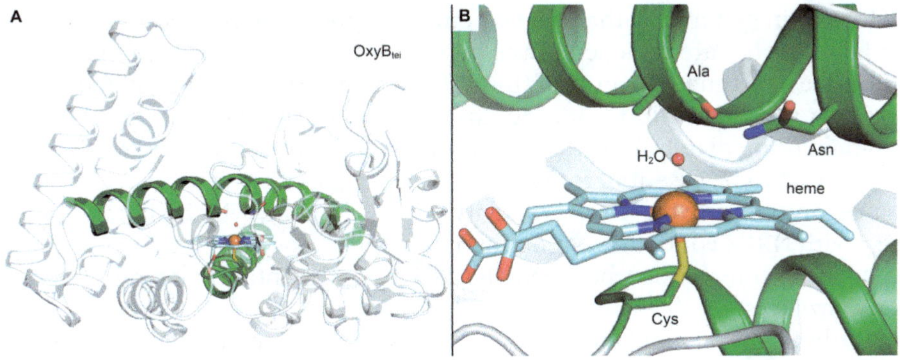

Figure 7.3 The X-ray crystal structure of OxyB from *Actinoplanes teichomyceticus* ($OxyB_{tei}$, PDB ID: 4TX3). This structure is similar to that of OxyB involved in vancomycin biosynthesis. (**A**) The full-length protein is shown as white ribbons, with the I- and L-helices colored dark green, the former spanning the length of the protein above the heme co-factor, and the latter lying underneath the heme co-factor and providing the proximal Cys ligand. (**B**) A close-up view of the heme co-factor in the resting state of the enzyme shows the distal water ligand, depicted as a red sphere. The heme co-factor is shown as sticks, with carbons colored light blue; nitrogens, blue; oxygens, red; and the heme iron as an orange sphere. The proximal cysteine ligand, as well as active-site alanine and asparagine residues that help stabilize the distal water ligand, are also shown as sticks, with sulfur colored yellow.

determined. These structures have confirmed the conservation of a large active site and the noted I-helix residues, as well as some other features that are hypothesized to be involved in protein–protein interactions with other domains in the NRPS pathway. However, the active site landscapes are otherwise not obviously conserved, and further studies are required to better understand exactly how these enzymes bind their peptide substrates.[56]

7.3.3 Mechanistic Proposals for Oxy Enzymes

To date, no rigorous mechanistic studies have been carried out to examine the mechanism by which the Oxy enzymes install the various cross-links in glycopeptides. The simple explanation for the absence of these kinds of studies is the complexity of the substrates on which Oxy enzymes act. Only a handful of enzymes have been shown to be active *in vitro* when supplied with various, mostly artificial, substrates. $OxyB_{van}$, the enzyme that installs the first bisaryl ether bond in vancomycin, was able to accept simplified hexa- and heptapeptide substrates, lacking the C-terminal Dpg in the former instance, and all modifications on the Tyr residues in both cases.[45,46] Similar success has been found using the teicoplanin enzymes on various simplified substrates, with $OxyB_{tei}$,[57] $OxyA_{tei}$,[58] and $OxyE_{tei}$[55] all forming bisaryl ether cross-links. To date, the enzymatic activity of a C–C-bond-forming Oxy enzyme has not been recapitulated *in vitro*.

One of the key findings of these experiments, both in the vancomycin and teicoplanin systems, is that the Oxy enzymes preferentially act upon peptide substrates that are attached to the peptidyl carrier protein (PCP) in the penultimate or ultimate module of the NRPS, which is responsible for adding the sixth or seventh amino acid. Moreover, a previously uncharacterized NRPS domain immediately following the last PCP-domain, the so-called X-domain, was found to be essential for $OxyA_{tei}$ and $OxyE_{tei}$ activity *in vitro*, while at the same time improving $OxyB_{tei}$ activity. The X-domain, which shares homology with NRPS condensation domains, has been postulated to play a role in mediating interactions between the PCP-bound peptide substrate and the Oxy enzymes.[59]

Without any detailed kinetic experiments, the precise mechanism by which aromatic cross-links are installed in glycopeptide antibiotics remains in the realm of hypothesis. A number of potential mechanisms can be imagined, and one of the first to be proposed involves a putative biradical generated by two sequential hydrogen atom abstractions.[46] Recombination of these radicals would then result in the various cross-links observed in glycopeptides. While this is certainly a possible mechanism, we prefer a single-radical mechanism such as that shown in Scheme 7.4. In this proposed scheme, the PCP-bound peptide substrate binds to the Oxy active site with the aryl rings to be cross-linked adjacent to one another. The first steps of our proposed catalytic cycle are akin to those of canonical CYPs.[60,61] Accordingly, upon substrate binding to the resting ferric state of the co-factor, which forms an out-of-plane high spin ferric heme, reduction to the ferrous form ensues (Scheme 7.4A,

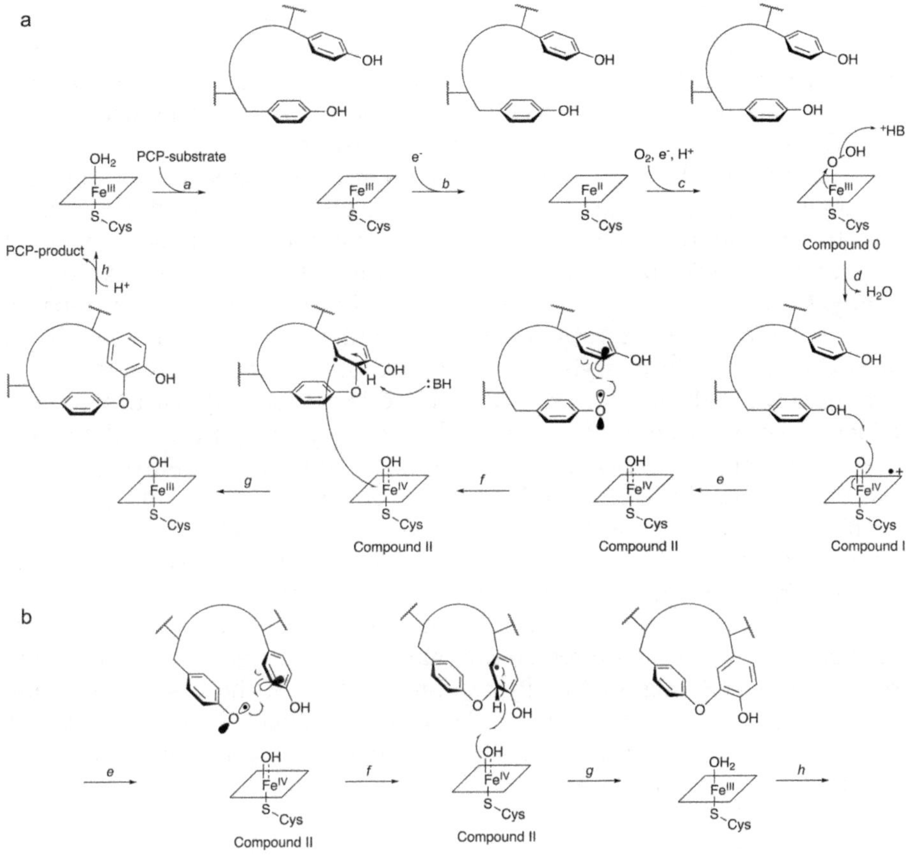

Scheme 7.4 Proposed reaction mechanisms for the Oxy-catalyzed bisaryl ether cross-links in vancomycin biosynthesis with (**A**) Compound II serving as an electron acceptor (and possibly a Brønsted base – see text), or (**B**) Compound II participating in H-atom abstraction, in a rebound-like process, to re-aromatize the cross-linked product. Note that steps *a*–*e* and step *h* in panel (B) are identical to those in panel (A).

steps *a* and *b*). The source of the electrons is typically a small redox protein, such as ferredoxin. Binding of O_2 concomitant with a second reduction step gives Compound 0, a ferric hydroperoxo intermediate (Scheme 7.4A, step *c*). Heterolytic O–O bond cleavage and the concurrent production of water give rise to the highly oxidative Compound I intermediate (Scheme 7.4A, step *d*), an Fe(IV)-oxo porphyrin cation radical species, which abstracts a hydrogen atom from the phenolic-OH of Tyr (Scheme 7.4A, step *e*). This generates a tyrosyl radical and Compound II, an Fe(IV)-hydroxy intermediate. Radical addition into the π-system of Hpg would lead to the formation of the ether cross-link between the Tyr and Hpg side chains (Scheme 7.4A, step *f*). Subsequent deprotonation of Hpg and electron transfer back to the heme allows for re-aromatization of the Hpg ring, completing the cross-linking reaction and returning the co-factor to the original Fe(III) oxidation state (Scheme 7.4A, step *g*). The deprotonation step could be carried out by an active site base, or alternatively and intriguingly, by Compound II, which has recently been shown to be fairly basic.[62–64] Finally, elimination of the product would reset the active site for another turnover (Scheme 7.4A, step *h*).

An alternative rebound-like mechanism may also be envisioned, in which Compound II abstracts an H-atom from the cross-linked radical intermediate to facilitate rearomatization and concurrently generate a molecule of water and the ferric state of the heme co-factor (Scheme 7.4B, steps *f* and *g*). Clearly, the orientation of the substrate would differ in this case compared to that shown in Scheme 7.4A. Mechanistic and structural studies will likely illuminate the details of aryl-ether cross-link formation in the future.

As mentioned above, an Oxy-catalyzed C–C-bond-forming reaction has not yet been reconstituted *in vitro*. Nonetheless, one can imagine a similar mechanism for the formation of biaryl Hpg–Dpg and Hpg–Trp cross-links (Scheme 7.5A and B). For these cross-links, the two aromatic side chains may form an off-center stacking interaction.[65,66] Following H-atom abstraction by Compound I from the Hpg phenol (Scheme 7.5, step *e*), overlap between the Hpg and Dpg/Trp π-systems would allow for C–C bond formation between the two aryl rings (Scheme 7.5, step *f*). Deprotonation of the upper ring and radical transfer to the heme would return the catalyst to its original Fe(III) oxidation state (Scheme 7.5, step *g*). Finally, tautomerization of the lower ring would deliver the cross-linked Hpg–Dpg and Hpg–Trp products (Scheme 7.5, step *g*). Again, alternative roles for Compound II, as discussed above, may come into play in these cross-links as well.

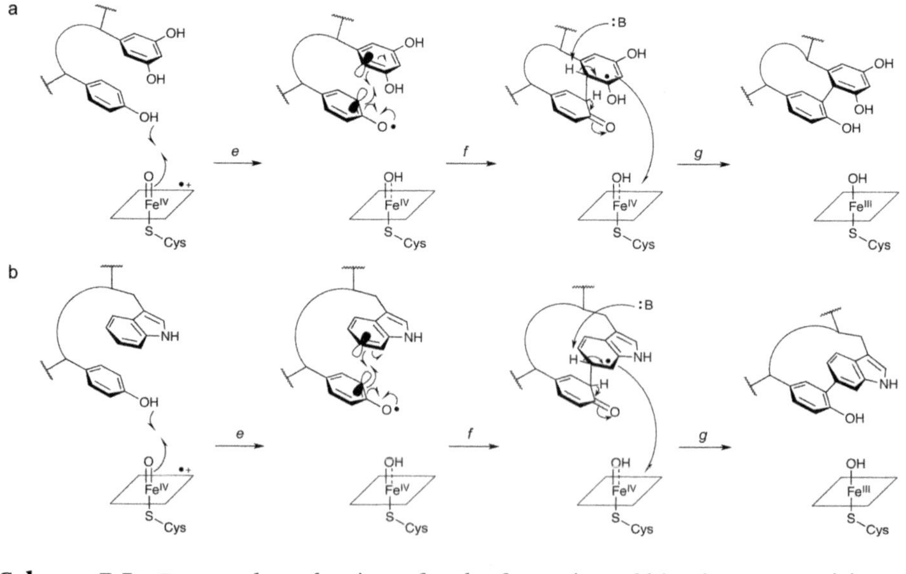

Scheme 7.5 Proposed mechanisms for the formation of biaryl Hpg–Dpg (**A**) and Hpg–Trp (**B**) cross-links in glycopeptide biosynthesis. Steps *a–d* are analogous to those in Scheme 7.4A.

While hypothetical, the foregoing discussion highlights that Oxy enzymes are a fascinating system in which to explore the structural and chemical features that underlie metalloenzyme-mediated peptide cross-links. Outstanding structural aspects include the interaction of Oxy enzymes with the X-PCP di-domain as well as the mode of substrate binding at the heme co-factor. Chemically, it remains puzzling as to why Nature would use highly oxidative Compound I intermediates, with an estimated reduction potential of 1.2–1.4 V, to oxidize phenolic groups with significantly lower radical reduction potentials (~0.8–0.9 V).[67–69] Finally, from a small molecule perspective, the discovery of new glycopeptides remains an attractive source of potentially new therapeutic molecules. These and other issues show that there is still a lot to learn regarding the chemistry, biosynthesis, and biology of glycopeptide antibiotics.

7.4 Radical SAM Enzymes Involved in Intramolecular RiPP Cross-links

The examples of metalloenzyme-catalyzed cross-links discussed so far have been studied for over half a century. Here, we examine a class of enzymes and natural products about which our understanding has rapidly expanded in the last two decades. The natural products are part of the ribosomally-synthesized and post-translationally modified peptides (RiPPs) and the enzymes belong to the radical SAM enzyme superfamily.

The explosion of genome sequences has revealed over 150 000 radical SAM enzymes in the genome database. Despite the wealth of sequence information, these enzymes remain sparsely studied, owing, no doubt, to their extreme oxygen sensitivity, necessitating strict anoxic conditions for protein preparation and assays. Radical SAM enzymes have been identified that play key roles in central metabolism, including the reduction of nucleic acids by class III ribonucleotide reductases and the synthesis of the vitamins thiamine and biotin.[70–72] The role of radical SAM enzymes in the production of RiPPs, however, has only begun to be explored.

RiPPs begin as precursor peptides that are encoded directly in the genome, usually in an operon or gene cluster along with enzymes responsible for the maturation and transport of the final product.[1] The precursor peptide typically consists of an N-terminal leader sequence followed by a core sequence. Additional elements, such as a 'follower' sequence C-terminal to the core, may also be present. The leader peptide is required for binding to the modification enzymes. The core sequence is modified by tailoring enzymes and, upon removal of the leader, secreted as the mature product. Cyclic RiPPs are common; they are formed usually *via* heterolytic chemistries involving intramolecular amide, ester, thioester, or thioether bond formation, among others. A few RiPPs, however, have been found to employ radical SAM enzymes to install cross-links involving unactivated, aliphatic carbons. The best studied of these systems are those involved in the biosynthesis of the redox co-factor pyrroloquinoline quinone (PQQ), sactipeptides such as the antibiotic subtilosin A, and the recently characterized streptide.[73–75]

7.4.1 PQQ

PQQ was initially characterized as the redox co-factor used by primary alcohol dehydrogenases found in methylotrophic bacteria, though since then it has been found widely in Nature.[76,77] Recent reviews can offer a better scope of the breadth of PQQ distribution and function in biological processes spanning all kingdoms.[78,79] A decade after the discovery of PQQ, the first steps in elucidating its biosynthesis were taken with the demonstration that it is generated from the amino acids Tyr and Glu.[80–82] Not long after, the *pqq* biosynthetic gene cluster was identified, and a gene, later named *pqqA*, was hypothesized as a possible precursor peptide, providing both the Tyr and Glu residues necessary for PQQ biosynthesis.[83,84] With the substrate identified, the question remained as to how the Tyr and Glu residues are transformed into the final product. Of particular interest was the carbon–carbon bond that is formed between the γ-carbon of Glu and the 3-position of the Tyr phenol. PqqE became the primary candidate for the formation of this particular cross-link when it was shown to be a radical SAM enzyme, and this hypothesis was recently confirmed *in vitro* (Scheme 7.6A).[85,86]

7.4.2 Sactipeptides

In 1985, a new cyclic peptide was isolated from *Bacillus subtilis*. This molecule, named subtilosin A, in addition to having antibiotic activity had an enigmatic and intriguing structure.[87] Nearly two decades later, the full NMR structure of subtilosin A revealed three unprecedented cross-links between

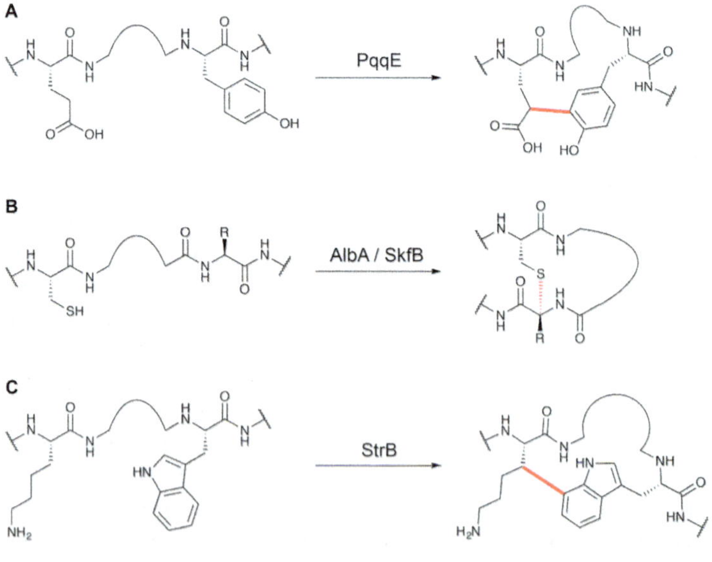

Scheme 7.6 Cross-links formed by radical SAM enzymes in the biosynthesis of PQQ (**A**), subtilosin A (**B**), and streptide (**C**).

the sulfur atoms of Cys residues and the α-carbons of two Phe and one Thr residue (Figure 7.1).[88] Since the discovery of subtilosin A, other sactipeptides ("sacti-" being derived from *s*ulfur to *a*lpha-*c*arbon) have been characterized, including the sporulation killing factor, SKF, also from *B. subtilis*, as well as various thuricins and thurincin H, from *Bacillus thuringiensis*.[89,90] In each instance, a radical SAM enzyme is responsible for installing these unique thioether bonds, and the mechanisms for AlbA and SkfB have been proposed by Marahiel and co-workers (Scheme 7.6B).[91,92]

7.4.3 Streptide

Within the last three years, a third class of radical SAM cross-linked RiPPs has been characterized. The representative new molecule, called streptide due to its discovery from streptococcal bacteria, contains a core peptide that is cyclized by a covalent bond between the β-carbon of lysine and the 7-carbon of the indole side chain of tryptophan (Scheme 7.6C).[75] Unlike PQQ and subtilosin A, there is no known biological function for streptide. However, some insights can be gleaned from how streptide production is regulated. Adjacent to the streptide biosynthetic gene cluster in *Streptococcus* spp. is a divergently expressed *shp-rgg* locus.[93,94] These genes, which encode a short hydrophobic peptide (SHP) and a Rgg type transcription factor, represent a common quorum sensing (QS) system in Streptococci. In agreement with this observation, streptide is produced at considerable quantities only when *S. thermophilus* is grown to high cell densities, a hallmark of QS-activated secondary metabolites. Furthermore, it has been shown that certain cyclic peptides are capable of inhibiting Rgg-type transcription factors, raising the possibility that streptide too could participate in modulation of QS-regulated processes.[95]

The biosynthesis of streptide requires expression of the precursor peptide, StrA; a radical SAM enzyme, StrB, which installs the Lys–Trp cross-link; and a transport/protease protein, StrC, which is ostensibly responsible for cleaving the leader peptide and transporting the mature streptide out of the cell. We will delve into streptide biosynthesis in greater detail, after a discussion of the general mechanisms underlying radical SAM enzymes.

7.4.4 Mechanisms of RiPP Cyclizations by Radical SAM Enzymes

The catalytic cycle of a typical radical SAM enzyme begins with a reduced $[4Fe–4S]^+$ cluster, which is usually bound at the N-terminus of a partial TIM barrel and, in most cases, ligated by a conserved CxxxCxxC motif. The three Cys residues in this motif ligate three of the four irons in the reduced Fe–S cluster. The fourth so-called unique Fe is open and serves as a binding site for SAM *via* its α-amine and carboxylate groups (Figure 7.4). Upon binding of SAM, the enzyme catalyzes homolytic cleavage of the bond between the methionine sulfur and the 5′-carbon of deoxyadenosine to form methionine,

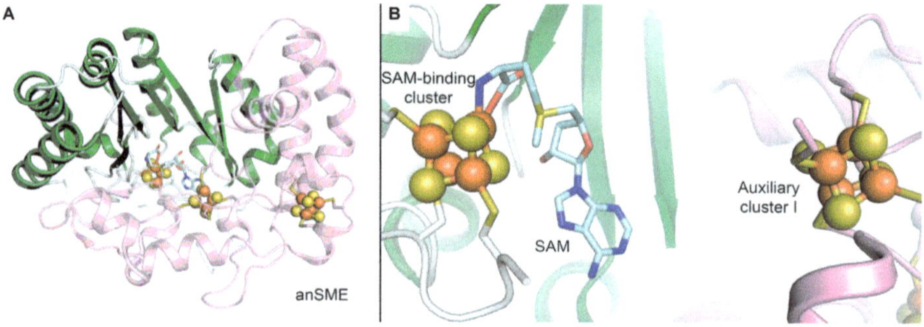

Figure 7.4 Structure of the SPASM-domain-containing radical SAM enzyme anSME involved in post-translational modification of a target protein. (**A**) The structure of anSME from *Clostridium perfringens* (PDB ID: 4K39), shown in a ribbon diagram. Two auxiliary [4Fe–4S] clusters can be seen in addition to the radical SAM cluster bound in the partial TIM barrel, which is rendered in dark green.[97] The C-terminal portion, including the SPASM domain, is colored light pink. The clusters are shown as ball-and-stick models, with iron colored orange and sulfur colored yellow. SAM is shown in stick representation, with carbons colored light blue; nitrogen, dark blue; and oxygen, red. (**B**) The active site of anSME contains a [4Fe–4S] cluster that is bound by the CxxxCxxC motif. SAM binds to the unique Fe within this cluster *via* its α-amine and carboxylate groups. The first auxiliary cluster can be seen at a distance of 16.9 Å from the SAM cluster.

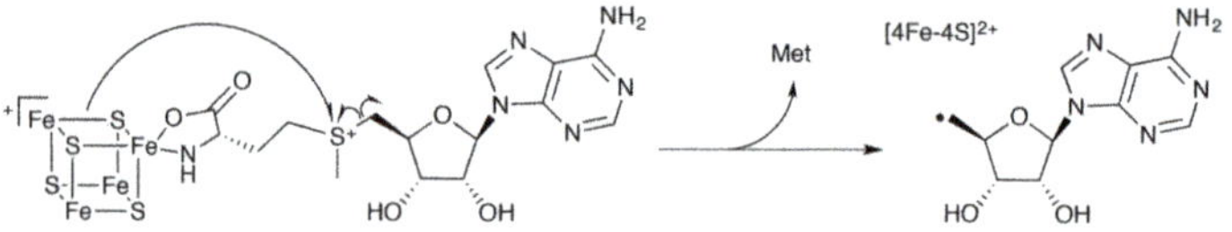

Scheme 7.7 Reductive homolytic cleavage of *S*-adenosylmethionine to form 5′-dA•, catalyzed by radical SAM enzymes.

a 5′-deoxyadenosyl primary radical (5′-dA•), and the oxidized $[4Fe–4S]^{2+}$ cluster (Scheme 7.7). The fact that this radical species is also produced by adenosylcobalamin-dependent enzymes has led some to argue that SAM is an evolutionary precursor of the much more complex corrinoid B12 co-factor.[70] Once the 5′-dA• has been generated, it is capable of abstracting a hydrogen atom from the substrate, which is positioned nearby, initiating the chemical transformation to the product. Following completion of the reaction, the resulting product radical may abstract a hydrogen atom from 5′-dA, which can regenerate SAM and $[4Fe–4S]^{+}$. Alternatively, depending on the reaction, the electron may be transferred to an auxiliary iron–sulfur cluster, or to an exogenous electron receptor. In the first instance, SAM is catalytic, while in the latter two cases, SAM is a co-substrate, with methionine and 5′-dA as reaction side products.[71,96]

In addition to the SAM-binding cluster, the RiPP-modifying radical SAM enzymes covered in this review also have other common features, notably a

C-terminal Fe–S cluster binding motif, known as a SPASM domain (for *s*ubtilosin A, *PQ*Q, *a*naerobic *s*ulfatase, and *m*ycofactocin), as well as an N-terminal *R*iPP precursor peptide *r*ecognition *e*lement, or RRE domain.[97,98] The C-terminal SPASM domain is widely found across radical SAM enzymes that catalyze various reactions. The primary purpose of the SPASM domain appears to be binding of up to two additional [4Fe–4S] clusters, also known as auxiliary clusters. These auxiliary clusters, in turn, are proposed to carry out a variety of functions, which include binding of the substrate and electron transfer.[96,97,99] They may play both of these roles in the intramolecular peptide cross-links that occur in sactipeptides, PQQ, and streptide, and it has been shown that knocking out the auxiliary cluster in StrB renders the enzyme catalytically inactive.[75] While not a RiPP-modifying enzyme, the anaerobic sulfatase maturating enzyme (anSME) also contains a SPASM domain and is the only structurally-characterized member of this subgroup (Figure 7.4). The crystal structure clearly shows two fully-ligated [4Fe–4S] clusters, consistent with biochemical and spectroscopic data.[100,101] The distance between the SAM cluster and the first and second auxiliary clusters was shown to be 16.9 and 26.7 Å, respectively, consistent with their involvement in electron transfer to and from the SAM cluster.

The N-terminal RRE domain has been implicated in binding the precursor peptide through recognition of its leader region. The RRE is usually fused to the N-terminus of the core TIM barrel fold, but can be found at the C-terminus or as a separately translated protein. In PQQ biosynthesis, it was found that the protein PqqD is a stand-alone RRE domain, and is required for the formation of the Glu–Tyr cross-link catalyzed by PqqE.[102] No crystal structure of an RRE domain in a radical SAM enzyme has yet been determined, and how this seemingly essential domain participates in substrate recognition and orientation for modification in the active-site remains poorly-understood.

As an example of how this class of radical SAM enzymes can install cross-links in RiPPs, we now return to the mechanism for Lys–Trp cross-link formation in streptide biosynthesis. Following translation, the precursor peptide StrA binds StrB, with an as-yet unknown mechanism of involvement by the RRE domain. Once bound, homolytic cleavage of SAM generates the active 5′-dA• (Scheme 7.8, step *a*), which abstracts a hydrogen atom from the β-carbon of an internal Lys residue (Scheme 7.8, step *b*). The resulting lysyl radical reacts with the π-system of Trp, forming a bond to the 7-carbon of the indole ring (Scheme 7.8, step *c*). Subsequent deprotonation at the 7-position of the indole and electron transfer to the auxiliary cluster gives the final Lys–Trp cross-link, as well as a reduced auxiliary cluster (Scheme 7.8, step *d*). Electron transfer from the auxiliary cluster to the radical SAM cluster could return StrB to the proper oxidation state, ready to carry out another round of turnover (Scheme 7.8, step *e*). This process may alternatively occur *via* small protein redox partners that bind to StrB. As with the Oxy enzymes, our knowledge regarding many aspects of this class of radical SAM enzymes is in its infancy and future studies will certainly provide additional mechanistic and structural insights.

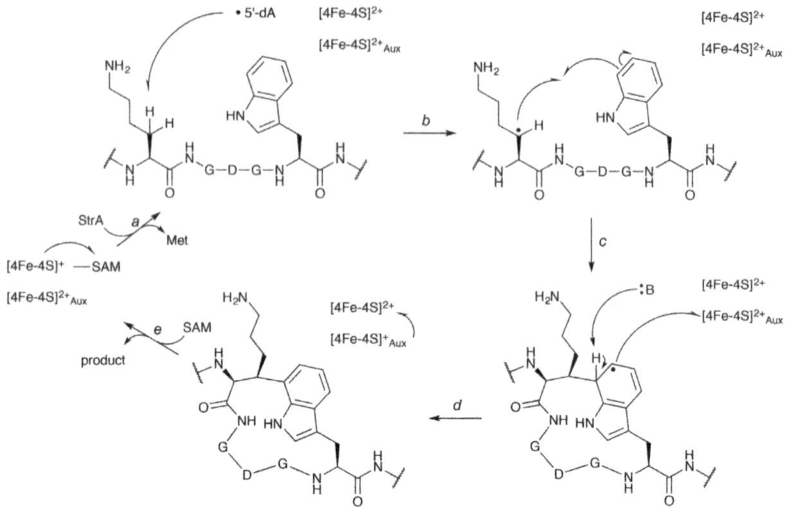

Scheme 7.8 Proposed mechanism for the StrB-catalyzed Lys–Trp cross-link formation on StrA during streptide biosynthesis.

7.5 Conclusions

Herein we have highlighted the mechanisms and structures of selected metalloenzymes that are involved in intramolecular peptide cross-linking reactions. The fact that the three enzyme classes discussed all employ different forms of an Fe co-factor is a testament to iron's versatility and utility in enzymatic systems. Notably, the key oxidizing intermediates are different in each of the three reactions. IPNS utilizes a ferric superoxide and an Fe(IV)-oxo intermediate to carry out two oxidative reactions. The Oxy enzymes probably use the well-known Compound I intermediate, an Fe(IV)-oxo porphyrin cation radical, while radical SAM enzymes catalyze reductive cleavage of SAM, mediated by $[4Fe–4S]^{+1}$, to generate the oxidizing 5′-dA•. The first two activate O_2 using Fe, while the radical SAM enzymes generate a powerful oxidant under anoxic conditions. These occurrences show that the involvement of iron-bearing metalloenzymes in peptide modification reactions is prevalent and that additional transformations along with their chemical mechanisms and structural underpinnings will likely be discovered in the future. At the same time, Nature has utilized other transition metals, especially Mn, Co, Ni, or Cu, to carry out reactions that are also performed by iron metalloenzymes; this is well-document in primary metabolic reactions. To name some examples among many, ribonucleotide reductases use a variety of metallo co-factors, containing Fe, Mn, or Co, and further, nitrogenases come in three varieties, bearing Fe, Mo, or V transition metals.[103–105] As such, we may also expect to find peptide-modifying metalloenzymes with alternative metals as investigations into NRP and RiPP biosyntheses continue and deepen in years to come.

Acknowledgements

We thank Princeton University start-up funds, the Edward C. Taylor 3rd Year Graduate Fellowship, and the Princeton Environmental Institute Innovative Research Award for support of our work.

References

1. P. G. Arnison, M. J. Bibb, G. Bierbaum, A. A. Bowers, T. S. Bugni, G. Bulaj, J. A. Camarero, D. J. Campopiano, G. L. Challis, J. Clardy, P. D. Cotter, D. J. Craik, M. Dawson, E. Dittmann, S. Donadio, P. C. Dorrestein, K.-D. Entian, M. A. Fischbach, J. S. Garavelli, U. Göransson, C. W. Gruber, D. H. Haft, T. K. Hemscheidt, C. Hertweck, C. Hill, A. R. Horswill, M. Jaspars, W. L. Kelly, J. P. Klinman, O. P. Kuipers, A. J. Link, W. Liu, M. A. Marahiel, D. A. Mitchell, G. N. Moll, B. S. Moore, R. Müller, S. K. Nair, I. F. Nes, G. E. Norris, B. M. Olivera, H. Onaka, M. L. Patchett, J. Piel, M. J. T. Reaney, S. Rebuffat, R. P. Ross, H.-G. Sahl, E. W. Schmidt, M. E. Selsted, K. Severinov, B. Shen, K. Sivonen, L. Smith, T. Stein, R. D. Süssmuth, J. R. Tagg, G.-L. Tang, A. W. Truman, J. C. Vederas, C. T. Walsh, J. D. Walton, S. C. Wenzel, J. M. Willey and W. A. van der Donk, *Nat. Prod. Rep.*, 2012, **30**, 108–160.
2. C. Walsh, *Antibiotics*, ASM Press, Washington, D.C., 2003.
3. A. Fleming, *Br. J. Exp. Pathol.*, 1929, **10**, 226–236.
4. R. B. Hamed, J. R. Gomez-Castellanos, L. Henry, C. Ducho, M. A. McDonough and C. J. Schofield, *Nat. Prod. Rep.*, 2012, **30**, 21–107.
5. E. B. Chain, *presented in Part at Nobel Lecture*, Stockholm, March 1946.
6. National Academy of Sciences (U.S.), *The Chemistry of Penicillin; Report on a Collaborative Investigation by American and British Chemists under the Joint Sponsorship of the Office of Scientific Research and Development, Washington, D.C., and the Medical Research Council, London. Comp. Under the Auspices of the National Academy of Sciences, Washington, D.C., Pursuant to a Contract with the Office of Scientific Research and Development*, ed. H. T. Clarke, J. R. Johnson and S. R. Robinson, Princeton Univ. Press, Princeton, 1949.
7. P. L. Roach, I. J. Clifton, C. M. H. Hensgens, N. Shibata, C. J. Schofield, J. Hajdu and J. E. Baldwin, *Nature*, 1997, **387**, 827–830.
8. H. R. V. Arnstein and D. Morris, *Biochem. J.*, 1960, **76**, 357–361.
9. P. A. Fawcett, P. B. Loder, T. J. Beesley and E. P. Abraham, *J. Gen. Microbiol.*, 1973, **79**, 293–309.
10. J. O'Sullivan, R. C. Bleaney, J. A. Huddleston and E. P. Abraham, *Biochem. J.*, 1979, **184**, 421–426.
11. T. Konomi, S. Herchen, J. E. Baldwin, M. Yoshida, N. A. Hunt and A. L. Demain, *Biochem. J.*, 1979, **184**, 427–430.
12. C. P. Pang, B. Chakravarti, R. M. Adlington, H. H. Ting, R. L. White, G. S. Jayatilake, J. E. Baldwin and E. P. Abraham, *Biochem. J.*, 1984, **222**, 789–795.

13. P. L. Roach, I. J. Clifton, V. Fülöp, K. Harlos, G. J. Barton, J. Hajdu, I. Andersson, C. J. Schofield and J. E. Baldwin, *Nature*, 1995, **375**, 700–704.
14. J. A. Hangasky, C. Y. Taabazuing, M. A. Valliere and M. J. Knapp, *Metallomics*, 2013, **5**, 287–301.
15. C. Krebs, D. Galonić Fujimori, C. T. Walsh and J. M. Bollinger, *Acc. Chem. Res.*, 2007, **40**, 484–492.
16. A. L. Feig and S. J. Lippard, *Chem. Rev.*, 1994, **94**, 759–805.
17. E. Tamanaha, B. Zhang, Y. Guo, W. Chang, E. W. Barr, G. Xing, J. St. Clair, S. Ye, F. Neese, J. M. Bollinger and C. Krebs, *J. Am. Chem. Soc.*, 2016, **138**, 8862–8874.
18. J. E. Baldwin and E. Abraham, *Nat. Prod. Rep.*, 1988, **5**, 129–145.
19. J. E. Baldwin and C. Schofield, in *The Chemistry of* β*-Lactams*, ed. M. I. Page, Springer, Netherlands, 1992, pp. 1–78.
20. A. M. Orville, V. J. Chen, A. Kriauciunas, M. R. Harpel, B. G. Fox, E. Munck and J. D. Lipscomb, *Biochemistry*, 1992, **31**, 4602–4612.
21. L. J. Ming, L. Que, A. Kriauciunas, C. A. Frolik and V. J. Chen, *Inorg. Chem.*, 1990, **29**, 1111–1112.
22. L. J. Ming, L. Que, A. Kriauciunas, C. A. Frolik and V. J. Chen, *Biochemistry*, 1991, **30**, 11653–11659.
23. R. A. Scott, S. Wang, M. K. Eidsness, A. Kriauciunas, C. A. Frolik and V. J. Chen, *Biochemistry*, 1992, **31**, 4596–4601.
24. F. Jiang, J. Peisach, L. J. Ming, L. Que and V. J. Chen, *Biochemistry*, 1991, **30**, 11437–11445.
25. W. A. van der Donk, C. Krebs and J. M. Bollinger Jr, *Curr. Opin. Struct. Biol.*, 2010, **20**, 673–683.
26. M. Lundberg and K. Morokuma, *J. Phys. Chem. B*, 2007, **111**, 9380–9389.
27. M. Lundberg, P. E. M. Siegbahn and K. Morokuma, *Biochemistry*, 2008, **47**, 1031–1042.
28. M. Lundberg, T. Kawatsu, T. Vreven, M. J. Frisch and K. Morokuma, *J. Chem. Theory Comput.*, 2009, **5**, 222–234.
29. N. I. Burzlaff, P. J. Rutledge, I. J. Clifton, C. M. H. Hensgens, M. Pickford, R. M. Adlington, P. L. Roach and J. E. Baldwin, *Nature*, 1999, **401**, 721–724.
30. R. S. Griffith, *Rev. Infect. Dis.*, 1981, **3**, S200–S204.
31. D. P. Levine, *Clin. Infect. Dis.*, 2006, **42**, S5–S12.
32. S. Bhooshan, J. Prasad, A. Dutta, V. Ke and C. Mukhopadhyay, *Int. J. Pharmacol. Pharm. Sci.*, 2016, **8**, 321–322.
33. M. H. Mccormick, J. M. Mcguire, G. E. Pittenger, R. C. Pittenger and W. M. Stark, *Antibiot. Annu.*, 1955, **3**, 606–611.
34. J. M. Mcguire, R. N. Wolfe and D. W. Ziegler, *Antibiot. Annu.*, 1955, **3**, 612–618.
35. M. A. Fischbach and C. T. Walsh, *Chem. Rev.*, 2006, **106**, 3468–3496.
36. M. Strieker, A. Tanović and M. A. Marahiel, *Curr. Opin. Struct. Biol.*, 2010, **20**, 234–240.
37. S. Pelzer, R. Süßmuth, D. Heckmann, J. Recktenwald, P. Huber, G. Jung and W. Wohlleben, *Antimicrob. Agents Chemother.*, 1999, **43**, 1565–1573.

38. A. A. van Wageningen, P. N. Kirkpatrick, D. H. Williams, B. R. Harris, J. K. Kershaw, N. J. Lennard, M. Jones, S. J. M. Jones and P. J. Solenberg, *Chem. Biol. (Oxford, U. K.)*, 1998, **5**, 155–162.
39. B. K. Hubbard and C. T. Walsh, *Angew. Chem., Int. Ed.*, 2003, **42**, 730–765.
40. K. C. Nicolaou, C. N. C. Boddy, S. Bräse and N. Winssinger, *Angew. Chem., Int. Ed.*, 1999, **38**, 2096–2152.
41. R. D. Süssmuth, S. Pelzer, G. Nicholson, T. Walk, W. Wohlleben and G. Jung, *Angew. Chem., Int. Ed.*, 1999, **38**, 1976–1979.
42. D. Bischoff, S. Pelzer, A. Höltzel, G. J. Nicholson, S. Stockert, W. Wohlleben, G. Jung and R. D. Süssmuth, *Angew. Chem., Int. Ed.*, 2001, **40**, 1693–1696.
43. D. Bischoff, S. Pelzer, B. Bister, G. J. Nicholson, S. Stockert, M. Schirle, W. Wohlleben, G. Jung and R. D. Süssmuth, *Angew. Chem., Int. Ed.*, 2001, **40**, 4688–4691.
44. D. Bischoff, B. Bister, M. Bertazzo, V. Pfeifer, E. Stegmann, G. J. Nicholson, S. Keller, S. Pelzer, W. Wohlleben and R. D. Süssmuth, *ChemBioChem*, 2005, **6**, 267–272.
45. K. Zerbe, K. Woithe, D. B. Li, F. Vitali, L. Bigler and J. A. Robinson, *Angew. Chem., Int. Ed.*, 2004, **43**, 6709–6713.
46. K. Woithe, N. Geib, K. Zerbe, D. B. Li, M. Heck, S. Fournier-Rousset, O. Meyer, F. Vitali, N. Matoba, K. Abou-Hadeed and J. A. Robinson, *J. Am. Chem. Soc.*, 2007, **129**, 6887–6895.
47. B. Nazari, C. C. Forneris, M. I. Gibson, K. Moon, K. R. Schramma and M. R. Seyedsayamdost, *MedChemComm*, 2017, **8**, 780–788.
48. K. Zerbe, O. Pylypenko, F. Vitali, W. Zhang, S. Rouset, M. Heck, J. W. Vrijbloed, D. Bischoff, B. Bister, R. D. Süssmuth, S. Pelzer, W. Wohlleben, J. A. Robinson and I. Schlichting, *J. Biol. Chem.*, 2002, **277**, 47476–47485.
49. D. C. Haines, D. R. Tomchick, M. Machius and J. A. Peterson, *Biochemistry*, 2001, **40**, 13456–13465.
50. J. Rittle and M. T. Green, *Science*, 2010, **330**, 933–937.
51. X. Wang, S. Peter, M. Kinne, M. Hofrichter and J. T. Groves, *J. Am. Chem. Soc.*, 2012, **134**, 12897–12900.
52. O. Pylypenko, F. Vitali, K. Zerbe, J. A. Robinson and I. Schlichting, *J. Biol. Chem.*, 2003, **278**, 46727–46733.
53. M. J. Cryle, J. Staaden and I. Schlichting, *Arch. Biochem. Biophys.*, 2011, **507**, 163–173.
54. K. Haslinger and M. J. Cryle, *FEBS Lett.*, 2016, **590**, 571–581.
55. M. Peschke, C. Brieke and M. J. Cryle, *Sci. Rep.*, 2016, **6**, 35584.
56. M. Peschke, M. Gonsior, R. D. Süssmuth and M. J. Cryle, *Curr. Opin. Struct. Biol.*, 2016, **41**, 46–53.
57. K. Haslinger, E. Maximowitsch, C. Brieke, A. Koch and M. J. Cryle, *ChemBioChem*, 2014, **15**, 2719–2728.
58. C. Brieke, M. Peschke, K. Haslinger and M. J. Cryle, *Angew. Chem., Int. Ed.*, 2015, **127**, 15941–15945.
59. K. Haslinger, M. Peschke, C. Brieke, E. Maximowitsch and M. J. Cryle, *Nature*, 2015, **521**, 105–109.

60. J. T. Groves, *Proc. Natl. Acad. Sci. U. S. A.*, 2003, **100**, 3569–3574.
61. X. Huang and J. T. Groves, *J. Biol. Inorg Chem.*, 2016, 1–23.
62. M. T. Green, J. H. Dawson and H. B. Gray, *Science*, 2004, **304**, 1653–1656.
63. T. H. Yosca, J. Rittle, C. M. Krest, E. L. Onderko, A. Silakov, J. C. Calixto, R. K. Behan and M. T. Green, *Science*, 2013, **342**, 825–829.
64. J. T. Groves, *Nat. Chem.*, 2014, **6**, 89–91.
65. C. A. Hunter and J. K. M. Sanders, *J. Am. Chem. Soc.*, 1990, **112**, 5525–5534.
66. C. R. Martinez and B. L. Iverson, *Chem. Sci.*, 2012, **3**, 2191–2201.
67. X. Wang, S. Peter, R. Ullrich, M. Hofrichter and J. T. Groves, *Angew. Chem., Int. Ed.*, 2013, **52**, 9238–9241.
68. X. Wang, R. Ullrich, M. Hofrichter and J. T. Groves, *Proc. Natl. Acad. Sci. U. S. A.*, 2015, **112**, 3686–3691.
69. C. Tommos, J. J. Skalicky, D. L. Pilloud, A. J. Wand and P. L. Dutton, *Biochemistry*, 1999, **38**, 9495–9507.
70. P. A. Frey, A. D. Hegeman and F. J. Ruzicka, *Crit. Rev. Biochem. Mol. Biol.*, 2008, **43**, 63–88.
71. S. J. Booker, *Curr. Opin. Chem. Biol.*, 2009, **13**, 58–73.
72. J. B. Broderick, B. R. Duffus, K. S. Duschene and E. M. Shepard, *Chem. Rev.*, 2014, **114**, 4229–4317.
73. L. Flühe and M. A. Marahiel, *Curr. Opin. Chem. Biol.*, 2013, **17**, 605–612.
74. J. P. Klinman and F. Bonnot, *Chem. Rev.*, 2014, **114**, 4343–4365.
75. K. R. Schramma, L. B. Bushin and M. R. Seyedsayamdost, *Nat. Chem.*, 2015, **7**, 431–437.
76. S. A. Salisbury, H. S. Forrest, W. B. T. Cruse and O. Kennard, *Nature*, 1979, **280**, 843–844.
77. J. A. Duine and J. A. Jongejan, in *Vitamins & Hormones*, ed. G. D. Aurbach and D. B. McCormick, Academic Press, 1989, vol. 45, pp. 223–262.
78. H. S. Misra, Y. S. Rajpurohit and N. P. Khairnar, *J. Biosci. (New Delhi, India)*, 2012, **37**, 313–325.
79. M. Akagawa, M. Nakano and K. Ikemoto, *Biosci., Biotechnol., Biochem.*, 2016, **80**, 13–22.
80. D. R. Houck, J. L. Hanners and C. J. Unkefer, *J. Am. Chem. Soc.*, 1988, **110**, 6920–6921.
81. D. R. Houck, J. L. Hanners and C. J. Unkefer, *J. Am. Chem. Soc.*, 1991, **113**, 3162–3166.
82. M. A. G. van Kleef and J. A. Duine, *FEBS Lett.*, 1988, **237**, 91–97.
83. N. Goosen, H. P. Horsman, R. G. Huinen and P. van de Putte, *J. Bacteriol.*, 1989, **171**, 447–455.
84. N. Goosen, R. G. Huinen and P. van de Putte, *J. Bacteriol.*, 1992, **174**, 1426–1427.
85. S. R. Wecksler, S. Stoll, H. Tran, O. T. Magnusson, S. Wu, D. King, R. D. Britt and J. P. Klinman, *Biochemistry*, 2009, **48**, 10151–10161.
86. I. Barr, J. A. Latham, A. T. Iavarone, T. Chantarojsiri, J. D. Hwang and J. P. Klinman, *J. Biol. Chem.*, 2016, **291**, 8877–8884.

87. K. Babasaki, T. Takao, Y. Shimonishi and K. Kurahashi, *J. Biochem.*, 1985, **98**, 585–603.
88. K. E. Kawulka, T. Sprules, C. M. Diaper, R. M. Whittal, R. T. McKay, P. Mercier, P. Zuber and J. C. Vederas, *Biochemistry*, 2004, **43**, 3385–3395.
89. M. C. Rea, C. S. Sit, E. Clayton, P. M. O'Connor, R. M. Whittal, J. Zheng, J. C. Vederas, R. P. Ross and C. Hill, *Proc. Natl. Acad. Sci. U. S. A.*, 2010, **107**, 9352–9357.
90. C. S. Sit, M. J. van Belkum, R. T. McKay, R. W. Worobo and J. C. Vederas, *Angew. Chem., Int. Ed.*, 2011, **50**, 8718–8721.
91. L. Flühe, T. A. Knappe, M. J. Gattner, A. Schäfer, O. Burghaus, U. Linne and M. A. Marahiel, *Nat. Chem. Biol.*, 2012, **8**, 350–357.
92. L. Flühe, O. Burghaus, B. M. Wieckowski, T. W. Giessen, U. Linne and M. A. Marahiel, *J. Am. Chem. Soc.*, 2013, **135**, 959–962.
93. M. Ibrahim, A. Guillot, F. Wessner, F. Algaron, C. Besset, P. Courtin, R. Gardan and V. Monnet, *J. Bacteriol.*, 2007, **189**, 8844–8854.
94. B. Fleuchot, C. Gitton, A. Guillot, J. Vidic, P. Nicolas, C. Besset, L. Fontaine, P. Hols, N. Leblond-Bourget, V. Monnet and R. Gardan, *Mol. Microbiol.*, 2011, **80**, 1102–1119.
95. V. Parashar, C. Aggarwal, M. J. Federle and M. B. Neiditch, *Proc. Natl. Acad. Sci. U. S. A.*, 2015, **112**, 5177–5182.
96. N. D. Lanz and S. J. Booker, *Biochim. Biophys. Acta, Mol. Cell Res.*, 2015, **1853**, 1316–1334.
97. T. A. J. Grell, P. J. Goldman and C. L. Drennan, *J. Biol. Chem.*, 2015, **290**, 3964–3971.
98. B. J. Burkhart, G. A. Hudson, K. L. Dunbar and D. A. Mitchell, *Nat. Chem. Biol.*, 2015, **11**, 564–570.
99. N. D. Lanz and S. J. Booker, *Biochim. Biophys. Acta, Proteins Proteomics*, 2012, **1824**, 1196–1212.
100. P. J. Goldman, T. L. Grove, L. A. Sites, M. I. McLaughlin, S. J. Booker and C. L. Drennan, *Proc. Natl. Acad. Sci. U. S. A.*, 2013, **110**, 8519–8524.
101. T. L. Grove, J. H. Ahlum, R. M. Qin, N. D. Lanz, M. I. Radle, C. Krebs and S. J. Booker, *Biochemistry*, 2013, **52**, 2874–2887.
102. I. Barr, J. A. Latham, A. T. Iavarone, T. Chantarojsiri, J. D. Hwang and J. P. Klinman, *J. Biol. Chem.*, 2016, **291**, 8877–8884.
103. J. Stubbe and W. A. van der Donk, *Chem. Rev.*, 1998, **98**, 705–762.
104. S. C. Lee and R. H. Holm, *Chem. Rev.*, 2004, **104**, 1135–1158.
105. B. M. Hoffman, D. Lukoyanov, Z.-Y. Yang, D. R. Dean and L. C. Seefeldt, *Chem. Rev.*, 2014, **114**, 4041–4062.

CHAPTER 8

Double-click Stapled Peptides for Inhibiting Protein–Protein Interactions

K. SHARMA[a], D. L. KUNCIW[a], W. XU[b], M. M. WIEDMANN[a], Y. WU[a], H. F. SORE[a], W. R. J. D. GALLOWAY[a], Y. H. LAU[a], L. S. ITZHAKI[b] AND D. R. SPRING*[a]

[a]University of Cambridge, Department of Chemistry, Lensfield Road, Cambridge, CB2 1EW, UK; [b]University of Cambridge, Department of Pharmacology, Tennis Court Road, Cambridge, CB2 1PD, UK
*E-mail: spring@ch.cam.ac.uk

8.1 Introduction

Protein–protein interactions describe the biochemical events during which a protein's activity and function are modulated by one or more separate interacting proteins. PPIs lead to measurable effects such as altering the kinetic properties of enzymes, allowing for substrate channeling, changing between active and inactive conformations, creating new binding sites, or serving as regulators in upstream or downstream events.[1] Complex networks of PPIs are intrinsic to most cellular functions of living organisms, and abnormal behavior of these proteins is often correlated with the commencement of various human diseases, like cancer.[2] Therefore, numerous studies have been carried out to assess the therapeutic potential of this target class and to discover novel PPI inhibitors.

Chemical Biology No. 6
Cyclic Peptides: From Bioorganic Synthesis to Applications
Edited by Jesko Koehnke, James Naismith and Wilfred A. van der Donk

Published by the Royal Society of Chemistry, www.rsc.org

PPI contact surfaces are often large (approximately 1500–3000 Å^2) compared to those of protein–small molecule interactions (approximately 300–1000 Å^2), and PPI interaction surfaces are usually flat and lack the grooves and pockets of proteins, which engage in small molecule binding.[3] Due to the aforementioned reasons, twenty years ago PPIs were considered to be "undruggable".[4] Over the past two decades, there has been a considerable amount of research in this area and a number of different approaches have been reported for inhibiting PPIs.[5] One such approach is peptidomimetics, which utilizes molecules designed to mimic the three-dimensional structure of peptides or proteins, while having the ability to interact with the biological target in the same way as the natural peptide or protein from which their structure was derived. While native peptides provide an accessible starting point for the design of peptidomimetics, peptides in isolation are poor contenders. On their own, short peptides often do not retain their native conformation, and hence binding capability, as there is a lack of structure inherently provided by the rest of the protein. Additionally, peptides are susceptible to rapid proteolysis and often suffer from poor cell permeability.

Many studies have focused on cyclic peptides as competitive inhibitors of PPIs as they are capable of selectively mimicking the protein contact.[6] There are many effective peptidomimetic strategies described in the literature.[7–9] In particular, stapled peptides, pioneered by Grubbs,[10] Verdine,[11] Walensky,[12] and Sawyer,[13] present a promising strategy to target "undruggable" therapeutic targets. Stapling as a form of macrocyclization refers to the process in which certain amino acid side chains of a peptide are covalently bonded for the stabilization of often, but not exclusively, α-helical structures in short peptide sequences.[14] In theory, if the resultant length, position, and characteristics of the staple are optimal, the problems with short peptides in isolation may be overcome and binding affinity restored. Stapling can reinforce α-helical secondary structures if the non-native amino acids are positioned specifically on the same face of the helix, such as i, $i+4$ and i, $i+7$[15] residues, or simply provide a means of macrocyclization for unstructured peptides. There are two general approaches: one-component and two-component stapling techniques. Whilst one-component stapling involves the direct linking of amino acid side chains, two-component stapling requires the use of a separate bi-functional linker to bridge the side chains of two non-proteogenic amino acids (Figure 8.1).

One-component stapling was the first to be proposed, starting with macrocyclization by lactamization *via* the incorporation of proteogenic amino acids Lys and Glu/Asp.[17] Seminal work on hydrocarbon stapling by Grubbs and Blackwell led to the use of ring-closing metathesis as a method of stapling utilizing olefin-bearing amino acids.[10] Other techniques include the reversible formation of disulfide bonds between two enantiomeric Cys residues,[18] the use of an alkyne-bearing side chain and an azide-containing side chain for macrocyclization by Cu(I)-catalyzed azide–alkyne cycloaddition (CuAAC),[19] and the formation of thioether bridges *via* a covalent linkage between Cys and α-bromo amide side chains.[20]

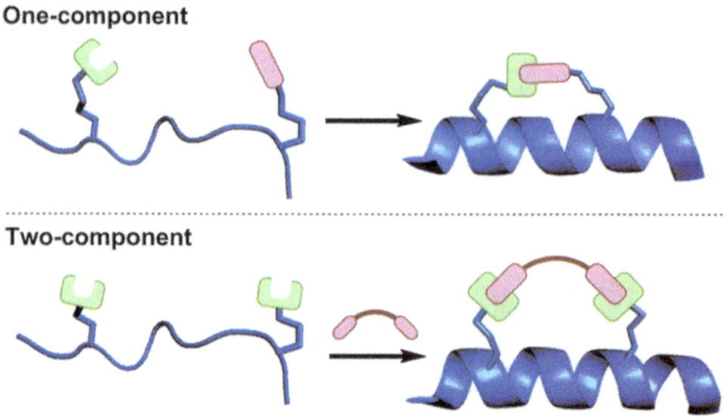

Figure 8.1 One- and two-component stapling techniques. Reproduced from ref. 16 with permission from The Royal Society of Chemistry.

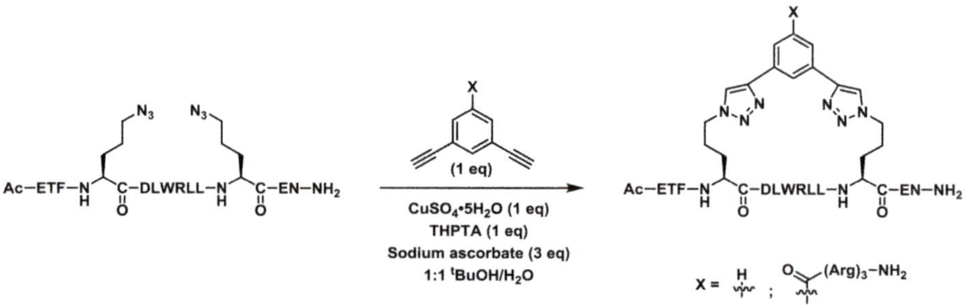

Scheme 8.1 Cu(I)-catalyzed double-click peptide stapling. Reproduced from ref. 16 with permission from The Royal Society of Chemistry.

Two-component stapling relies on two steps: the intermolecular coupling of the bifunctional linker and peptide followed by an intramolecular coupling to complete the cyclization. An alternative, competing pathway, which is not possible in one-component stapling, results in double addition of the linker forming a linear peptide. Advances in this area have led to coping mechanisms such as conformational preorganization, varying the reacting side chain positions, and dilutions.[16] Despite this complication, there are many advantages of two-component stapling, such as the ability to incorporate diverse staple linkages efficiently without the need to synthesize complex side chain-bearing amino acids for use in solid phase peptide synthesis (spps).

The Spring Group have devised a two-component stapling strategy termed the "double-click" approach to peptide stapling (Scheme 8.1).[21] The approach is based on the robust nature of CuAAC, the archetypal click reaction reported by Sharpless[22] and Meldal[23] and based on the original 1,3-dipolar cycloaddition developed by Huisgen.[24] This technique calls for the incorporation of two non-proteogenic azido amino acids into the peptide sequence of interest. Dialkynyl linkers then allow the formation of bistriazole-containing macrocyclized peptides.[25] A large variety of stapled peptides can be synthesized by reacting the linear peptide with a library of dialkynyl linkers in a divergent manner. Furthermore, the required azido amino acids are easy to synthesize and are compatible with Fmoc chemistry and can be incorporated into the peptide chain *via* regular Merrifield spps.[26] In addition, the azide functionality allows chemospecific reactions and is tolerant to many other functional groups.

The remainder of this chapter will focus on the work of the Spring Group in the area of two-component "double-click" stapling, specifically:

- Efficient synthesis of Fmoc-protected azido amino acids.
- Optimization of the peptide sequence and use of functionalized staple linkages to modulate the cellular activity of stapled peptides.
- Metal-free strain-promoted peptide stapling.
- The application of double stapling in targeting the substrate-recognition domain of tankyrase to antagonize Wnt signaling, and the transcription factor HNF1β/Importin α PPI, by using constrained non-α-helical peptide inhibitors.

It is important to note that the stapling approach developed by the Spring group has been used not only to constrain near-native peptidomimetics into α-helices as in the p53/MDM2 PPI, but also to provide the benefits of macrocyclization to unstructured peptidomimetics to target Tankyrase/Wnt and the HNF1β/Importin α PPI.

8.2 Non-proteogenic Amino Acid Synthesis

Non-proteogenic amino acids containing azide groups are known to be useful in the design of synthetic peptides and proteins as a biorthogonal handle allowing for further functionalization *via* Staudinger ligation or CuAAC reactions. Azido amino acids have been used previously for the synthesis of peptidomimetics including triazole-containing macrocycles *via* CuAAC.[27]

Since each coupling during spps requires multiple equivalents of pure amino acid, we sought to create a route to azido amino acids that was short, scalable, and high yielding, starting from inexpensive, commercially available compounds. Although there are existing literature procedures[27–30] for the synthesis of azido amino acids, we decided to use copper-catalyzed diazotransfer chemistry to design a more straightforward and atom-efficient pathway.

To this end, amine bearing precursors, such as Fmoc-Dap[†]-OH (**3**), Fmoc-Dab[‡]-OH (**4**), and Fmoc-Orn[§]-OH (**8**), were prepared. Primary amines **3** and **4** were synthesized *via* Hofmann rearrangement of readily available

---

[†]Dap = 2,3-diaminopropionic acid.
[‡]Dab = 2,4-diaminobutyric acid.
[§]Orn = Ornithine.

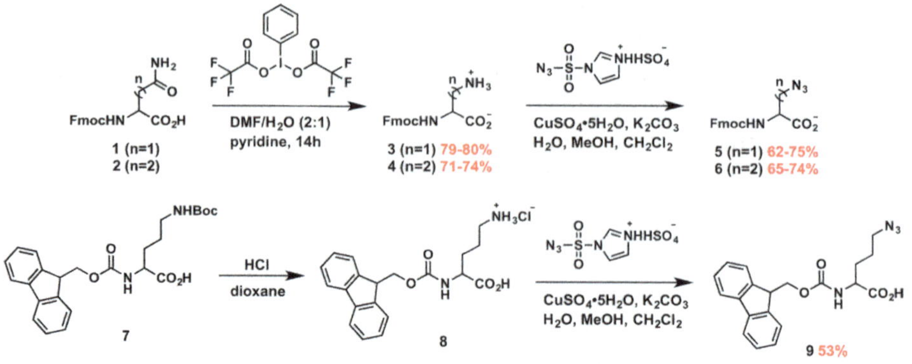

Scheme 8.2 Synthesis of Fmoc-protected azido amino acids. Reproduced from Y. H. Lau and D. R. Spring, *Synlett*, 2011, 1917–1919[32] with permission from Thieme; Copyright © 2011 Georg Thieme Verlag KG.

amino acids **1** and **2** (Scheme 8.2). Fmoc-Orn-OH (**8**) was synthesized *via* the acid deprotection of Fmoc-Orn(Boc)-OH (**7**). Imidazole-1-sulfonyl azide hydrochloride was chosen as the diazotransfer reagent as it is relatively less explosive than triflyl azide, its synthesis is scalable, and it is stable over the long term when stored at 4 °C.[31]

From the primary amines and the diazotransfer reagent, a route was developed using a biphasic solvent mixture of H_2O, MeOH, and CH_2Cl_2 adjusted to pH 9 with potassium carbonate to afford the azido amino acid **9** *via* CuAAC in relatively high yields. The synthesis can be performed on multigram scale with >98% chromatographic purity following an aqueous workup.[32] This work serves as the sturdy foundation of all our experiments.

8.3 Peptide Sequence Optimization and Use of Functionalized Staple Linkages for Modulating the Cellular Activity of Stapled Peptides

By applying the double-click stapling concept to the p53/MDM2 interaction, a validated target for anticancer therapeutics, it was shown that Pro-27 replacement by the staple is the most suitable position (**P1**, Figure 8.2a)[¶] and Orn(N_3) is the ideal side chain length to attain optimal binding affinity (3.21 ± 0.38 nM) and cellular activity, when stapled with linker **10** (**P1–10**).[33] Substituting Orn(N_3) with Aha and Lys(N_3) as the side chain led to a decrease in the binding affinity of the stapled peptide to 10.5 ± 0.76 nM (**P9–10**) and 9.63 ± 0.87 nM (**P10–10**), respectively. Moreover, it was observed that activity could be induced in an otherwise impermeable p53 stapled peptide by incorporating a tri-arginine motif (**11**) on the staple linkage, rather than adjusting the peptide sequence (Figure 8.2a).[34] A series of linear aliphatic staples (**12–14**)

¶Aha = Azidohomoalanine; Lys = Lysine.

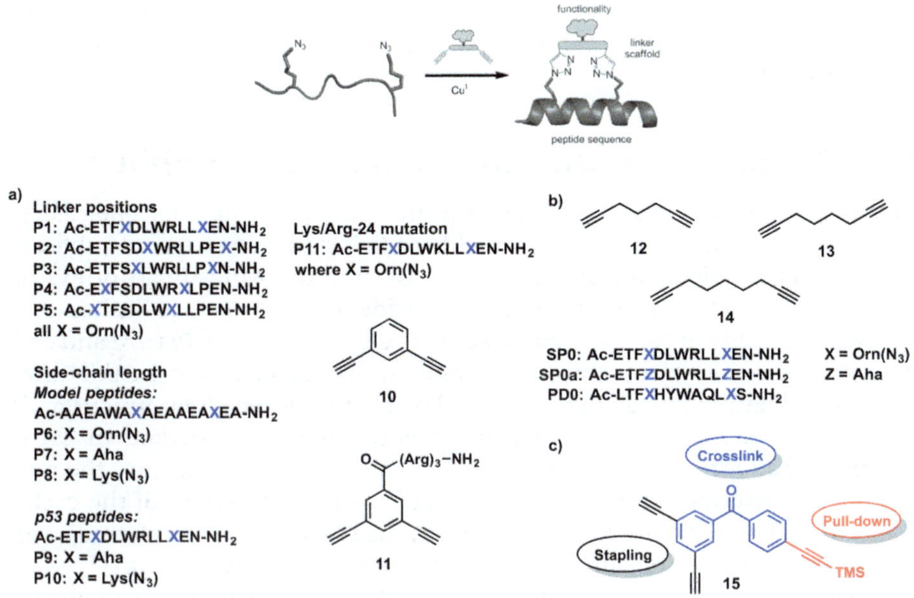

Figure 8.2 (a) Diazido peptides for stapling with dialkyne linkers **10** and **11** (reproduced from ref. 33 with permission from The Royal Society of Chemistry). (b) Linear aliphatic dialkyne linkers (from Y. H. Lau, P. de Andrade, G. J. McKenzie, A. R. Venkitaraman and D. R. Spring, Linear Aliphatic Dialkynes as Alternative Linkers for Double-Click Stapling of p53-Derived Peptides,[25] John Wiley and Sons, Copyright © 2014 by John Wiley & Sons, Inc.). (c) A multifunctional linker for peptide stapling and photoaffinity labelling (from Yuteng Wu, Lasse B. Olsen, Yu Heng Lau, Claus Hatt Jensen, Maxim Rossmann, Ysobel R. Baker, Hannah F. Sore, Súil Collins, David R. Spring, Development of a Multifunctional Benzophenone Linker for Peptide Stapling and Photoaffinity Labelling, John Wiley and Sons,[36] Copyright © 2016 by John Wiley & Sons, Inc.).

that complement the aromatic linkers were also developed.[25] Optimization of the combination of staple and sequence suggested that the aliphatic scaffolds can lead to enhanced binding *in vitro* and superior p53 activation in cells when combined with a phage-display-derived[35] sequence **PD0** (Figure 8.2b). The results suggest that different staple linkages can lead to very different peptide bioactivity in cells.

A benzophenone moiety was also incorporated into our dialkyne linker for photo-cross-linking. This novel multifunctional linker, **15**, serves as both a peptide stapling reagent and a photoaffinity probe with pulldown potential (Figure 8.2c).[36] The linker **15** can be accessed in four steps and the TMS group was conveniently removed under click conditions to reveal the terminal alkyne. Subsequent reaction with a biotinylated azide demonstrated its pulldown capability. As a proof of concept, this methodology was applied to a p53-dervied peptide, which was effectively cross-linked with MDM2 after

UV irradiation. Current work is underway to extend this strategy to MDM2 labelling and pulldown in cell lysates or live cells, as well as applying it to study other PPIs.

8.4 Metal-free Strain-promoted Peptide Stapling

In 2003, Carolyn Bertozzi founded the field of bioorthogonal chemistry, involving reactions that can occur inside of living systems without interfering with native biochemical processes.[37] Despite the widespread utility of biorthogonal CuAAC reactions, their use inside living systems is limited due to the cytotoxicity of the Cu(I) catalyst involved (Figure 8.3a). *In vitro* and cell culture studies have demonstrated that metals like copper have the potential to cause oxidative damage to the cell and disrupt critical cellular functions.[38] Therefore, a copper-free click reaction was developed by Bertozzi in 2004, by utilizing a high-energy strained cyclooctyne molecule to increase the rate of reaction without the need for a catalyst.[39] As the ring strain of the cyclic molecule drives the click reaction forward, the reaction is also referred to as a strain-promoted azide–alkyne cycloaddition (SPAAC) (Figure 8.3b).[39]

Since then, a number of strained molecules have been synthesized and applied to undergo metal-free click reactions *in vitro* and *in vivo*.[40] In 2002, Orita and co-workers[41,42] reported the synthesis of strained Sondheimer–Wong diyne[43] **16** by following a one-pot double-elimination protocol (Scheme 8.3).

This protocol was utilized by the Spring group to develop a metal-free double-click peptide stapling methodology (Scheme 8.4).[44] The double-click stapling of a p53-based diazidopeptide **17** with linker **16** in 1 : 1 $^tBuOH/H_2O$ gave stapled peptide **18** in 60% yield.

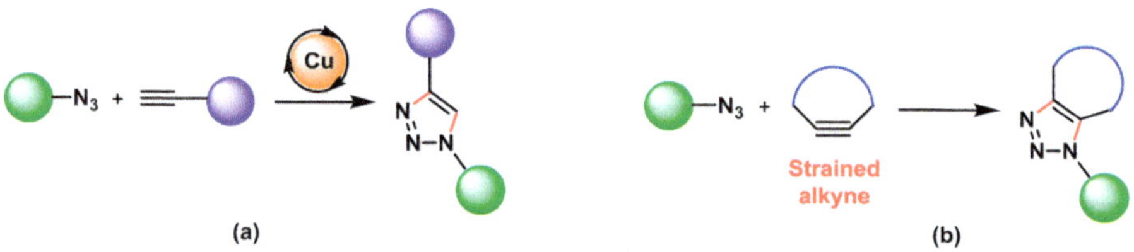

Figure 8.3 (a) General schematic of the Cu(I)-catalyzed click reaction by Sharpless and Meldal. (b) General schematic of the copper-free click reaction developed by Bertozzi.

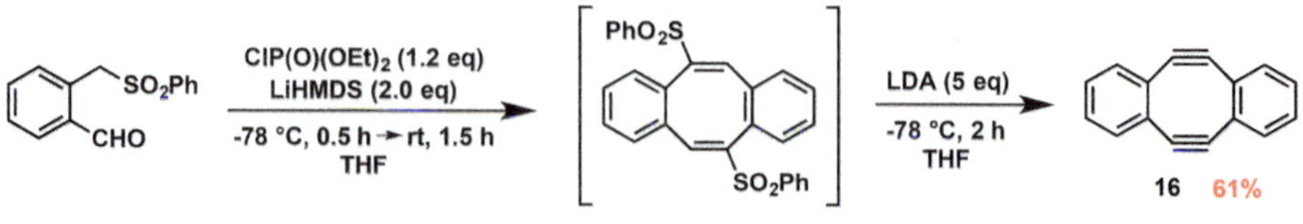

Scheme 8.3 Orita's synthesis of a Sondheimer–Wong diyne. Reprinted with permission from ref. 42. Copyright (2014) American Chemical Society.

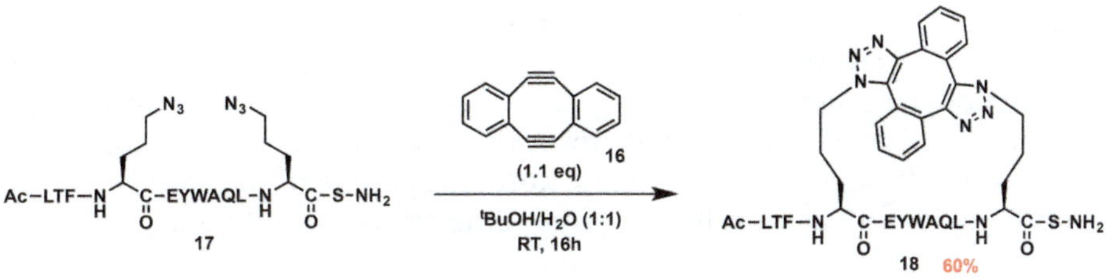

Scheme 8.4 Metal-free double-click peptide stapling.

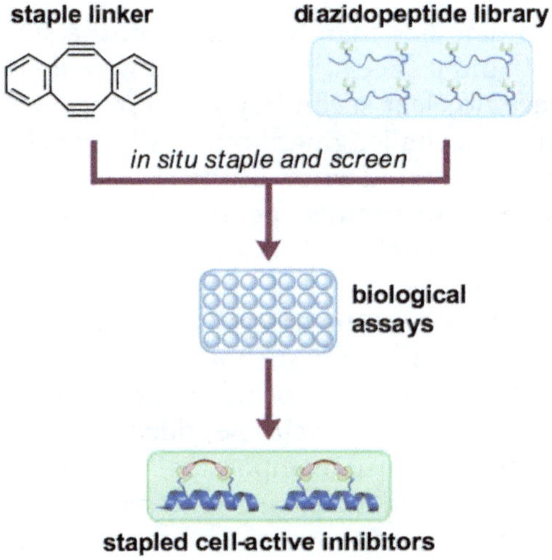

Figure 8.4 *In situ* strategy combining stapling and a primary biological assay in a single step.[44]

HPLC analysis of the reaction mixture indicated that along with the major product **18**, there were other minor by-products with the same mass. These were suggested to be non-interchanging conformations of *syn* and *anti* regioisomers of the stapled peptide. Later on, X-ray crystallography studies on the MDM2-bound stapled peptide demonstrated the major isomer **18** to be the *anti*-regioisomer, which also confirmed the α-helical conformation of the stapled peptide. The stapled peptide **18** was found to be a potent helical inhibitor of the p53–MDM2 interaction. This methodology was extended to staple multiple MDM2-binding peptides in parallel, directly in the culture medium of a primary cell-based 96-well assay (Figure 8.4).[44] This *in situ* screening process led to the rapid selection of an optimal candidate with nanomolar binding affinity and enhanced proteolytic stability. This technique provides a faster way of screening a large peptide library avoiding the need to perform a separate stapling reaction for each peptide variant.

Two limitations of this stapling methodology are the poor water solubility of the linker leading to its precipitation during stapling, and the lack of functional groups to which other motifs could be attached. Hence, current work is underway towards the synthesis of different heteroatom-substituted variants of the Sondheimer–Wong diyne **16**, with the aim of improving its solubility and providing a handle for further reactivity. Functionalization of the linker will hopefully also impart novel properties to the stapled peptide modifying its activity in cells and its binding affinity with MDM2.

8.5 Constrained Macrocyclic Non-α-helical Peptide Inhibitors

Peptide stapling to date has mainly focused on the stabilization of α-helical peptides[11,12,45–47] and β-sheets.[48–50] Generating mimetics of short peptide sequences that have no clear secondary structure prior to association with their binding partner is challenging,[51] but has been achieved through macrocyclization and hence stabilization of the short peptides to give so-called constrained peptides. Constrained peptides can be synthesized through head-to-tail, head-to-side chain, side chain-to-tail or side chain-to-side chain cyclization (Figure 8.5).[52]

Numerous examples of such macrocyclic peptides with high potency and cell permeability exist in nature,[53] such as cyclosporin A,[52,54] antimicrobial polymyxins[55] and the hormone oxytocin.[56] In addition to mono-macrocyclic peptides, naturally occurring bicyclic peptides[57–59] have inspired the synthesis of synthetic bicyclic drug compounds.[60,61] Macrocyclization of these peptide sequences may be less efficient due to high conformational flexibility compared to α-helical peptides and there are no structure-based rules to follow (such as placing unnatural amino acids at i, $i + 4$ or $i + 7$ positions for stapling the same face of a helical turn),[1,2] as every irregularly structured peptide is unique in its secondary structure when bound to the target protein. It is also difficult to predict the optimum linkage length for cross-linking such peptides, since computational approaches become less accurate in comparison to predictions for stapling α-helical peptides. Nevertheless, synthetic stapling or macrocyclization allows the peptides to be constrained in their bioactive conformation resulting in a lower entropic loss upon binding.[62] Chemical modification of these irregularly structured peptides using

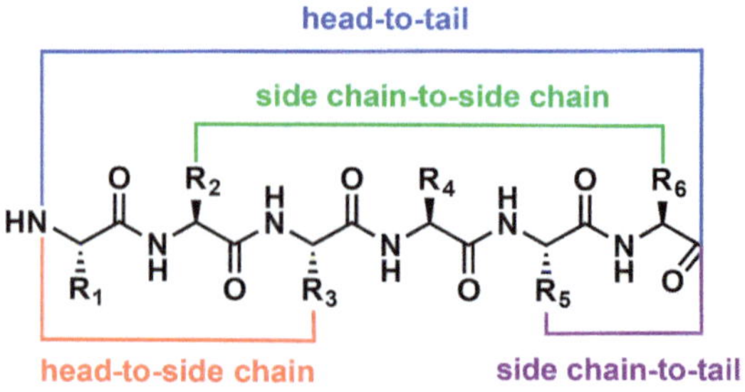

Figure 8.5 Peptide cyclization strategies.

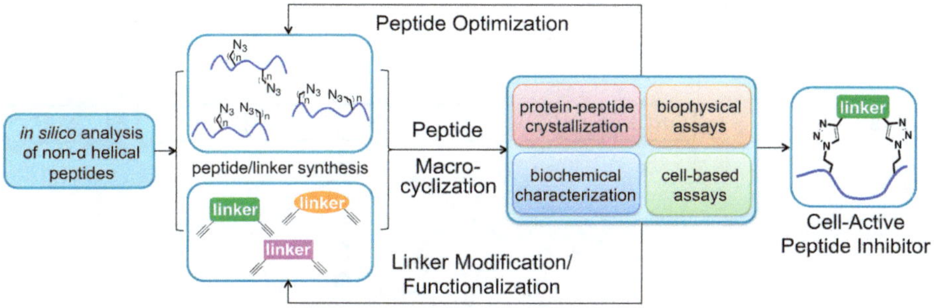

Figure 8.6 Design of constrained peptide inhibitors with a non-α-helical structure.

side chain-to-side chain double-click macrocyclization can potentially help to increase peptide stability towards proteasome-mediated degradation in cells, as well as their cell-penetrating capability. Using *in silico* analysis to scan through a peptide sequence for potential macrocyclization positions can minimize the number of peptides and linkers to be synthesized, and following a rational approach to improve the macrocyclic peptides using the information gained from protein crystallography and screening assays can efficiently provide potential peptide inhibitors that are protease-resistant, highly selective and bioactive in cells (Figure 8.6).

8.5.1 Design of Macrocyclic Peptide Inhibitors to Target the Substrate-recognition Domain of Tankyrase and Antagonize Wnt Signaling

Tankyrase (TNKS) is an ankyrin repeat-containing protein with a catalytic poly(ADP-ribose) polymerase (PARP) domain (Figure 8.7).[63,64] The ankyrin-repeat domain, known as an ankyrin repeat cluster (ARC), is responsible for substrate recognition,[65] whereas the PARP activity of TNKS proteins plays a key role in controlling the axin level, a concentration-determining component of the β-catenin destruction complex in Wnt signaling.[66–68] It has been an attractive therapeutic target for regulating β-catenin in many Wnt-dependent cancers, such as colorectal cancer, gastric cancer, breast cancer, and hepatocellular carcinoma, where accumulation of β-catenin is often observed.[69–77] Small molecules developed for PARP inhibition of TNKS proteins prevent the PARylation and thereby degradation of axin *via* the ubiquitin-proteasome system, and axin in turn promotes the activity of the destruction complex, which phosphorylates β-catenin for degradation. However, potential drug resistance and target-specificity of these small molecule inhibitors remain challenges in this field, as many other members of the PARP family share homology in the PARP domain.[78–80] Peptide inhibitors that instead target the substrate-recognition domain of TNKS represent an alternative approach to intervene in Wnt signaling. An unstructured motif, REAGDGEE, recognized by the TNKS ARC domain was determined by Guettler *et al.*[81] and provided

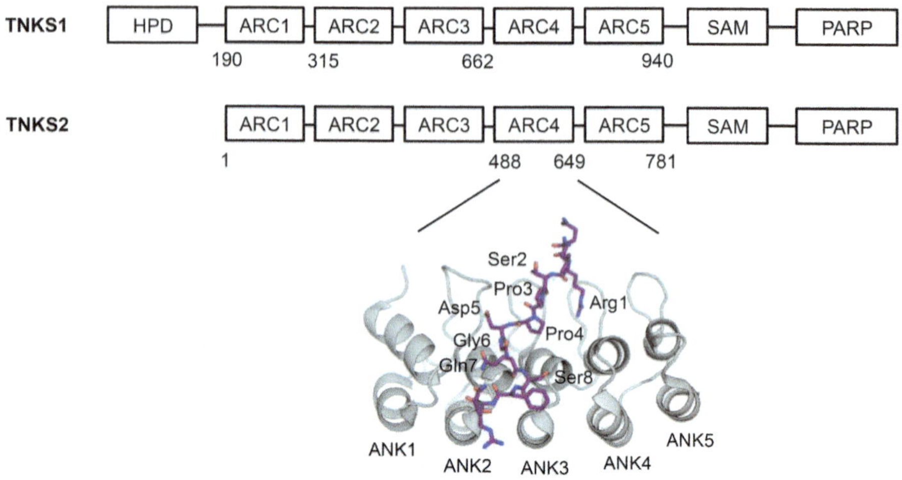

Figure 8.7 Domain architecture of TNKS1 and TNKS2, comprising a homopolymeric run of histidine, proline and serine (HPS) residues, the ankyrin repeat cluster (ARC), a sterile alpha motif (SAM), and catalytic PARP domains. The structure of human TNKS2 ARC4 (grey cartoon) is shown in complex with a substrate peptide LPHLQ**RSPPDGQS**FRS (purple; PDB ID: 3TWR);[81] for clarity, only the central part of the peptide (in bold) is labelled. Adapted from Wenshu Xu, Yu Heng Lau, Gerhard Fischer, Yaw Sing Tan, Anasuya Chattopadhyay, Marc de la Roche, Marko Hyvönen, Chandra Verma, David R. Spring, and Laura S. Itzhaki, Macrocyclized Extended Peptides: Inhibiting the Substrate- Recognition Domain of Tankyrase, *The Journal of American Chemical Society*, 2017, **139**, 2245–2256,[82] © 2017 American Chemical Society. Published under the terms of the CC BY 4.0 licence, https://creativecommons.org/licenses/by/4.0/.

the initial sequence basis for developing TNKS-specific peptide inhibitors to disrupt the TNKS–axin interaction in order to abolish the subsequent axin PARylation. In this work, the two-component double-click strategy has been applied to macrocyclize the TNKS-binding peptides and lock their conformation in the active form, which is an extended, non-helical structure. The macrocyclized peptides showed enhanced binding affinities, proteolytic stability and cell permeability compared with the linear peptide, and one exhibited dose-dependent inhibition of Wnt signaling in cells.

One challenge for the initial macrocyclized peptide design was to determine the cross-linking positions in the sequence as the TNKS-binding peptides are non-helical and therefore the conventional rules for stapling α-helical peptides do not apply here. Computational alanine scanning was first carried out to assess which positions would be amenable to replacement by an azido-functionalized unnatural amino acid (UAA) in the consensus peptide (REAGDGEE, pep1 thereafter) without compromising the binding interaction with the ARC domain of TNKS (Figure 8.8a). Based on the *in silico* analysis, the first panel of peptides were synthesized, each with a pair of azido-functionalized UAAs (Figure 8.8b and d) at different positions in the

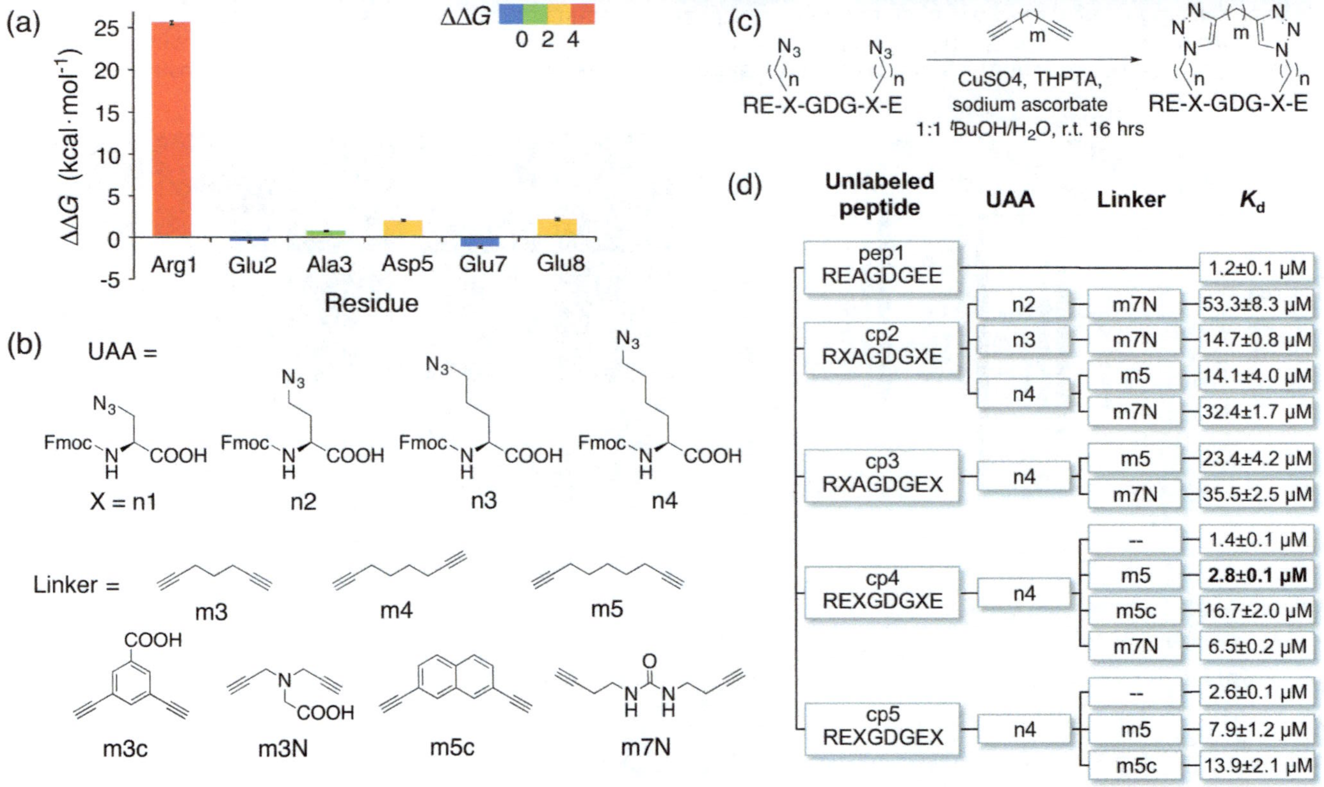

Figure 8.8 (a) Computational alanine scanning of consensus TNKS-binding peptide (REAGDGEE) residues. $\Delta\Delta G$ for Ala3 was obtained by mutating it to glycine. Hot, warm, cool and cold spots are colored red, yellow, green and blue, respectively. (b) Unnatural amino acids (UAAs) and linkers used in the macrocyclized peptides. (c) Reaction scheme of the double-click chemistry on the extended TNKS-binding peptides. (d) The first panel of macrocyclized peptides; K_d values were obtain from competitive fluorescence polarization (FP). Adapted from Wenshu Xu, Yu Heng Lau, Gerhard Fischer, Yaw Sing Tan, Anasuya Chattopadhyay, Marc de la Roche, Marko Hyvönen, Chandra Verma, David R. Spring, and Laura S. Itzhaki, Macrocyclized Extended Peptides: Inhibiting the Substrate- Recognition Domain of Tankyrase, *The Journal of American Chemical Society*, 2017, **139**, 2245–2256,[82] © 2017 American Chemical Society. Published under the terms of the CC BY 4.0 licence, https://creativecommons.org/licenses/by/4.0/.

(a)

Peptide	K_d (μM) TNKS2 ARC4
pep1	0.78 ± 0.06
cp4n1m3	n.d.
cp4n4m5*	2.82 ± 0.10
cp4n3m5	1.91 ± 0.05
cp4n2m5	1.29 ± 0.04
cp4n3m4	1.11 ± 0.03
cp4n2m4	0.81 ± 0.02
cp4n2m3	0.60 ± 0.01
cp4n2m3c	**0.44 ± 0.01**

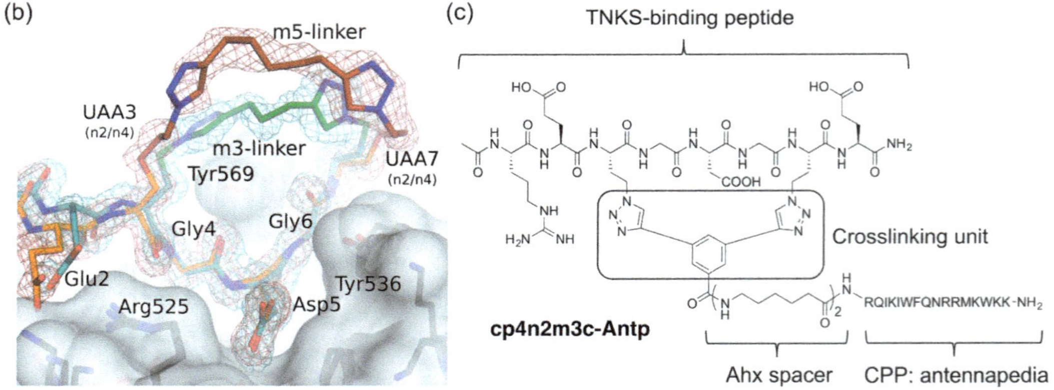

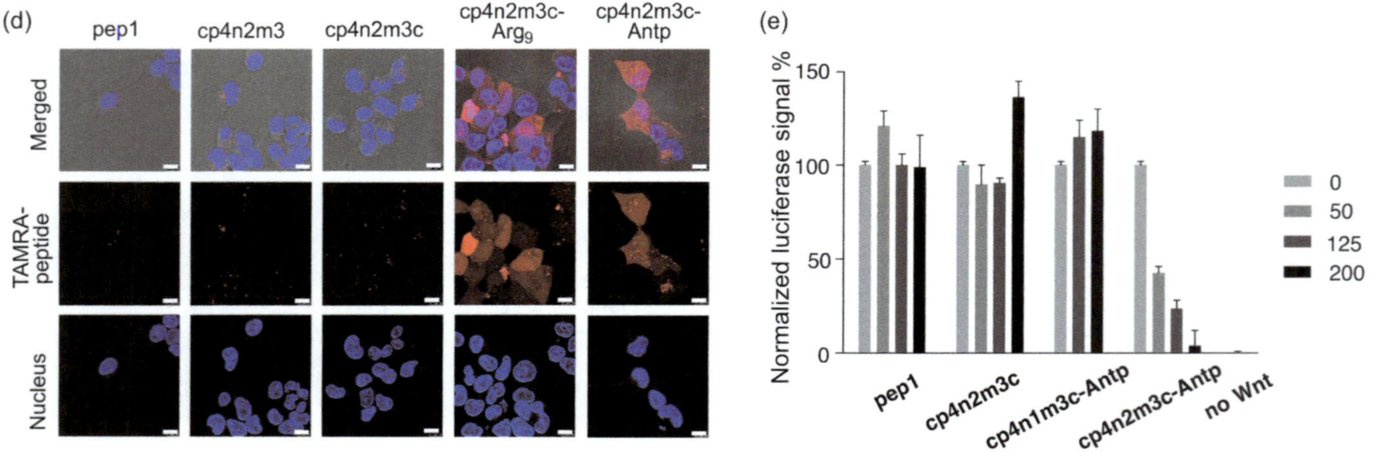

Figure 8.9 (a) A selected list of the second panel of TNKS-binding peptides and their binding affinities to TNKS2 ARC4 measured from FP assays. All peptides were TAMRA-labelled except for cp4n4m5 (*non-labelled), the K_d of which was listed as a comparison. (b) Overlay of two crystal structures of the peptide–TNKS2 ARC4 complex in the macrocycle region of the peptides, cp4n2m3

sequence. The successful double-click reaction (Figure 8.8c) was confirmed by IR spectroscopy and high-resolution mass spectrometry, where only a single product was present where the two azide groups on the peptide reacted with one dialkynyl linker. From the first round of screening using a competitive fluorescence polarization (FP) assay, the sequence containing UAAs at positions 3 and 7 (cp4: REXGDGXE, where X stands for the UAA) was found to provide the highest binding affinity among the series (Figure 8.8d), and therefore this sequence was used for further optimizing the lengths of the UAA side chain, as well as the dialkynyl linker (Figure 8.8b).

To rationally improve the macrocyclic peptide for binding to the ARC domain of TNKS, the crystal structure of TNKS2 ARC4 in complex with a tight-binding macrocyclic peptide, cp4n4m5, was solved at 1.35 Å resolution (Figure 8.9b). The crystal structure suggested that the flexibility of the cross-linking allowed the peptide to adopt the same non-helical bioactive conformation as the linear peptide, though the lengths of the linker as well as the side chains of the UAAs could be shortened to provide more constraint and reduce the entropic cost during the binding. A second panel of macrocyclic peptides were then synthesized based on the optimal sequence cp4, in which the lengths of the side chain in the azido-functionalized UAAs and the dialkynyl linker were varied (Figure 8.9a and 8.8b). Two peptides, cp4n2m3 and cp4n2m3c, with shorter and more constrained cross-linking moieties were shown to bind more strongly than cp4n4m5 and also the linear pep1 to TNKS2 ARC4 in an FP assay (Figure 8.9a) and in isothermal titration calorimetry (ITC). The crystal structure of TNKS2 ARC4 in complex with the tight-binding cp4n2m3 (PBD ID: 5BXO) was subsequently solved at 1.33 Å resolution, confirming the design rationale that the smaller macrocycle size reduced the flexibility in the peptide and improved the binding affinity (Figure 8.9b). Similar to a stapled α-helical peptide, these macrocyclic peptides were shown to be much more resistant to proteolytic degradation than the linear peptide when treated with AspN peptidase, which cleaves between Gly–Asp of the sequence.

(peptide: cyan, linker: green) and cp4n4m5 (peptide: orange, linker: red). PBD IDs: 5BXO and 5BXU.[82] (c) Chemical structure of the active peptide cp4n2m3c-Antp. (d) Confocal images of U2OS cells after incubation with 10 μM of TAMRA-labelled peptides for 4 hours. Bar represents 10 μm. The linker of cp4n2m3c-Arg_9 contains a polyarginine CPP and that of cp4n2m3c-Antp contains a penetratin peptide. (e) Dual-luciferase reporter assay showing the luciferase signal corresponding to the β-catenin level in Wnt3a-activated HEK 293T cells treated with a selection of unlabeled peptides. Macrocyclized CPP-conjugated cp4n2m3c-Antp showed dose-dependent inhibition of luciferase activity. Adapted from Wenshu Xu, Yu Heng Lau, Gerhard Fischer, Yaw Sing Tan, Anasuya Chattopadhyay, Marc de la Roche, Marko Hyvönen, Chandra Verma, David R. Spring, and Laura S. Itzhaki, Macrocyclized Extended Peptides: Inhibiting the Substrate- Recognition Domain of Tankyrase, *The Journal of American Chemical Society*, 2017, **139**, 2245–2256,[82]

The cell-penetrating capabilities of the macrocyclic peptides were enhanced by coupling the linker to a selection of cell-penetrating peptides (CPPs); this did not disrupt the peptide–protein interaction because the cross-linking unit was pointing away from the protein surface according to the two crystal structures (Figure 8.9b) The CPP-conjugated macrocyclized peptides had significant cellular uptake even at 10 μM (Figure 8.9d) compared to the unconjugated macrocyclic peptides and the linear pep1. The unlabeled CPP-conjugated macrocyclic peptides did not exhibit any cytotoxicity at 100 μM. Treating HEK 293T cells in Wnt-activated conditions showed that cp4n2m3c-Antp, a macrocyclic TNKS-binding peptide conjugated with the penetratin sequence (Figure 8.9c), antagonized Wnt signaling by decreasing the β-catenin level in a dose-dependent manner (Figure 8.9e). This is the first example of a peptide inhibitor that directly targets the TNKS–axin binding interaction to antagonize Wnt signaling instead of blocking the catalytic PARP activity of TNKS. This proof-of-concept method provides guidance to develop future peptide inhibitors for TNKS in treating Wnt/β-catenin-dependent cancers, and may overcome issues related to target-specificity and off-target effects of small molecule PARP inhibitors.[82]

8.5.2 Development of Cell-permeable, Non-helical, Constrained Peptides to Target a Key Protein–Protein Interaction in Ovarian Cancer

Ovarian clear cell carcinoma (CCC) is a subtype of ovarian cancer.[83,84] Prognosis for patients with advanced stage or relapsed disease is poor due to intrinsic resistance to platinum-based chemotherapy and the lack of targeted therapies for CCC.[85,86] The development of novel, targeted therapeutics for CCC is therefore of high priority. Transcription factor HNF1β is overexpressed in most, if not all, CCC cases[87,88] and proliferation in CCC cell lines drops upon HNF1β knockdown, caused by induced apoptotic cell death in CCC cells.[89] Drugs targeting HNF1β have yet to be developed.

Human HNF1β protein consists of 557 amino acids and a nuclear localization signal (NLS), which directs the nuclear import of the protein by interacting with structured binding sites on the nuclear import protein Importin α,[90] is contained within the DNA-binding domain of HNF1β.[91–93]

Transcription factors are generally considered inferior drug targets,[94] but are highly attractive as key regulators of gene expression.[95–97] Interestingly, 49% of transcription factor sequences consist of intrinsically disordered domains (IDDs).[98] Therapeutic targeting of the nuclear import of transcription factors provides a strategy for inhibiting transcription factor function, since activity depends on successful localization to the nucleus for transcription to take place.[99,100] Competitive inhibition of the HNF1β NLS–Importin α PPI by a constrained peptide inhibitor that is based on the HNF1β NLS sequence may interfere with HNF1β transcription factor function as shown in Figure 8.10. The self-inhibitory domain of Importin α binds to Importin β to free up the NLS-binding sites on Importin α. The constrained peptide

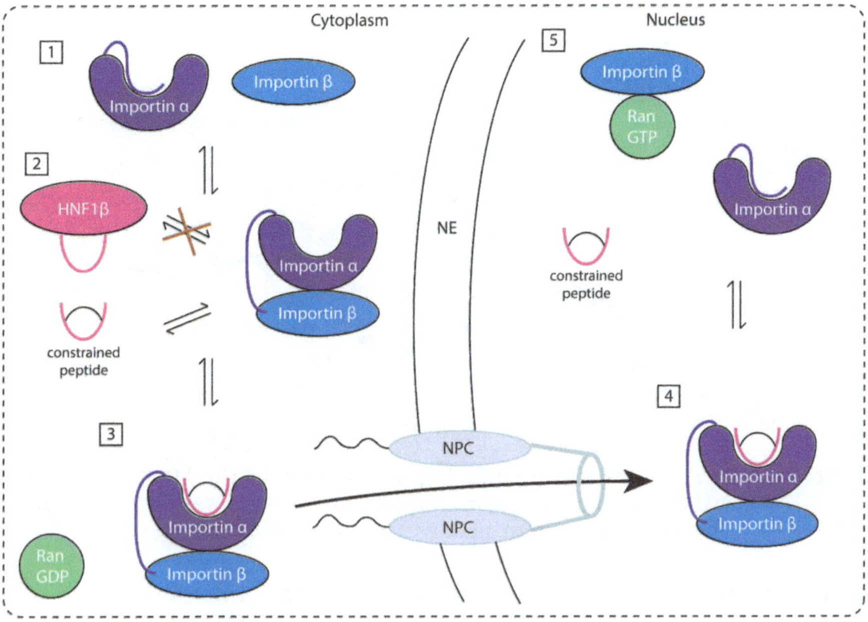

Figure 8.10 Targeting scheme of the nuclear import of transcription factor HNF1β with a constrained peptide inhibitor based on the NLS sequence. From Mareike M. Wiedmann, Yaw Sing Tan, Yuteng Wu, Shintaro Aibara, Wenshu Xu, Hannah F. Sore, Chandra S. Verma, Laura Itzhaki, Murray Stewart, James D. Brenton, David R. Spring, Development of Cell-Permeable, Non-Helical Constrained Peptides to Target a Key Protein–Protein Interaction in Ovarian Cancer,[101] © 2016 The Authors. Published by Wiley-VCH Verlag GmbH & Co. KGaA. Published under the terms of the CC BY 4.0 licence, https://creativecommons.org/licenses/by/4.0/.

competes with the HNF1β protein for Importin α binding, which is in complex with Importin β in the cytoplasm. The trimer is imported through the nuclear pore complex (NPC) in the nuclear envelope (NE) into the nucleus, where it is released by RanGTP binding to Importin β.

The HNF1β NLS peptide, which is located in a flexible surface loop in between the two DNA-binding domains, was stapled to stabilize the NLS in its binding conformation. The crystal structure of the mImportin α1 ΔIBB with the HNF1β NLS peptide was determined and the HNF1β NLS peptide sequence was used as a starting point for constrained peptide design (Figure 8.11).

Three independently performed molecular dynamics (MD) simulations revealed the most important binding interactions of the HNF1β NLS peptide with two flanking residues on either side (1-TN**KKMRRNR**FK-11) with the mImportin α1 protein. Two discrepancies between the obtained crystal structure and the MD simulation results were identified. In the crystal structure, the backbone carbonyl group of Thr1 interacts with Arg238 on Importin α,

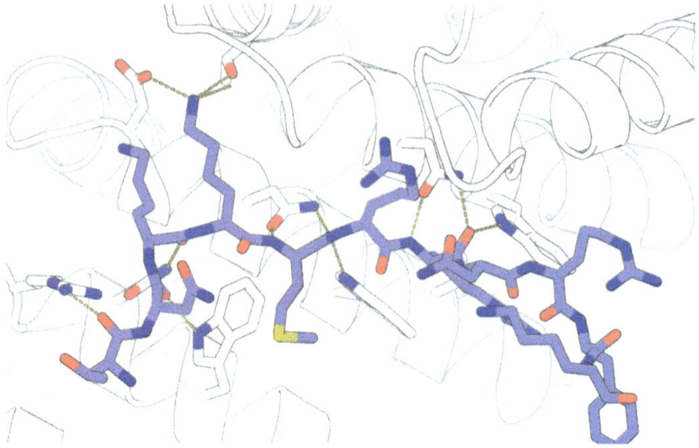

Figure 8.11 Crystal structure depicting interactions of the HNF1β NLS peptide with mImportin α1 with tryptophan stacking. Generated by Dr Shintaro Aibara and reprinted with kind permission.[93]

whereas in two out of three MD simulations, this interaction is lost and H-bonding with Asp270 is observed instead (Figure 8.12A and B). Secondly, Arg9 forms a salt bridge with Glu465 on Importin α in the crystal structure, whereas a salt bridge with Glu107 was observed instead in all three MD simulations (Figure 8.12C and D). The results highlight the importance of using MD simulations to eliminate the effect of crystal packing and non-physiological contacts with neighboring proteins in the crystal, and to restore the protein structure in solution.[102] In summary, the results suggest that residues Thr1 and Arg9 should be retained to maintain the binding potency of the constrained peptides.

The contribution of each HNF1β NLS residue to the binding was then assessed by binding free energy decomposition.[103] Residues Phe10 and Lys11 contribute almost 0 kcal mol^{-1} to the total binding free energy suggesting that they can be removed from the peptide with minimal disruption of the overall binding. Hence, the peptides are based on the following peptide sequence: 1-**TNKKMRRNRNR**-9. Computational alanine scanning on structures extracted from the MD simulations was then used to determine suitable stapling locations. Each residue was mutated to alanine and the difference in the binding free energy between mutant and wild-type was calculated, allowing the identification of residues that have little or negative contribution to the binding. The analysis revealed that Thr1, Lys4, Arg6, Arg7 and Arg9 are the most important residues for binding, whereas Asn2, Asn8, Phe10 and Lys11 only make negligible contributions to the binding. Since the staple is preferably placed on residues with side chains that have little or negative contribution to the binding, the staple may hence be placed at residues Asn2 and Asn8, Phe10, or Lys11 which contribute almost 0 kcal mol^{-1} to the total binding free energy. Based on the obtained crystal structure (Figure 8.11), the staple is preferentially placed at residues Asn2 and Asn8 as

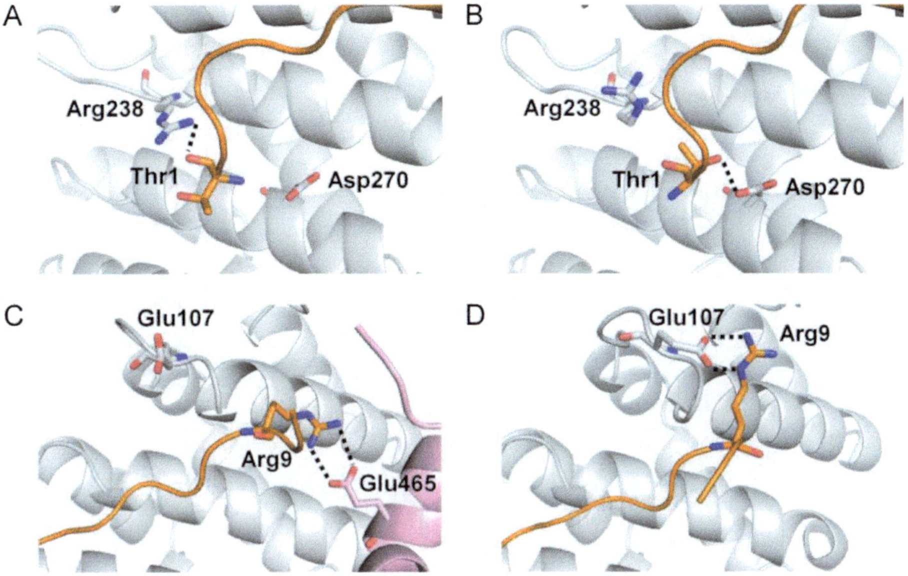

Figure 8.12 Binding interactions of the HNF1β NLS peptide (orange) with mImportin α1 (white) determined from X-ray crystallography and MD simulations. Trajectory structures shown are snapshots taken from the end of the simulations. (A) Backbone carbonyl oxygen of Thr1 hydrogen bonds with the side chain of Arg238 in the crystal structure. (B) Side chain of Thr1 hydrogen bonds with the side chain of Asp270 in the MD simulations. (C) Arg9 forms a salt bridge with Glu465 from a neighboring protein chain (pink) in the crystal structure. (D) Arg9 forms a salt bridge with Glu107 in the MD simulations. From Mareike M. Wiedmann, Yaw Sing Tan, Yuteng Wu, Shintaro Aibara, Wenshu Xu, Hannah F. Sore, Chandra S. Verma, Laura Itzhaki, Murray Stewart, James D. Brenton, David R. Spring, Development of Cell-Permeable, Non-Helical Constrained Peptides to Target a Key Protein–Protein Interaction in Ovarian Cancer,[101]

their side chains point towards each other in the bound peptide, to minimize constraint of the constrained peptide's bound conformation by the staple. The peptide sequence is therefore TAMRA-Ahx-^{227}**TX$_n$KKMRRX$_n$R**^{235}, with X_n referring to non-proteogenic azido amino acids used for stapling.

All linear peptides are synthesized using Fmoc-spps and then stapled using Cu(I)-click chemistry with three symmetrical dialkynyl linkers (Figure 8.13A and B). A direct fluorescence polarization assay is used to determine the binding constants. The dye, in this case TAMRA-5, reports a lower mobility of the dye-labelled ligand as it binds to the much heavier and hence less mobile protein, resulting in a higher degree of fluorescence polarization.[104] The linear HNF1β NLS peptide sequence (Ac-**TNKKMRRNRFK**-NH_2) binds mImportin α1 with a K_d of 1.7 μM. TAMRA-5 alone binds mImportin α1 with a

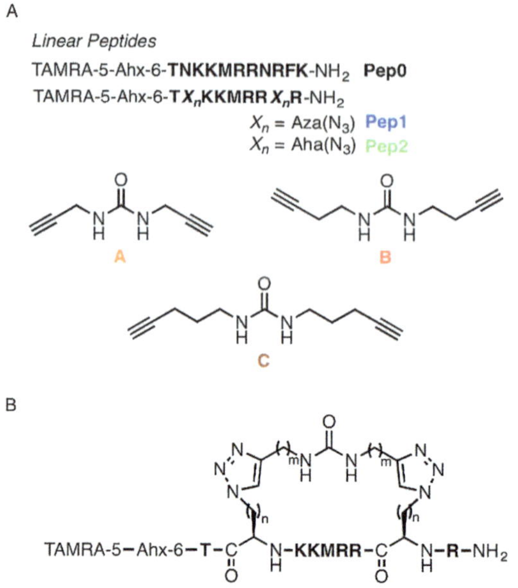

C

Peptide	K_d in μM
Pep0	5.32 ± 0.30
Pep1	2.06 ± 0.07
Pep2	6.84 ± 0.26
Pep1A	95.9 ± 27.3
Pep1B	48.2 ± 4.88
Pep1C	55.9 ± 2.43
Pep2A	4.54 ± 0.03

Figure 8.13 (A) Synthesized peptide sequences containing azido amino acids and linkers A–C. (B) General structure of bis-triazole constrained peptides with $n = 1$ or 2 and $m = 1$–3. (C) Direct FP assay binding affinities for (constrained) peptides in μM. From Mareike M. Wiedmann, Yaw Sing Tan, Yuteng Wu, Shintaro Aibara, Wenshu Xu, Hannah F. Sore, Chandra S. Verma, Laura Itzhaki, Murray Stewart, James D. Brenton, David R. Spring, Development of Cell-Permeable, Non-Helical Constrained Peptides to Target a Key Protein–Protein Interaction in Ovarian Cancer,[101]

K_d of approximately 125 μM, and is a useful control to establish if binding takes place *via* the peptide motif or is attributed to the dye. Compared to the linear wild-type peptide Pep0 there was a 2.5-fold improvement in the binding of Pep1. Constrained peptides Pep1A–Pep1C follow a rough trend with the binding affinity increasing with increasing linker length (Figure 8.13C). Constrained Pep2A bound with slightly improved binding affinity compared to linear peptide Pep0 and bound more tightly than its unconstrained precursor Pep2. This confirmed that an entropically-driven gain in binding affinity was achieved for Pep2A.

The cell permeability of the synthesized TAMRA-labelled linear and constrained peptides was assessed in the normal ovarian cell line IOSE4 using live-cell fluorescence microscopy. Pep1 and Pep2 displayed good cell permeability, which was retained upon stabilization (Pep1B, Pep2A). In conclusion, this work represents the first example of constraining an NLS peptide to target the nuclear import pathway. Further structural information on transcription factor HNF1β binding to Importin α is required for the future design of Importin α isoform-selective inhibitors.

Acknowledgements

The Spring lab acknowledges support from the European Research Council under the European Union's Seventh Framework Programme (FP7/2007–2013)/ERC grant agreement no [279337/DOS]. In addition, the group research was supported by grants from the Engineering and Physical Sciences Research Council, Biotechnology and Biological Sciences Research Council, Medical Research Council, Royal Society and Welcome Trust. LSI and WX acknowledge support from the Medical Research Council and the Medical Research Foundation.

References

1. E. M. Phizicky and S. Fields, *Microbiol. Rev.*, 1995, **59**, 94–123.
2. T. L. Nero, C. J. Morton, J. K. Holien, J. Wielens and M. W. Parker, *Nat. Rev. Cancer*, 2014, **14**, 248–262.
3. J. A. Wells and C. L. McClendon, *Nature*, 2007, **450**, 1001–1009.
4. G. L. Verdine and L. D. Walensky, *Clin. Cancer Res.*, 2007, **13**, 7264.
5. H. Yin and A. D. Hamilton, *Angew. Chem., Int. Ed.*, 2005, **44**, 4130–4163.
6. B. N. Bullock, A. L. Jochim and P. S. Arora, *J. Am. Chem. Soc.*, 2011, **133**, 14220–14223.
7. R. M. J. Liskamp, D. T. S. Rijkers, J. A. W. Kruijtzer and J. Kemmink, *ChemBioChem*, 2011, **12**, 1626–1653.
8. V. Azzarito, K. Long, N. S. Murphy and A. J. Wilson, *Nat. Chem.*, 2013, **5**, 161–173.
9. A. Grauer and B. König, *Eur. J. Org. Chem.*, 2009, **2009**, 5099–5111.
10. H. E. Blackwell and R. H. Grubbs, *Angew. Chem., Int. Ed.*, 1998, **37**, 3281–3284.
11. C. E. Schafmeister, J. Po and G. L. Verdine, *J. Am. Chem. Soc.*, 2000, **122**, 5891–5892.
12. L. D. Walensky, A. L. Kung, I. Escher, T. J. Malia, S. Barbuto, R. D. Wright, G. Wagner, G. L. Verdine and S. J. Korsmeyer, *Science*, 2004, **305**, 1466–1470.
13. Y. S. Chang, B. Graves, V. Guerlavais, C. Tovar, K. Packman, K.-H. To, K. A. Olson, K. Kesavan, P. Gangurde, A. Mukherjee, T. Baker, K. Darlak, C. Elkin, Z. Filipovic, F. Z. Qureshi, H. Cai, P. Berry, E. Feyfant, X. E. Shi, J. Horstick, D. A. Annis, A. M. Manning, N. Fotouhi, H. Nash, L. T. Vassilev and T. K. Sawyer, *Proc. Natl. Acad. Sci.*, 2013, **110**, E3445–E3454.
14. L. D. Walensky and G. H. Bird, *J. Med. Chem.*, 2014, **57**, 6275–6288.
15. O. Torres, D. Yüksel, M. Bernardina, K. Kumar and D. Bong, *ChemBioChem*, 2008, **9**, 1701–1705.
16. Y. H. Lau, P. de Andrade, Y. Wu and D. R. Spring, *Chem. Soc. Rev.*, 2015, **44**, 91–102.
17. A. M. Felix, E. P. Heimer, C.-T. Wang, T. J. Lambros, A. Fournier, T. F. Mowles, S. Maines, R. M. Campbell, B. B. Wegrzynski, V. Toome, D. Fry and V. S. Madison, *Int. J. Pept. Protein Res.*, 1988, **32**, 441–454.

18. D. Y. Jackson, D. S. King, J. Chmielewski, S. Singh and P. G. Schultz, *J. Am. Chem. Soc.*, 1991, **113**, 9391–9392.
19. J. E. Moses and A. D. Moorhouse, *Chem. Soc. Rev.*, 2007, **36**, 1249–1262.
20. F. M. Brunel and P. E. Dawson, *Chem. Commun.*, 2005, 2552–2554.
21. Y. H. Lau, Y. Wu, P. de Andrade, W. R. J. D. Galloway and D. R. Spring, *Nat. Protoc.*, 2015, **10**, 585–594.
22. V. V. Rostovtsev, L. G. Green, V. V. Fokin and K. B. Sharpless, *Angew. Chem., Int. Ed.*, 2002, **41**, 2596–2599.
23. C. W. Tornøe, C. Christensen and M. Meldal, *J. Org. Chem.*, 2002, **67**, 3057–3064.
24. R. Huisgen, *Proc. Chem. Soc.*, 1961, 357–396.
25. Y. H. Lau, P. de Andrade, G. J. McKenzie, A. R. Venkitaraman and D. R. Spring, *ChemBioChem*, 2014, **15**, 2680–2683.
26. R. B. Merrifield, *J. Am. Chem. Soc.*, 1963, **85**, 2149–2154.
27. M. Roice, I. Johannsen and M. Meldal, *QSAR Comb. Sci.*, 2004, **23**, 662–673.
28. S. I. van Kasteren, H. B. Kramer, H. H. Jensen, S. J. Campbell, J. Kirkpatrick, N. J. Oldham, D. C. Anthony and B. G. Davis, *Nature*, 2007, **446**, 1105–1109.
29. K. L. Kiick, E. Saxon, D. A. Tirrell and C. R. Bertozzi, *Proc. Natl. Acad. Sci.*, 2002, **99**, 19–24.
30. N. Miller, G. M. Williams and M. A. Brimble, *Org. Lett.*, 2009, **11**, 2409–2412.
31. E. D. Goddard-Borger and R. V. Stick, *Org. Lett.*, 2007, **9**, 3797–3800.
32. Y. H. Lau and D. R. Spring, *Synlett*, 2011, 1917–1919.
33. Y. H. Lau, P. de Andrade, N. Skold, G. J. McKenzie, A. R. Venkitaraman, C. Verma, D. P. Lane and D. R. Spring, *Org. Biomol. Chem.*, 2014, **12**, 4074–4077.
34. Y. H. Lau, P. de Andrade, S.-T. Quah, M. Rossmann, L. Laraia, N. Skold, T. J. Sum, P. J. E. Rowling, T. L. Joseph, C. Verma, M. Hyvonen, L. S. Itzhaki, A. R. Venkitaraman, C. J. Brown, D. P. Lane and D. R. Spring, *Chem. Sci.*, 2014, **5**, 1804–1809.
35. B. Hu, D. M. Gilkes and J. Chen, *Cancer Res.*, 2007, **67**, 8810.
36. Y. Wu, L. B. Olsen, Y. H. Lau, C. H. Jensen, M. Rossmann, Y. R. Baker, H. F. Sore, S. Collins and D. R. Spring, *ChemBioChem*, 2016, **17**, 689–692.
37. E. M. Sletten and C. R. Bertozzi, *Angew. Chem., Int. Ed.*, 2009, **48**, 6974–6998.
38. D. C. Kennedy, C. S. McKay, M. C. B. Legault, D. C. Danielson, J. A. Blake, A. F. Pegoraro, A. Stolow, Z. Mester and J. P. Pezacki, *J. Am. Chem. Soc.*, 2011, **133**, 17993–18001.
39. N. J. Agard, J. A. Prescher and C. R. Bertozzi, *J. Am. Chem. Soc.*, 2004, **126**, 15046–15047.
40. E. M. Sletten and C. R. Bertozzi, *Acc. Chem. Res.*, 2011, **44**, 666–676.
41. A. Orita, D. Hasegawa, T. Nakano and J. Otera, *Chem.–Eur. J.*, 2002, **8**, 2000–2004.

42. F. Xu, L. Peng, K. Shinohara, T. Morita, S. Yoshida, T. Hosoya, A. Orita and J. Otera, *J. Org. Chem.*, 2014, **79**, 11592–11608.
43. H. N. C. Wong, P. J. Garratt and F. Sondheimer, *J. Am. Chem. Soc.*, 1974, **96**, 5604–5605.
44. Y. H. Lau, Y. Wu, M. Rossmann, B. X. Tan, P. de Andrade, Y. S. Tan, C. Verma, G. J. McKenzie, A. R. Venkitaraman, M. Hyvönen and D. R. Spring, *Angew. Chem., Int. Ed.*, 2015, **54**, 15410–15413.
45. M. M. Madden, A. Muppidi, Z. Li, X. Li, J. Chen and Q. Lin, *Bioorg. Med. Chem. Lett.*, 2011, **21**, 1472–1475.
46. M. L. Stewart, E. Fire, A. E. Keating and L. D. Walensky, *Nat. Chem. Biol.*, 2010, **6**, 595–601.
47. K. Pitter, F. Bernal, J. LaBelle and L. D. Walensky, 2008, **446**, 387–408.
48. O. Khakshoor and J. S. Nowick, *Org. Lett.*, 2009, **11**, 3000–3003.
49. A. A. Virgilio and J. A. Ellman, *J. Am. Chem. Soc.*, 1994, **116**, 11580–11581.
50. A. A. Virgilio, S. C. Schürer and J. A. Ellman, *Tetrahedron Lett.*, 1996, **37**, 6961–6964.
51. D. E. Scott, A. R. Bayly, C. Abell and J. Skidmore, *Nat. Rev. Drug Discovery*, 2016, **15**, 533–550.
52. T. A. Hill, N. E. Shepherd, F. Diness and D. P. Fairlie, *Angew. Chem., Int. Ed. Engl.*, 2014, **53**, 13020–13041.
53. D. J. Craik, *Science*, 2006, **311**, 1563–1564.
54. D. Faulds, K. L. Goa and P. Benfield, *Drugs*, 1993, **45**, 953–1040.
55. T. Velkov, P. E. Thompson, R. L. Nation and J. Li, *J. Med. Chem.*, 2010, **53**, 1898–1916.
56. J. P. Rose, C.-K. Wu, C.-D. Hsiao, E. Breslow and B.-C. Wang, *Nat. Struct. Mol. Biol.*, 1996, **3**, 163–169.
57. S. D. Jolad, J. J. Hoffman, S. J. Torrance, R. M. Wiedhopf, J. R. Cole, S. K. Arora, R. B. Bates, R. L. Gargiulo and G. R. Kriek, *J. Am. Chem. Soc.*, 1977, **99**, 8040–8044.
58. M. Zalacaín, E. Zaera, D. Vázquez and A. Jiménez, *FEBS Lett.*, 1982, **148**, 95–97.
59. D. A. Bushnell, P. Cramer and R. D. Kornberg, *Proc. Natl. Acad. Sci. U. S. A.*, 2002, **99**, 1218–1222.
60. A. Angelini, L. Cendron, S. Chen, J. Touati, G. Winter, G. Zanotti and C. Heinis, *ACS Chem. Biol.*, 2012, **7**, 817–821.
61. S. Chen, J. Morales-Sanfrutos, A. Angelini, B. Cutting and C. Heinis, *ChemBioChem*, 2012, **13**, 1032–1038.
62. J. S. McMurray, P. K. Mandal, W. S. Liao, J. Klostergaard and F. M. Robertson, *JAKSTAT*, 2012, **1**, 263–347.
63. H. Seimiya and S. Smith, *J. Biol. Chem.*, 2002, **277**, 14116–14126.
64. S. Smith, I. Giriat, A. Schmitt and T. de Lange, *Science*, 1998, **282**, 1484–1487.
65. J. I. Sbodio and N.-W. Chi, *J. Biol. Chem.*, 2002, **277**, 31887–31892.
66. S.-M. A. Huang, Y. M. Mishina, S. Liu, A. Cheung, F. Stegmeier, G. A. Michaud, O. Charlat, E. Wiellette, Y. Zhang, S. Wiessner, M. Hild, X. Shi,

C. J. Wilson, C. Mickanin, V. Myer, A. Fazal, R. Tomlinson, F. Serluca, W. Shao, H. Cheng, M. Shultz, C. Rau, M. Schirle, J. Schlegl, S. Ghidelli, S. Fawell, C. Lu, D. Curtis, M. W. Kirschner, C. Lengauer, P. M. Finan, J. A. Tallarico, T. Bouwmeester, J. A. Porter, A. Bauer and F. Cong, *Nature*, 2009, **461**, 614–620.
67. S. Ikeda, S. Kishida, H. Yamamoto, H. Murai, S. Koyama and A. Kikuchi, *EMBO J.*, 1998, **17**, 1371–1384.
68. A. Salic, E. Lee, L. Mayer and M. W. Kirschner, *Mol. Cell*, 2000, **5**, 523–532.
69. O. Aguilera, M. F. Fraga, E. Ballestar, M. F. Paz, M. Herranz, J. Espada, J. M. Garcia, A. Munoz, M. Esteller and J. M. Gonzalez-Sancho, *Oncogene*, 2006, **25**, 4116–4121.
70. J. N. Anastas and R. T. Moon, *Nat. Rev. Cancer*, 2013, **13**, 11–26.
71. K. Breuhahn, T. Longerich and P. Schirmacher, *Oncogene*, 2006, **25**, 3787–3800.
72. H. Clevers and R. Nusse, *Cell*, 2012, **149**, 1192–1205.
73. G. J. Klarmann, A. Decker and W. L. Farrar, *Epigenetics*, 2008, **3**, 59–63.
74. M. Kurayoshi, N. Oue, H. Yamamoto, M. Kishida, A. Inoue, T. Asahara, W. Yasui and A. Kikuchi, *Cancer Res.*, 2006, **66**, 10439–10448.
75. P. Laurent-Puig and J. Zucman-Rossi, *Oncogene*, 2006, **25**, 3778–3786.
76. S. Y. Lin, W. Xia, J. C. Wang, K. Y. Kwong, B. Spohn, Y. Wen, R. G. Pestell and M. C. Hung, *Proc. Natl. Acad. Sci. U. S. A.*, 2000, **97**, 4262–4266.
77. P. J. Morin, A. B. Sparks, V. Korinek, N. Barker, H. Clevers, B. Vogelstein and K. W. Kinzler, *Science*, 1997, **275**, 1787–1790.
78. S. L. Edwards, R. Brough, C. J. Lord, R. Natrajan, R. Vatcheva, D. A. Levine, J. Boyd, J. S. Reis-Filho and A. Ashworth, *Nature*, 2008, **451**, 1111–1115.
79. C. J. Lord and A. Ashworth, *Curr. Opin. Pharmacol.*, 2008, **8**, 363–369.
80. M. Rouleau, A. Patel, M. J. Hendzel, S. H. Kaufmann and G. G. Poirier, *Nat. Rev. Cancer*, 2010, **10**, 293–301.
81. S. Guettler, J. LaRose, E. Petsalaki, G. Gish, A. Scotter, T. Pawson, R. Rottapel and F. Sicheri, *Cell*, 2011, **147**, 1340–1354.
82. W. Xu, Y. H. Lau, G. Fischer, Y. S. Tan, A. Chattopadhyay, M. de la Roche, M. Hyvönen, C. Verma, D. R. Spring and L. S. Itzhaki, *J. Am. Chem. Soc.*, 2017, **139**, 2245–2256.
83. S. Vaughan, J. I. Coward, R. C. Bast Jr, A. Berchuck, J. S. Berek, J. D. Brenton, G. Coukos, C. C. Crum, R. Drapkin, D. Etemadmoghadam, M. Friedlander, H. Gabra, S. B. Kaye, C. J. Lord, E. Lengyel, D. A. Levine, I. A. McNeish, U. Menon, G. B. Mills, K. P. Nephew, A. M. Oza, A. K. Sood, E. A. Stronach, H. Walczak, D. D. Bowtell and F. R. Balkwill, *Nat. Rev. Cancer*, 2011, **11**, 719–725.
84. H. Kobayashi, K. Sumimoto, N. Moniwa, M. Imai, K. Takakura, T. Kuromaki, E. Morioka, K. Arisawa and T. Terao, *Int. J. Gynecol. Cancer*, 2007, **17**, 37–43.
85. D. S. P. Tan and S. Kaye, *J. Clin. Pathol.*, 2007, **60**, 355–360.
86. D. S. Tan, R. E. Miller and S. B. Kaye, *Br. J. Cancer*, 2013, **108**, 1553–1559.
87. N. Kato, S. Sasou and T. Motoyama, *Mod. Pathol.*, 2006, **19**, 83–89.

88. K. Yamaguchi, M. Mandai, T. Oura, N. Matsumura, J. Hamanishi, T. Baba, S. Matsui, S. K. Murphy and I. Konishi, *Oncogene*, 2010, **29**, 1741–1752.
89. A. Tsuchiya, M. Sakamoto, J. Yasuda, M. Chuma, T. Ohta, M. Ohki, T. Yasugi, Y. Taketani and S. Hirohashi, *Am. J. Pathol.*, 2003, **163**, 2503–2512.
90. D. Kalderon, W. D. Richardson, A. F. Markham and A. E. Smith, *Nature*, 1984, **311**, 33–38.
91. G. Wu, S. Bohn and G. U. Ryffel, *Eur. J. Biochem.*, 2004, **271**, 3715–3728.
92. S. Bohn, H. Thomas, G. Turan, S. Ellard, C. Bingham, A. T. Hattersley and G. U. Ryffel, *J. Am. Soc. Nephrol.*, 2003, **14**, 2033–2041.
93. M. M. Wiedmann, S. Aibara, D. R. Spring, M. Stewart and J. D. Brenton, *J. Struct. Biol.*, 2016, **195**, 273–281.
94. P. Imming, C. Sinning and A. Meyer, *Nat. Rev. Drug Discovery*, 2006, **5**, 821–834.
95. A. G. Papavassiliou and P. A. Konstantinopoulos, *J. Am. Med. Assoc.*, 2011, **305**, 2349–2350.
96. M. N. Patel, M. D. Halling-Brown, J. E. Tym, P. Workman and B. Al-Lazikani, *Nat. Rev. Drug Discovery*, 2013, **12**, 35–50.
97. F. Fontaine, J. Overman and M. François, *Cell Regener.*, 2015, **4**, 1–12.
98. Y. Minezaki, K. Homma, A. R. Kinjo and K. Nishikawa, *J. Mol. Biol.*, 2006, **359**, 1137–1149.
99. P. Ferrigno and P. A. Silver, *Oncogene*, 1999, **18**, 6129–6134.
100. A. Komeili and E. K. O'Shea, *Curr. Opin. Cell Biol.*, 2000, **12**, 355–360.
101. M. M. Wiedmann, Y. S. Tan, Y. Wu, S. Aibara, W. Xu, H. F. Sore, C. S. Verma, L. Itzhaki, M. Stewart, J. D. Brenton and D. R. Spring, *Angew. Chem., Int. Ed.*, 2017, **56**, 524–529.
102. T. Terada and A. Kidera, *J. Phys. Chem. B*, 2012, **116**, 6810–6818.
103. H. Gohlke, C. Kiel and D. A. Case, *J. Mol. Biol.*, 2003, **330**, 891–913.
104. J. R. Unruh, G. Gokulrangan, G. H. Lushington, C. K. Johnson and G. S. Wilson, *Biophys. J.*, 2005, **88**, 3455–3465.

CHAPTER 9

Libraries of Head-to-tail Peptides

ANDREW T. BALL AND ALI TAVASSOLI*

Chemistry, University of Southampton, Southampton, SO17 1BJ, UK
*E-mail: ali1@soton.ac.uk

9.1 Introduction

An emerging challenge within contemporary drug discovery is the lack of effective ligands for many targets that have significant disease relevance, but are typically considered 'undruggable' by way of the existing small molecule paradigm.[1] Peptide-based drugs have been proposed as a solution for challenging targets, such as protein–protein interactions (PPIs), which often have extended binding interfaces and lack well defined small molecule pockets. Peptides often have beneficial characteristics over many small molecule drugs, such as lower toxicity profiles and enhanced selectivity. Conversely, they are also substrates for many proteases in the digestive system and therefore are often prone to short half-lives *in vivo* and poor oral bioavailability.[2,3] Macrocyclic peptides offer a solution to some of the common problems with peptide drugs, as they not only possess enhanced resistance to many peptidases, but often also exhibit improved activity over their linear counterparts by virtue of their conformational pre-organisation.[4,5]

Head-to-tail (or homodetic) cyclic peptides are a class of peptide macrocycles formed by the direct connection of the N- and C-termini *via* an amide bond. While these peptides share the conformational rigidity and

Chemical Biology No. 6
Cyclic Peptides: From Bioorganic Synthesis to Applications
Edited by Jesko Koehnke, James Naismith and Wilfred A. van der Donk

Published by the Royal Society of Chemistry, www.rsc.org

resistance to endopeptidase activity common to other types of macrocyclic peptides, the lack of exposed termini also confers exopeptidase resistance.

The head-to-tail cyclic peptide motif is of particular interest amongst macrocycles as it is commonly exhibited in many bioactive natural products (Figure 9.1). The often-cited example of a successful macrocyclic peptide drug is cyclosporine A – a head-to-tail cyclic natural product that exhibits immunosuppressant activity. The same structural element is also present in many antimicrobial peptides, such as gramicidin S, as well as some bicyclic peptides, such as sunflower trypsin inhibitor-1 (SFTI-1).

Bioactive linear peptides have also benefitted in terms of affinity and bioavailability from modification with head-to-tail cyclization. For instance, the integrin recognition sequence RGD was found to be more potent when incorporated as part of the cyclic peptide c(RGDfV).[6]

Whilst head-to-tail cyclic peptides show some potential for use in drug discovery, thus far most of the currently marketed drugs in this class are derived from existing natural products, with the development of novel structures being a relatively under-explored field.[7,8] The emerging interest in these compounds has fueled a demand for high-throughput screening platforms and accompanying libraries of head-to-tail peptides to utilize with them.

An important feature of many of the bioactive cyclic peptide examples is the presence of non-ribosomal amino acid moieties. Of particular note is the presence of *N*-methylated amides, which contribute towards cell permeability, and potentially the activity of the peptide. However, a systematic approach to *N*-methylation across all peptide structures is so far not currently established, and so in many cases the effect of this modification has been approached empirically *via* library screening.[9,10] It therefore follows that approaches that incorporate non-canonical elements into cyclic peptide libraries are of significant interest.

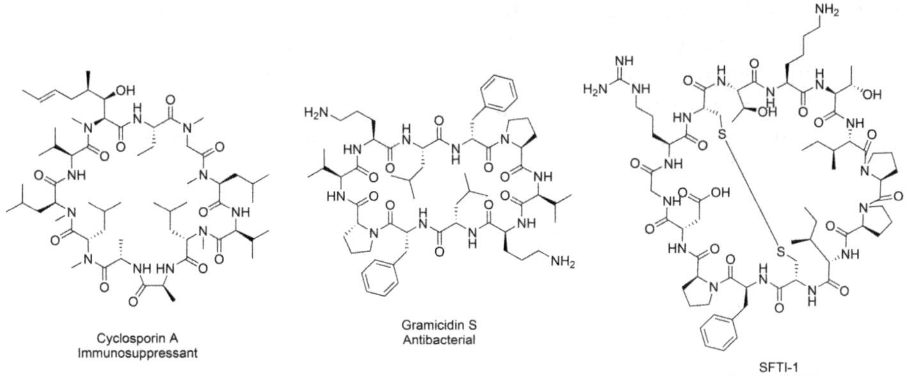

Figure 9.1 Examples of head-to-tail cyclic peptide natural products and their respective bioactivity.

The strategies for library generation discussed in this chapter theoretically allow for almost unlimited diversity of cyclic peptides; however, in practice the ability to resolve hits from the rest of the library becomes the limiting factor in determining its size. With increasing library complexity, the means by which a hit can be distinguished reliably becomes more challenging, and as such all approaches will be considered in the context of their deconvolution strategy. A general theme when discussing these methods of library generation is that those which offer more reliable deconvolution can reach larger sizes.

This chapter will present current methods for library development of head-to-tail cyclic peptides, which will be broadly divided into synthetic and genetically derived categories. Each approach will be discussed in terms of the preparation, application, deconvolution and functional diversity of the libraries generated.

9.2 Chemically Synthesized Libraries

9.2.1 Synthesis and Deconvolution of Diverse Linear Peptide Libraries

The synthesis of diverse peptide libraries has been possible since the development of solid-phase peptide synthesis (SPPS), which allows for controlled, step-wise extension of the peptide sequence *via* successive coupling and deprotection reactions, building from a solid support.[11] Peptide library synthesis has since developed into a mature field that brings together SPPS and combinatorial chemistry methods.[12–14]

A key advantage of a chemical synthetic approach toward peptide libraries is the high compound diversity, since amino acids with non-proteinogenic side chains and even non-amino acid moieties may be introduced, assuming that they are compatible with the SPPS workflow. Synthetic libraries can therefore cover a large region of chemical space; however, it should be noted that many of these building blocks are significantly more expensive than the 20 proteinogenic amino acid set, which may limit their use within a practical context.

The simplest approach to library synthesis is the parallel synthesis of peptides, where a unique position in a microplate well or on a pin is treated with different amino acids during each step. Peptides are then screened either in separate reaction wells or on a solid support. Peptide microarrays in particular utilize this solid support format, whereby peptides are site-specifically immobilized on a surface to which a single application of assay reagents may be applied, enabling rapid screening.[15] Importantly, the spatial isolation of peptides in these formats allows for extremely simple hit identification, as positive results can be characterized according to their respective positions. The simplicity of deconvolution, however, comes at the cost of small library size, as the number of positions exponentially increases and becomes rapidly unmanageable with the addition of more variable sites and amino acid derivatives in the sequence.

A convenient method of ensuring larger libraries is split-and-pool synthesis. At each coupling stage the solid support is divided into separate portions (splitting), to each of which is coupled a different amino acid. This is followed by recombination of these portions (pooling) into one mixture. The process is repeated until the desired chain length is reached, generating a highly diverse set of linear peptide sequences. The synthesis has the advantage of a more straightforward workflow than that of parallel synthesis, but as the peptides are present as a mixture, more advanced deconvolution strategies must be employed. Following screening, hits can be identified by mass spectrometry, with the connectivity of the amino acids in the sequence deduced either from sequential Edman degradation or tandem mass spectrometry techniques such as MS/MS.

These approaches are effective for the generation of linear peptide libraries, and there exist multiple commercial routes for the acquisition of peptide libraries prepared in this manner. The general combinatorial methods described and the chemistry of amino acid coupling are also applied in the same manner for all cyclic peptide libraries; however, additional considerations must be made to accommodate the cyclization step – particularly in the case of head-to-tail peptides.

9.2.2 Head-to-tail Cyclization of Peptide Libraries

Head-to-tail cyclization requires the formation of an amide bond between the N- and C-terminus of the peptide. In typical SPPS, the C-terminus of the peptide is tethered to the solid support and therefore must be cleaved prior to cyclization. This precludes the possibility of screening these peptides on a solid support and identifying hits by physically isolating active beads. A further complicating factor is the competing reactivity of carboxylic acid or amine-containing side chains with each of the termini under cyclization conditions.

The synthetic solutions to these issues are not insurmountable, with many well-established methods in the literature available for cyclizing peptides.[16] Typically, side chain reactivity is addressed through inclusion of protecting groups that are resistant to the resin cleavage conditions, either with orthogonal reactivity or by utilizing milder cleavages with more labile resins. The protected linear peptide is then cyclized under high dilution to favor intramolecular cyclization over competing processes, such as oligomerization. Following this, the side chains of the cyclized product may then be deprotected. This approach is, however, more suitable for the synthesis of low numbers of individual compounds; the inclusion of additional steps and purification processes adds complexity to the workflow, which during library synthesis can be difficult to monitor and optimize due to the diversity of compounds. To this end, chemistry has been developed specifically with the simplification of the head-to-tail cyclization step during library synthesis in mind.

The primary strategy developed to address this issue is the tethering of an amino acid side chain to the resin, rather than the typical C-terminus,

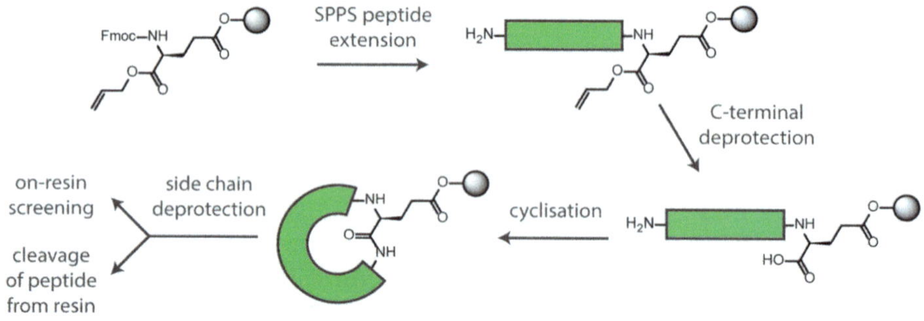

Figure 9.2 On-resin cyclization chemistry using a side chain-tethered glutamic acid residue.

allowing for on-resin cyclization. Performing this step on a solid phase, rather than in solution, is convenient for library synthesis as the cyclization reagents can be washed off the solid support for simple purification.[10,17,18] In addition, the isolation of peptides on separate resin sites also provides a 'pseudo-dilution' effect that favors intramolecular reactions.

The linear peptide is tethered to the resin *via* alternative positions to the C-terminus, typically with a reactive side chain (Asp/Glu, Asn/Gln, Tyr or Lys are commonly used). The C-terminal carboxylic acid is orthogonally protected (allyl or benzyl). After completion of the linear sequence, the C-terminus is then deprotected and the peptide is cyclized on-resin using standard coupling reagents. Side chain protecting groups may then be removed and the peptide can be screened on the solid support, or cleaved from the resin if required (Figure 9.2).

Alternatively, an additional step may be circumvented by simultaneously cyclizing and cleaving the peptide from the resin. This has been achieved by employing 'safety-catch' linkers, whereby the peptide is tethered to the resin by a linker that is inert under SPPS conditions but can be selectively activated following synthesis, initiating cyclization (Figure 9.3).[19–21] This approach generally removes side chain protecting groups prior to safety-catch activation, which has the benefit of accomplishing almost all steps on the solid phase, but may result in side chains such as Lys or Arg reacting instead of the N-terminus during ring closure. In the case of the synthesis of a tyrocidine A library of 192 members, the β-turn induced by a proline residue ensured correct orientation of the amine terminus to achieve selectivity over an exposed ornithine residue.[22]

Tyrocidine A was a scaffold of interest for generating libraries *via* enzymatic cyclization. This strategy made use of a linker that mimics phosphopantetheine, a substrate of the TycC terminal thioesterase (TE) domain, which simultaneously cleaves and cyclizes the resin-bound peptide in a process analogous to that of the natural biosynthetic pathway. This approach was used to generate an *in vitro* library of analogues from synthetic precursors assembled by SPPS in 96-well microplates.[23]

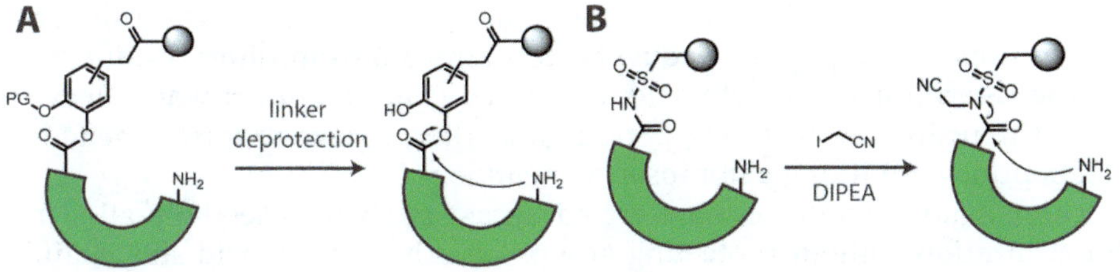

Figure 9.3 Two examples of safety-catch linkers. Approach (A) utilizes a caged-group approach, by protecting the otherwise labile leaving group with either tBu or Bz, whereas (B) requires an activation reagent to initiate cyclization.

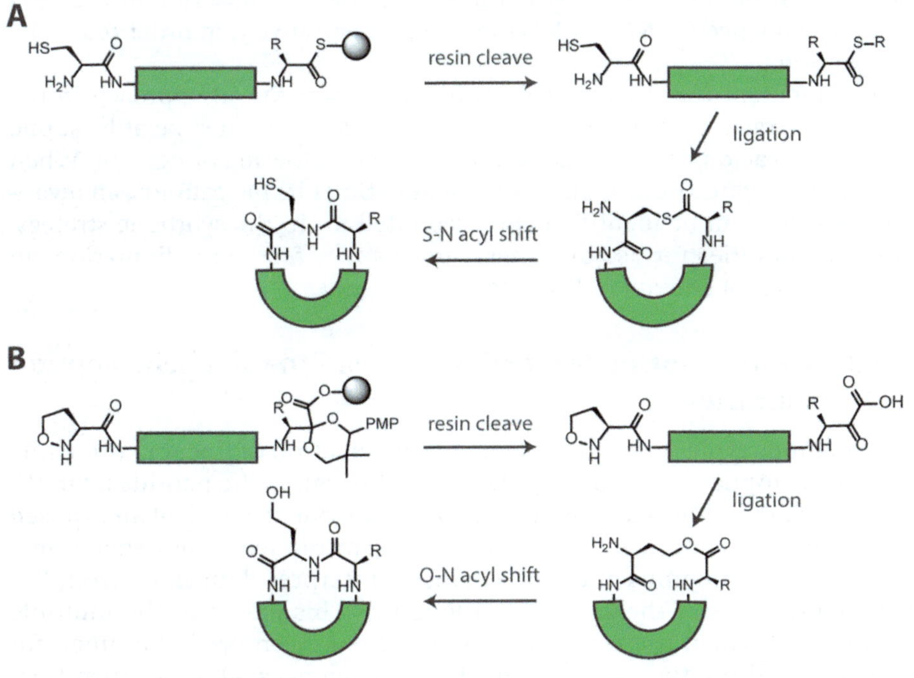

Figure 9.4 Cyclic peptide ligation by (A) native chemical ligation and (B) KAHA ligation.

Finally, ligation-based strategies can be employed to cyclize a peptide that has the appropriate 'sticky-ends' on the N- and C-terminus. Native chemical ligation (NCL) has been used extensively to attach peptide fragments to produce longer sequences that would otherwise be difficult to access *via* SPPS (Figure 9.4A).[24] This approach also finds application in head-to-tail cyclization by incorporating the N-terminal cysteine and C-terminal thioester on the same peptide, for both on-resin and solution phase cyclizations.[25,26]

The α-keto-acid hydroxylamine (KAHA) ligation has also been adapted for use in cyclic peptide synthesis (Figure 9.4B), where a recent iteration of this technique used substrates more suitable for library synthesis. The N-terminus of the peptide sequence contains a 5-oxaproline residue and on the C-terminus an α-keto-acid, generated from the linker when cleaved from the resin. The resulting peptide may then cyclize to form a head-to-tail peptide containing a homoserine residue.[27]

These ligation strategies have the advantage of chemoselectivity, allowing for cyclization without protecting groups on other amino acid side chains. Extensively protected peptides have problematic solubility and cyclization rates and as such these strategies may present convenient methods for reliable production of some of the larger cyclic peptides for libraries. In addition, these ligations occur spontaneously under the correct pH and so minimal purification is required to remove additional reagents. A key requirement, however, is the inclusion of either a cysteine or a homoserine in the final peptide sequence for NCL or KAHA ligation, respectively, in order to enable the ligation.

It is apparent that there currently is no one-size-fits-all approach in the synthetic strategy for cyclization. In many cases, there must be at least one fixed amino acid position to allow for side chain tethering or ligation. When generating peptide libraries derived from existing hits or scaffolds, an invariable position can be identified and accounted for in the synthetic strategy. However, libraries for use in *de novo* screens may face some limitations in that they cannot contain fully randomized peptides.

9.2.3 Deconvolution Strategies for Head-to-tail Cyclic Peptide Libraries

The methods typically used for sequence elucidation of linear peptide hits cannot be applied as simplistically for head-to-tail cyclic peptides. Firstly, Edman degradation sequencing is not possible due to a lack of an exposed N-terminus. Secondly, fragments arising from tandem mass spectrometry are more numerous and challenging to interpret than those from linear counterparts of the same chain length.[28] This is due to the multiple positions at which the ring can be opened – a given cyclic hexamer, for instance, will produce six different linear sequences which in turn fragment further.

A result of this limitation is that historically, many cyclic peptide libraries have been prepared by parallel synthesis to ensure reliable hit determination. A key development in this field, however, is the one bead two compound (OBTC) strategy, whereby a resin bead contains both a displayed cyclic peptide and an accompanying linear 'coding' peptide that can be sequenced by conventional means, ultimately circumventing the difficulties associated with cyclic peptide analysis. A resin bead is segregated between inner and outer layers, which can be primed independently with linear or cyclic precursors, respectively. The cyclic precursor is side chain-tethered to the resin, whereas the linear precursor is coupled to the resin on the C-terminus. The peptide sequences are then coupled to both layers concurrently by split and pool synthesis. Finally, the C-terminal protecting group on the cyclic peptide

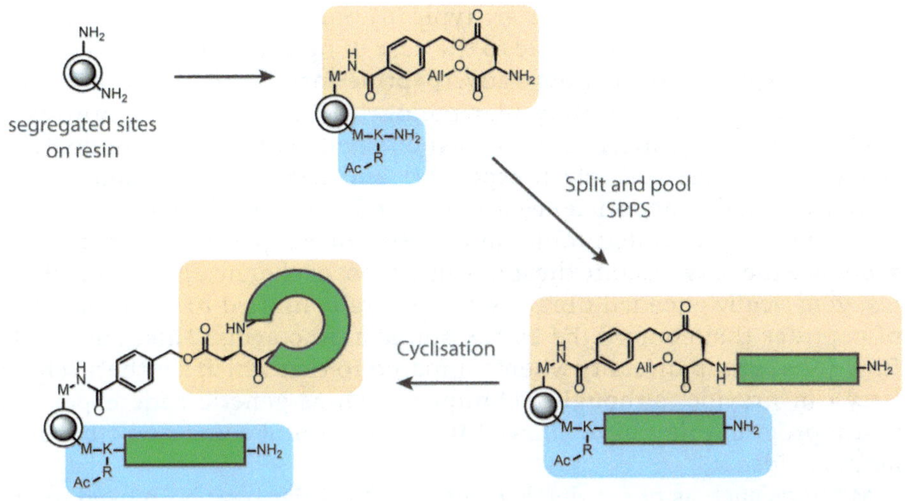

Figure 9.5 One bead two compound library assembly. The final cyclic peptide may be retained on the bead or cleaved prior to screening. The linear peptide contains elements, such as an extra R residue, to aid with MS analysis.

precursor is cleaved and head-to-tail cyclization occurs only on the displayed outer layer of the bead. The cyclic peptide can be retained on the bead for screening or alternatively may be cleaved off selectively for solution-based assays. Once hits are determined, the linear partner can then be recovered for sequencing (Figure 9.5).[29,30]

The comparative ease of deconvolution through OBTC opens up the possibility for much larger chemically synthesized cyclic peptide libraries than those discussed thus far. This method has been used to generate and screen a library containing up to 5 variable positions with 26 amino acids (1.19×10^7 compounds) against the extracellular domain of prolactin, identifying head-to-tail cyclic peptide ligands with low micromolar affinity.[31]

9.3 Genetically Derived Libraries

An alternative to producing chemically synthesized peptide libraries is to utilize ribosomal biosynthesis from living organisms, using DNA sequences that may be transcribed and translated to form a peptide. A DNA library may be formed using a degenerate oligonucleotide that randomizes the DNA sequence encoding the target peptide, which in turn leads to random codons that are translated into different amino acids incorporated into the peptides being expressed, resulting in a peptide library. Translation and transcription of the genetic code into the peptide library can be achieved by either introducing DNA *via* vectors (plasmids, retroviruses *etc.*) into live organisms, such as *E. coli*, or by *in vitro* techniques such as *in vitro* transcription/translation (IVTT), which are mixtures containing purified expression components.

Genetically derived libraries are typically much larger than their chemical counterparts, primarily due to the ease of deconvolution. DNA or RNA that corresponds to each derivative peptide may be amplified and then sequenced by standard molecular biology techniques. The synthesis of DNA sequences themselves is well established, with custom randomized sequences being relatively inexpensive and accessible *via* commercial sources. Finally, the efficiency and specificity of the ribosomal machinery allows for accurate production of the correct peptides regardless of sequence identity. Despite these advantages over chemically derived libraries, genetically encoded libraries are generally limited to the production of peptides that contain the 20 ribosomal amino acids. This reduces the functional group diversity when compared to that which synthetic chemistry can provide, although techniques such as genetic code expansion and reprogramming have allowed incorporation of some non-ribosomal amino acids.

Methods such as phage display and mRNA display have been used to prepare large libraries consisting of over 10^{12} members of linear, and cyclic peptides. The power of these techniques is in the simple deconvolution of hits *via* the display tags attached to each peptide (mRNA, or a phage), but as these tags are bound to the peptide terminus, head-to-tail cyclization is not possible, and these peptides are cyclized *via* their side chains (*e.g.* disulfide bonds). There are, however, a handful of methods for the production of genetically encoded libraries of head-to-tail cyclic peptides.

9.3.1 SICLOPPS

Inteins (internal protein) are protein domains that spontaneously splice out of their surrounding polypeptide during maturation (Figure 9.6A). Upon splicing, the remaining extein (external protein) fragments are ligated *via* a native peptide bond. This natural mechanism for forming an N-to-C connection has been exploited in a technique termed SICLOPPS (Split-Intein Circular Ligation of Proteins and Peptides), whereby the intein and extein sequences derived from *Synechocystis* sp. (*Ssp*) PCC6803 DnaE are inverted, such that the extein is positioned in between fragments of the original intein that has been split (Figure 9.6B).[32,33] Upon association of the split components of the intein, the complex splices, leaving the extein as a head-to-tail cyclized peptide product. By introducing a randomized DNA sequence within the extein-encoding region, a diverse library can be generated.[34–36]

Peptides produced by SICLOPPS must contain either a cysteine or serine residue as the first amino acid to ensure splicing, but there are no other sequence constraints, or requirements for externally administered reagents to enact cyclization. SICLOPPS plasmid libraries are transformed into cells (one member of the plasmid library per cell) and the cellular machinery transcribes and translates the SICLOPPS gene to give inteins that splice, yielding

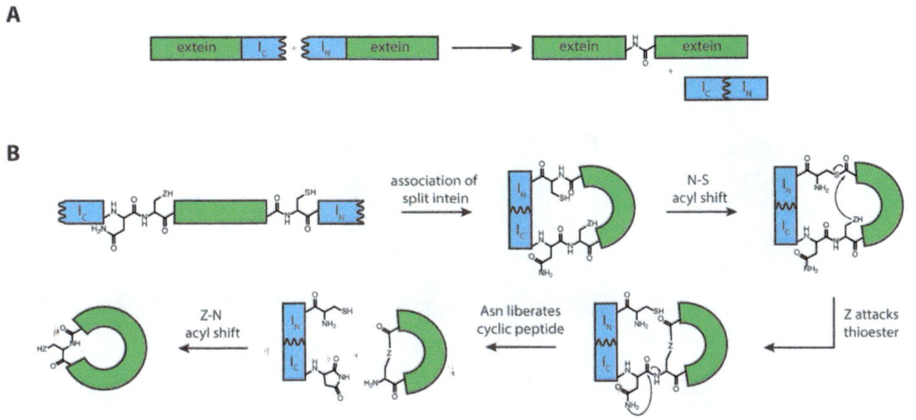

Figure 9.6 Inteins and SICLOPPS. (A) Association of the I_C and I_N components in the natural *Ssp* intein. Upon association, the intein complex is spliced, leaving the ligated extein sequences. (B) Mechanism of SICLOPPS splicing. The inteins are positioned on either side of an extein sequence, which in SICLOPPS encodes for the desired cyclic peptide. Z = O or S.

cyclic peptides in cells. Consequently, SICLOPPS libraries are used in conjunction with phenotypic cellular screens.[37,38]

In a frequently used iteration, SICLOPPS has been combined with a bacterial reverse two-hybrid system (RTHS)[39,40] to provide a high-throughput method of screening and hit identification against a variety of protein–protein interactions.[40–43] The RTHS is grown on selective media that inhibits cell growth unless the reporter gene cassette is expressed, which is prevented by the association of a given PPI (Figure 9.7A). When a cyclic peptide inhibitor is present, the expression of reporter genes is permitted resulting in the formation of a bacterial colony (Figure 9.7B). The compartmentalization of each unique plasmid – and therefore cyclic peptide – within a bacterium means that surviving colonies, derived from a single transformant, express (several copies of) a single active member of the library. The DNA sequence corresponding to this can then be isolated *via* straightforward plasmid extraction followed by DNA sequencing to determine the identity of the hit, overall enabling rapid hit deconvolution (Figure 9.7C).

Inteins are capable of ligating exteins of a variety of lengths – from small cyclic peptides, to the cyclization of proteins such as GFP.[44] It therefore follows that the theoretical size of a SICLOPPS DNA-encoded library is not explicitly constrained by the length of the peptide sequence. In practice, however, the screening platform and subsequent deconvolution process determines library size. As SICLOPPS libraries are generated within live host organisms, the maximum transformation efficiency of the host becomes the limiting factor when incorporating a library. In *E. coli*, for example, this equates to about 10^8 library members. As a result, SICLOPPS libraries are typically designed

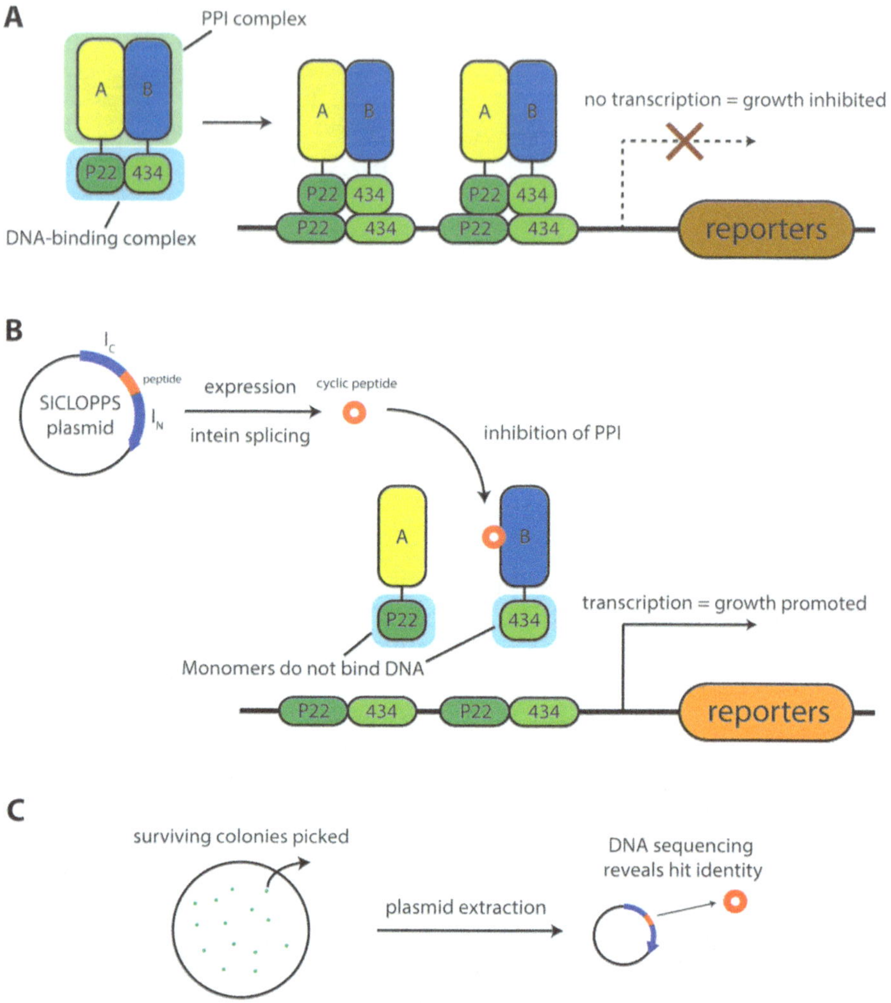

Figure 9.7 SICLOPPS and the bacterial RTHS. (A) RTHS without an inhibitor present. The two proteins in a PPI are each expressed as fusions with P22 or 434, respectively. Upon complex formation, the P22/434 association allows for DNA binding. The protein complex binds to a bacterial cassette and suppresses transcription of downstream reporter genes that are required for survival. (B) RTHS combined with SICLOPPS screening. The split intein protein is expressed from a library of SICLOPPS plasmids, generating cyclic peptides. In the event of inhibition of the PPI, the suppressive DNA binding complex cannot form, allowing expression of reporter genes that aid survival. (C) Deconvolution and hit identification from SICLOPPS screening in the RTHS.

to contain 5 or 6 variable amino acid positions (corresponding to a library size of 3.2×10^6 or 6.4×10^7 members, respectively) to ensure that the library can be fully represented in the screen. Although libraries containing a larger number of randomized positions are certainly possible, it should be noted that only a fraction of the library would be assessed in the screen in these cases.

There are several other examples of combining a SICLOPPS library with phenotypic assays in bacteria. In one example, inhibitors of the protease ClpXP were identified using an assay that linked GFP degradation to ClpXP activity. Hits were then determined using fluorescence-assisted cell sorting (FACS) to identify cells containing GFP stabilized by inhibition of the protease, which were subsequently analyzed by DNA sequencing of the associated SICLOPPS plasmid.[45]

The capacity for SICLOPPS screening in eukaryotic organisms has not been explored as extensively as in bacteria, but there are a handful of examples that demonstrate its potential in these more complex cellular systems. SICLOPPS inteins can be successfully expressed and spliced within human cell lines and yeast; however, in these cases the lower transformation efficiencies of the host limit the library size.[46–48] Examples include identification of cyclic peptides that reduce the toxicity of α-synuclein by utilizing SICLOPPS in yeast,[48] and using human B cells as a host for an SX_4 library of cyclic peptides (1.6×10^5 members) to identify inhibitors of the interleukin-4 (IL-4) signaling pathway.[46]

SICLOPPS libraries currently use *Ssp* inteins, which have variable splicing efficiency depending on the amino acids surrounding the splice junctions. The resulting slower splicing or prevention of splicing will result in some sequences being incompletely processed instead of producing cyclic peptides.[49] These limitations have been recently overcome by using the faster splicing and more residue-tolerant *Npu* split-intein system (derived from *Nostoc punctiforme*) in SICLOPPS.[50] Another interesting development involved the use of orthogonal AARS/$tRNA_{CUA}$ pairs with SICLOPPS to introduce unnatural amino acids into the cyclic peptide libraries. The combined platform was used to identify cyclic peptide inhibitors of HIV protease containing *p*-benzoylphenylalanine or *O*-methyltyrosine as the 21st amino acid building block.[51] It should be noted that a cyclic hexamer identified by SICLOPPS has been successfully converted to a small molecule protein–protein inhibitor, which has shown activity in cells and *in vivo*.[52,53]

9.3.2 Genetically Encoded Cyclic Peptide Library Production *In vitro*

In addition to the chemical and cell-based approaches detailed above for the production of cyclic peptide libraries, there are examples of *in vitro* production of genetically encoded cyclic peptide libraries by mRNA display. This approach uses *in vitro* transcription/translation (IVTT) mixtures, which are either derived from cell lysates, or assembled from recombined components of the transcription/translation machinery (PURE system).[54] This has the benefit of allowing the inclusion of multiple non-natural amino acids into the library by replacing one or more tRNA from the IVTT mixture with variants that have been "mischarged" with a non-natural amino acid (Figure 9.8), a process termed genetic code reprogramming. The most widespread method for this uses a promiscuous ribozyme termed a flexizyme,[55] which charges a diverse range of non-proteinogenic amino acids with varying functionalities onto tRNA *via* this method.[56] This system, pioneered by Suga and colleagues, is named flexible *in vitro* translation (FIT). This approach has been used for the incorporation of a diverse range of functionalities and structures into peptide libraries, and can be used to introduce multiple unnatural amino acids at once.

FIT has been adapted to enable the production of head-to-tail cyclic peptide libraries by insertion of glycolic acid (^{HO}G) in the peptide sequence. A "C-P-(^{HO}G)" residue triplet undergoes a spontaneous N–S acyl shift, followed

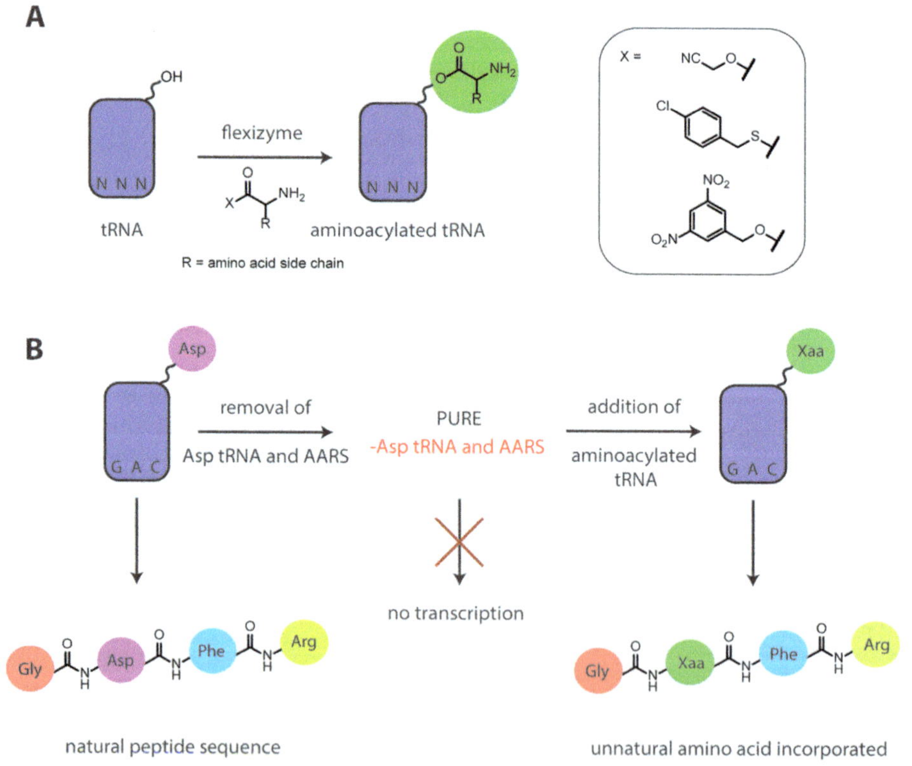

Figure 9.8 The FIT system. (A) Flexizymes are used to charge tRNA with synthetic amino acids containing activated acyl groups indicated by X. (B) Incorporation of unnatural amino acids using the PURE IVTT system. Asp recoding is used as an example.

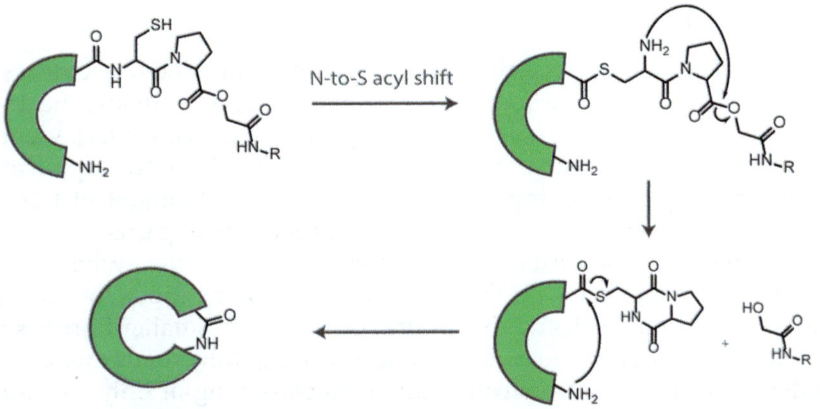

Figure 9.9 Peptide cyclization initiated by rearrangement of the C-P-(^{HO}G) residue triplet.

by formation of a diketopiperazine thioester, which in the presence of an N-terminal amine will cyclize to form the desired cyclic peptide (Figure 9.9). This approach was used to express the natural products SFTI-1 and rhesus theta defensin-1 (RTD-1), as well as analogues that included *N*-methylated and γ-amino acids.[57]

The screening method for these cyclic peptides is an *in vitro* assay performed in microplates, where multiple DNA sequences – and their corresponding cyclic peptides following IVTT – are present per well. The activity in each well is then evaluated and those that contain hits are deconvoluted by cycles of limiting-dilution PCR and assay repetition until the molecule of interest is isolated. This process was demonstrated through the screening of a library of SFTI-1 mutants using three variable amino acid positions ($19^3 = 6859$ compounds).[57]

FIT reaches its full potential when combined with the mRNA display platform Random non-standard Peptide Integrated Discovery (RaPID), whereby the library size can reach $>10^{12}$ members, from which hits can be easily deconvoluted *via* reverse transcription of the affixed mRNA tags.[58] Unfortunately, the nature of head-to-tail cyclic peptide formation precludes mRNA display as these are typically bound to the C-terminus of the peptide, and therefore the screening and identification of hits is less straightforward. Given that the simplicity of deconvolution is the limiting stage of large library screening, FIT-derived head-to-tail cyclic peptide libraries are generally designed to be smaller than what is theoretically possible using fully randomized DNA templates – the upper limit is considered to be 10^5 compounds.[59] Despite this, FIT libraries utilize a more efficient and accessible method of deconvolution than their chemically synthesized counterparts, whilst also offering scope for unnatural amino acid incorporation that is not possible through SICLOPPS.

9.4 Conclusion

Cyclic peptides are of key interest when considering peptide-based drugs, as they intrinsically possess properties that can potentially overcome the disadvantages of linear peptides (such as stability and cell permeability). Identifying cyclic peptide hits against novel targets typically calls for the application of high-throughput screening programs, and the development of libraries and screening platforms are key elements in accomplishing this.

The main constraint pertaining to libraries of head-to-tail cyclic peptides over their linear equivalents is that the deconvolution methods for synthetic libraries are less optimal, and this limitation results in smaller library sizes. A consequence for *de novo* screens is that the set of fully randomized cyclic peptides to be assayed is typically limited to consisting of only 5–6 amino acids. Compared with the tighter binding hits derived from techniques like phage or mRNA display (10–14mers), these smaller peptides will have fewer contacts with their target proteins, which often results in weaker binding; however, the smaller ring sizes will be more readily translated into small molecules.[52] Nonetheless, the novel inhibitors identified from both synthetic and genetic libraries described have provided valuable chemical tools, which inevitably require further optimization for effective use as therapeutic agents. The application of these methods to identify hits for novel targets, including those often considered challenging, such as PPIs and transcription factors, has been well established.

Chemically synthesized and genetically derived cyclic peptides each have their own advantages and disadvantages, and the optimal technique to utilize is dependent on the resources available and the stage of screening. SICLOPPS libraries are simple to generate and generally only require standard molecular biology resources; however, this comes at a cost of diminished coverage of chemical space within the library. Often these are best suited toward early stage screening, using randomized peptides against novel targets in order to discover hits. Techniques like SICLOPPS have been extensively used to identify leads or tool compounds that target otherwise intractable targets.[38] Chemical synthesis on the other hand offers unparalleled diversity, but generating large libraries often requires automation equipment or specialized deconvolution methods, thereby presenting a high barrier of entry for many academic labs. Smaller scale parallel library synthesis is a more common technique that uses relatively accessible equipment, although this is usually applied to optimizing or developing SARs from existing scaffolds. Libraries synthesized *via* techniques like OBTC do, however, bridge the gap between the library sizes derived from genetic and synthetic means.

The methodologies detailed here have been developed to reliably generate libraries of head-to-tail cyclic peptides in response to the emerging interest in this class of compound. The use of these libraries coupled with high-throughput screening forms a strong basis from which new inhibitors may be developed. Importantly, these approaches span a variety of disciplines, covering fields within both chemistry and biology. Consequently, there are multiple entry points for those interested in the field. As interest in 'difficult' drug targets increases, it is anticipated that these library generation methods will be of great utility when developing new tool and therapeutic molecules.

References

1. A. L. Hopkins and C. R. Groom, *Nat. Rev. Drug Discovery*, 2002, **1**, 727–730.
2. P. Vlieghe, V. Lisowski, J. Martinez and M. Khrestchatisky, *Drug Discovery Today*, 2010, **15**, 40–56.
3. K. Fosgerau and T. Hoffmann, *Drug Discovery Today*, 2015, **20**, 122–128.
4. P. K. Madala, J. D. A. Tyndall, T. Nall and D. P. Fairlie, *Chem. Rev.*, 2010, **110**, PR1–PR31.
5. E. M. Driggers, S. P. Hale, J. Lee and N. K. Terrett, *Nat. Rev. Drug Discovery*, 2008, **7**, 608–624.
6. M. Gurrath, G. Muller, H. Kessler, M. Aumailley and R. Timpl, *Eur. J. Biochem.*, 1992, **210**, 911–921.
7. F. Giordanetto and J. Kihlberg, *J. Med. Chem.*, 2014, **57**, 278–295.
8. A. M. White and D. J. Craik, *Expert Opin. Drug Discovery*, 2016, **11**, 1151–1163.
9. P. Thansandote, R. M. Harris, H. L. Dexter, G. L. Simpson, S. Pal, R. J. Upton and K. Valko, *Bioorg. Med. Chem.*, 2014, **23**, 322–327.
10. T. R. White, C. M. Renzelman, A. C. Rand, T. Rezai, C. M. McEwen, V. M. Gelev, R. A. Turner, R. G. Linington, S. S. F. Leung, A. S. Kalgutkar, J. N. Bauman, Y. Zhang, S. Liras, D. A. Price, A. M. Mathiowetz, M. P. Jacobson and R. S. Lokey, *Nat. Chem. Biol.*, 2011, **7**, 810–817.
11. R. B. Merrifield, *J. Am. Chem. Soc.*, 1963, **85**, 2149–2154.
12. R. a. Houghten, C. Pinilla, S. E. Blondelle, J. R. Appel, C. T. Dooley and J. H. Cuervo, *Nature*, 1991, **354**, 84–86.
13. M. A. Gallop, R. W. Barrett, W. J. Dower, S. P. Fodor and E. M. Gordon, *J. Med. Chem.*, 1994, **37**, 1233–1251.
14. K. S. Lam, M. Lebl and V. Krchnák, *Chem. Rev.*, 1997, **97**, 411–448.
15. H. Andresen and F. F. Bier, *Methods Mol. Biol.*, 2009, **509**, 123–134.
16. C. J. White and A. K. Yudin, *Nat. Chem.*, 2011, **3**, 509–524.
17. M. Gonçalves, K. Estieu-Gionnet, G. Laïn, M. Bayle, N. Betz and G. Déléris, *Tetrahedron*, 2005, **61**, 7789–7795.
18. M. C. Alcaro, G. Sabatino, J. Uziel, M. Chelli, M. Ginanneschi, P. Rovero and A. M. Papini, *J. Pept. Sci.*, 2004, **10**, 218–228.
19. L. Yang and G. Morriello, *Tetrahedron Lett.*, 1999, **40**, 8197–8200.
20. J. Ravn, G. T. Bourne and M. L. Smythe, *J. Pept. Sci.*, 2005, **11**, 572–578.
21. G. T. Bourne, J. L. Nielson, J. F. Coughlan, P. Darwen, M. R. Campitelli, D. A. Horton, A. Rhümann, S. G. Love, T. T. Tran and M. L. Smythe, *Peptide Synthesis and Applications*, Humana Press, Totowa, NJ, 2005, pp. 151–166.

22. C. Qin, X. Bu, X. Wu and Z. Guo, *J. Comb. Chem.*, 2003, **5**, 353–355.
23. R. M. Kohli, C. T. Walsh and M. D. Burkart, *Nature*, 2002, **418**, 658–661.
24. P. Thapa, R.-Y. Zhang, V. Menon and J.-P. Bingham, *Molecules*, 2014, **19**, 14461–14483.
25. J. A. Camarero, G. J. Cotton, A. Adeva and T. W. Muir, *J. Pept. Res.*, 2009, **51**, 303–316.
26. J. Tulla-Puche and G. Barany, *J. Org. Chem.*, 2004, **69**, 4101–4107.
27. F. Rohrbacher, G. Deniau, A. Luther and J. W. Bode, *Chem. Sci.*, 2015, **6**, 4889–4896.
28. J. E. Redman, K. M. Wilcoxen and M. R. Ghadiri, *J. Comb. Chem.*, 2003, **5**, 33–40.
29. S. H. Joo, Q. Xiao, Y. Ling, B. Gopishetty and D. Pei, *J. Am. Chem. Soc.*, 2006, **128**, 13000–13009.
30. Y.-U. Kwon and T. Kodadek, *Chem. Commun. (Camb.)*, 2008, 5704–5706.
31. T. Liu, S. H. Joo, J. L. Voorhees, C. L. Brooks and D. Pei, *Bioorg. Med. Chem.*, 2009, **17**, 1026–1033.
32. H. Wu, Z. Hu and X. Q. Liu, *Proc. Natl. Acad. Sci. U. S. A.*, 1998, **95**, 9226–9231.
33. C. P. Scott, E. Abel-Santos, M. Wall, D. C. Wahnon and S. J. Benkovic, *Proc. Natl. Acad. Sci. U. S. A.*, 1999, **96**, 13638–13643.
34. C. P. Scott, E. Abel-Santos, A. D. Jones and S. J. Benkovic, *Chem. Biol.*, 2001, **8**, 801–815.
35. A. Tavassoli and S. J. Benkovic, *Nat. Protoc.*, 2007, **2**, 1126–1133.
36. E. L. Osher and A. Tavassoli, *Methods in Molecular Biology (Clifton, N.J.)*, 2017, vol. 1495, pp. 27–39.
37. K. R. Lennard and A. Tavassoli, *Chem.–Eur. J.*, 2014, **20**, 10608–10614.
38. A. Tavassoli, *Curr. Opin. Chem. Biol.*, 2017, **38**, 30–35.
39. A. R. Horswill, S. N. Savinov and S. J. Benkovic, *Proc. Natl. Acad. Sci. U. S. A.*, 2004, **101**, 15591–15596.
40. A. Tavassoli and S. J. Benkovic, *Angew. Chem., Int. Ed.*, 2005, **44**, 2760–2763.
41. A. Tavassoli, Q. Lu, J. Gam, H. Pan, S. J. Benkovic and S. N. Cohen, *ACS Chem. Biol.*, 2008, **3**, 757–764.
42. E. Miranda, I. K. Nordgren, A. L. Male, C. E. Lawrence, F. Hoakwie, F. Cuda, W. Court, K. R. Fox, P. a Townsend, G. K. Packham, S. a Eccles and A. Tavassoli, *J. Am. Chem. Soc.*, 2013, **135**, 10418–10425.
43. C. N. Birts, S. K. Nijjar, C. A. Mardle, F. Hoakwie, P. J. Duriez, J. P. Blaydes and A. Tavassoli, *Chem. Sci.*, 2013, **4**, 3046.
44. H. Iwai, A. Lingel and A. Pluckthun, *J. Biol. Chem.*, 2001, **276**, 16548–16554.
45. L. Cheng, T. A. Naumann, A. R. Horswill, S.-J. Hong, B. J. Venters, J. W. Tomsho, S. J. Benkovic and K. C. Keiler, *Protein Sci.*, 2007, **16**, 1535–1542.
46. T. M. Kinsella, C. T. Ohashi, A. G. Harder, G. C. Yam, W. Li, B. Peelle, E. S. Pali, M. K. Bennett, S. M. Molineaux, D. A. Anderson, E. S. Masuda and D. G. Payan, *J. Biol. Chem.*, 2002, **277**, 37512–37518.
47. I. N. Mistry and A. Tavassoli, *ACS Synth. Biol.*, 2017, **6**, 518–527.

48. J. A. Kritzer, S. Hamamichi, J. M. McCaffery, S. Santagata, T. A. Naumann, K. A. Caldwell, G. A. Caldwell and S. Lindquist, *Nat. Chem. Biol.*, 2009, **5**, 655–663.
49. H. Iwai, S. Züger, J. Jin and P. H. Tam, *FEBS Lett.*, 2006, **580**, 1853–1858.
50. J. E. Townend and A. Tavassoli, *ACS Chem. Biol.*, 2016, **11**, 1624–1630.
51. T. S. Young, D. D. Young, I. Ahmad, J. M. Louis, S. J. Benkovic and P. G. Schultz, *Proc. Natl. Acad. Sci. U. S. A.*, 2011, **108**, 11052–11056.
52. I. B. Spurr, C. N. Birts, F. Cuda, S. J. Benkovic, J. P. Blaydes and A. Tavassoli, *ChemBioChem*, 2012, **13**, 1628–1634.
53. D. J. Asby, F. Cuda, M. Beyaert, F. D. Houghton, F. R. Cagampang and A. Tavassoli, *Chem. Biol.*, 2015, **22**, 838–848.
54. Y. Shimizu, A. Inoue, Y. Tomari, T. Suzuki, T. Yokogawa, K. Nishikawa and T. Ueda, *Nat. Biotechnol.*, 2001, **19**, 751–755.
55. H. Saito, D. Kourouklis and H. Suga, *EMBO J.*, 2001, **20**, 1797–1806.
56. Y. Goto, T. Katoh and H. Suga, *Nat. Protoc.*, 2011, **6**, 779–790.
57. T. Kawakami, A. Ohta, M. Ohuchi, H. Ashigai, H. Murakami and H. Suga, *Nat. Chem. Biol.*, 2009, **5**, 888–890.
58. T. Passioura and H. Suga, *Chem. Commun.*, 2017, **53**, 1931–1940.
59. T. Katoh, Y. Goto, M. S. Reza and H. Suga, *Chem. Commun.*, 2011, **47**, 9946.

CHAPTER 10

An Introduction to Bacterial Lasso Peptides

JULIAN D. HEGEMANN*[a,b] AND MOHAMED A. MARAHIEL*[a,b]

[a]Department of Chemistry, Biochemistry, Philipps-University Marburg, Hans-Meerwein-Strasse 4, Marburg, Germany; [b]LOEWE-Center for Synthetic Microbiology, D-35032, Marburg, Germany
*E-mail: marahiel@staff.uni-marburg.de; jdhegemann@googlemail.com

10.1 An Introduction to Bacterial Lasso Peptides

Amongst RiPPs, lasso peptides comprise an intriguing class of cyclic peptides with a unique fold in Nature.[1–3] Fundamentally, they consist of a macrolactam ring that is threaded by the C-terminal linear peptide tail. The ring itself is formed by condensation of the N-terminal α-amino group with a carboxyl side chain. The carboxyl group used for this isopeptide bond formation typically belongs to an aspartate or glutamate residue, which is located at position 7, 8 or 9 of the peptide scaffold.[3,4]

As their threaded structures are reminiscent of lariat knots, these natural products were named lasso peptides. So far inaccessible *via* chemical synthesis, the lasso fold is merely stabilized by steric interactions of bulky amino acid side chains positioned in the C-terminal region above and below the ring.[1–3] These residues are therefore often referred to as plugs or plug amino acids. The size needed for a side chain to be suitable as a plug directly depends on the size of the present macrolactam ring. While the small seven-residue rings can utilize every amino acid bigger than serine

Chemical Biology No. 6
Cyclic Peptides: From Bioorganic Synthesis to Applications
Edited by Jesko Koehnke, James Naismith and Wilfred A. van der Donk

Published by the Royal Society of Chemistry, www.rsc.org

for stabilizing the lasso fold,[4] larger nine-residue rings can only use the largest amino acids (*e.g.* phenylalanine, tyrosine, tryptophan or arginine) for maintenance of their fold.[5–8]

All lasso peptides can be differentiated into three classes.[1–3,9] Numbered based on the order of their discovery, they are distinguished by the presence and number of disulfide bonds. Class I lasso peptides contain a total of four cysteines that form two disulfide bridges. One of these cysteines is always found at the first position of the peptide. Class II lasso peptides, on the other hand, completely lack disulfide bonds, while the so far only member of class III features a single disulfide bond. In further contrast to class I lasso peptides, the class III lasso peptide, BI-32169, has no cysteine at position 1.[9,10] Though the disulfide bridges in class I and III lasso peptides constitute an additional covalent stabilization of their structures, these compounds also contain suitable plug amino acids and therefore remain in their fold upon reductive opening of the disulfide bonds.[11] The majority of investigated lasso peptides belong to class II. Therefore, any lasso peptide that is discussed in this chapter should be assumed to belong to class II, unless stated otherwise.

To allow an overview of the diversity found in the lasso peptide framework, a selection of lasso peptides from all available three-dimensional structures is depicted in Figure 10.1.

Lasso peptides are still a rather young family of natural products and their first representative, anantin, was discovered in 1991.[3,18] From this point onward until 2008, all further lasso peptide isolations were driven by chance and only compounds identified in activity-based approaches were reported.[3] After the description of anantin, it took eight years until the first sequence of a lasso peptide biosynthetic gene cluster was published. This cluster, found on a plasmid of *Escherichia coli* AY25, is responsible for the production of the lasso peptide microcin J25 (MccJ25) and represents the best investigated lasso peptide system.[3,19] The cluster consists of four genes, organized in two operons. The precursor peptide-encoding gene *mcjA* has its own promoter, while the three genes *mcjB*, *mcjC* and *mcjD* all share another promoter and are oriented in the opposite coding direction to *mcjA*. Remarkably, this is the only known lasso peptide biosynthetic gene cluster organized in this way, as in all other reported clusters the genes involved are situated in a single operon with a single promoter.[2,3] Still, the MccJ25 system shows all the basic principles every lasso peptide biosynthetic machinery has in common. The gene *mcjA* encodes a short precursor peptide (58 residues), which in turn is processed by the enzymes McjB and McjC into the mature lasso peptide (21 residues).[1–3,20] Furthermore, McjD is an ATP-binding cassette (ABC) transporter, which exports the antimicrobial MccJ25 and thereby confers immunity to the producing organism.[21] Thus, unlike homologs of McjA, McjB and McjC, a D protein is not essential for biosynthesis and therefore many lasso peptide producing gene clusters are devoid of a gene coding for a dedicated lasso peptide export protein.[2,3,22–24]

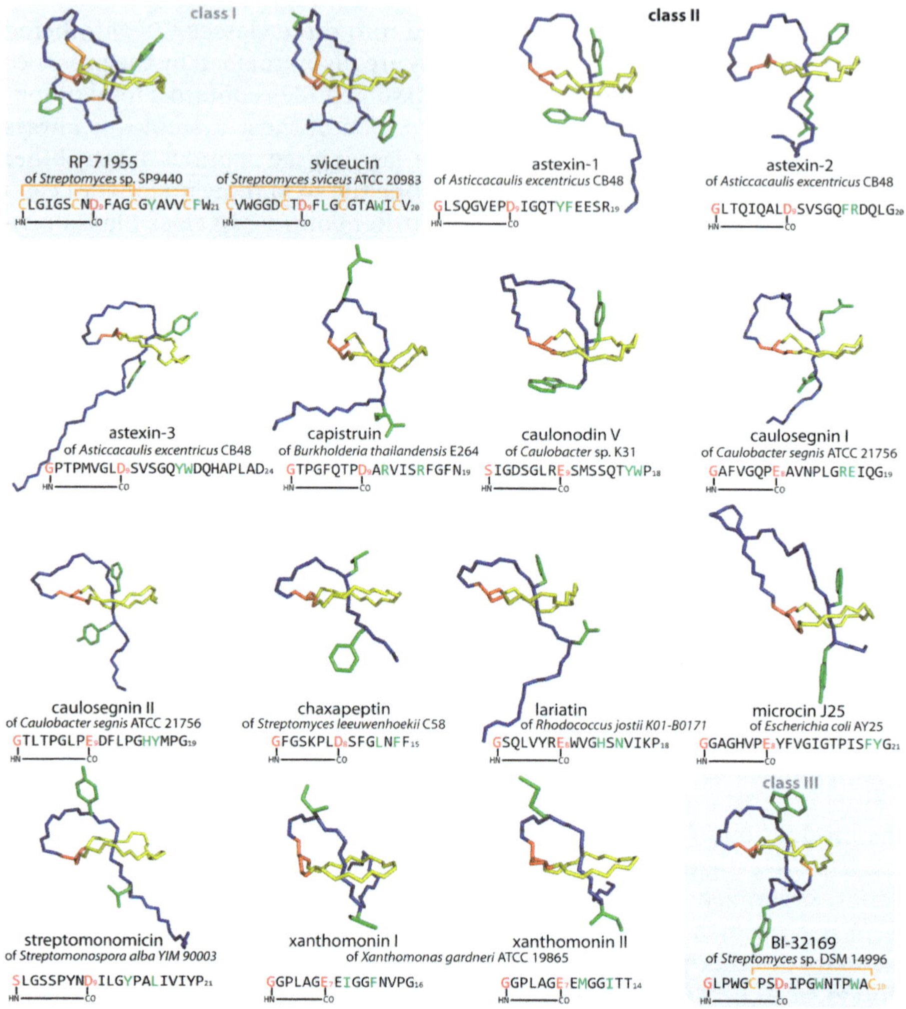

Figure 10.1 Overview of reported and accessible lasso peptide structures. The common color code used depicts the macrolactam rings in yellow, the tail in blue and the ring-forming aspartate or glutamate residues in red. Additionally, plug residues are highlighted in green and disulfide bridges in orange. The corresponding PDB accession codes are: 1RPB (RP 71955),[12] 2LS1 (sviceucin),[11] 2M37 (astexin-1),[6] 2N6U (astexin-2),[13] 2N6V (astexin-3),[13] 2MLJ (caulonodin V),[7] 2LX6 (caulosegnin I),[5] 5D9E (caulosegnin II),[8] 2N5C (chaxapeptin),[14] 1Q71 (microcin J25),[15] 2MW3 (streptomonomicin),[16] 4NAG (xanthomonin I),[4] 2MFV (xanthomonin II)[4] and 3NJW (BI-32169).[10] The BRMB number of capistruin is 20014.[17]

The exact mechanism of lasso peptide maturation is still not completely elucidated, due to prevalent problems with the heterologous production and purification of the processing enzymes. Nevertheless, in a recent study some *in vitro* results were obtained for McjB and McjC.[20,25] The most important results from these investigations were that apparently McjB and McjC are interdependent and that McjB requires ATP for its activity. The interdependency was first suggested by the observation that neither McjB nor McjC could act on its own, while single enzyme activities could be probed when either enzyme was incubated with an inactive variant of the other one.[20] Generally, the C proteins are homologous to aspartate synthetases and act in a similar manner, *i.e.* instead of an ammonia molecule, they transfer the N-terminal α-amino group to the side chain of an aspartate or glutamate. Before the transfer can occur, these side chains need to be activated by ATP-dependent adenylation of the carboxylic acid moieties. The B proteins are ATP-dependent cysteine proteases that cleave off the leader sequence of the precursor peptide and thereby free the N-terminal α-amino group of the core peptide for the C protein-mediated macrocyclization. Closer inspection revealed that these proteases, in particular, hydrolyze ATP into ADP and inorganic phosphate, as adenosine 5′-(β,χ-imido)triphosphate (an ATP analogue with a non-hydrolyzable bond between ADP and the last phosphate group) renders McjB inactive.[20] In a similar manner, α,β-methyleneadenosine 5′-triphosphate (another ATP analogue with a non-hydrolyzable bond between AMP and the pyrophosphate group) inhibits McjC activity.[20] While the ATP-dependency of McjC was anticipated due to its homology to asparagine synthetases, the ATP-dependency of McjB was quite surprising. As the proteolytic activity should not require ATP-consumption, an additional activity is proposed for McjB. Namely, it is thought that the ATP-consumption facilitates a further chaperone-like activity that mediates the prefolding of the core peptide before the macrocyclization.[1–3,20,23] This is necessary, as the same factors that sterically maintain the lasso fold after maturation of course also prohibit the threading to occur post-cyclization. A proposed mechanism for MccJ25 biosynthesis that is derived from this experimental data is shown in Figure 10.2. It should be noted that recent studies have suggested that the ATP-dependency of the B protein might not be present in all lasso peptide biosynthetic machineries.[26,27]

Interestingly, there are other lasso peptide biosynthesis systems that feature an additional small open reading frame (ORF) that was shown to be necessary for precursor peptide processing *in vivo*. First described for the gene cluster of lariatin,[28] additional examples have been reported recently (chaxapeptin,[14] lassomycin,[29] paeninodin,[30] streptomonomycin[16] and sviceucin (which is a class I lasso peptide)[11]). What was first thought to be a novel enzyme in the context of lasso peptide maturation, turned out, upon closer bioinformatic investigation, to be just a small protein homologous to the N-terminal domain of McjB that is subsequently followed by

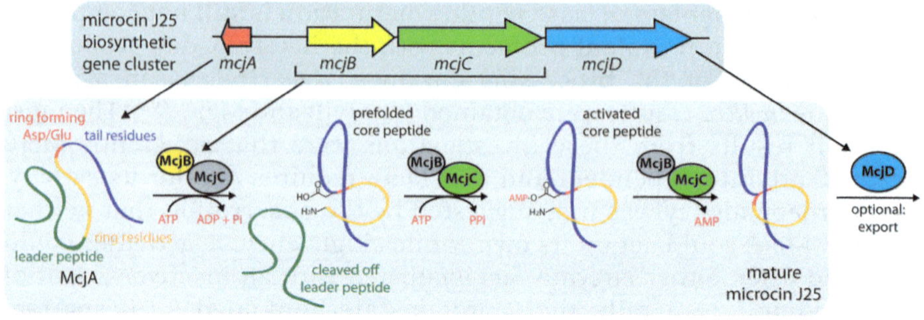

Figure 10.2 Proposed mechanism for the biosynthesis of the lasso peptide MccJ25.[1–3,20,23] Initially, the B protein cleaves off the leader peptide and prefolding of the core peptide occurs (possibly in an ATP-dependent manner). Then, the C protein activates the carboxylic acid side chain *via* adenylation and subsequently mediates the macrocyclization yielding the mature lasso peptide. The optional export of the mature compound by the D protein is also depicted.

an ORF coding for a homolog of the C-terminal portion of a full-length B protein.[3,11,14,23,24,30,31] Hence, these two enzymes are referred to as split B proteins, and therefore were named B1 and B2. While the larger B2 proteins contain the catalytic triads and thus will also mediate the proteolytic cleavage of the leader peptides, the smaller B1 proteins feature the conserved N-terminal motif of full-length B proteins. Although the concrete function of this motif has not been elucidated yet, it was suggested that it contains a site needed for interaction with ATP and therefore would mediate the proposed chaperone-like activity. This is further suggested by the fact that B1 proteins show homology to PqqD.[1,3,30–33] PqqD is another small protein of similar size to B1 proteins that has no catalytic activity on its own, but was shown to be essential for the biosynthesis of the RiPP pyrroloquinoline quinone (PQQ). Hence, this B protein region and the corresponding B1 proteins could fulfill the same or at least a similar role in lasso peptide biosynthesis. Intriguingly, a recent study inspected all known RiPP systems bioinformatically and found that these PqqD-like domains or proteins are present in a large number of them.[31] Further experiments revealed that these domains show a high affinity to the leader regions of the corresponding precursor peptides.[31] Hence, it was suggested that these domains are crucial for recognition of the leader sequences and that they mediate the interactions between precursor peptides and processing enzymes.[31,33] Thus, the term RiPP recognition element (RRE) has been introduced for these domains and proteins.[31] More recent studies have put the general ATP-dependency of B proteins and the interdependency of B and C proteins in doubt. During investigations of lasso peptide–protein fusions, it was shown that co-expression of the astexin-1 B protein and the corresponding precursor peptide fused to the N-terminus of a leucine zipper protein led to complete removal of the leader peptide even in the absence of a C protein.[26] In another study it was shown that proteolysis of the precursor peptide could be achieved *in vitro* with only the B1 and B2 proteins of the paeninodin system present.[27] Furthermore, the proteolysis happened in the absence of ATP and the reaction rate was not increased by ATP addition.[27] Future studies need to shed more light on these matters and will help to figure out if the MccJ25 system could be a special example with a slightly different underlying biosynthetic mechanism.

In addition to the MccJ25 gene cluster (which features a so far unique architecture) and the fact that there are operons encoding split B proteins, many more variations of lasso peptide biosynthetic gene clusters have been reported.[2,3,7,22–24,30] Fundamentally, proteo- and actinobacterial lasso peptide producing gene clusters can be distinguished by the order of the ORFs in their operons. While proteobacteria seem to have basic A B C organizations, actinobacteria switch the order of the processing enzyme encoding genes to A C B.[2,3,7,11,22–24] In the case of a split B protein, the ORFs encoding B1 and B2 are always adjacent to each other and always in the order B1 B2. Further differentiation is achieved by the presence or absence of accessory genes. An ORF encoding an McjD homolog is often found in clusters that produce antimicrobially active lasso peptides. Proteobacterial clusters lacking a D protein homolog normally contain no additional genes in the lasso peptide operons, but seem to exclusively appear next to another highly conserved cluster facing in the opposite direction. This neighboring cluster contains up to four genes, which encode proteins annotated as σ-factors, anti-σ-factors, TonB-dependent receptors and peptidases.[5–7,22–24] The occurrence of these operons next to the lasso peptide gene clusters does not appear to be coincidental, as studies have shown that the putative peptidases are indeed highly specific lasso peptide isopeptidases.[13,24,34,35] These serine proteases cleave only the isopeptide bond in their substrate lasso peptides and thereby linearize them. Additionally, they have a high affinity for the lasso peptides of the adjacent biosynthetic gene clusters and act with a very high selectivity for their targets.[13,24,34] The crystal structures of the astexin-2/3 isopeptidase (AtxE2) and of the sphingopyxin I isopeptidase (SpI-IsoP) have been elucidated, showing that these enzymes have a prolyl-oligopeptidase-like structure featuring a unique broken β-propeller fold of their recognition domain.[34,35] The co-crystal structure of astexin-3 and AtxE2 furthermore showed that the protease recognized the fold of the lasso peptide substrate rather than specific residues in its sequence.[35]

Other putative clusters were identified *via* genome mining approaches that also encode novel types of proteins, whose functions are not yet known.[3,23,30] Possibly at least some of them could act as tailoring enzymes that introduce additional chemical modifications to the lasso scaffolds. One likely example for this is the putative *O*-methyltransferase found in the lassomycin cluster.[29] Though not experimentally proven, the fact that lassomycin features a methyl ester group instead of the usually free C-terminus suggests that this modification was introduced enzymatically. Another recent example is the kinase found in the paeninodin gene cluster, which phosphorylates the side chain of the C-terminal serine of the precursor peptide.[30,36] This modified precursor peptide is then processed by the B1, B2 and C proteins into a phosphorylated lasso peptide. Even though the paeninodin cluster is so far the only characterized lasso peptide operon from a firmicute, a number of homologous clusters identified in *Bacillus* and *Paenibacillus* spp., as well as some putative proteobacterial clusters, have been reported that also feature a gene encoding a homologous kinase, which was shown to have a similar function.[23,30,36] Interestingly, the identified close-by precursor genes exclusively encode peptides with C-terminal serines, which further supports this notion.[30,36] To give an overview of the diversity in lasso peptide biosynthetic gene clusters, examples of all types of reported, functional operons are shown in Figure 10.3.

As stated earlier, lasso peptide discovery was limited to activity-driven compound isolations until 2008. In this specific year, a genome mining-based approach was employed for the first time for the isolation of a before unknown lasso peptide, namely capistruin, from *Burkholderia thailandensis* E264.[17] From this point onward, genome mining-based approaches became the predominant method for isolating new representatives of this RiPP family and today the lasso peptides discovered this way far outnumber the ones isolated by other approaches.[3]

Generally, there are two major genome mining methods that have been commonly employed in recent years, both having certain advantages. One is the B protein-centric genome mining approach, which utilizes the fact that there are no known close homologs to the B proteins outside of lasso peptide biosynthetic gene clusters.[3,23] Hence, a simple search for homologs

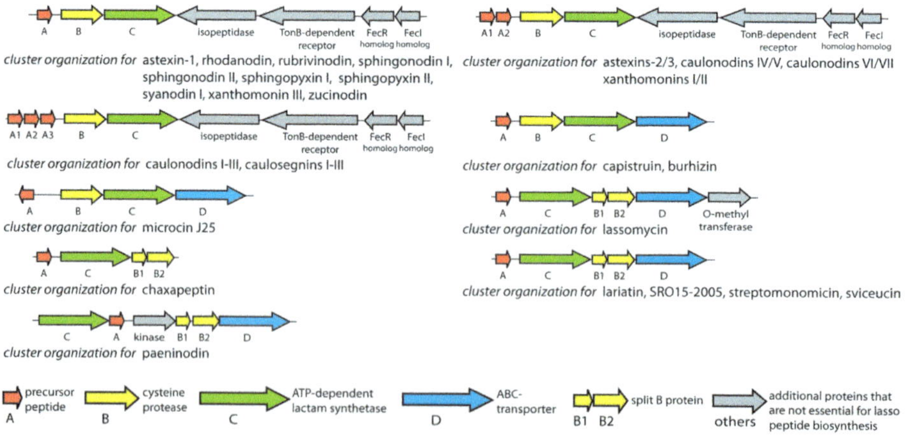

Figure 10.3 Schematic overview of all reported lasso peptide biosynthetic gene clusters that have been experimentally shown to be functional.

of a B protein sequence is performed and followed up by a manual search for C protein homologs and suitable precursor peptide-encoding genes in close proximity on the genome. Alternatively, an algorithm has been developed to analyze genomic data banks for the presence of small ORFs encoding suitable lasso peptide precursors.[2,22] Again, the identification of a potential hit is followed by checking for the presence of genes encoding potential processing enzymes in close genomic proximity. While in the precursor-centric approach the identification and annotation of the small ORFs is automated, this often has to be done manually in the other approach. Still, the B-protein centric approach has the advantage that it also allows the identification of putative lasso peptide biosynthetic gene clusters with unusual precursor peptides, which would be overlooked by the algorithm, as these precursor peptides would not fit into the characteristics screened for.[7,16,30] Figure 10.4 shows a graphic comparison of both genome mining approaches.

In addition, there is a rather elegant way to screen for the production of a lasso peptide in a specific organism. While this method was developed to scan for the general presence of secondary metabolites, it was also successfully employed to identify two novel lasso peptides.[37] In short, colony-MALDI MS is combined with automatic fragmentation of major MS peaks. Based on the fragments, possible peptide sequences are predicted and used as a basis

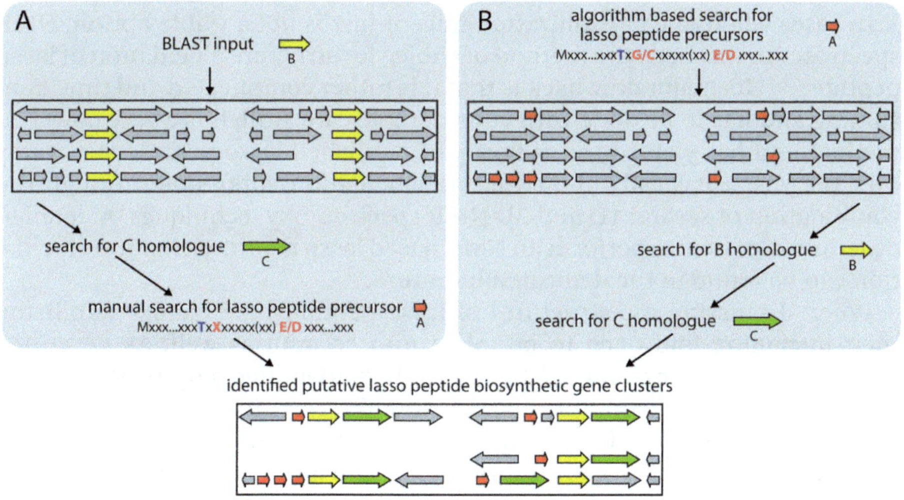

Figure 10.4 A graphical overview of the (A) B protein-centric[23] and (B) precursor-centric[22] genome mining approaches. The latter illustrates the criteria used in the original publication on the precursor-centric genome mining approach (threonine at position -2, glycine or cysteine at position 1 and a glutamate or aspartate in a suitable distance to it).[22] For subsequent genome mining studies, this algorithm could be updated to take most recent findings into account, *e.g.* position 1 residues different to glycine and cysteine.

for a bioinformatic-facilitated screen for a potential precursor peptide-encoding gene in the corresponding genome and thereby for identifying the complete biosynthetic gene cluster.

10.2 Investigation of Lasso Peptide Structures

As stated above, the core motif of lasso peptide structures is the threaded macrolactam ring, which is stabilized by the plug residues in the tail region. Their three-dimensional structures can be elucidated by NMR spectroscopy and crystallographic methods.[3,4,8,10,38]

The low number of currently available lasso peptide crystal structures may indicate a low success rate for such endeavors.[4,8,10,34] As their size is somewhere in between small organic molecules and small proteins, the reported instances of lasso peptide crystallizations do not necessarily follow a typical approach for either compound class. While sufficiently water soluble lasso peptides can be directly applied in protein crystallization screens using a near saturated stock solution in water,[4,34] stock solutions of more hydrophobic compounds are prepared in organic solvents or water–solvent mixtures.[8,10] In the latter case, the dilution of the solvent stock under crystallization conditions and possible hygroscopy (*e.g.* when using DMSO) will decrease the overall solvent content and thereby the solubility of a hydrophobic lasso peptide in the mixture, and thus can facilitate crystallization.

In cases where the crystallization fails or just is not a viable option, NMR spectroscopy has been the method of choice for structure elucidation of lasso peptides.[3,38] Its major drawback is that it is rather complicated and time-consuming and has to be done with considerable care. Still, it is a very powerful method and has yielded the majority of published lasso peptide structures over the years. Basically, the structure elucidation is possible through the combination of several 1D and 2D NMR spectroscopy techniques. A detailed description of how to perform an NMR based lasso peptide structure elucidation can be found in the dedicated literature.[38]

When discussing the structures of lasso peptides, the factors stabilizing their intriguing fold need to be taken into account as well. As described above, the fold is maintained by sterically demanding plug residues positioned in the tail region above and below the ring. In class I and III, further stabilization is achieved by the presence of two or one disulfide bond, respectively. The steric stabilization alone is usually sufficient to stabilize class II lasso peptides under standard conditions, even though they lack any disulfide bridges. Still, it is possible to discriminate between heat-stable and heat-sensitive lasso peptides, depending on their behavior at elevated temperatures.[4–8,13,23,24,30,34] Heat-stable lasso peptides will resist denaturing in solution even at very high temperatures. MccJ25 was reported to even withstand autoclaving at 120 °C without losing any of its antimicrobial activity.[39] On the other hand, heat-sensitive lasso peptides will easily unthread into branched-cyclic peptides when the temperature is increased.[4–8,23,24,30,34,40] The temperature where the unthreading starts can be different for each heat-sensitive lasso peptide, but as a general definition lasso peptides that readily unthread at 95 °C can be described as heat-sensitive. Studies have shown that thermal stability can be dependent on the amino acids used as plugs, with the general trend that the more spacious the side chain, the higher the likelihood of a thermally stable fold.[4,6–8,40] Therefore, the change of a plug residue with a bigger or smaller amino acid can also change how a lasso peptide behaves at elevated temperatures. Nonetheless, in some cases, even the use of the biggest available canonical amino acids cannot confer thermal stability to the lasso peptide, which is mostly observed for compounds with large nine-residue macrolactam rings.[7,8] On the other hand, for the small seven-residue macrolactam rings found in some lasso peptides, every amino acid bigger than serine is suitable to maintain a heat-stable lasso fold.[4] In a recent study, it was further demonstrated that not only the nature of the plug residues and the ring size are important for the thermal stability of a lasso peptide, but that also residues located inside the macrolactam ring can have a crucial effect on the fold stability.[8]

An advantage of dealing with a thermally sensitive lasso peptide is that this characteristic can be exploited to confirm the lasso topology without needing to rely on a complete structure elucidation.[3–8,23,30,34] Generally, the unthreading of a lasso peptide at higher temperatures can be monitored *via* LC-MS. After incubation at a sufficiently high temperature, a new peak will appear in the chromatogram with a different retention time to the lasso peptide, but with identical mass. This observation suggests a non-chemical conversion of topologies, which in this case is the thermally induced unthreading of the lasso into a branched-cyclic peptide. This can be confirmed by utilizing carboxypeptidase Y, a protease that sequentially cleaves off amino acids from the C-terminus of a peptide chain.[3–8,11,23,30,34] When incubated with a lasso peptide, this protease will cleave off no or only very few amino acids below the lower plug. This can be explained by the fact that in a lasso fold, the macrolactam ring will shield a large portion of the C-terminus, thereby rendering it inaccessible for proteolytic digestion. The C-terminus of a branched-cyclic peptide, on the other hand, lacks this kind of protection and hence can be degraded much further by carboxypeptidase Y, in some cases down to the macrolactam ring. Figure 10.5 elaborates on the procedure to confirm the topology of a heat-sensitive lasso peptide by thermal and carboxypeptidase Y treatment.

In addition to carboxypeptidase Y treatment, recent investigations have also shown that ion mobility-mass spectrometry (IM-MS) can be used to efficiently discriminate between a lasso peptide and its branched-cyclic analogue.[30,41] Using sulfolane and formic acid to achieve high charge states (+3 or +4), the repulsion of the positive charges in the molecule will cause the branched-cyclic peptide to adopt a less compact conformation and therefore have a longer drift time. On the other hand, the rigidity and

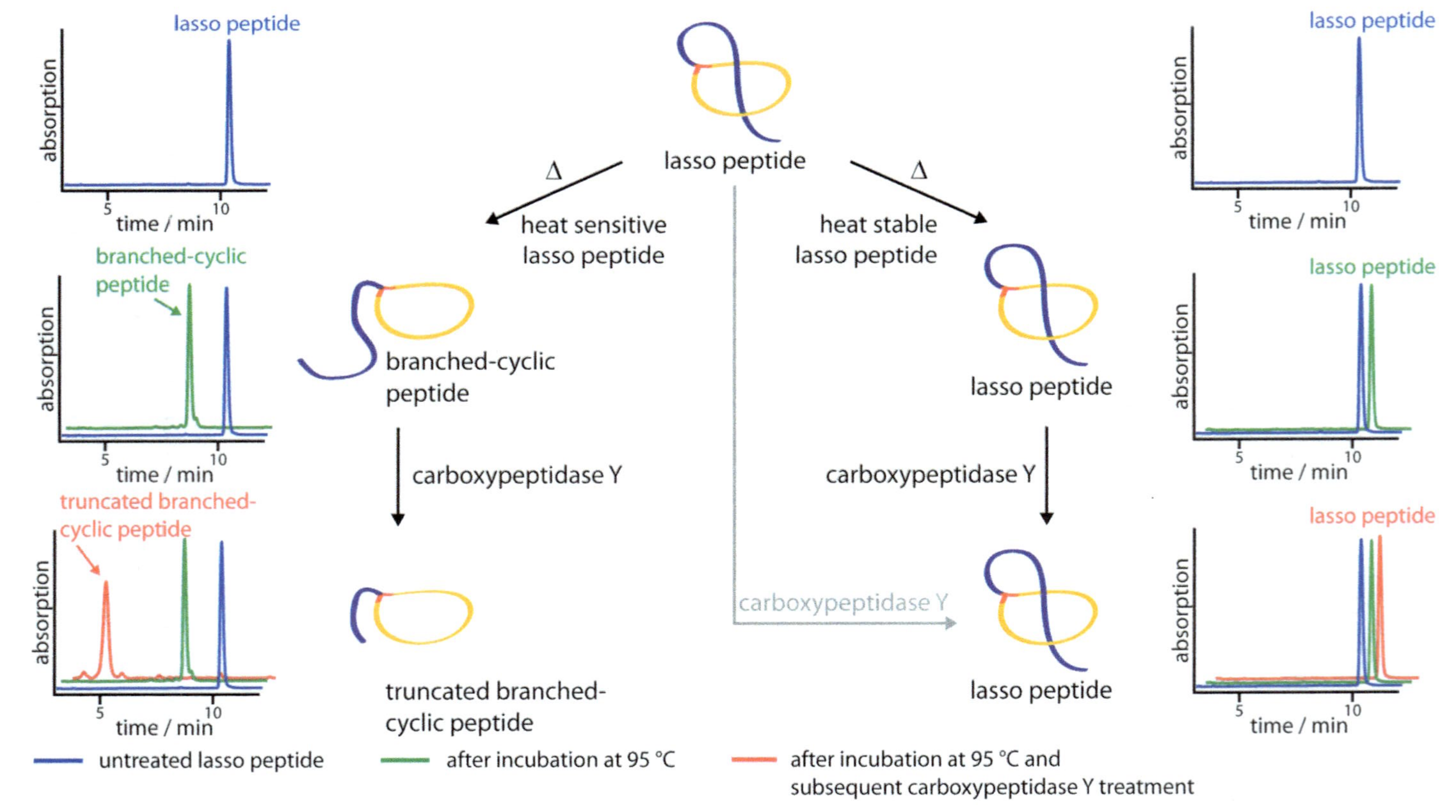

Figure 10.5 Schematic depiction of the differences between heat-stable and heat-sensitive lasso peptides when employed in a combined thermal and carboxypeptidase Y stability assay. On the left, the heat-sensitive peptide unthreads during incubation at 95 °C and is subsequently degraded by carboxypeptidase Y. On the right, the heat-stable lasso peptide remains unaltered by both thermal and subsequent carboxypeptidase Y treatment. At the far edges, exemplary chromatograms showing the LC analysis after each step are shown.

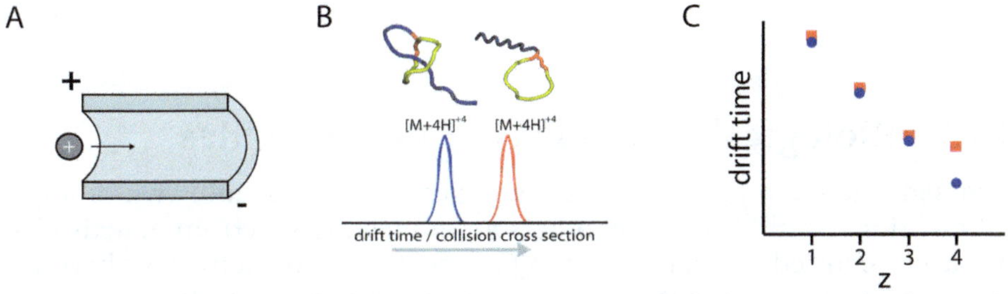

Figure 10.6 Using IM-MS to differentiate between a lasso and a branched-cyclic peptide with identical amino acid sequences. (A) Schematic depiction of the principle of IM. A positively charged ion moves through the drift tube in the direction of opposite charge. As the drift tube is filled with an inert buffer gas, the ion will get slowed down by collisions with the gas molecules. The larger the collision cross-section of the ion, the stronger the effect and hence the longer the time it needs to pass through the drift tube. (B) The different collision cross-sections of a lasso peptide and its branched-cyclic analogue at high charge states can be observed by the differences in their drift times.[30,41] (C) An exemplary graph depicting the correlation of the positive charge state *z* and the drift time of a lasso peptide (blue circles) and its branched-cyclic analogue (red squares). In this example, the lasso and branched-cyclic peptides behave in a very similar manner up to a charge state of +3. At the higher charge state of +4, their collision cross-sections and in turn their drift times change significantly and thus differentiation of the topologies becomes possible by IM-MS.[41]

steric stabilization of the lasso peptide trap the tail in a very confined conformation and hence the structure remains compact even at such high charge states. Both the shorter drift time and the approximately linear correlation of the collision cross-sections and charge states of a lasso peptide allow its differentiation from a branched-cyclic peptide with identical primary structure. Figure 10.6 shows an example of this to illustrate the methodology.[41]

As a final way to obtain information about lasso peptide structures, mutational studies can be performed.[4–7,30,34,40] This can be especially useful for determining the plug residues in a lasso peptide, whose three-dimensional structure is unknown, but whose lasso topology was confirmed by the methods described above. Generally, a simple alanine scan of all residues of the tail region that could act as a plug is often sufficient for this. In cases where the lower plug amino acid is replaced by alanine, the fold cannot be maintained after biosynthesis and readily unthreads inside the cell and the resulting branched-cyclic peptide is subsequently degraded by proteases.[4–7] In these cases, heterologous production often yields no or only trace amounts of the corresponding unthreaded peptide and truncations thereof. In some cases, it was observed that bulky residues located after the actual lower plug can still somewhat maintain the lasso fold after the plug is replaced with alanine.[4,40] Here, testing of double or even triple alanine substitutions can help to clarify the results and enable the identification of the correct plug residue after all.

10.3 Biological Functions of Lasso Peptides

Although there are plenty of lasso peptides with so far unknown biological functions,[3–5,7,22–24] for some others an interesting spectrum of activities has been reported in the literature.[3] The most common activities observed are of an antimicrobial nature directed against a narrow spectrum of other, often related or ecologically associated, bacteria.[3,11,14,16,17,29,39,43–47] Examples of this are capistruin, lariatin, lassomycin, MccJ25, propeptin, siamycins (which are class I lasso peptides), streptomonomycin and sviceucin (also a class I lasso peptide). Amongst these, the actual targets are only known in a few cases, namely capistruin, lassomycin and MccJ25.[29,48–54] While the primary target of capistruin and MccJ25 is the Gram-negative RNA-polymerase,[48–51] lassomycin binds to the ClpC ATPase of *Mycobacterium tuberculosis* and thereby uncouples ClpC-mediated ATP-hydrolysis from ClpP-dependent proteolysis, which is essential for the viability of *M. tuberculosis*.[29] Additionally, a secondary mode of action was reported for MccJ25, where it affects the respiratory chain and in the course of this triggers superoxide formation, although the concrete mechanism behind this is not fully understood yet.[52–54]

In addition to these antimicrobial activities, screens for compounds of potential therapeutic interest have revealed lasso peptides that can inhibit human enzymes or bind to human receptors as well as compounds that show anticancer properties.[3] Examples among the inhibitors are propeptin[46] (inhibits the prolyl endopeptidase) or siamyicins[45] (class I lasso peptides that inhibit the myosin light chain kinase). Additionally, anantin, BI-32169 (a class III lasso peptide) and RES-701-type lasso peptides have been demonstrated to act as receptor antagonists for the atrial natriuretic factor,[18] glucagon[55] and endothelin type-B receptors,[56,57] respectively. Anti-HIV activity was furthermore described for the siamycins[44] and efficacy against invasion by the human cell cancer line A549 was observed for the closely related sungsanpin and chaxapeptin lasso peptides.[14,58]

As mentioned above, there are still numerous lasso peptides, especially ones discovered in genome mining studies, for which no biological activities have been observed so far.[3–5,7,22–24,30] Hence, their actual roles in Nature remain enigmatic, but recent studies potentially provided the first stepping stone toward resolving this question. In particular, thorough genome mining in proteobacteria revealed a subtype of lasso peptide biosynthetic gene clusters that always appear adjacent to another highly conserved operon.[2–7,22–24,34] These operons face in the opposite direction of the lasso peptide clusters and typically encode homologs to peptidases, TonB-dependent receptors and σ-/anti-σ factor pairs. The fact that the co-occurrence with the lasso peptide biosynthetic gene clusters is not coincidental was recently shown in a study that demonstrated the peptidases encoded close to the astexin biosynthetic gene clusters in *Asticcacaulis excentricus* CB48 to be functional lasso peptide isopeptidases.[24] These enzymes selectively cleave the isopeptide bonds of the lasso peptides produced by the neighboring clusters and were shown to lack cross-activities for the respective other lasso peptides from this organism. This behavior was also observed for a lasso peptide–isopeptidase pair found in *Sphingopyxis alaskensis* RB2256.[34] Interestingly, the lasso fold seems to play a crucial role in substrate recognition as on the one hand branched-cyclic analogues and other lasso peptides are not hydrolyzed, while on the other hand most point-mutations in the native lasso peptide substrates are readily tolerated.[13,24,34] This hypothesis is supported by the co-crystal structure of astexin-3 with its isopeptidase (AtxE2), which lacks any extensive specific lasso peptide side chain-protease contacts.[35]

The very high specificity of these serine proteases combined with their large size (~70–80 kDa) suggests that the targeted degradation of the respective lasso peptides is of importance for the cell. Therefore, it has been hypothesized that these lasso peptides might fulfill siderophore-like functions[24] or are used as signaling molecules by the producing strains.[3] The presence of genes encoding TonB-dependent receptors and regulatory proteins in the isopeptidase operons further suggests both dedicated uptake and regulatory mechanisms that are dependent on lasso peptides. In this regard, future studies of these interesting enzymes and the clusters surrounding them might help to better understand the functions Nature has devised for the lasso peptides in question.

10.4 Lasso Peptides as Scaffolds for Drug Development

A hallmark of lasso peptides or RiPPs in general is the promiscuity of their processing enzymes toward the amino acid sequences of their precursor peptides.[1,4–8,13,25,34,40,42,59–63] For lasso peptides, numerous studies have shown that their core peptide sequences are amenable to exchanges of most residues, with the exception of those that are involved in the actual ring formation or whose substitution would cause a destabilization of the overall fold (*e.g.* replacement of a plug residue with a much smaller amino acid). In two parallel studies, it was additionally shown that single non-canonical amino acids can be incorporated into lasso peptides *via* the well-established routes of amber codon suppression or the use of expression strains that are auxotrophic for single amino acids and thus allow introduction of structurally similar residues under certain conditions.[62,63] Based on all of these findings, lasso peptides appear to be attractive scaffolds for the incorporation of small bioactive peptide sequences.[42,61]

It was generally observed that most single substitutions have at least a slight detrimental effect on the lasso peptide yield, with some even completely abolishing production. Therefore, the number of changes incorporated in

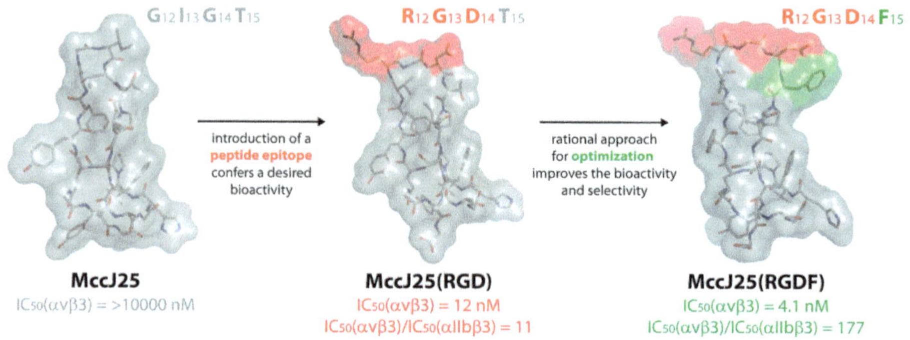

Figure 10.7 An example of epitope grafting utilizing the lasso peptide MccJ25. First, the RGD peptide epitope is introduced to target the αvβ3 integrin receptor.[42] Then, the epitope is modified based on rational means to improve affinity and selectivity for the desired target.[61]

one single scaffold can be limited, but at the moment the prediction of what is tolerated and what is not is not possible. This is nonetheless not too much of an issue for the use of lasso peptides as epitope grafting scaffolds in the context of drug development, as the basic lasso topology allows only the use of small, medicinally relevant peptide epitopes to begin with. Generally, the most suitable region of a lasso peptide to incorporate a bioactive peptide epitope is the loop, as this is both the most exposed region as well as one of the most rigid parts of the lasso scaffold. In addition to this, the processing enzymes often show a high promiscuity regarding substitutions in the central portion of the loops.

The feasibility of this concept was shown for MccJ25 with regard to first introducing a peptide epitope to target the αvβ3 integrin receptor and then optimizing the affinity and selectivity for this target by rational means.[42,61] A graphical summary of these studies is shown in Figure 10.7.

Finally, another very interesting application of the lasso peptide biosynthetic machinery was recently shown. In the corresponding study, the end of the gene coding for the astexin-1 precursor peptide was fused to the start of a gene encoding either a leucine zipper protein or GFP with a short GSSG-linker region in between.[26] These precursor peptide–protein fusions were co-expressed with the corresponding astexin-1 processing enzymes. Remarkably, the biosynthetic enzymes did recognize the leader peptide regions on the N-terminal part of the proteins and correctly processed them into the lasso fold, yielding proteins capped N-terminally with a lasso peptide region.[26] Though this was only a proof-of-concept study, the authors state that it could yield ways to establish high-throughput screening techniques for lasso peptides,[26] which would allow directed evolution-driven identification of lasso peptide sequences that can interact with specific targets instead of merely relying on rational approaches in this matter. Such a technique could also be helpful for optimizing a lasso peptide drug lead to a compound that can actually be considered for clinical trials.

References

1. P. G. Arnison, M. J. Bibb, G. Bierbaum, A. A. Bowers, T. S. Bugni, G. Bulaj, J. A. Camarero, D. J. Campopiano, G. L. Challis, J. Clardy, P. D. Cotter, D. J. Craik, M. Dawson, E. Dittmann, S. Donadio, P. C. Dorrestein, K. D. Entian, M. A. Fischbach, J. S. Garavelli, U. Goransson, C. W. Gruber, D. H. Haft, T. K. Hemscheidt, C. Hertweck, C. Hill, A. R. Horswill, M. Jaspars, W. L. Kelly, J. P. Klinman, O. P. Kuipers, A. J. Link, W. Liu, M. A. Marahiel, D. A. Mitchell, G. N. Moll, B. S. Moore, R. Muller, S. K. Nair, I. F. Nes, G. E. Norris, B. M. Olivera, H. Onaka, M. L. Patchett, J. Piel, M. J. Reaney, S. Rebuffat, R. P. Ross, H. G. Sahl, E. W. Schmidt, M. E. Selsted, K. Severinov, B. Shen, K. Sivonen, L. Smith, T. Stein, R. D. Sussmuth, J. R. Tagg, G. L. Tang, A. W. Truman, J. C. Vederas, C. T. Walsh, J. D. Walton, S. C. Wenzel, J. M. Willey and W. A. van der Donk, *Nat. Prod. Rep.*, 2013, **30**, 108.
2. M. O. Maksimov and A. J. Link, *J. Ind. Microbiol. Biotechnol.*, 2014, **41**, 333.
3. J. D. Hegemann, M. Zimmermann, X. Xie and M. A. Marahiel, *Acc. Chem. Res.*, 2015, **48**, 1909.
4. J. D. Hegemann, M. Zimmermann, S. Zhu, H. Steuber, K. Harms, X. Xie and M. A. Marahiel, *Angew. Chem., Int. Ed.*, 2014, **53**, 2230.
5. J. D. Hegemann, M. Zimmermann, X. Xie and M. A. Marahiel, *J. Am. Chem. Soc.*, 2013, **135**, 210.
6. M. Zimmermann, J. D. Hegemann, X. Xie and M. A. Marahiel, *Chem. Biol.*, 2013, **20**, 558.
7. M. Zimmermann, J. D. Hegemann, X. Xie and M. A. Marahiel, *Chem. Sci.*, 2014, **5**, 4032.
8. J. D. Hegemann, C. D. Fage, S. Zhu, K. Harms, F. S. Di Leva, E. Novellino, L. Marinelli and M. A. Marahiel, *Mol. BioSyst.*, 2016, **12**, 1106.
9. T. A. Knappe, U. Linne, X. Xie and M. A. Marahiel, *FEBS Lett.*, 2010, **584**, 785.
10. H. Nar, A. Schmid, C. Puder and O. Potterat, *ChemMedChem*, 2010, **5**, 1689.
11. Y. Li, R. Ducasse, S. Zirah, A. Blond, C. Goulard, E. Lescop, C. Giraud, A. Hartke, E. Guittet, J. L. Pernodet and S. Rebuffat, *ACS Chem. Biol.*, 2015, **10**, 2641.
12. D. Frechet, J. D. Guitton, F. Herman, D. Faucher, G. Helynck, B. Monegier du Sorbier, J. P. Ridoux, E. James-Surcouf and M. Vuilhorgne, *Biochemistry*, 1994, **33**, 42.
13. M. O. Maksimov, J. D. Koos, C. Zong, B. Lisko and A. J. Link, *J. Biol. Chem.*, 2015, **290**, 30806.
14. S. S. Elsayed, F. Trusch, H. Deng, A. Raab, I. Prokes, K. Busarakam, J. A. Asenjo, B. A. Andrews, P. van West, A. T. Bull, M. Goodfellow, Y. Yi, R. Ebel, M. Jaspars and M. E. Rateb, *J. Org. Chem.*, 2015, **80**, 10252.
15. K. J. Rosengren, R. J. Clark, N. L. Daly, U. Goransson, A. Jones and D. J. Craik, *J. Am. Chem. Soc.*, 2003, **125**, 12464.
16. M. Metelev, J. I. Tietz, J. O. Melby, P. M. Blair, L. Zhu, I. Livnat, K. Severinov and D. A. Mitchell, *Chem. Biol.*, 2015, **22**, 241.

17. T. A. Knappe, U. Linne, S. Zirah, S. Rebuffat, X. Xie and M. A. Marahiel, *J. Am. Chem. Soc.*, 2008, **130**, 11446.
18. W. Weber, W. Fischli, E. Hochuli, E. Kupfer and E. K. Weibel, *J. Antibiot.*, 1991, **44**, 164.
19. J. O. Solbiati, M. Ciaccio, R. N. Farias, J. E. Gonzalez-Pastor, F. Moreno and R. A. Salomon, *J. Bacteriol.*, 1999, **181**, 2659.
20. K. P. Yan, Y. Li, S. Zirah, C. Goulard, T. A. Knappe, M. A. Marahiel and S. Rebuffat, *ChemBioChem*, 2012, **13**, 1046.
21. H. G. Choudhury, Z. Tong, I. Mathavan, Y. Li, S. Iwata, S. Zirah, S. Rebuffat, H. W. van Veen and K. Beis, *Proc. Natl. Acad. Sci. U. S. A.*, 2014, **111**, 9145.
22. M. O. Maksimov, I. Pelczer and A. J. Link, *Proc. Natl. Acad. Sci. U. S. A.*, 2012, **109**, 15223.
23. J. D. Hegemann, M. Zimmermann, S. Zhu, D. Klug and M. A. Marahiel, *Biopolymers*, 2013, **100**, 527.
24. M. O. Maksimov and A. J. Link, *J. Am. Chem. Soc.*, 2013, **135**, 12038.
25. R. Ducasse, K. P. Yan, C. Goulard, A. Blond, Y. Li, E. Lescop, E. Guittet, S. Rebuffat and S. Zirah, *ChemBioChem*, 2012, **13**, 371.
26. C. Zong, M. O. Maksimov and A. J. Link, *ACS Chem. Biol.*, 2016, **11**, 61.
27. S. Zhu, C. D. Fage, J. D. Hegemann, A. Mielcarek, D. Yan, U. Linne and M. A. Marahiel, *Sci. Rep.*, 2016, **6**, 35604.
28. J. Inokoshi, M. Matsuhama, M. Miyake, H. Ikeda and H. Tomoda, *Appl. Microbiol. Biotechnol.*, 2012, **95**, 451.
29. E. Gavrish, C. S. Sit, S. Cao, O. Kandror, A. Spoering, A. Peoples, L. Ling, A. Fetterman, D. Hughes, A. Bissell, H. Torrey, T. Akopian, A. Mueller, S. Epstein, A. Goldberg, J. Clardy and K. Lewis, *Chem. Biol.*, 2014, **21**, 509.
30. S. Zhu, J. D. Hegemann, C. D. Fage, M. Zimmermann, X. Xie, U. Linne and M. A. Marahiel, *J. Biol. Chem.*, 2016, **291**, 13662.
31. B. J. Burkhart, G. A. Hudson, K. L. Dunbar and D. A. Mitchell, *Nat. Chem. Biol.*, 2015, **11**, 564.
32. J. A. Latham, A. T. Iavarone, I. Barr, P. V. Juthani and J. P. Klinman, *J. Biol. Chem.*, 2015, **290**, 12908.
33. B. M. Wieckowski, J. D. Hegemann, A. Mielcarek, L. Boss, O. Burghaus and M. A. Marahiel, *FEBS Lett.*, 2015, **589**, 1802.
34. C. D. Fage, J. D. Hegemann, A. J. Nebel, R. M. Steinbach, S. Zhu, U. Linne, K. Harms, G. Bange and M. A. Marahiel, *Angew. Chem., Int. Ed.*, 2016, **55**, 12717.
35. J. R. Chekan, J. D. Koos, C. Zong, M. O. Maksimov, A. J. Link and S. K. Nair, *J. Am. Chem. Soc.*, 2016, **138**, 16452.
36. S. Zhu, C. D. Fage, J. D. Hegemann, D. Yan and M. A. Marahiel, *FEBS Lett.*, 2016, **590**, 3323.
37. R. D. Kersten, Y. L. Yang, Y. Xu, P. Cimermancic, S. J. Nam, W. Fenical, M. A. Fischbach, B. S. Moore and P. C. Dorrestein, *Nat. Chem. Biol.*, 2011, **7**, 794.
38. X. Xie and M. A. Marahiel, *ChemBioChem*, 2012, **13**, 621.

39. R. A. Salomon and R. N. Farias, *J. Bacteriol.*, 1992, **174**, 7428.
40. T. A. Knappe, U. Linne, L. Robbel and M. A. Marahiel, *Chem. Biol.*, 2009, **16**, 1290.
41. K. Jeanne Dit Fouque, C. Afonso, S. Zirah, J. D. Hegemann, M. Zimmermann, M. A. Marahiel, S. Rebuffat and H. Lavanant, *Anal. Chem.*, 2015, **87**, 1166.
42. T. A. Knappe, F. Manzenrieder, C. Mas-Moruno, U. Linne, F. Sasse, H. Kessler, X. Xie and M. A. Marahiel, *Angew. Chem.*, 2011, **50**, 8714.
43. O. Potterat, H. Stephan, J. W. Metzger, V. Gnau, H. Zähner and G. Jung, *Liebigs Ann. Chem.*, 1994, 741.
44. M. Tsunakawa, S. L. Hu, Y. Hoshino, D. J. Detlefson, S. E. Hill, T. Furumai, R. J. White, M. Nishio, K. Kawano and S. Yamamoto, *et al.*, *J. Antibiot.*, 1995, **48**, 433.
45. K. Yano, S. Toki, S. Nakanishi, K. Ochiai, K. Ando, M. Yoshida, Y. Matsuda and M. Yamasaki, *Bioorg. Med. Chem.*, 1996, **4**, 115.
46. K. Kimura, F. Kanou, H. Takahashi, Y. Esumi, M. Uramoto and M. Yoshihama, *J. Antibiot.*, 1997, **50**, 373.
47. M. Iwatsuki, H. Tomoda, R. Uchida, H. Gouda, S. Hirono and S. Omura, *J. Am. Chem. Soc.*, 2006, **128**, 7486.
48. M. A. Delgado, M. R. Rintoul, R. N. Farias and R. A. Salomon, *J. Bacteriol.*, 2001, **183**, 4543.
49. J. Yuzenkova, M. Delgado, S. Nechaev, D. Savalia, V. Epshtein, I. Artsimovitch, R. A. Mooney, R. Landick, R. N. Farias, R. Salomon and K. Severinov, *J. Biol. Chem.*, 2002, **277**, 50867.
50. K. Adelman, J. Yuzenkova, A. La Porta, N. Zenkin, J. Lee, J. T. Lis, S. Borukhov, M. D. Wang and K. Severinov, *Mol. Cell*, 2004, **14**, 753.
51. K. Kuznedelov, E. Semenova, T. A. Knappe, D. Mukhamedyarov, A. Srivastava, S. Chatterjee, R. H. Ebright, M. A. Marahiel and K. Severinov, *J. Mol. Biol.*, 2011, **412**, 842.
52. M. R. Rintoul, B. F. de Arcuri, R. A. Salomon, R. N. Farias and R. D. Morero, *FEMS Microbiol. Lett.*, 2001, **204**, 265.
53. A. Bellomio, P. A. Vincent, B. F. de Arcuri, R. N. Farias and R. D. Morero, *J. Bacteriol.*, 2007, **189**, 4180.
54. M. C. Chalon, A. Bellomio, J. O. Solbiati, R. D. Morero, R. N. Farias and P. A. Vincent, *FEMS Microbiol. Lett.*, 2009, **300**, 90.
55. O. Potterat, K. Wagner, G. Gemmecker, J. Mack, C. Puder, R. Vettermann and R. Streicher, *J. Nat. Prod.*, 2004, **67**, 1528.
56. Y. Morishita, S. Chiba, E. Tsukuda, T. Tanaka, T. Ogawa, M. Yamasaki, M. Yoshida, I. Kawamoto and Y. Matsuda, *J. Antibiot.*, 1994, **47**, 269.
57. T. Ogawa, K. Ochiai, T. Tanaka, E. Tsukuda, S. Chiba, K. Yano, M. Yamasaki, M. Yoshida and Y. Matsuda, *J. Antibiot.*, 1995, **48**, 1213.
58. S. Um, Y. J. Kim, H. Kwon, H. Wen, S. H. Kim, H. C. Kwon, S. Park, J. Shin and D. C. Oh, *J. Nat. Prod.*, 2013, **76**, 873.
59. O. Pavlova, J. Mukhopadhyay, E. Sineva, R. H. Ebright and K. Severinov, *J. Biol. Chem.*, 2008, **283**, 25589.

60. S. J. Pan and A. J. Link, *J. Am. Chem. Soc.*, 2011, **133**, 5016.
61. J. D. Hegemann, M. De Simone, M. Zimmermann, T. A. Knappe, X. Xie, F. S. Di Leva, L. Marinelli, E. Novellino, S. Zahler, H. Kessler and M. A. Marahiel, *J. Med. Chem.*, 2014, **57**, 5829.
62. R. S. Al Toma, A. Kuthning, M. P. Exner, A. Denisiuk, J. Ziegler, N. Budisa and R. D. Sussmuth, *ChemBioChem*, 2015, **16**, 503.
63. F. J. Piscotta, J. M. Tharp, W. R. Liu and A. J. Link, *Chem. Commun.*, 2015, **51**, 409.

CHAPTER 11

Biological Synthesis and Affinity-based Selection of Small Macrocyclic Peptide Ligands

TOBY PASSIOURA, YUKI GOTO, TAKAYUKI KATOH AND HIROAKI SUGA*

Department of Chemistry, Graduate School of Science, The University of Tokyo, 7-3-1 Hongo, Bunkyo-ku, 113-0033 Tokyo, Japan
*E-mail: hsuga@rcast.u-tokyo.ac.jp

11.1 Introduction

Comparatively small peptides, such as animal hormones and venom components, display many of the features of good drugs, including strong activity and high selectivity. Moreover, and unlike small molecule drugs, such peptides are capable of blocking protein–protein interactions, making them potentially applicable to targeting the so-called "undruggable" proteome.[1] However, the peptidic nature of such molecules, in addition to their relatively high mass as compared to small molecule drugs, results in poor pharmacokinetics and they exhibit rapid cleavage by peptidases and/or clearance from the blood combined with poor membrane permeability and oral availability. This has, consequently, limited the clinical development of such molecules to date.

Chemical Biology No. 6
Cyclic Peptides: From Bioorganic Synthesis to Applications
Edited by Jesko Koehnke, James Naismith and Wilfred A. van der Donk

Published by the Royal Society of Chemistry, www.rsc.org

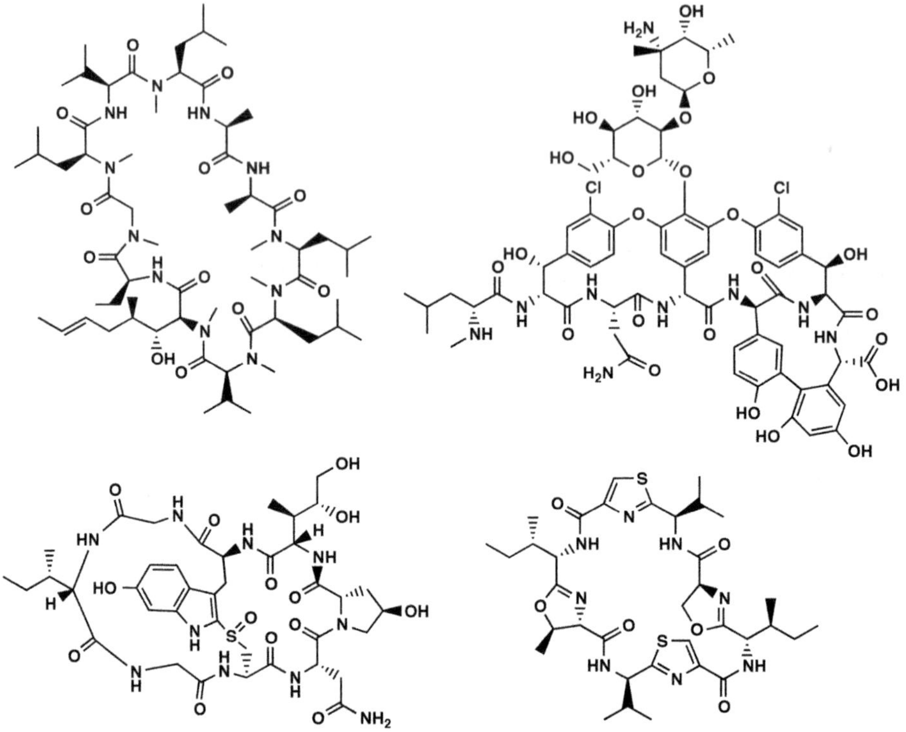

Figure 11.1 Representative natural macrocyclic peptides. Clockwise from top left: the immunosuppressant cyclosporine, the antibiotic vancomycin, the cyanobacterial secondary metabolite patellamide A, and the mushroom toxin α-amanitin.

Differing both in size and structure from such hormones and venom components is a diverse class of bioactive macrocyclic peptides found as secondary metabolites in fungi and prokaryotes (Figure 11.1). These molecules are typically smaller than 2000 Da and, in addition to their macrocyclic structures, they generally exhibit a range of unusual structural features, including D-stereochemistry, backbone *N*-methylation, and non-canonical side chains. Like larger and more canonical bioactive peptides, however, such compounds are often highly potent (the mushroom toxin α-amanitin, for example, has an oral LD_{50} of ~100 μg kg^{-1} in humans) and appear to retain the ability to disrupt protein–protein interactions.[2,3] Several of these compounds have been developed into clinically useful drugs, either in their natural form or following structural optimization, including the immunosuppressant cyclosporine, the antibiotics vancomycin, daptomycin and actinomycin D, and the antifungal echinocandins.[4–7]

Notably, many of these microbial secondary metabolites overcome the apparent limitations of peptides as drugs, and exhibit oral availability, resistance to peptidases and acceptable pharmacokinetics generally. This appears to be a direct consequence of their non-canonical structures: macrocyclization and *N*-methylation confer both peptidase resistance and the ability to cross biological membranes, whilst the presence of unusual side chains and D-stereochemistry are also critical for bioactivity, possibly due to improvements in both peptidase resistance and target affinity.[6,8–11]

The ability of small, macrocyclic, non-canonical peptides to inhibit protein–protein interactions, whilst exhibiting oral availability and acceptable pharmacokinetics, makes them an appealing class for drug discovery and development. However, all of the clinically relevant compounds in this class discovered to date have been derived from natural products, and are generally cytotoxic/antibiotic, presumably as a result of evolutionary pressure for host defenses. Thus, the screening of natural product cyclic peptides for other disease indications seems unlikely to be a generally effective strategy.

As an alternative to the isolation of macrocyclic peptide ligands from Nature, a number of strategies have been developed. Combinatorial chemistry-type approaches have had some success, but are limited to the screening of libraries containing a few million compounds for technical reasons.[12] However, because peptides can be synthesized ribosomally through translation, a number of alternative techniques based on genetically encoded libraries have been developed. These techniques allow the synthesis and screening of very diverse (on the order of 10^{13} or more) libraries of macrocyclic peptides through translational synthesis and affinity selection against a disease-related target of interest, and typically produce ligands with high target affinity and selectivity. Moreover, by combining such techniques with genetic code reprogramming techniques (which allow the synthesis of non-canonical peptides), compounds with structural similarities to microbial secondary metabolites can be isolated. The present chapter describes the development and application of such techniques.

11.2 Selection of Cyclic Peptides from Libraries Composed of Canonical Amino Acids

11.2.1 Head-to-tail Peptide Cyclization Using Split-inteins (SICLOPPS)

Inteins are internal protein regions with self-splicing activity that can excise themselves and ligate the remaining (N- and C-terminal flanking) regions of the protein, termed exteins.[13,14] The splicing reaction is initiated by an N to S/O acyl shift to form a (thio)ester at a conserved Cys/Ser/Thr residue at the N-terminus of the intein (Figure 11.2a). Then, the N-terminal extein is transferred to the side chain of a Cys/Ser/Thr residue at the C-terminus of the extein through trans(thio)esterification. Finally, the amide bond between the intein and the C-terminal extein is cleaved *via* cyclization of a C-terminal Asn of the intein, followed by spontaneous S/O to N acyl rearrangement to give a new amide bond connecting the two exteins.

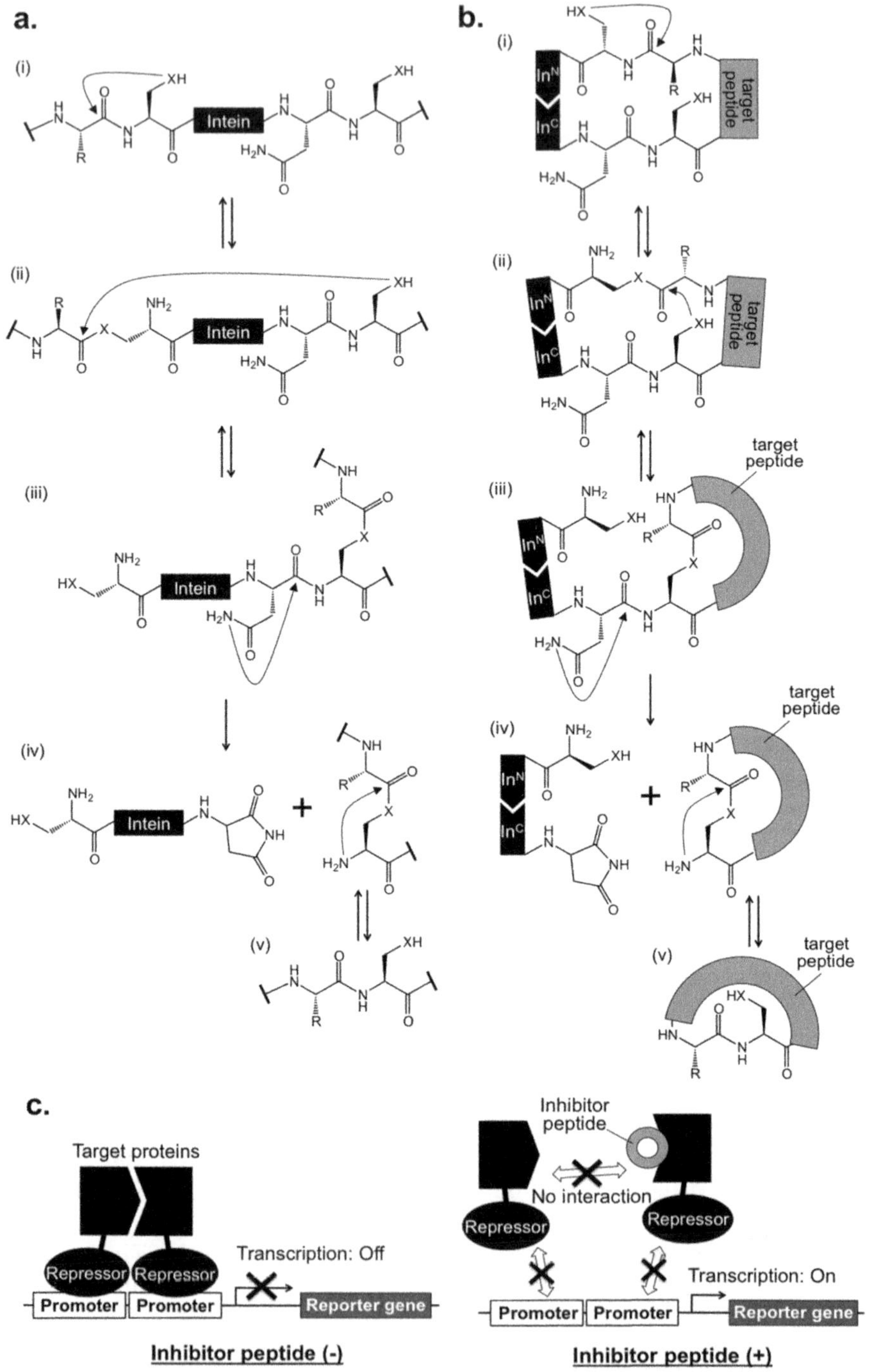

Figure 11.2 Scheme of head-to-tail peptide cyclization by SICLOPPS technology. (a) The mechanism of protein self-splicing mediated by an intein.

SICLOPPS (split-intein circular ligation of peptides and proteins) is a method for the synthesis of cyclic peptides in cells utilizing split inteins (Figure 11.2b).[15] Two intein domains, designated the N-terminal (In^N) and C-terminal (In^C) domains, are engineered into an expression construct such that they flank the target peptide, resulting in an In^C–target-peptide–In^N polypeptide. When expressed, the split In^C and In^N domains associate and reconstitute the intein activity to cleave the target peptide at the N- and C-termini, and ligate them to yield a head-to-tail macrocyclic peptide. In SICLOPPS, the DnaE split intein derived from a catalytic subunit of DNA polymerase III of *Synechocystis* sp. PCC6803 has generally been used. A limitation of this method is that the split intein requires a Cys, Ser or Thr residue at the N-terminus of the target peptide, where Cys is generally more favorable than Ser and Thr. Moreover, the C-terminal residue of the target peptide should have a sterically non-demanding side chain for efficient ligation, and thus, approximately 30% of peptides with variation of the C-terminal residue may not be cyclized efficiently by the *Synechocystis* DnaE split intein.[16,17] To improve the cyclization efficiency, Townend and Tavassoli have recently devised a new split intein with higher tolerance for variations in the C-terminal amino acid and a faster splicing rate by modifying the *Nostoc punctiforme* (*Npu*) DnaE split intein.[18] Although the *Npu* split intein showed a high toxicity in host *E. coli* cells, the toxicity could be diminished by use of an SsrA tag to target the spliced *Npu* intein to the ClpXP machinery, a protease complex that degrades SsrA-tagged substrates.

SICLOPPS can be applied to the construction of macrocyclic peptide libraries by randomizing the sequence of the target peptide. By introducing a plasmid vector encoding In^C–random-peptide–In^N into appropriate host cells, such as *E. coli*, yeast or human cells, cyclic peptide libraries can be expressed in the cells, and utilized for selection of active macrocyclic peptides.[19] Generally, a peptide library with 5–7 randomized amino acids with an N-terminal invariant Cys is used for selection. Although the theoretical diversity of such

---

(i) N to S/O acyl shift at the N-terminus of an intein. (ii) Trans(thio) esterification between an N-terminal extein and the side chain of a Cys/Ser/Thr at the C-terminus of an extein. (iii) Cleavage of the amide bond between an intein and a C-terminal extein caused by cyclization of the C-terminal Asn of an intein. (iv) Formation of a new amide bond between the two exteins *via* S/O-to-N acyl shift. X indicates S or O. (b) Cyclization of a target peptide by means of a split intein. The intein is divided into N-terminal (In^N) and C-terminal (In^C) domains, and linked to the C- and N-termini of the target peptide, respectively. The reactions shown in (i)–(iv) occur in a similar manner as for the intact intein. X indicates S or O. (c) A reverse two-hybrid system combined with a cyclic peptide library produced by SICLOPPS technology. Without an inhibitor peptide, expression of a reporter gene is repressed by two repressor proteins conjugated to the target proteins. When the interaction between the two target proteins is inhibited by a macrocyclic peptide in the library, the repressor is released from the promoter and the reporter gene is expressed.

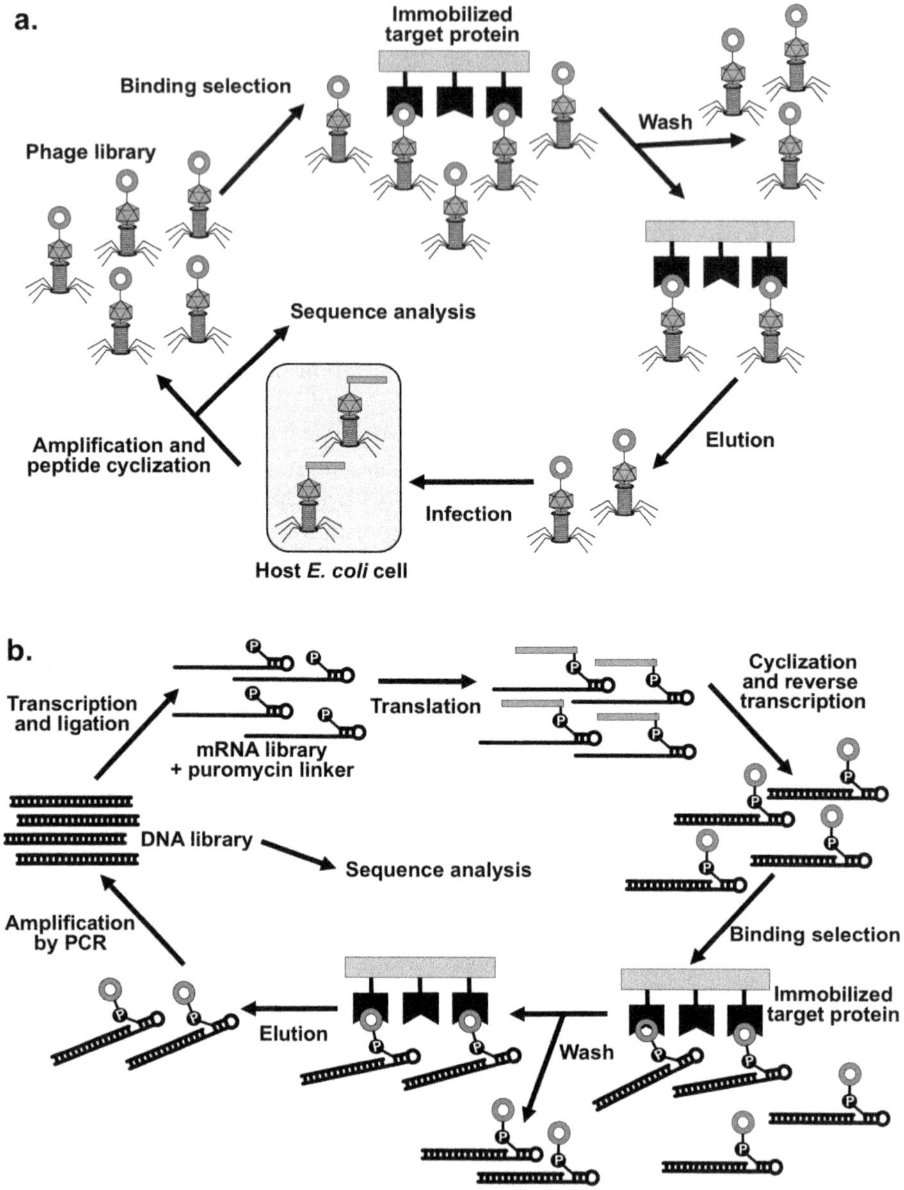

Figure 11.3 Schematic depiction of phage display and mRNA display. (a) Scheme of phage display combined with a macrocyclic peptide library. The peptide library on the surface of phages is cyclized using an appropriate chemical reagent, and then mixed with immobilized target protein. The phages with target-binding peptides are selectively recovered and infected into host bacterial cells for amplification. After several rounds of affinity selection, the sequences of the target-binding peptides are analyzed by sequencing of the encapsulated phage DNA. (b) Scheme of mRNA display combined with a macrocyclic peptide library. A random

a peptide library is around 10^6–10^9, in practice it is limited by the transformation efficiency of the vector (typically around 10^8). For the detection of active macrocyclic peptides, a reverse two-hybrid system has been utilized in combination with SICLOPPS technology.

To discover inhibitors of specific protein–protein interactions, the two target proteins are conjugated to repressor proteins, which regulate the expression of a downstream reporter gene essential for cell growth (Figure 11.2c). When the two proteins interact, the fused repressor regions bind to the promoter, the reporter gene is not expressed and thus no cell growth is observed. If a macrocyclic peptide expressed using SICLOPPS inhibits the interaction of the two target proteins, the repressor region fails to bind to the promoter and the reporter gene will be expressed to induce cell growth. Thereafter, the SICLOPPS vector can be recovered from the surviving cells in order to determine the sequence of the active peptides. The first application of this system was a screen for inhibitors of heterodimerization of ribonucleotide reductase subunits using a peptide library with 5 randomized amino acids and a fixed N-terminal Cys. The obtained peptides showed moderate inhibitory activities in the μM range.[20] To date, similar approaches have been applied to screen for inhibitors of several specific protein–protein interactions, such as homodimerization of AICAR transformylase/IMP cyclohydrolase (ATIC)[21] or interaction of the HIV Gag protein with TSG101 (tumor susceptibility gene 101).[22]

11.2.2 Phage/Phagemid Display

Phage display is a screening method for identification of peptides or proteins with affinity to a given target using a random peptide/protein library displayed on the surface of bacteriophages, typically fused to a coat protein (Figure 11.3a).[23] A library of phages is panned against target molecules immobilized on a carrier, such as resin or magnetic beads. By washing away unbound phages, the fraction that selectively binds to the target molecule is recovered. The recovered phages are then used to infect host bacterial cells as a means of amplification, and the resulting enriched phage library can be used for a subsequent round of affinity selection. Through iteration of the selection cycle, phages with high binding affinity to the target molecule are enriched. Finally, the sequence of the binding peptide/protein can be obtained by analyzing the encapsulated phage DNA. For efficient enrichment of strong binders, the unbound fraction is necessarily washed out

DNA library is transcribed into an mRNA library, followed by puromycin-linker ligation. Subsequently, random peptides are translated and covalently linked to the mRNA *via* the puromycin linker. After cyclization of the peptide and reverse transcription of the mRNA to form mRNA–cDNA hybrids, the library is panned against immobilized target proteins to selectively recover the target-binding peptides. Finally, the bound fraction is amplified by PCR and the sequence of the DNA library is analyzed.

under stringent conditions, such as high temperature and/or extreme pH. Thus, bacteriophage species for expression of peptide/protein library should be carefully chosen. Filamentous phages such as M13, f1 and fd are typically used due to their high stability under harsh conditions. Normally, a library consisting of ~10^9–10^{10} different species in theory (if all peptides are correctly displayed) can be screened in a single experimental cycle. To date, phage display has been applied for developing various antibody fragments,[24–27] as well as short peptides.

Phagemid display is a variant of phage display that utilizes phagemids for expression of phages.[28,29] Phagemids are plasmids that have an origin of replication (ori) for double-stranded DNA, as well as the f1 origin required for the replication of single-stranded DNA. Phagemids can be amplified as a double-stranded plasmid and easily introduced into host bacteria by simple transformation methods. The single-stranded DNA gene is then replicated and packed into a viral particle, typically that of a filamentous phage such as M13. As transformation of phagemids is more efficient than transfection of the phage itself, a higher diversity of libraries can be obtained. An additional advantage of phagemid display is that the expression level of the library can be easily controlled.

For the construction of cyclic peptide libraries in phage/phagemid display format, the simplest way is through introduction of a disulfide bond between two Cys residues assigned at the N- and C-termini of the random region, which is represented by CX_nC (with X indicating any amino acid, and n indicating the number of such randomized amino acid residues). Although introduction of a disulfide bond contributes to a higher observed binding affinity of the recovered peptides against the target molecules, such disulfide bonds are usually unstable under reducing conditions.

To improve the stability of the macrocycle, various methods for linking multiple Cys residues through non-reducible bonds have been developed. For example, Taki and co-workers recently demonstrated a gp10-based thioetherification ($10BASE_d$-T) method that can be applied for the linkage of two Cys thiols,[30] in which *N,N′*-[1,2-ethanediyl-oxy-2,1-ethanediyl]bis(2-bromoacetamide) (EBB) is mixed with peptides displayed on T7 phage to form a crown ether-like macrocyclic peptide library (Figure 11.4a). From a library consisting of approximately 1.5×10^9 members, a peptide against the N-terminal domain of Hsp90 with a K_D value of 1.7 μM was obtained. Derda and co-workers have also developed several thiol reactive linkers, such as decafluoro-diphenylsulfone,[31] dichloro-oxime derivatives,[32] and bis(allenamide) derivatives[33] (Figure 11.4b–d). Heinis and co-workers developed a method for linking three Cys residues through thioether bonds using 1,3,5-tris-(bromomethyl)benzene (TBMB) (Figure 11.4e).[34] In this approach, peptides containing a CX_nCX_nC sequence were reacted with TBMB to form a bicyclic structure composed of three thioether bonds. 1,3,5-Triacryloyl-1,3,5-triazinane (TATA) and *N,N′,N″*-(benzene-1,3,5-triyl)-tris(2-bromoacetamide) (TBAB) are similar aromatic linkers for connecting three Cys residues by thioether bonds (Figure 11.4f and g). As an example of the use of these linkers,

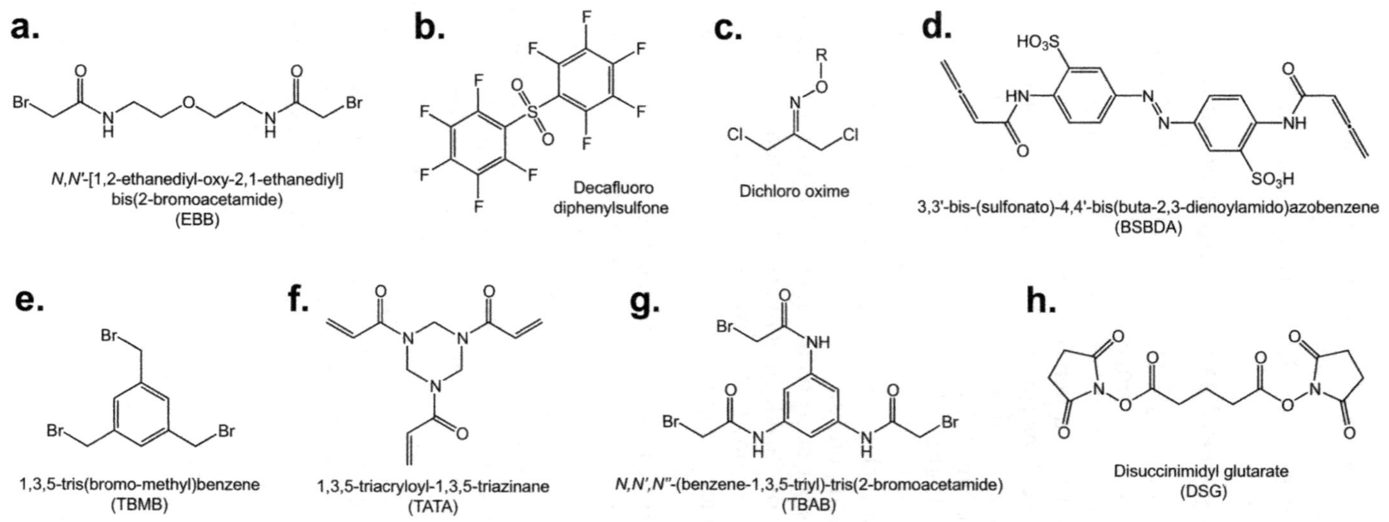

Figure 11.4 Representative chemical reagents used for peptide cyclization. (a) *N,N'*-[1,2-ethanediyl-oxy-2,1-ethanediyl]bis(2-bromoacetamide) (EBB), (b) decafluoro diphenylsulfone, (c) a dichloro oxime derivative, (d) 3,3′-bis-(sulfonato)-4,4′-bis(buta-2,3-dienoylamido)azobenzene (BSBDA), (e) 1,3,5-tris(bromo-methyl)benzene (TBMB), (f) 1,3,5-triacryloyl-1,3,5-triazinane (TATA), (g) *N,N′,N″*-(benzene-1,3,5-triyl)-tris(2-bromoacetamide) (TBAB), and (h) disuccinimidyl glutarate (DSG). All reagents except for (h) are used for linkages between two Cys residues by thioether bond formation. (h) is used for linkages between the N-terminal amine and the side chain of a Lys residue.

Bertoldo *et al.* used TBMB, TATA and TBAB in parallel for phage selection of bicyclic peptides against β-catenin.[35] In these experiments, three libraries based on the ACX_6CX_6CG sequence were prepared, and independently reacted with TBMB, TATA and TBAB. Selections using all three libraries resulted in successful enrichment of peptides with affinity for s-catenin.

11.2.3 mRNA Display, cDNA Display and Ribosome Display

In mRNA display,[36,37] cDNA display[38] and ribosome display,[39] peptide libraries are prepared by *in vitro* translation, as compared to SICLOPPS and phage libraries in which translation occurs in cells. In these methods, a random mRNA library is translated into a peptide library using a cell-free reconstituted translation system, such as the PURE system,[40,41] or using cell extract expression systems, such as rabbit reticulocyte lysate or *E. coli* S30 fraction. One of the advantages of mRNA/cDNA/ribosome display over SICLOPPS and phage display is that higher library diversities are achievable, because the total number of peptides is not limited by the transformation or infection efficiency. In addition, relatively stringent selection conditions, such as extreme pH and/or high temperatures, can be employed.

Typical mRNA libraries for such techniques consist of a Shine–Dalgarno (in prokaryotic systems) or Kozak (in eukaryotic systems) region, a start codon (AUG), a random region, a stop codon and a 3′-teminal spacer. As required, the random sequence can be flanked by appropriate tag and spacer sequences.

In addition, some form of linkage between the mRNA/cDNA (genotype) and peptide (phenotype) is required in order to identify the sequence of specific binding peptides. In mRNA display, the peptide and mRNA are linked by a covalent bond through a puromycin linker. Puromycin is an analogue of the 3′ tyrosyl adenosine moiety of Tyr-tRNA, in which the ester bond between tyrosine and adenosine is substituted with an amide bond. Upon ribosomal stalling at a pause site, such as an oligo dA region or UAG stop codon (the latter being particularly effective when release factor-1 is removed from the cell-free translation system, *e.g.*, the RaPID system, *vide infra*),[42,43] a puromycin moiety linked to the 3′ end of the mRNA enters the A site of the ribosome, and is ligated onto the C-terminus of the peptide. A similar method that links the N-terminus of the peptide and the 5′ end of the mRNA has also been reported.[44] Such peptide–mRNA conjugates are then reverse-transcribed into peptide–mRNA–cDNA conjugates, and mixed with an immobilized target molecule to pan for specific peptides with target affinity (Figure 11.3b). The bound fraction is recovered and amplified by PCR, and the sequences of the recovered peptides can be determined by analyzing the DNA sequence. By repeating this selection cycle 4–10 times, strong binders can generally be obtained. cDNA display is a modified version of mRNA display in which the peptide library is covalently linked to the cDNA.[38] In this method, a branched puromycin linker that has a reverse-transcription site is ligated to the mRNA to make the resulting reverse-transcribed cDNA covalently linked to the peptide. In the case of ribosome display, in contrast, the peptide library is not covalently linked to the mRNA nor cDNA. Instead, a stable peptide–ribosome–mRNA complex is formed by means of ribosomal stalling induced through the absence of a stop codon in the mRNA. Antibiotics that cause ribosomal stalling or fusion of a C-terminal ricin A sequence that inactivates translation can also be used for forming stable peptide–ribosome–mRNA complexes.[45] Since the C-terminal region of the nascent peptide in the peptide–ribosome–mRNA complex is buried in the ribosomal tunnel, at least 23–30 amino acids should be added to the C-terminus in order to display the random region outside of the ribosome. An advantage of ribosome display relative to comparable methods is that not only mRNA libraries but also DNA libraries can be used if a transcription–translation coupled system is employed, since preparation of mRNA with a linker moiety is not required.

For cyclization of peptide libraries used for mRNA, cDNA or ribosome display, disulfide bonds between multiple Cys residues can be easily introduced as described for phage display above, although they exhibit the same drawbacks with respect to reduction of the final product. Similarly, cyclization *via* thioether bond formation through the use of benzyl halides is also possible. For instance, mRNA display-based peptide libraries containing N- and C-terminal Cys residues closed by α,α-dibromoxylene have been reported.[46] Roberts and co-workers utilized disuccinimidyl glutarate (DSG) to post-translationally ligate the N-terminal amine and an ε-amine of a downstream Lys by forming two amide bonds (Figure 11.4h).[47] By using an mRNA display platform, they successfully found inhibitor peptides against the Gail signal transduction protein.

11.3 Broadening Library Chemical Diversity

The examples provided in the previous sections demonstrate the utility of techniques involving selection from high diversity peptide libraries for the identification of relatively small ligands with moderate to very high affinities for a protein target of interest. However, the chemical diversity of the ligands identified using such approaches is limited by their reliance on the 20 canonical amino acids as monomers for peptide synthesis. As discussed in the Introduction, bioactive microbial secondary metabolite cyclic peptides nearly invariably contain non-canonical structural features (D-stereochemistry, *N*-methylation, unusual side chains, *etc.*). Moreover, these modifications appear to be critical for bioactivity, and impart characteristics such as biostability and bioavailability, which are critical for potential drug development.[6,8–10,48,49] Thus, the application of such "canonical amino acid only" techniques to the identification of microbial secondary metabolite-like compounds (with the potential for favorable pharmacokinetics) is problematic.

The natural biosynthesis of microbial secondary metabolite cyclic peptides generally occurs through one of two routes: (i) synthesis by non-ribosomal peptide synthetases (NRPSs), or (ii) translation followed by extensive enzymatic post-translational modification to produce so-called ribosomally synthesized and post-translationally modified peptides (RIPPs). NRPSs are large, multi-subunit enzyme complexes, composed of modular functional domains, each of which catalyzes a specific reaction.[50] Molecular engineering of NRPSs is possible, and the products can, in principle, be screened by phage display.[51–53] However, the generation of high or very high diversity peptide libraries through NRPS engineering remains extremely challenging, and we are not aware of any studies demonstrating the utility of this approach for the screening of such libraries.

In contrast to the products of NRPSs, RIPPs are synthesized ribosomally prior to modification by one or more specific enzymes.[54] This can lead to the synthesis of peptides with a surprisingly diverse set of non-canonical structural features (including macrocyclization) as evidenced by compounds such as the patellamides or amanitins (Figure 11.1). The mRNA-dependent synthetic route of RIPPs means that (unlike peptides synthesized by NRPSs) the final structure can be determined from the mRNA sequence alone, provided that the activity and substrate specificity of the modifying enzyme(s) are well-understood.[55] Furthermore, post-translational processing of peptides containing non-canonical amino acids has been shown to produce analogues of natural RIPPs with increased potency, suggesting that the combination of post-translational modification enzymes with genetic code reprogramming techniques (*vide infra*) may by applicable to the synthesis and screening of very high diversity libraries with unique chemistries.[56] However, to the best of our knowledge, no such studies have been reported to date.

To summarize, display techniques involving peptides derived only from the 20 canonical amino acids are insufficient for the identification of microbial secondary metabolite-like macrocyclic peptides, since they do not allow for sufficient structural diversity. As tools for the synthesis of non-canonical peptide libraries for screening, both NRPSs and RIPP enzymes have some promise, particularly the latter since they are mRNA-dependent and therefore compatible with display screening and genetic code reprogramming techniques, *vide infra*, but neither is currently applicable to the synthesis and screening of very high diversity libraries. As such, alternative techniques allowing for the synthesis of peptides containing non-canonical moieties in an mRNA-dependent fashion are required.

11.3.1 Genetic Code Expansion

The fidelity of peptide translation from an mRNA template is controlled at two levels: (i) the specific base-pairing interaction of the mRNA codon with the tRNA anti-codon, and (ii) the specific aminoacylation of each tRNA with its cognate amino acid as catalyzed by a specific aminoacyl-tRNA synthetase (AARS). In the case of (i), engineering of the mRNA sequence (mutagenesis) has, of course, been widely used in many areas of molecular biology as a means of substituting one canonical amino acid for another (*e.g.* alanine scanning). In such cases, while both the mRNA and peptide sequences are altered, the genetic code itself remains unchanged (*i.e.* each codon still specifies the same amino acid). By contrast, engineering of (ii) through the synthesis of "mis-aminoacylated" tRNAs allows for reprogramming of the genetic code itself, such that a given codon can be reassigned to an amino acid of choice. In this way, peptides containing diverse, non-canonical residues can be synthesized ribosomally in an mRNA-dependent manner.

In essence, genetic code reprogramming techniques comprise two basic components: (i) a technique for the synthesis of non-canonical aminoacylated tRNAs (AA-tRNAs), and (ii) a technique for the translation of these AA-tRNAs into peptides. The earliest studies in this area employed chemo-enzymatic synthesis for the generation of AA-tRNAs in combination with a simple cell lysate-based translation reaction.[57,58] This involved synthesis of the non-canonical amino acid to be incorporated as a dinucleotide amino acid ester, which was then ligated to a tRNA body lacking the last two ribonucleotides using an RNA ligase. The purified AA-tRNA was then translated into peptides using a bacterial cell lysate system. In order to avoid competition from endogenous AA-tRNAs, the tRNA for reprogramming was designed to translate at the UAG (amber) stop codon (so-called nonsense suppression), so that peptides containing up to 21 different amino acids could be synthesized. Such an approach may be considered to be genetic code expansion, rather than genetic code reprogramming *per se*, since although it allows the translation of one (or sometimes more) non-canonical amino acid it does not allow for true reprogramming of the genetic code. Importantly, however, this approach demonstrated that diverse amino acids could be employed as substrates for ribosomal polypeptide formation. Subsequent developments in the field have included the use of 4 base codons (derived from relatively rare codons) which allow translation at non-stop codons, and the development of orthogonal tRNA/amino acid/AARS systems for non-canonical AA-tRNA synthesis, which allow for genetic code reprogramming in living cells.[59–64]

However, the reliance on rare or stop codons in the suppression techniques described above limits the number of non-canonical amino acids that can be incorporated into a single peptide. Moreover, because these techniques employ cellular translation systems (either lysates or in living cells), competition between the non-canonical AA-tRNA and endogenous canonical AA-tRNAs or release factors is unavoidable. This can greatly decrease the efficiency of reprogramming, particularly in the case of poorly translated amino acids such as *N*-methylated or D-amino acids.[57,58,65,66]

11.3.2 Genetic Code Reprogramming in Reconstituted Translation Systems

To circumvent these limitations, fully-reconstituted translation systems have been developed to replace cellular translation systems. In such systems, all of the individual components required for translation are individually purified from bacteria and then recombined to produce a system that is essentially free of all other cellular components.[40,67,68] Whilst laborious to produce initially, the great advantage of such systems is that the concentration of

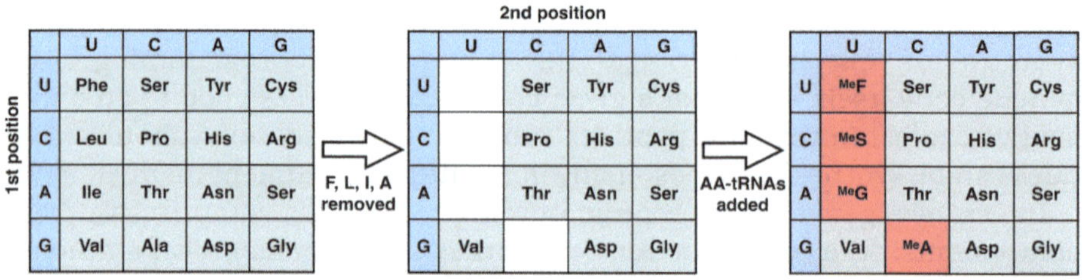

Figure 11.5 Genetic code reprogramming in reconstituted translation reactions. The codon table for NNU codons (N = A, G, C or U) for a reaction containing all of the required elements for canonical translation is show on the left (third position nucleotide not shown). Removal of Phe (F), Leu (L), Ile (I) and Ala (A) and/or their cognate AARSs "vacates" their respective codons (middle table) making them available for the introduction of non-canonical residues. Addition of tRNAs (pre-aminoacylated with non-canonical amino acids) with anti-codons complementary to the vacant codons, in this case α-*N*-methyl Phe (^{Me}F), Ser (^{Me}S) Gly (^{Me}G) and Ala (^{Me}A), results in the reprogramming of these codons to non-canonical residues.

individual components can be altered, or they can be omitted altogether. In this way, codons for canonical amino acids can be "vacated" though omission of the relevant amino acid and/or its cognate AARS, making the codon available for reprogramming using an appropriate AA-tRNA (Figure 11.5). In the first report of this approach, AA-tRNAs were synthesized using the chemo-enzymatic approach and Asn, Val and Thr codons were reprogrammed to three amino acids with non-canonical side chains, thereby creating a novel genetic code.[67]

Subsequent studies using similar translation systems have demonstrated that the bacterial translation machinery is highly tolerant of variations in amino acid structure, and will accept a diverse range of non-canonical amino acid substrates (*vide infra*).[69–71] In general, however, most subsequent studies have not used the chemo-enzymatic approach for AA-tRNA synthesis, due to the relative laboriousness of this technique, and it has been replaced with fully enzymatic methods. These either employ endogenous protein AARSs, taking advantage of their limited promiscuity with respect to amino acid substrates, or promiscuous ribozyme AARSs (flexizymes), and are discussed in detail below.

11.3.3 Enzymatic Aminoacylation by Natural AARSs

As alternatives to chemo-enzymatic AA-tRNA synthesis, techniques based on the use of aminoacylating catalysts have been developed for approaches involving extensive, *in vitro*, genetic code reprogramming in fully reconstituted systems. The first of these simply uses endogenous bacterial AARS proteins, and takes advantage of the fact that while these enzymes display high amino acid substrate selectivity with respect to canonical amino acids (*e.g.* FRS does not use Tyr as a substrate and *vice versa*), they do display limited promiscuity for close analogues of their respective substrates (*e.g.* many AARSs will accept analogues of their cognate amino acid substrate in which a single H has been replaced with F).[72] Indeed, it has been demonstrated that at least 90 such non-canonical amino acids function as substrates for natural AARSs, and of these more than half are also efficiently translated by the ribosome.[72,73] Using such an approach, the synthesis of peptides containing as many as 13 non-canonical amino acids (and 2 canonical ones) has been reported; to our knowledge, this is the most heavily reprogrammed genetic code designed to date.[46]

However, the use of natural AARSs for *in vitro* genetic code reprogramming still has significant limitations. Most notably, this technique is limited to close analogues of the canonical amino acids, since more structurally dissimilar amino acids are poor substrates for the AARSs. In fact, even some very close analogues of canonical amino acids are poor substrates for the natural AARSs precluding their use (*e.g.* whilst the 4,4-difluoro analogue of Glu is a substrate for aminoacylation, the 3,3-difluoro analogue is not).[72] In general, non-canonical moieties in the side chains appear to be tolerated better than modifications of the peptide backbone. With respect to the synthesis of microbial secondary metabolite-like molecules, a particular limitation is that α-*N*-methyl amino acids appear to be very poor substrates for natural AARSs (only α-*N*-methyl-His and Asp exhibited both acylation and translation, and even this was relatively weak).[72]

Several techniques have been developed to circumvent some of the limitations of AARS-mediated genetic code reprogramming. For example, dehydroalanine-containing peptides, which cannot be directly incorporated into peptides, can be synthesized by translation of selenolysine (using the KRS) followed by oxidative elimination.[74,75] With respect to α-*N*-methyl-amino acids, the relatively poor aminoacylation of these by natural AARSs can be bypassed by performing enzymatic aminoacylation in a separate reaction, chemically *N*-methylating the α-amine and then introducing the methylated, aminoacylated tRNAs into the translation reaction.[76] In this way, efficient incorporation of *N*-methyl Val, Leu and Thr has been reported, although translation of other residues still appears to be problematic.

The great advantages of AARS-mediated genetic code reprogramming are that it is fast, relatively facile and requires little synthetic organic chemistry since many of the amino acids to be used can be obtained from commercial suppliers. However, the dependence upon the natural AARSs involves significant limitations, particularly with respect to the diversity of amino acids that can be employed. Additionally, the reliance on AARSs may increase the potential for "mis-reads" back to the canonical amino acid, since trace canonical amino acids in the reaction are likely to be a more efficient substrate for the AARS than its non-canonical analogue. Nonetheless, AARS-mediated genetic code reprogramming is an effective and relatively simple technique for the synthesis of non-canonical peptides, and is compatible with display screening techniques (*vide infra*).

11.3.4 Aminoacylation of tRNAs Catalyzed by Flexizymes

In addition to the chemo-enzymatic and AARS-mediated approaches to the synthesis of non-canonical AA-tRNAs, a third technique based on aminoacylating ribozymes (flexizymes) has been developed. These short (~45 nucleotide) RNAs were originally identified through studies of the "RNA world" hypothesis investigating the possible functions of ribozymes before the development of proteinaceous life, and were subsequently adapted for genetic code reprogramming techniques.[42,43] The amino acid substrates for flexizyme-mediated aminoacylation must be activated with an appropriate leaving group on the carboxyl moiety prior to tRNA ligation. This limitation aside, they are remarkably promiscuous with respect to both tRNA and amino acid substrates, making them ideal for the aminoacylation of diverse amino acids onto tRNAs.[77,78]

At present, three flexizymes are in general use for genetic code reprogramming applications: enhanced flexizyme (eFx), dinitro flexizyme (dFx) and amino flexizyme (aFx). All three of these recognize the universal CCA 3′ end of the tRNA substrate, with no dependence upon the tRNA body or anti-codon, meaning that they can be used for the aminoacylation of essentially any tRNA. By contrast, each flexizyme recognizes different moieties in the amino acid substrate, giving them some (very limited) degree of substrate specificity. However, their activities have been engineered so as to be complementary, such that between them almost any molecule with a carboxylic acid can be ligated onto a tRNA acceptor.

The eFx, which has the most closely related sequence to the prototype of the flexizymes, called Fx3,[78,79] recognizes an aromatic (or hydrophobic) moiety in the side chain of an amino acid substrate activated through a cyanomethyl ester (CME), or the aromatic group present in a chlorobenzyl thioester (CBT) (Figure 11.6).[78,79] The dFx recognizes the aromatic group in a dinitrobenzyl ester (DBE) with no dependence on the side chain or α-amine, and aFx similarly recognizes the aromatic moiety in an (aminoethyl)amidocarboxybenzyl thioester (ABT) in a manner independent of the structure of the amino acid itself (Figure 11.6).[77,78] Using these three catalysts, almost any amino acid (or indeed, any molecule with a carboxylic acid) can be ligated onto any tRNA, providing that it is soluble in aqueous 20% DMSO. For example, α-amino acids with both canonical and non-canonical side chains,[80–83] β-amino acids,[84] α-*N*-alkylated amino acids,[84–88] D-amino acids,[84,89,90] and even short peptides[84] have all been shown to be efficient substrates for flexizyme-catalyzed aminoacylation.

As in genetic code reprogramming techniques employing chemo-enzymatically synthesized AA-tRNAs, flexizyme-mediated acylation is performed prior to translation, with the resulting AA-tRNAs simply purified by ethanol precipitation and then added to the translation reaction. Such a combination of flexizyme-mediated AA-tRNA synthesis with a reconstituted translation system is known as flexible *in vitro* translation (FIT). Compared to AARS-mediated translation, FIT can be used to incorporate amino

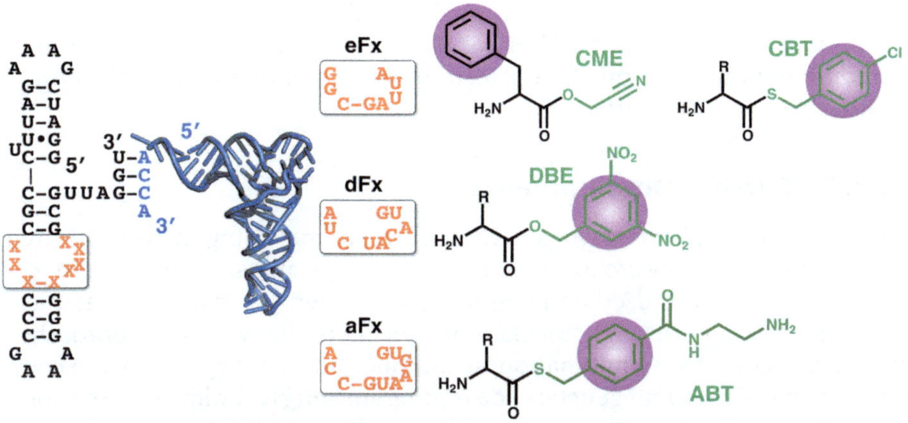

Figure 11.6 Flexizymes used for genetic code reprogramming. The conserved sequence of the three predominantly used flexizymes (eFx, dFx and aFx) is shown on the left in black, with the variable amino acid substrate binding pocket indicated in red. Hybridization of the flexizyme 3′ end to the tRNA (blue) 3′ end is shown. The center and right show the sequence of the flexizyme amino acid substrate binding pocket (red) and the structure of the amino acid substrate for each of eFx, dFx and aFx. Leaving groups are indicated in green, with the aromatic moiety recognized by the flexizyme highlighted in pink. CME = cyanomethyl ester, CBT = chlorobenzyl thioester, DBE = dinitrobenzyl ester and ABT = (aminoethyl)amidocarboxybenzyl thioester.

acids with substantially more diverse structures, since it is not limited to analogues of canonical amino acids, and the AARSs for the codons to be reprogrammed can be omitted from the reaction, suppressing mis-reading caused by trace amounts of the cognate canonical amino acid. Compared to chemo-enzymatic AA-tRNA synthesis, FIT requires only facile organic synthesis of the amino acid substrates, making it substantially less laborious. For these reasons, FIT has been applied to the translation of peptides containing many more different amino acids than either of the two aforementioned techniques, with hundreds of different amino acids incorporated to date. These include non-canonical α-amino acids,[80,91–94] α-hydroxy acids,[81,83] β-amino acids,[84] α-*N*-alkylated and acetylated amino acids,[85–88] and D-amino acids,[89,95,96] leading to the ribosomal synthesis of diverse products including peptoids, polyesters and a number of pseudo-natural peptides.[80,87,90,97]

A particularly useful application of FIT has been the synthesis of structurally diverse cyclic peptides, allowing the synthesis of macrocyclic peptide libraries. This is achieved through the inclusion of two reactive moieties which can couple, forming a non-reducible (in contrast to disulfide bonds) macrocyclic structure. Examples of this approach include cyclization through Huisgen 1,3-dipolar cycloaddition, Michael addition, native chemical ligation and oxidative coupling.[80,85,97,98] By far the most widely used technique, however, involves the incorporation of an N-terminal chloroacetyl group, which spontaneously undergoes a nucleophilic substitution reaction with the thiol of a Cys residue.[90,99–101] It has also been demonstrated that through the combination of several of these techniques, bicyclic peptides with defined architectures can be synthesized.

11.3.5 Further Developments

The first reports of extensive genetic code reprogramming in reconstituted translation reactions were more than a decade ago, and since then these systems have been optimized and engineered in order to expand the chemical structures available through translational synthesis. Many of these approaches may be thought of as "combination" techniques, in that they involve the combination of two different genetic code reprogramming techniques, or the combination of a genetic code reprogramming technique with post-translational chemical modification, such as the chemical cross-linking used to produce macrocyclic structures for phage display (*vide supra*). Indeed, for AARS-mediated reprogramming approaches, such cross-linking (through the use of dibromoxylene) is, to the best of our knowledge, the only approach that has been demonstrated to yield non-reducible macrocyclic structures, since pairs of spontaneously self-reactive amino acids compatible with protein AARSs have not been described.[46] A similar approach has also been described in combination with FIT reprogramming, in this case using a tribromoxylene in combination with genetically reprogrammed chloroacetyl/thioether cyclization to yield a tri-cyclic peptide with a defined architecture.[102] Combination techniques involving the combined use of both FIT and AARS-mediated genetic code reprogramming have also been described. For example, four non-canonical amino acids were incorporated using an AARS-mediated strategy, with cyclization effected using a FIT-mediated chloroacetyl/thioether approach.[103]

Recent technical developments have further broadened the scope of *in vitro* genetic code reprogramming techniques, by improving the translational efficiency of particular amino acids/peptide motifs. For example, tRNA engineering has been shown to improve the translation of certain non-canonical amino acids, and to allow the synthesis of peptides containing multiple consecutive D-residues.[96,104,105]

Other recent developments have altered the nature of the reprogramming itself. For example, through the use of multiple initiator tRNAs with differing anti-codons, it has been shown that peptide libraries composed of molecules with differing cyclization chemistries can be synthesized.[106] A second example is the demonstration that by removing all of the canonical tRNAs from the translation reaction and replacing them with *in vitro* transcribed variants, it is possible to "split" groups of codons assigned to the same canonical amino acid, such that the canonical amino acid remains translated at one (or more) codon, but a non-canonical amino acid is translated at another.[91] For example, the Val GUG codon can be retained by inclusion of *in vitro* transcribed Val tRNA with a CAC anti-codon, but the Val GUC codon can be reprogrammed to a non-canonical residue by inclusion of a (flexizyme generated) non-canonical AA-tRNA with a GAC anti-codon (Figure 11.7). This may be of particular

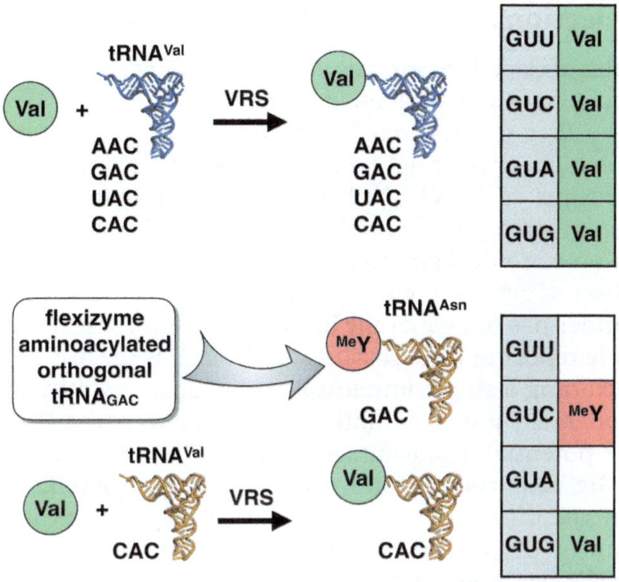

Figure 11.7 Reprogrammed division of codon "boxes". Upper panel: during canonical translation of the four natural Val codons, the four Val tRNAs ($tRNA^{Val}$) with the relevant anti-codons (5′ to 3′ AAC, GAC, UAC and CAC) are amino acylated with Val by the Val tRNA synthetase (VRS) leading to the assignment of Val at each codon. Lower panel: by removing the endogenous tRNAs from the reaction and replacing them with a single, *in vitro* transcribed, analogue bearing a CAC anti-codon, it is possible to specify Val at only one of the natural codons (in this case, the GUG codon). This "vacates" the other codons for reprogramming using pre-aminoacylated, orthologous tRNAs (in this case, an Asn tRNA aminoacylated with an α-*N*-methyl Tyr amino acid (^{Me}Y) prepared using a flexizyme), allowing the translation of both a canonical and non-canonical residue at different codons. Endogenous tRNAs are shown in blue, and *in vitro* transcribed tRNAs are shown in orange.

significance to affinity selection-based techniques for the isolation of macrocyclic peptide ligands from genetically reprogrammed libraries (*vide infra*), since it allows, in principle, the synthesis and screening of libraries containing more than 20 amino acids, thereby increasing the accessible chemical diversity.

11.4 Genetically Engineered Selections of Target-binding Macrocyclic Peptides

As discussed above, genetic code manipulation techniques enable mRNA-templated synthesis of non-canonical macrocyclic peptides that mimic naturally occurring peptide secondary metabolites. Thus, the combination of such techniques with cellular or *in vitro* selection methods greatly facilitates the discovery of drug-like macrocyclic peptides exhibiting high binding affinities for specific targets.

11.4.1 Selections Involving Genetic Code Expansion

Early examples of selection campaigns for peptides with non-canonical amino acids utilized genetic code expansion strategies rather than reprogramming methods. The first reports may be considered as proof-of-concept experiments in which linear peptides bearing a biotin-containing amino acid were selected by means of mRNA display.[107] Thereafter, many examples utilizing different display strategies for *in vitro* selection of linear peptides or proteins containing non-canonical amino acids have been reported.[108–111] By contrast, the application of genetic code expansion methods to the selection of macrocyclic peptides has not generally been attempted. To the best of our knowledge, the only report of a successful selection thus far is a SICLOPPS-based selection including a single unnatural amino acid, resulting in the identification of moderately potent inhibitors ($IC_{50} \sim 1\ \mu M$) of the HIV protease.[112]

One of the potential disadvantages in using genetic code expansion techniques for the construction and selection of macrocyclic peptide libraries would be the difficulty encountered in strategies for the simultaneous incorporation of the diverse non-canonical structures found in naturally occurring cyclic peptides, limiting the chemical diversity. One possible way to alleviate this disadvantage may be ribosomal synthesis of macrocyclic organo-peptide hybrids that have been demonstrated by Fasan *et al.*[113–116] In this strategy, small organic precursors are post-translationally introduced onto non-canonical amino acids through specific functional groups incorporated *via* the amber codon. In combination with intein constructs, this approach can be used to induce macrocyclization through the reaction of a nucleophilic non-canonical side chain with the thioether intermediate formed during intein splicing, yielding non-canonical macrocyclic peptides that mimic natural products. By "feeding" genetically engineered bacteria an appropriate non-canonical amino acid, this approach can be used for the *in vivo* synthesis of non-canonical macrocyclic peptides that mimic natural products, and is thus, in principle, applicable to SICLOPPS-based selections. This may allow the discovery of novel organo-peptide hybrids with potent bioactivities in the future.

The appeal of such approaches notwithstanding, an alternative, and more proven strategy for the discovery of artificial cyclic peptide mimics of naturally occurring peptides with multiple non-canonical residues or backbone modifications is the integration of *in vitro* selection methods with genetic code reprogramming technologies. Such combined strategies based on ARS-mediated or flexizyme-mediated genetic code reprogramming are discussed in turn below.

11.4.2 Selections Involving ARS-mediated Genetic Code Reprogramming

The combination of ARS-mediated genetic code reprogramming with mRNA display has been demonstrated to be an effective strategy for the identification of non-canonical macrocyclic peptides with strong affinity to a desired protein target. For example, an mRNA-displayed peptide library consisting of 10 randomized amino acids flanked by two cysteine residues was ribosomally constructed by means of a reprogrammed genetic code involving 8 canonical and 12 non-canonical amino acids, which were charged onto appropriate endogenous tRNAs by the 20 canonical ARSs.[46] The linear peptide library was then subjected to chemical macrocyclization using dibromoxylene to link two cysteine residues. *In vitro* selection experiments based on this library isolated potent inhibitors of the human plasma protease thrombin, which showed binding and inhibition constants in the low nanomolar range.

ARS-mediated genetic code reprogramming has also been applied to the construction and screening of an artificial lanthipeptide library. Seebeck *et al.* incorporated a dehydroalanine into an mRNA-displayed peptide library *via* selenolysine as described above, and subsequently conducted an intramolecular Michael-type reaction with a cysteine to form a lanthionine residue forming a macrocycle.[117] This artificial lanthipeptide library was subjected to *in vitro* selection against the bacterial target sortase A. Unfortunately, the resulting peptides only had moderately strong binding affinities (low micromolar range) for the target and did not show inhibition. Nonetheless, this report demonstrates the potential of this combination strategy for the discovery of natural product-like peptides from genetically reprogrammed libraries.

11.4.3 Selections Involving FIT-mediated Genetic Code Reprogramming

An advantage of FIT-mediated genetic code reprogramming over ARS-mediated strategies is that non-canonical amino acids not analogous to proteinogenic amino acids can be used. This advantage allows for the construction of thioether-closed macrocyclic peptides containing modified backbones or mechanism-based warhead residues.

Taking advantage of the ability of the FIT system to translate *N*-methylated, D-, and *N*-chloroacetylated amino acids (which can induce spontaneous macrocyclization with a downstream cysteine as described above), a macrocyclic *N*-methylated peptide library was constructed and screened for inhibitors of the E6AP ubiquitin ligase by means of a modified mRNA display method.[99] The resulting macrocyclic E6AP inhibitors exhibited extremely high affinity for the target, with dissociation constants in the picomolar range, and inhibited E6AP-catalyzed ubiquitination by disrupting the protein–protein interaction between E6AP and the E2 ubiquitin-conjugating enzyme. It should be noted that the inhibitors obtained here were arguably more like peptidic natural products than any other non-natural macrocyclic peptide inhibitors discovered to date – *e.g.* one of them was a 13-mer macrocyclic peptide containing one D-amino acid as well as four *N*-methylated amino acids.

As a means of directing such screening processes to identify inhibitors of a specific class of enzymes, selection using a macrocyclic peptide library bearing a non-canonical amino acid targeting the active site of the enzyme of interest has also been reported.[92] In this case, FIT-mediated genetic code reprogramming enabled the ribosomal incorporation of trifluoroacetyl-lysine (K^{Tfa}), a known mechanism-based warhead of the NAD-dependent deacetylase (sirtuin) family. Warhead residue-containing inhibitors selected from this library exhibited high potency ($IC_{50} \sim 100$ nM) and a high degree of isoform specificity for sirtuin 2 over sirtuins 1 and 3. The crystal structure of sirtuin 2 in complex with a representative inhibitor revealed that the inhibitor induces a dynamic structural change in sirtuin 2, illustrating the potential of macrocyclic peptides as isoform-specific ligands.

The combination strategy of FIT with mRNA display, as described in the two examples above, has been named *ra*ndom non-standard *p*eptides *i*ntegrated *d*iscovery (RaPID). In addition to the examples already discussed, RaPID screening has been used to discover many bioactive macrocyclic peptides that demonstrate strong and selective affinity for diverse therapeutic targets.[92,99–101,118–125] The targets of the peptides obtained by the RaPID system include not only enzymes but also membrane proteins such as receptors and transporters. The modes of action of the peptides identified are also diverse with examples of activity *via* enzymatic inhibition by binding to the active site, inhibition of protein–protein interaction, or allosteric inhibition all observed. Some such peptide ligands have also been used as cocrystallization chaperones of intractable transmembrane proteins in order to facilitate crystallization and phase determination.[101,118,119] Additionally, this system was used to develop agonists of the tyrosine kinase receptor cMet – homodimers of the cMet-binding peptides identified through RaPID screening induced dimerization and activation of cMet on the cell membrane.[120] These examples have clearly demonstrated the potential of the RaPID system as a platform technology enabling development of non-canonical cyclic peptide ligands targeting diverse proteins of interest.

Recently, several co-crystal structures of RaPID-identified macrocyclic peptides in complex with their protein targets have been reported, revealing at a molecular level how the macrocyclic scaffolds contribute not only to target interaction but also to the rigidity of their diverse tertiary structures. Among the nine protein structures with thioether-macrocyclic peptide ligands isolated by the RaPID system that have been reported to date, we discuss here three representatives (Figure 11.8). First, a K^{Tfa}-containing cyclic peptide, S2iL5, inhibiting sirtuin 2 adopts a ring-shaped structure.[126] Multiple direct and water-mediated hydrogen bonds stabilize the scaffold with the guanidinium group of R8 flipped within the cyclic scaffold, while a number of intramolecular hydrophobic and hydrophilic interactions orient $K^{Tfa}7$ to extend outside of the ring and occupy the active site of sirtuin 2 (Figure 11.8, panels a and b). Alternatively, thioether-closed macrocyclic peptides can adopt unique conformations that mimic the specific secondary structures found in proteins. One such example is the 17-mer macrocyclic peptide MaL6, that binds to a multidrug and toxic compound extrusion (MATE) transporter and stabilizes its outward-open conformation.[118] The co-crystal structure illustrated compact turn motifs connecting two β-strands in MaL6, which are stabilized by several intramolecular hydrogen bonds (Figure 11.8,

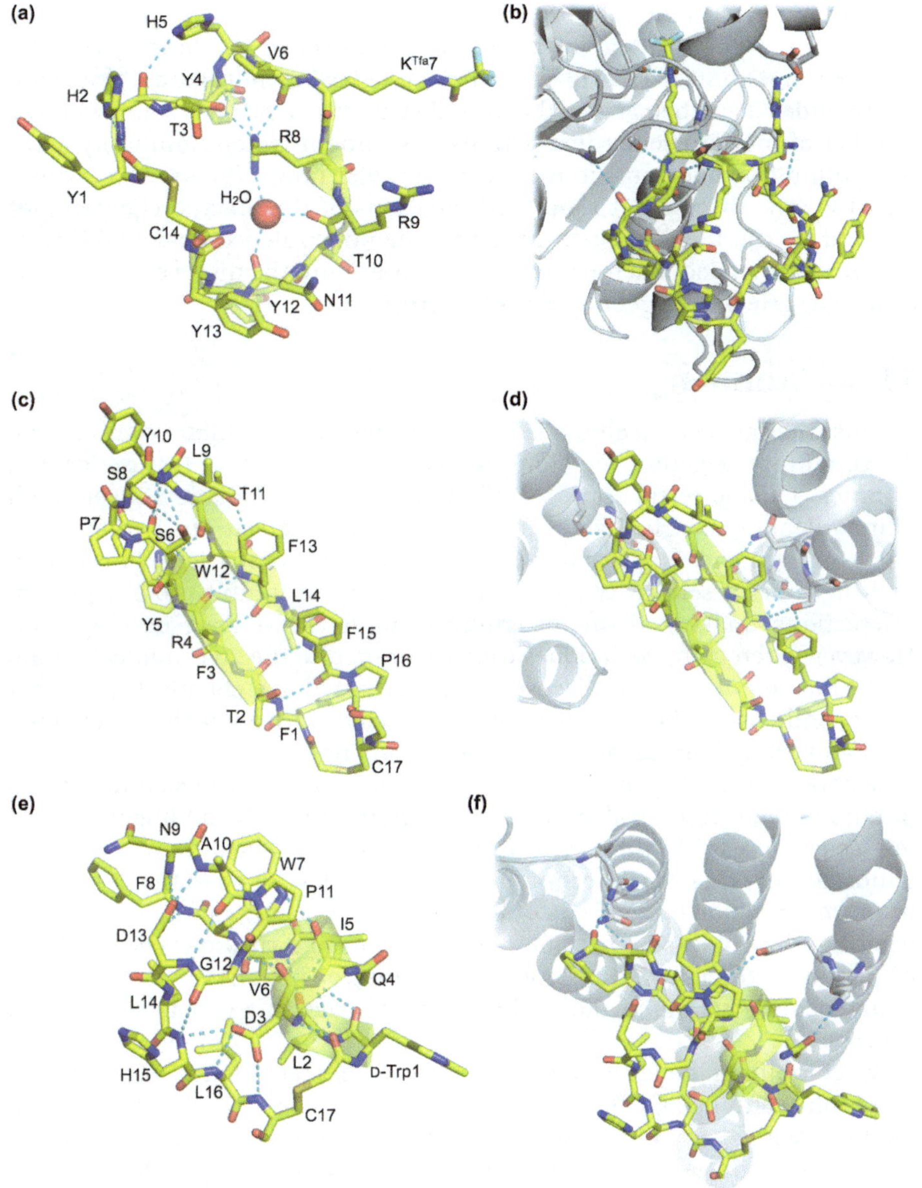

Figure 11.8 Crystal structures of macrocyclic peptide ligand–protein complexes. The peptide ligands are shown with the following element color code; carbon in yellow, nitrogen in blue, oxygen in red, and sulfur in orange. (a and b) Ring shape conformation of the S2iL5 peptide in complex with SIRT2 (PDB ID: 4L3O). (c and d) Compact strand–turn–strand conformation of the MaL6 peptide in complex with a MATE transporter (PDB ID: 3WBN). (e and f) Partial α-helical conformation of the aCAP peptide in complex with CmABCB1 (PDB ID: 3WMG). Intramolecular and intermolecular hydrogen bonds are shown as cyan dashed lines in panels a, c and e, and panels b, d and f, respectively.

panels c and d). In another example, a 17-mer macrocyclic peptide having one D-amino acid, aCAP, acting as a surface clamp of an ATP-binding cassette transporter (CmABCB1) involves a partial α-helical conformation.[119] As with the peptides described above, the overall structure of aCAP is stabilized by a number of intramolecular interactions, resulting in potent inhibitory activities (inhibition constants in the low nanomolar range) (Figure 11.8, panels e and f). The diverse but rigid conformations of the macrocyclic peptides isolated by the RaPID system have thus demonstrated the adaptability of genetically encoded macrocyclic peptide libraries to the discovery of ligands against a variety of target proteins of interest.

11.5 Summary

Because they are genetically encoded, translated peptide libraries are amenable to affinity selection screening of very high diversity (>10^{12} molecules) libraries. This powerful approach allows for the rapid isolation of very high affinity ligands to virtually any protein target of interest. The peptides isolated through such approaches commonly exhibit strong target inhibition and high target selectivity and have the ability to inhibit protein–protein interactions, making them intriguing candidates for drug development. However, when using techniques that are limited to the 20 canonical amino acids, the peptide ligands discovered are restricted to a chemical space that is not conducive to the identification of molecules with good pharmacokinetics, hampering their further clinical development.

Genetic code reprogramming techniques allow for the translational incorporation of non-canonical amino acids, thereby greatly expanding the chemical space accessible through ribosomal synthesis. Since the first reports almost 30 years ago, such techniques have been developed to the point that, in *in vitro* reactions, hundreds of different amino acids can be used as the basic building blocks for peptide synthesis. This in turn allows the synthesis of peptides containing numerous non-canonical moieties (*N*-methylation, D-stereochemistry, non-canonical side chains, *etc.*) in a genetically defined process.

By combining genetic code reprogramming techniques with affinity selection techniques, it is possible to identify macrocyclic non-canonical peptides with many of the structural features of drug-like microbial secondary metabolites. Such compounds exhibit very high target affinities and selectivities, and greatly improved peptidase resistance and cell membrane permeability relative to canonical peptides, making them ideal candidates for drug development.

Currently used techniques in this field already allow the rapid identification of drug-like macrocyclic peptides, and we anticipate that over the coming years a number of clinically useful drugs will be developed from molecules identified through such screens. However, further improvements to genetic code reprogramming and screening techniques can still be made. We expect that the development of synthetic technologies based on ribosomal translation will continue apace, with novel amino acid building blocks further expanding the available chemistries. In addition, engineering of the translation system itself and/or the development of chemical or enzymatic post-translational modification will allow for the synthesis of highly diverse peptidic structures.

Acknowledgements

This work was supported by the Japan Science and Technology Agency (JST) Core Research for Evolutional Science and Technology (CREST) for JPMJCR12L2, the Platform Project for Supporting in Drug Discovery and Life Science Research (Platform for Drug Discovery, Informatics, and Structural Life Science) from the Ministry of Education, Culture, Sports, and Science (MEXT) and the Japan Agency for Medical Research and development (AMED), AMED Basic Science and Platform Technology Program for Innovative Biological Medicine, to H.S., and Japan Society for Promotion of Science KAKENHI grants to Y.G. (15K12739, 16H01131, and 17H04762), T.K (26560429) and T.P. (15K16558).

References

1. A. L. Hopkins and C. R. Groom, *Nat. Rev. Drug Discovery*, 2002, **1**, 727–730.
2. T. Wieland, *Int. J. Pept. Protein Res.*, 1983, **22**, 257–276.
3. M. Gao, K. Cheng and H. Yin, *Biopolymers*, 2015, **104**, 310–316.
4. N. Takahashi, T. Hayano and M. Suzuki, *Nature*, 1989, **337**, 473–475.
5. G. Fischer, B. Wittmann-Liebold, K. Lang, T. Kiefhaber and F. X. Schmid, *Nature*, 1989, **337**, 476–478.
6. E. M. Driggers, S. P. Hale, J. Lee and N. K. Terrett, *Nat. Rev. Drug Discovery*, 2008, **7**, 608–624.
7. D. S. Perlin, *Future Microbiol.*, 2011, **6**, 441–457.
8. J. G. Beck, J. Chatterjee, B. Laufer, M. U. Kiran, A. O. Frank, S. Neubauer, O. Ovadia, S. Greenberg, C. Gilon, A. Hoffman and H. Kessler, *J. Am. Chem. Soc.*, 2012, **134**, 12125–12133.
9. O. Ovadia, S. Greenberg, J. Chatterjee, B. Laufer, F. Opperer, H. Kessler, C. Gilon and A. Hoffman, *Mol. Pharm.*, 2011, **8**, 479–487.
10. E. Biron, J. Chatterjee, O. Ovadia, D. Langenegger, J. Brueggen, D. Hoyer, H. A. Schmid, R. Jelinek, C. Gilon, A. Hoffman and H. Kessler, *Angew. Chem., Int. Ed. Engl.*, 2008, **47**, 2595–2599.
11. C. L. Ahlbach, K. W. Lexa, A. T. Bockus, V. Chen, P. Crews, M. P. Jacobson and R. S. Lokey, *Future Med. Chem.*, 2015, **7**, 2121–2130.
12. R. Liu, X. Li, W. Xiao and K. S. Lam, *Adv. Drug Delivery Rev.*, 2017, **110–111**, 13–37.
13. P. Kane, C. Yamashiro, D. Wolczyk, N. Neff, M. Goebl and T. Stevens, *Science*, 1990, **250**, 651–657.
14. G. Volkmann and H. D. Mootz, *Cell. Mol. Life Sci.*, 2013, **70**, 1185–1206.

15. C. P. Scott, E. Abel-Santos, M. Wall, D. C. Wahnon and S. J. Benkovic, *Proc. Natl. Acad. Sci.*, 1999, **96**, 13638–13643.
16. C. P. Scott, E. Abel-Santos, A. D. Jones and S. J. Benkovic, *Chem. Biol.*, 2001, **8**, 801–815.
17. J. A. Kritzer, S. Hamamichi, J. M. McCaffery, S. Santagata, T. A. Naumann, K. A. Caldwell, G. A. Caldwell and S. Lindquist, *Nat. Chem. Biol.*, 2009, **5**, 655–663.
18. J. E. Townend and A. Tavassoli, *ACS Chem. Biol.*, 2016, **11**, 1624–1630.
19. A. R. Horswill and S. J. Benkovic, *Curr. Protoc. Protein Sci.*, 2006, **46**, 19.15.1–19.15.19.
20. A. R. Horswill, S. N. Savinov and S. J. Benkovic, *Proc. Natl. Acad. Sci. U. S. A.*, 2004, **101**, 15591–15596.
21. A. Tavassoli and S. J. Benkovic, *Angew. Chem., Int. Ed. Engl.*, 2005, **44**, 2760–2763.
22. A. Tavassoli, Q. Lu, J. Gam, H. Pan, S. J. Benkovic and S. N. Cohen, *ACS Chem. Biol.*, 2008, **3**, 757–764.
23. G. P. Smith, *Science*, 1985, **228**, 1315–1317.
24. T. Clackson, H. R. Hoogenboom, A. D. Griffiths and G. Winter, *Nature*, 1991, **352**, 624–628.
25. G. Winter, A. D. Griffiths, R. E. Hawkins and H. R. Hoogenboom, *Annu. Rev. Immunol.*, 1994, **12**, 433–455.
26. T. J. Vaughan, A. J. Williams, K. Pritchard, J. K. Osbourn, A. R. Pope, J. C. Earnshaw, J. McCafferty, R. A. Hodits, J. Wilton and K. S. Johnson, *Nat. Biotechnol.*, 1996, **14**, 309–314.
27. M. Hust, A. Frenzel, T. Schirrmann and S. Dubel, *Methods Mol. Biol.*, 2014, **1101**, 305–320.
28. H. Qi, H. Lu, H. J. Qiu, V. Petrenko and A. Liu, *J. Mol. Biol.*, 2012, **417**, 129–143.
29. A. Fagerlund, A. H. Myrset and M. A. Kulseth, *Methods Mol. Biol.*, 2014, **1088**, 19–33.
30. K. Fukunaga, T. Hatanaka, Y. Ito, M. Minami and M. Taki, *Chem. Commun. (Camb.)*, 2014, **50**, 3921–3923.
31. S. Kalhor-Monfared, M. R. Jafari, J. T. Patterson, P. I. Kitov, J. J. Dwyer, J. M. Nuss and R. Derda, *Chem. Sci.*, 2016, **7**, 3785–3790.
32. S. Ng and R. Derda, *Org. Biomol. Chem.*, 2016, **14**, 5539–5545.
33. M. R. Jafari, J. Lakusta, R. J. Lundgren and R. Derda, *Bioconjugate Chem.*, 2016, **27**, 509–514.
34. C. Heinis, T. Rutherford, S. Freund and G. Winter, *Nat. Chem. Biol.*, 2009, **5**, 502–507.
35. D. Bertoldo, M. M. Khan, P. Dessen, W. Held, J. Huelsken and C. Heinis, *ChemMedChem*, 2016, **11**, 834–839.
36. R. W. Roberts and J. W. Szostak, *Proc. Natl. Acad. Sci. U. S. A.*, 1997, **94**, 12297–12302.
37. N. Nemoto, E. MiyamotoSato, Y. Husimi and H. Yanagawa, *FEBS Lett.*, 1997, **414**, 405–408.

38. J. Yamaguchi, M. Naimuddin, M. Biyani, T. Sasaki, M. Machida, T. Kubo, T. Funatsu, Y. Husimi and N. Nemoto, *Nucleic Acids Res.*, 2009, **37**, e108.
39. L. C. Mattheakis, R. R. Bhatt and W. J. Dower, *Proc. Natl. Acad. Sci. U. S. A.*, 1994, **91**, 9022–9026.
40. Y. Shimizu, A. Inoue, Y. Tomari, T. Suzuki, T. Yokogawa, K. Nishikawa and T. Ueda, *Nat. Biotechnol.*, 2001, **19**, 751–755.
41. T. Kanamori, Y. Fujino and T. Ueda, *Biochim. Biophys. Acta*, 2014, **1844**, 1925–1932.
42. Y. Goto and H. Suga, *Methods Mol. Biol.*, 2012, **848**, 465–478.
43. J. Morimoto, Y. Hayashi, K. Iwasaki and H. Suga, *Acc. Chem. Res.*, 2011, **44**, 1359–1368.
44. S. Ueno, H. Arai, M. Suzuk and Y. Husimi, *Int. J. Biol. Sci.*, 2007, **3**, 365–374.
45. S. Fujita, J. M. Zhou and K. Taira, *Methods Mol Biol*, 2007, **352**, 221–236.
46. Y. V. Schlippe, M. C. Hartman, K. Josephson and J. W. Szostak, *J. Am. Chem. Soc.*, 2012, **134**, 10469–10477.
47. S. W. Millward, T. T. Takahashi and R. W. Roberts, *J. Am. Chem. Soc.*, 2005, **127**, 14142–14143.
48. T. Rezai, B. Yu, G. L. Millhauser, M. P. Jacobson and R. S. Lokey, *J. Am. Chem. Soc.*, 2006, **128**, 2510–2511.
49. T. R. White, C. M. Renzelman, A. C. Rand, T. Rezai, C. M. McEwen, V. M. Gelev, R. A. Turner, R. G. Linington, S. S. Leung, A. S. Kalgutkar, J. N. Bauman, Y. Zhang, S. Liras, D. A. Price, A. M. Mathiowetz, M. P. Jacobson and R. S. Lokey, *Nat. Chem. Biol.*, 2011, **7**, 810–817.
50. M. Strieker, A. Tanovic and M. A. Marahiel, *Curr. Opin. Struct. Biol.*, 2010, **20**, 234–240.
51. V. Siewers, R. San-Bento and J. Nielsen, *Biotechnol. Bioeng.*, 2010, **106**, 841–844.
52. B. Wilkinson and J. Micklefield, *Nat. Chem. Biol.*, 2007, **3**, 379–386.
53. J. Yin, F. Liu, M. Schinke, C. Daly and C. T. Walsh, *J. Am. Chem. Soc.*, 2004, **126**, 13570–13571.
54. M. A. Ortega and W. A. van der Donk, *Cell Chem. Biol.*, 2016, **23**, 31–44.
55. Y. Goto, Y. Ito, Y. Kato, S. Tsunoda and H. Suga, *Chem. Biol.*, 2014, **21**, 766–774.
56. P. J. Knerr, T. J. Oman, C. V. Garcia De Gonzalo, T. J. Lupoli, S. Walker and W. A. van der Donk, *ACS Chem. Biol.*, 2012, **7**, 1791–1795.
57. J. A. Ellman, D. Mendel and P. G. Schultz, *Science*, 1992, **255**, 197–200.
58. C. J. Noren, S. J. Anthony-Cahill, M. C. Griffith and P. G. Schultz, *Science*, 1989, **244**, 182–188.
59. T. Hohsaka, M. Fukushima and M. Sisido, *Nucleic Acids Res. Suppl.*, 2002, 201–202.
60. T. Hohsaka, N. Muranaka, C. Komiyama, K. Matsui, S. Takaura, R. Abe, H. Murakami and M. Sisido, *FEBS Lett.*, 2004, **560**, 173–177.

61. H. Taira, M. Fukushima, T. Hohsaka and M. Sisido, *J. Biosci. Bioeng.*, 2005, **99**, 473–476.
62. L. Davis and J. W. Chin, *Nat. Rev. Mol. Cell Biol.*, 2012, **13**, 168–182.
63. H. Neumann, K. Wang, L. Davis, M. Garcia-Alai and J. W. Chin, *Nature*, 2010, **464**, 441–444.
64. C. C. Liu and P. G. Schultz, *Annu. Rev. Biochem.*, 2010, **79**, 413–444.
65. J. D. Bain, E. S. Diala, C. G. Glabe, D. A. Wacker, M. H. Lyttle, T. A. Dix and A. R. Chamberlin, *Biochemistry (Mosc.)*, 1991, **30**, 5411–5421.
66. J. D. Bain, D. A. Wacker, E. E. Kuo and A. R. Chamberlin, *Tetrahedron*, 1991, **47**, 2389–2400.
67. A. C. Forster, Z. Tan, M. N. Nalam, H. Lin, H. Qu, V. W. Cornish and S. C. Blacklow, *Proc. Natl. Acad. Sci. U. S. A.*, 2003, **100**, 6353–6357.
68. Y. Goto, T. Katoh and H. Suga, *Nat. Protoc.*, 2011, **6**, 779–790.
69. Z. Tan, S. C. Blacklow, V. W. Cornish and A. C. Forster, *Methods*, 2005, **36**, 279–290.
70. B. Zhang, Z. Tan, L. G. Dickson, M. N. Nalam, V. W. Cornish and A. C. Forster, *J. Am. Chem. Soc.*, 2007, **129**, 11316–11317.
71. M. Y. Pavlov, R. E. Watts, Z. Tan, V. W. Cornish, M. Ehrenberg and A. C. Forster, *Proc. Natl. Acad. Sci. U. S. A.*, 2009, **106**, 50–54.
72. M. C. Hartman, K. Josephson and J. W. Szostak, *Proc. Natl. Acad. Sci. U. S. A.*, 2006, **103**, 4356–4361.
73. M. C. Hartman, K. Josephson, C. W. Lin and J. W. Szostak, *PLoS One*, 2007, **2**, e972.
74. F. P. Seebeck and J. W. Szostak, *J. Am. Chem. Soc.*, 2006, **128**, 7150–7151.
75. F. P. Seebeck, A. Ricardo and J. W. Szostak, *Chem. Commun. (Camb.)*, 2011, **47**, 6141–6143.
76. A. O. Subtelny, M. C. Hartman and J. W. Szostak, *J. Am. Chem. Soc.*, 2008, **130**, 6131–6136.
77. N. Niwa, Y. Yamagishi, H. Murakami and H. Suga, *Bioorg. Med. Chem. Lett.*, 2009, **19**, 3892–3894.
78. H. Murakami, A. Ohta, H. Ashigai and H. Suga, *Nat. Methods*, 2006, **3**, 357–359.
79. H. Murakami, H. Saito and H. Suga, *Chem. Biol.*, 2003, **10**, 655–662.
80. Y. Goto, K. Iwasaki, K. Torikai, H. Murakami and H. Suga, *Chem. Commun. (Camb.)*, 2009, 3419–3421.
81. A. Ohta, H. Murakami and H. Suga, *ChemBioChem*, 2008, **9**, 2773–2778.
82. T. J. Kang and H. Suga, *Biochem. Cell Biol.*, 2008, **86**, 92–99.
83. A. Ohta, H. Murakami, E. Higashimura and H. Suga, *Chem. Biol.*, 2007, **14**, 1315–1322.
84. Y. Goto and H. Suga, *J. Am. Chem. Soc.*, 2009, **131**, 5040–5041.
85. Y. Yamagishi, H. Ashigai, Y. Goto, H. Murakami and H. Suga, *ChemBioChem*, 2009, **10**, 1469–1472.
86. T. Kawakami, H. Murakami and H. Suga, *Chem. Biol.*, 2008, **15**, 32–42.
87. T. Kawakami, H. Murakami and H. Suga, *J. Am. Chem. Soc.*, 2008, **130**, 16861–16863.

88. Y. Goto, H. Ashigai, Y. Sako, H. Murakami and H. Suga, *Nucleic Acids Symp. Ser. (Oxf.)*, 2006, 293–294.
89. T. Fujino, Y. Goto, H. Suga and H. Murakami, *J. Am. Chem. Soc.*, 2013, **135**, 1830–1837.
90. Y. Goto, A. Ohta, Y. Sako, Y. Yamagishi, H. Murakami and H. Suga, *ACS Chem. Biol.*, 2008, **3**, 120–129.
91. Y. Iwane, A. Hitomi, H. Murakami, T. Katoh, Y. Goto and H. Suga, *Nat. Chem.*, 2016, **8**, 317–325.
92. J. Morimoto, Y. Hayashi and H. Suga, *Angew. Chem., Int. Ed. Engl.*, 2012, **51**, 3423–3427.
93. K. Ojemalm, T. Higuchi, P. Lara, E. Lindahl, H. Suga and G. von Heijne, *Proc. Natl. Acad. Sci. U. S. A.*, 2016, **113**, 10559–10564.
94. K. Ojemalm, T. Higuchi, Y. Jiang, U. Langel, I. Nilsson, S. H. White, H. Suga and G. von Heijne, *Proc. Natl. Acad. Sci. U. S. A.*, 2011, **108**, E359–E364.
95. Y. Goto, H. Murakami and H. Suga, *RNA*, 2008, **14**, 1390–1398.
96. T. Katoh, K. Tajima and H. Suga, *Cell Chem. Biol.*, 2017, **24**, 46–54.
97. T. Kawakami, A. Ohta, M. Ohuchi, H. Ashigai, H. Murakami and H. Suga, *Nat. Chem. Biol.*, 2009, **5**, 888–890.
98. Y. Sako, J. Morimoto, H. Murakami and H. Suga, *J. Am. Chem. Soc.*, 2008, **130**, 7232–7234.
99. Y. Yamagishi, I. Shoji, S. Miyagawa, T. Kawakami, T. Katoh, Y. Goto and H. Suga, *Chem. Biol.*, 2011, **18**, 1562–1570.
100. Y. Hayashi, J. Morimoto and H. Suga, *ACS Chem. Biol.*, 2012, **7**, 607–613.
101. C. J. Hipolito, Y. Tanaka, T. Katoh, O. Nureki and H. Suga, *Molecules*, 2013, **18**, 10514–10530.
102. N. K. Bashiruddin, M. Nagano and H. Suga, *Bioorg. Chem.*, 2015, **61**, 45–50.
103. T. Kawakami, T. Ishizawa, T. Fujino, P. C. Reid, H. Suga and H. Murakami, *ACS Chem. Biol.*, 2013, **8**, 1205–1214.
104. K. W. Ieong, M. Y. Pavlov, M. Kwiatkowski, M. Ehrenberg and A. C. Forster, *RNA*, 2014, **20**, 632–643.
105. J. Wang, M. Kwiatkowski, M. Y. Pavlov, M. Ehrenberg and A. C. Forster, *ACS Chem. Biol.*, 2014, **9**, 1303–1311.
106. Y. Goto, M. Iseki, A. Hitomi, H. Murakami and H. Suga, *ACS Chem. Biol.*, 2013, **8**, 2630–2634.
107. A. Frankel and R. W. Roberts, *RNA*, 2003, **9**, 780–786.
108. C. C. Liu, A. V. Mack, E. M. Brustad, J. H. Mills, D. Groff, V. V. Smider and P. G. Schultz, *J. Am. Chem. Soc.*, 2009, **131**, 9616–9617.
109. C. C. Liu, A. V. Mack, M. L. Tsao, J. H. Mills, H. S. Lee, H. Choe, M. Farzan, P. G. Schultz and V. V. Smider, *Proc. Natl. Acad. Sci. U. S. A.*, 2008, **105**, 17688–17693.
110. M. Liu, S. Tada, M. Ito, H. Abe and Y. Ito, *Chem. Commun. (Camb.)*, 2012, **48**, 11871–11873.
111. N. Muranaka, T. Hohsaka and M. Sisido, *Nucleic Acids Res.*, 2006, **34**, e7.

112. T. S. Young, D. D. Young, I. Ahmad, J. M. Louis, S. J. Benkovic and P. G. Schultz, *Proc. Natl. Acad. Sci. U. S. A.*, 2011, **108**, 11052–11056.
113. J. M. Smith, F. Vitali, S. A. Archer and R. Fasan, *Angew. Chem., Int. Ed. Engl.*, 2011, **50**, 5075–5080.
114. M. Satyanarayana, F. Vitali, J. R. Frost and R. Fasan, *Chem. Commun. (Camb.)*, 2012, **48**, 1461–1463.
115. J. R. Frost, F. Vitali, N. T. Jacob, M. D. Brown and R. Fasan, *ChemBioChem*, 2013, **14**, 147–160.
116. J. R. Frost, N. T. Jacob, L. J. Papa, A. E. Owens and R. Fasan, *ACS Chem. Biol.*, 2015, **10**, 1805–1816.
117. F. T. Hofmann, J. W. Szostak and F. P. Seebeck, *J. Am. Chem. Soc.*, 2012, **134**, 8038–8041.
118. Y. Tanaka, C. J. Hipolito, A. D. Maturana, K. Ito, T. Kuroda, T. Higuchi, T. Katoh, H. E. Kato, M. Hattori, K. Kumazaki, T. Tsukazaki, R. Ishitani, H. Suga and O. Nureki, *Nature*, 2013, **496**, 247–251.
119. A. Kodan, T. Yamaguchi, T. Nakatsu, K. Sakiyama, C. J. Hipolito, A. Fujioka, R. Hirokane, K. Ikeguchi, B. Watanabe, J. Hiratake, Y. Kimura, H. Suga, K. Ueda and H. Kato, *Proc. Natl. Acad. Sci. U. S. A.*, 2014, **111**, 4049–4054.
120. K. Ito, K. Sakai, Y. Suzuki, N. Ozawa, T. Hatta, T. Natsume, K. Matsumoto and H. Suga, *Nat. Commun.*, 2015, **6**, 6373.
121. K. Iwasaki, Y. Goto, T. Katoh, T. Yamashita, S. Kaneko and H. Suga, *J. Mol. Evol.*, 2015, **81**, 210–217.
122. Y. Matsunaga, N. K. Bashiruddin, Y. Kitago, J. Takagi and H. Suga, *Cell Chem. Biol.*, 2016, **23**, 1341–1350.
123. S. A. Jongkees, S. Caner, C. Tysoe, G. D. Brayer, S. G. Withers and H. Suga, *Cell Chem. Biol.*, 2017, **24**, 381–390.
124. H. Yu, P. Dranchak, Z. Li, R. MacArthur, M. S. Munson, N. Mehzabeen, N. J. Baird, K. P. Battalie, D. Ross, S. Lovell, C. K. Carlow, H. Suga and J. Inglese, *Nat. Commun.*, 2017, **8**, 14932.
125. A. Kawamura, M. Munzel, T. Kojima, C. Yapp, B. Bhushan, Y. Goto, A. Tumber, T. Katoh, O. N. King, T. Passioura, L. J. Walport, S. B. Hatch, S. Madden, S. Muller, P. E. Brennan, R. Chowdhury, R. J. Hopkinson, H. Suga and C. J. Schofield, *Nat. Commun.*, 2017, **8**, 14773.
126. K. Yamagata, Y. Goto, H. Nishimasu, J. Morimoto, R. Ishitani, N. Dohmae, N. Takeda, R. Nagai, I. Komuro, H. Suga and O. Nureki, *Structure*, 2014, **22**, 345–352.

CHAPTER 12

Mass Spectrometric Analysis of Cyclic Peptides

YULIN QI AND DIETRICH A. VOLMER*

Institute of Bioanalytical Chemistry, Saarland University, D-66123 Saarbrücken, Germany
*E-mail: Dietrich.Volmer@mx.uni-saarland.de

12.1 Classification of Cyclic Peptides

The cyclic peptide backbone is linked *via* amide bonds; this cyclic arrangement facilitates the formation of internal hydrogen bonds and disulfide bridges to generate a rigid backbone conformation. This feature makes them attractive scaffolds for protein-based drugs because cyclic peptides are able to survive in the human digestive tract.[1] On the other hand, this "rigid" trait also makes cyclic peptides extremely resistant to enzymatic digestion for proteomics studies as compared to their linear forms.[2,3]

Normally, cyclic peptides are classified according to the different types of bonds that comprise the ring structure:

1. Homodetic cyclic peptides are peptides formed by amino acids *via* amide bonds between an α-carboxyl group of one residue and an α-amino group of another amino acid (Figure 12.1); an example of this type of cyclic peptide is cyclosporine.

Chemical Biology No. 6
Cyclic Peptides: From Bioorganic Synthesis to Applications
Edited by Jesko Koehnke, James Naismith and Wilfred A. van der Donk

Published by the Royal Society of Chemistry, www.rsc.org

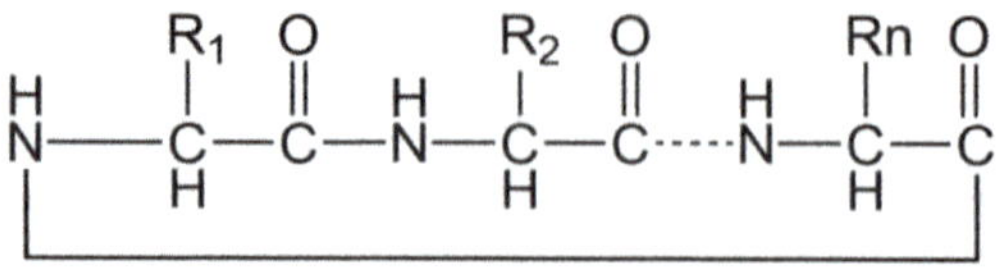

Figure 12.1 Chemical structure of homodetic cyclic peptides.

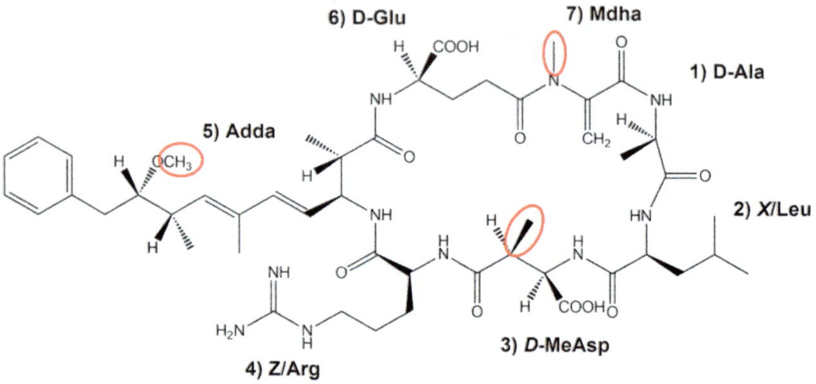

Figure 12.2 The structure of microcystin MC-LR and possible demethylation sites; the numbering scheme for the amino acids is also shown. Reproduced from *Journal of The American Society for Mass Spectrometry*, Detailed Study of Cyanobacterial Microcystins Using High Performance Tandem Mass Spectrometry, **25**, 2014, 1253–1262, Y. Qi,[77] © American Society for Mass Spectrometry 2014. With permission from Springer.

2. Cyclic isopeptides contain at least one non-α-amide bond at the side chain of one residue to the α-carboxyl group of another residue, as is seen in microcystins (Figure 12.2).
3. Cyclic depsipeptides are composed of one or more lactone ester linkages (–C(O)OR–) in the amides, *e.g.* didemnin B (Figure 12.3a).
4. Bicyclic peptides contain a bridging group between two side chains, *e.g.* amanin (Figure 12.3b).

12.2 Nomenclature

In tandem mass spectrometry, peptide ions are dissociated and the structures of the molecules are deduced from the fragment ions. For improved structural interpretation and unambiguous assignment of peptide fragments, an easy-to-use and unambiguous descriptor system is important. The commonly accepted nomenclature for labelling peptide fragments (Figure 12.4) was initially proposed by Roepstorff and Fohlman in 1984[4] and later modified by Biemann;[5] the Biemann nomenclature is used throughout this book chapter. Briefly, the method applies letters to indicate the particular bonds of the amino acid residue and subscripts indicate the position

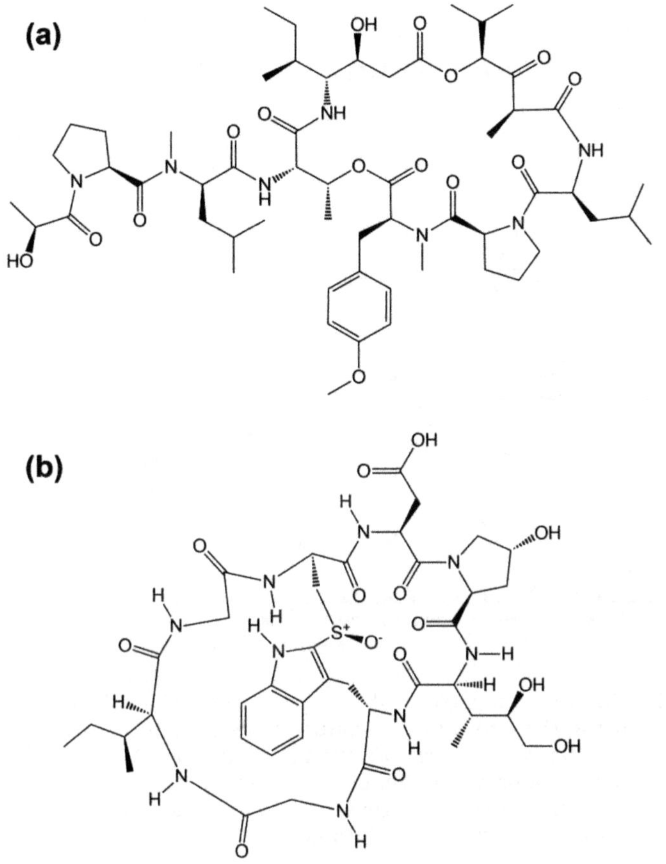

Figure 12.3 Structure of didemnin B (a) and amanin (b).

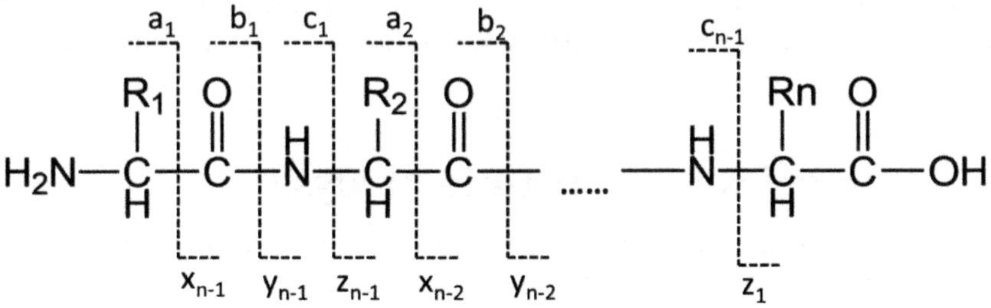

Figure 12.4 Nomenclature for peptide/protein fragmentations.

within the peptide chain. This nomenclature scheme is not directly applicable to cyclic peptides, however, as they possess no N- and C-terminals. As a result, new systems have been proposed to label the product ions from cyclic peptides.[6,7] For example, Gross *et al.* utilized a four-part descriptor of the general formula x_{nJZ},[7] where x is the ion designator and n is the number

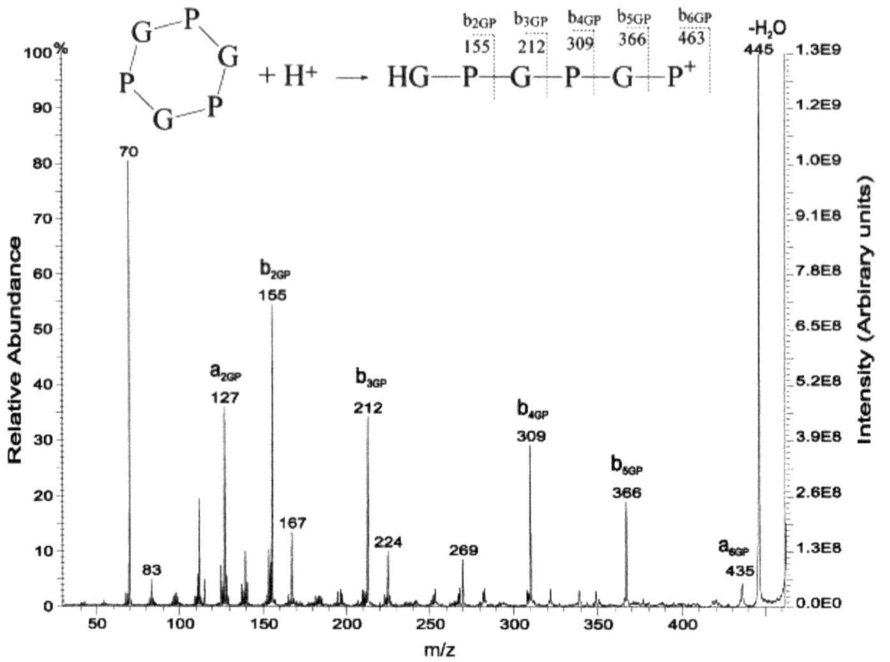

Figure 12.5 CID mass spectrum of FAB-produced $[M + H]^+$ of cyclo(Pro–Gly)$_3$ using the Gross nomenclature system. Reproduced from *Journal of The American Society for Mass Spectrometry*, A Nomenclature System for Labeling Cyclic Peptide Fragments, **10**, 1999, 360–363, L. C. M. Ngoka,[7] Copyright © 1999 American Society for Mass Spectrometry. With permission from Springer.

of amino acid residues within the ion, the same as in the Biemann system. The subscripts J and Z are the one-letter codes representing the two amino acid residues where cleavage occurs to form the decomposed linear ion (Figure 12.5).

12.3 Strategies for Structural Analysis

Classic peptide sequencing *via* Edman degradation[8] is not preferred for cyclic peptides because the N-termini are concealed in the cyclic structures. Moreover, cyclic peptides usually contain uncommon amino acids synthesized by microorganisms. Therefore, mass spectrometry (MS) and nuclear magnetic resonance (NMR) are the most powerful approaches for cyclic peptide analysis.[9] Specialized NMR techniques utilizing the Nuclear Overhauser effect (NOE) are useful for peptides,[10] which measure the amide bonds along the peptide and allow assignment of the connectivities of amino acid residues. Moreover, the NOE can also be exploited in 2-dimensional NMR experiments for fast recording of peptide structural information.[10] However, NMR requires relatively large amounts (milligrams) of analyte from purified

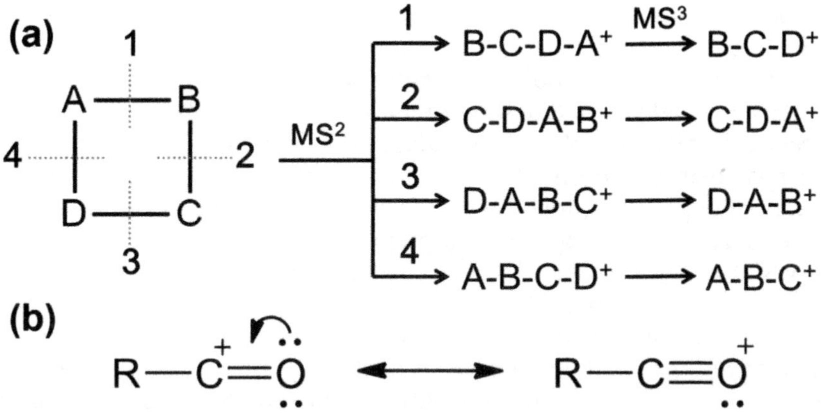

Figure 12.6 (a) Ring opening and fragmentation of a cyclic tetrapeptide and (b) the acylium ion formed from the scission of the peptide *N*-acyl bond.

or concentrated samples and only reveals average structural features of the bulk mixtures, while information on low abundant compounds cannot be determined. MS, on the other hand, readily provides complementary information to NMR, due to its ability to detect much smaller amounts of sample (micrograms or less) and generate structural information *via* tandem mass spectrometry (MS^n). In theory, MS is able to sequence any peptide and protein with no limitation on the molecular mass; however, as the size increases, data processing becomes more challenging because of limitations of the computational algorithms. In recent years, novel MS^n techniques have been developed and computational power has strongly increased; thus, MS methods have increased in popularity for peptide sequencing.[11]

During mass spectral experiments, sequential peptide backbone cleavages occur *via* MS^n to produce smaller peptide fragments, and hence, the sequence information can be deduced from the cleavages and losses of consecutive amino acids along the peptide chain.[12] Unfortunately, for the cyclic structure, the first MS^n cleavage on the peptide backbone only opens the cyclic ring; afterwards, various linear peptides with the same mass-to-charge ratio (m/z), but different linear sequences, are formed. Consequently, at least two cleavages are required for the backbone of the cyclic peptide to release a neutral segment (Figure 12.6a). The principle of MS approaches must, therefore, be to separate the secondary peptide fragments from the MS^2 step and conduct MS^3 on the various open ring peptides, for interpretation and assignment of the sequence information.

12.4 Ionization Methods

Fast atom bombardment (FAB) was the first ionization technique for systematic MS analysis of peptides. FAB is a relatively soft ionization technique, in which the sample is mixed with a viscous matrix (*e.g.* glycerol) and bombarded by a high-energy beam of atoms. During the FAB process, intact protonated or deprotonated molecules are produced.[13,14] FAB greatly reduced thermally induced rearrangement in the gas phase and therefore constituted a milestone for ionizing large, polar molecules such as peptides, proteins and carbohydrates. In addition, FAB was amenable to analysis of cyclic peptides with subsequent MS^n.[9] Gross *et al.* postulated the structure of *Helminthosporium carbonum* toxin based on FAB-MS^n experiments.[15] Tomer *et al.* investigated several cyclic peptides, including dimers and tetramers of tri- and tetrapeptide sequences, and proposed a strategy for sequencing cyclic peptides *via* FAB-MS^n.[16] Eckart *et al.* emphasized the difficulties of sequencing cyclic peptides using acylium ions (Figure 12.6b); that is, the absolute position of an amino acid residue cannot be easily recognized as for linear peptides, and this is indeed the most challenging aspect for cyclic peptide sequencing.[17]

Subsequently, electrospray ionization (ESI) emerged as the most powerful method for producing intact ions from large and complex species in solution.[18] ESI is a soft ionization technique, which permitted MS analysis of various fragile and polar molecules commonly encountered in biological systems. ESI was invented by Fenn in the late 1980s and the significance of ESI-MS was recognized with the Nobel Prize in 2002, together with another cutting edge ionization technique – matrix assisted laser desorption ionization (MALDI).[19]

ESI can be operated in both positive and negative ionization mode, as shown in Figure 12.7. During ESI operation, the sample solution is loaded into a capillary tube held at a high electric potential of several kV. The strong electric field between the capillary and the counter electrode induces charge accumulation at the liquid surface located at the tip of the capillary, which releases small charged droplets. With the evaporation of the solvent from the droplet, the radius of the droplet decreases and the charge density further increases. When the droplet radius reaches the Rayleigh limit, the Coulombic repulsion exceeds the surface tension and leads to a "Coulombic explosion", releasing smaller sized offspring droplets.[20,21] This process continues until the solvent is completely evaporated and ions are released into the gas phase.[22,23] The distinct advantage of ESI over other ionization techniques is the ability to form multiply-charged ions, where the high charge state allows for analysis of heavy molecules such as peptides, proteins, petroleum, polymers, organometallics, and other organic compounds.[24–27] Furthermore, the series of multiply-charged ions for each molecule greatly supports the measurement for large biopolymers.

Compared to proteins, peptides are much easier to measure because they are readily soluble. Large proteins are usually digested by an endo-protease such as trypsin, and the generated peptides are separated by high-performance liquid chromatography (HPLC) prior to ESI-MS analysis. The structures of the peptides are then deduced from the measured *m*/*z* values of the fragments generated by MS^n. In proteomics research, fragment ion spectra are compared to databases of existing proteins, to determine the sequence and modifications of fragments.[28]

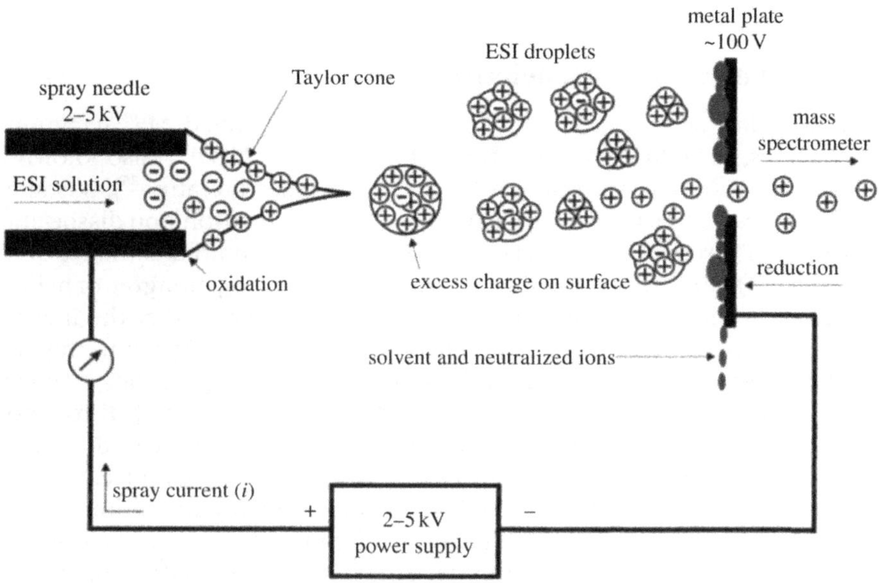

Figure 12.7 Schematic principle of electrospray ionization (ESI). Reproduced with permission from ref. 22. Cech N. B., Enke C. G. Reproduced from N. B. Cech and G. G. Enke, Practical Implications of Some Recent Studies in Electrospray Ionization Fundamentals, *Mass Spectrom. Rev.*, John Wiley and Sons. © 2001 John Wiley & Sons, Inc.

In addition to multiply-charged ions from solution phase analytes, the sample can also be ionized from a matrix crystal, to produce singly-charged ions *via* matrix-assisted laser desorption/ionization (MALDI). This technique was initially developed by Karas and Hillenkamp.[29,30] Tanaka applied a similar method to ionize large proteins and polymers, with masses up to 100 000 Da.[31] The exact MALDI mechanism is still not fully understood due to its complexity. Briefly, MALDI combines two processes: desorption and ionization. Samples are first mixed with an excess of matrix (aromatic compounds) and irradiated with an UV laser. The matrix molecules absorb laser energy, followed by ablation of the upper matrix layers. Vaporized gas phase mixtures of neutrals and ionized matrix molecules, as well as neutral sample molecules, are then formed, followed by proton transfer which leads to the formation of charged analyte species. MALDI is a soft ionization technique that is particularly well suited to non-volatile species such as biomolecules and polymers. The ions generated by MALDI are typically singly-charged species, with a better tolerance for salts than electrospray ionization; usually less sample preparation is required as well. However, singly-charged peptides often produce relatively simple tandem mass spectra, which are dominated by preferred low-energy pathways.[32,33]

12.5 Fragmentation Methods

12.5.1 Threshold Dissociations

Threshold ion activation techniques are routinely applied MS^n techniques for peptides, including collision-induced dissociation (CID)[34] (also known as collisionally-activated dissociation (CAD) in the older literature[35]), sustained off-resonance dissociation (SORI–CID)[36] and infrared multiphoton dissociation (IRMPD).[37] These methods are based on excitation of vibrational modes of the molecules.[38] During the process, a neutral gas such as nitrogen, argon, or helium, is present in the collision cell of the mass spectrometer, so that the inelastic collision process between the gas and the ions converts energy into the ions' internal energy. A similar effect can be achieved by heating the ions by means of an infrared laser. The vibrational energy is redistributed rapidly throughout the entire ion before fragmentation occurs. Accordingly, the cleaved bonds in threshold-based methods are mostly the weakest bonds in the molecule.

When a CID experiment is applied to a peptide, the fragmentation starts at a backbone amide bond and produces *b* and *y* ions (Figure 12.4) along with side chain losses of water, ammonia, and CO_2. Unfortunately, post-translational modifications (PTMs) of peptides also dissociate easily before backbone cleavage. As a result, important structural information such as glycosylation, phosphorylation or sulfation is lost during the CID process. Generally, the inability to measure PTMs is a major limitation of all the vibrational excitation methods, as in most cases the essential biological function of biopolymers is represented by their PTMs.

12.5.2 Ion–Electron Dissociations (ExD)

The prototype of ion–electron reaction in MS was reported by Cody and Freiser in 1979 and termed 'electron impact excitation of ions from organics' (EIEIO): it was initially applied to the analysis of substituted benzene radicals.[39] Unfortunately, it was not until the 1990s that the method was further explored for ion–electron reactions. In 1996, McLafferty and co-workers rediscovered the technique by irradiating multiply-charged proteins using ultraviolet (UV) light.[40] In these experiments, unusual *c*/*z* fragments (Figure 12.4) were observed from proteins, indicating that the UV light triggered an ion fragmentation pathway, which was different from conventional CID techniques. Subsequently, in 1998, Zubarev *et al.* utilized Fourier transform ion cyclotron resonance mass spectrometry (FTICR-MS) for ion–electron reactions and illustrated that the charge-reduced ion species and *c*/*z* fragments were actually produced by secondary electrons generated by the UV photons hitting the metal surface of the mass spectrometer.[41] This discovery led to the first modern ExD technique; *viz.*, electron capture dissociation (ECD).[42,43]

The ECD reaction process is quite different from CID and greatly facilitated peptide and protein analysis,[44,45] including PTM analysis,[46] top-down sequencing,[47] differentiation of peptide isomers,[48] and even protein–protein interactions.[49] Because ion–electron dissociations produce distinct fragment patterns, which are not generated by vibrational excitation methods, research on ECD and development of other ExD techniques has intensified during the last decade. By manipulating the electrons' energy, various ion–electron dissociation methods have been developed to broaden their application ranges; *e.g.*, electron transfer dissociation (ETD) for implementation on mass analyzers other than FTICR-MS,[50] hot-ECD for secondary fragmentation of peptides,[51] electron-induced dissociation (EID) for singly-charged peptides,[52] electron detachment dissociation (EDD) and negative ion electron capture dissociation (niECD) for analysis of negatively-charged peptides.[53,54]

12.5.3 MALDI-related Methods

Post-source decay (PSD) is a method specific to MALDI time-of-flight (TOF) mass spectrometry, where product ions (commonly *a*, *b*, *y*, *z* and *d* ions) are produced from metastable transitions or CID in the flight tube prior to entering the TOF reflector.[55] PSD has proved to be a valuable method for primary structure analysis in peptide sequencing, due to its high sensitivity and tolerance to the inhomogeneities of the MALDI sampling, together with additional structural information obtained from the PSD process.[56]

An alternative approach for sequencing biomolecules with MALDI is so-called in-source decay (ISD) fragmentation.[57] In ISD, the fragmentation process is initialized by the laser in the hot MALDI plume during sample ionization. In peptide applications, ISD induces N–Cα bond cleavage *via* hydrogen transfer from the matrix to the peptide backbone, and leads to a *c*/*z* ion series,[58] similar to ECD experiments. In addition, fragment ion series such as *a*, *b*, and *y* ions have also been reported, depending on the matrix used and the analyte itself.[59,60] The ISD experiment can be performed in TOF[57] and other MS instruments, such as FTICR[61] and orbitrap;[62] however, its major drawbacks are the precursor ion selection and the intense matrix cluster background in the lower *m*/*z* range.

Both MALDI-PSD and ISD are valuable techniques for sequencing of peptides and proteins, because of the ability of MALDI to handle small quantities of complex samples. However, so far, they are complementary methods rather than a substitute for ESI-MS/MS techniques, as on-line separation techniques and automation are difficult due to the solid phase nature of the MALDI experiment.

12.6 Application of Tandem Mass Spectrometry to Cyclic Peptides

The application of MS^n methods to fast peptide sequencing has been most beneficial for structural identification of linear peptides.[63] Characterization of cyclic peptides, on the other hand, is challenging because their termini are not well-defined. Consequently, the amino acid sequence cannot be assigned in an orderly fashion. Furthermore, two bond cleavages are required for the cyclic peptides for the loss of amino acid residues, which makes the correct sequence assignment difficult because of the ambiguities encountered during the ring opening process. For these reasons, analysis of cyclic peptides requires more elaborate, higher dimension MS^n data acquisition techniques,[64,65] which are summarized in the following sections.

12.6.1 General Procedure

Following ionization, protonated cyclic peptides can be characterized stepwise in the mass spectrometer using sequential MS^n. In 1982, Gross and co-workers successfully sequenced an unknown cyclic peptide using CID.[15] Later, the same group developed a general strategy for sequencing of cyclic peptides using multiple MS^n steps.[16] This procedure requires selection of one of the primary acylium ions from the first generation MS^2 product ions, and subjecting it to consecutive CID experiments. In the first MS/MS spectrum (MS^2), product ions that originated from the protonated molecules after the loss of common amino acid residues must be selected as precursor ions for higher order MS^n experiments.[64] This method is easy to operate without the need for chemical derivatization. Moreover, the strategy avoids the interpretation of complicated mass spectra produced by high-energy CID.

Eckart *et al.* also described an alternative MS^n approach, by separating the sequencing procedure into two steps: analysis of the linear connectivity of the amino acid residues in the first step, followed by determination of the orientation in step two.[17] However, this method exhibits limitations for the separation of overlapping dipeptide and tripeptide fragment isomers. The multiple stage MS^n strategy by Gross *et al.* circumvents this limitation for isomer fragments during the MS^2 experiment of protonated cyclic peptides, and has therefore been utilized more often in structural elucidation studies of cyclic peptides.

12.6.2 Metal Complexation

One useful approach to increase the fragmentation levels from MS^n experiments is the use of metal complexation.[66] Due to their structural characteristics, cyclic peptides can readily bind metal ions within their cyclic cavities through carbonyl oxygen atoms and side chain groups. Multistep MS^n of these chelate complexes can generate enhanced ion abundances of the diagnostic fragments and thus offer improved insights into structure and biological function.

Ngoka *et al.* studied the fragmentation behavior of two cyclic metal–peptide complexes, cyclo(D-Trp–D-Asp–Pro–D-Val–Leu) and cyclo(D-Trp–D-Asp–Pro–D-Ile–Leu), and found that the metal–peptides appear to initialize 'charge remote' fragmentation because successive CID cleavages did not change the position of the metal ion.[67] They also successfully applied the same methodology to metal depsipeptides, including didemnin B, beauvericin, and enniatin B1.[68] Subsequently, Cotter and co-workers studied sodium complexes of nine cyclic hexapeptides using CID.[69] Hue *et al.* characterized the sodium complexes of surfactin, which is an octapeptide.[70] Brodbelt's group systematically investigated the CID fragmentation behavior of protonated and metal-cationized peptides formed by ESI.[66] The size of the cyclic peptides ranged from six to twelve residues with various metals (Ag^+, Mn^{2+}, Ni^{2+}, Pb^{2+}, Co^{2+}, K^+, Na^+, Li^+, Sr^{2+}, and Ca^{2+}). It was shown that metal–peptides provided superior and complementary information for the identification of cyclic peptides as compared to the data obtained from protonated peptides (Figure 12.8).

12.6.3 Ion–Electron Dissociation (ExD) for Cyclic Peptides

During the last decade, ExD techniques, in particular electron capture dissociation (ECD), have significantly extended the field of tandem mass spectrometry, for providing complementary structural information to CID fragmentation.[44] ECD has also been applied to cyclic peptides. For example, O'Connor and co-workers studied the ECD fragmentation of small, doubly-charged cyclic peptides,[71] and observed unexpected amino acid residue losses. This was unusual, because the fragmentation of cyclic peptides requires

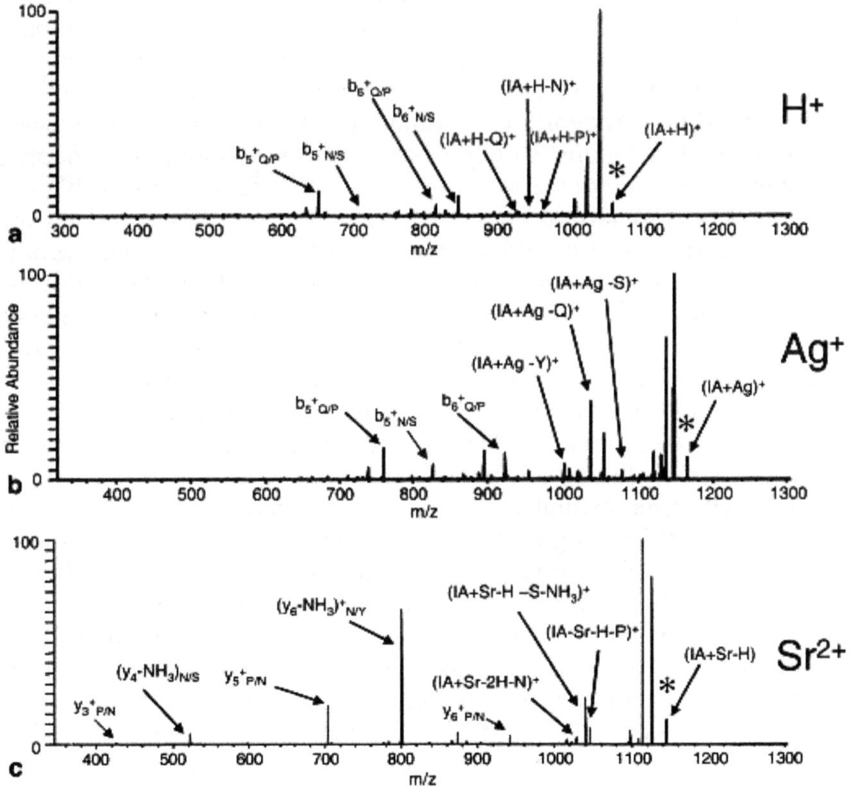

Figure 12.8 MS^2 spectra for complexes of (a) (iturin A + H)$^+$, (b) (iturin A + Ag)$^+$, and (c) (iturin A + Sr–H)$^+$. IA = iturin A. An asterisk identifies the precursor ion. Reproduced from *Journal of The American Society for Mass Spectrometry*, MSn Characterization of Protonated Cyclic Peptides and Metal Complexes, **15**, 2004, 1039–1054, S. M. Williams and J. S. Brodbelt.[66] Copyright © 2004 American Society for Mass Spectrometry. With permission from Springer.

a single electron capture to trigger two or more cleavage reactions of the peptide backbone. To explain this phenomenon, a "free radical cascade mechanism" was proposed:[71] the nonergodic cleavage of the cyclic chain generates a distonic ion, and the radical site can propagate along the peptide backbone, causing extensive secondary backbone cleavages. Since then, ExD has proved extremely useful for the analysis of cyclic peptides. This section summarizes several recent ExD applications for the characterization of cyclic peptides.

Microcystins (MCs) are a group of cyanotoxins produced by cyanobacterial genera in natural water systems.[72,73] MCs are cyclic peptides containing seven amino acid units (Figure 12.2), with the general structure cyclo(D-Ala^1–X^2–D-MeAsp^3–Z^4–Adda^5–D-Glu^6–Mdha^7), where X and Z are variable amino acids, D-MeAsp is D-*erythro*-β-methyl aspartic acid, Adda is an unusual amino acid (2S,3S,8S,9S-3-amino-9-methoxy-2,6,8-trimethyl-10-phenyldeca-4E,6E-dienoic acid), and Mdha is *N*-methyldehydroalanine. In mass spectrometric analysis, the combination of LC-MS and CID is the most commonly applied approach for quick identification of MC variants.[74,75] In addition to deducing the chemical structures, it is also important to examine the exact biological action mechanisms of MC toxins and their metal binding complexes, as it has been discovered that expression of MCs is controlled by levels of metal ions taken up from the ecosystem.[76] Unfortunately, CID has failed in this mission, as threshold methods preferentially cleave the weakest bond, which happens to be the metal binding to the MC molecules. To overcome this problem, Qi *et al.* applied ECD to several MCs and their metal complexes.[77] The experiment showed that ECD provided more detailed, complementary product ions as compared to CID. For example, extensive side chain losses from the amino acid residues were observed, which can serve as diagnostic markers for the variable amino acid units of MCs. More importantly, ECD enabled the accurate location of the metal binding sites of MCs for the first time.

Non-ribosomal peptides (NPs) are a class of peptide secondary metabolites produced by microorganisms. They are important natural products for use as clinical drugs, including antibiotics, antitumor agents and immunosuppressants.[78] NPs often exhibit cyclic structures, carrying modifications such as *N*-methyl and *N*-formyl groups, acylation, glycosylation, and heterocyclic ring formation (oxazoline, thiazole, and thiazoline). This structural diversity accounts for their biological activities and potential for pharmaceutical drug action.[79] Sequencing NPs is challenging because of the cyclization and many other building blocks present in the compounds. Wills *et al.* applied CID and EID to characterize actinomycin D, a non-ribosomal peptide used in the treatment of various cancers.[80] Their data demonstrated that CID only provided limited sequence information, which was in agreement with previous reports.[81] In contrast, EID was shown to be superior for structural identification, because it generated a larger degree of fragmentation, which was not restricted to *b*/*y* ion formation through preferential cleavages in CID (Figure 12.9). Moreover, it was shown that by changing the charge carriers in EID, a better fitting metal cation (Na^+) directed fragmentation more effectively through the entire ring structure.

Eptifibatide is a potent antagonist for glycoprotein IIb/IIIa and has been widely utilized in clinical practices to inhibit platelet aggregation.[82]

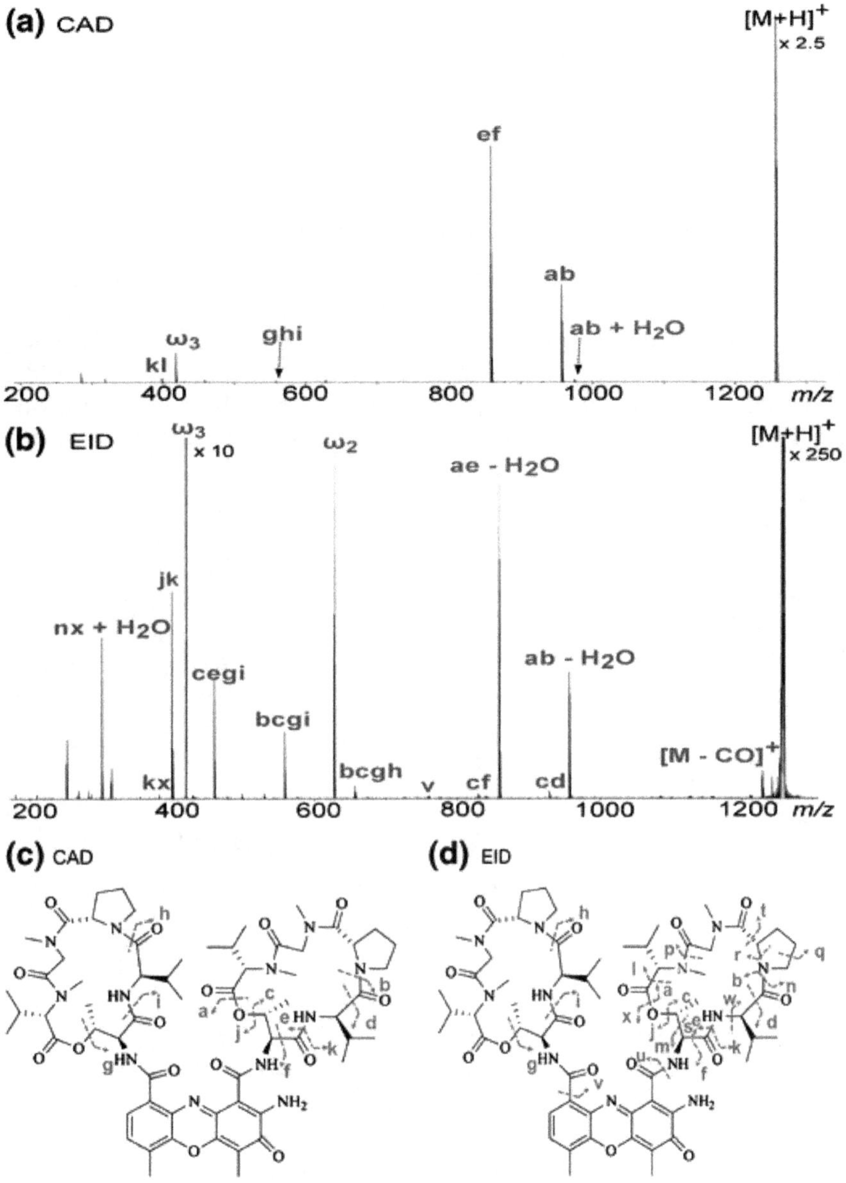

Figure 12.9 CID (a) and EID (b) mass spectra of protonated actinomycin D (*m*/*z* 1255.63), with illustrations of the fragments assigned from CID (c) and EID (d). Reproduced from *Journal of the American Society for Mass Spectrometry*, Structural Characterization of Actinomycin D Using Multiple Ion Isolation and Electron Induced Dissociation, **25**, 2013, 186–195, R. H. Wills and P. B. O'Connor,[80] © American Society for Mass Spectrometry 2013. With permission from Springer; a full list of peak assignments can be found in the supplementary material of ref. 80.

Interestingly, its cyclic backbone structure is held together by a disulfide bond, which cannot be cleaved efficiently using CID.[83] Duan *et al.* demonstrated that CID of eptifibatide failed to produce sequence-informative fragmentation patterns, whereas ETD induced an initial ring scission of the disulfide bond, thus generating simpler sequence-informative mass spectra for the disulfide-bridged cyclic peptide.[84]

12.6.4 Post-source Decay and In-source Decay

Cyclic peptide sequencing is not as straightforward as for linear species due to the lack of N- and C-termini in the closed ring. PSD and ISD spectra feature more abundant product ions for the characterization of cyclic peptides. Schilling *et al.* investigated a series of synthetic cyclic peptides by MALDI-PSD mass spectrometry,[85] and proved that the fragmentation pattern depended on the amino acid sequence (Figure 12.10). Erhard *et al.*

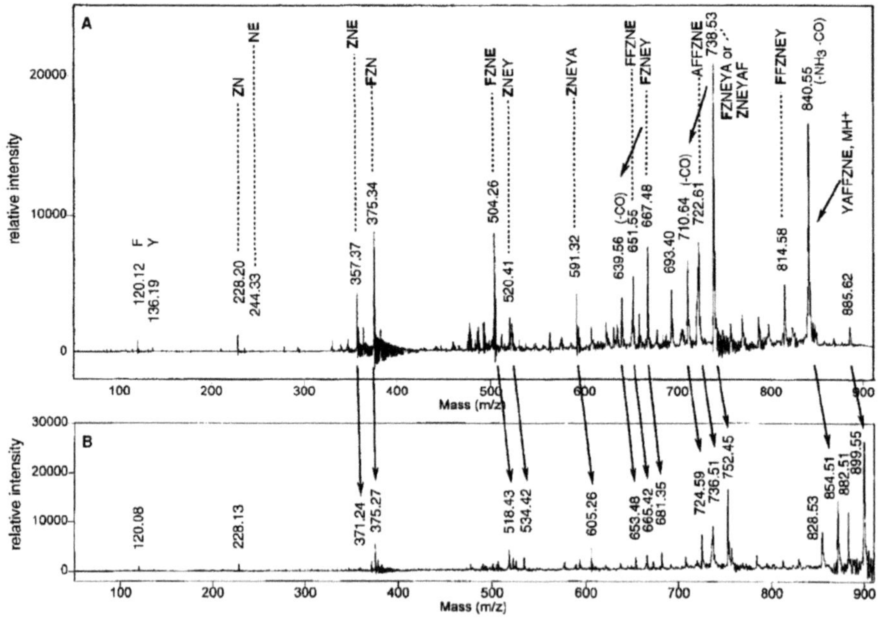

Figure 12.10 (A) MALDI-PSD spectrum of the peptide [cyclo(Asn–Glu–Tyr–Ala–Phe–Phe–Aca)]. Fragment ions are annotated using the one letter code of amino acids. Arrows indicate satellite ion formation *via* neutral losses. (B) MALDI-PSD spectrum of the same peptide after acid-catalyzed esterification with MeOH. Arrows between the spectra connect the corresponding fragment ions. Reproduced with permission from B. Schilling, W. Wang, J. S. McMurray and K. F. Medzihradszky, Fragmentation and Sequencing of Cyclic Peptides by Matrix-Assisted Laser Desorption/Ionization Post-Source Decay Mass Spectrometry, *Rapid Commun. Mass Spectrom.*,[85] John Wiley and Sons. Copyright © 1999 John Wiley & Sons, Ltd.

reported the use of MALDI-PSD to observe additional structural information for cyclic peptides, which were formed by metabolism of cyanobacteria. Unique mass fingerprints (Figure 12.11) were obtained for a variety of microorganisms and were used to differentiate their toxicities.[86] Recently, Asakawa *et al.* investigated the potential fragmentation pathways of MALDI-based PSD and ISD analyses using a model cyclic peptide (Arg–Gly–Asp–D-Phe–Val) and 1,5-diaminonaphthalene (1,5-DAN) as a matrix.[87] The study shows that ISD produced a hydrogen-abundant radical $[M + 2H]^{+\bullet}$ after N–Cα bond cleavage, and subsequently reacted with the matrix to give $[M + 3H]^+$ and $[M + H + matrix]^+$ species. PSD of the $[M + H + matrix]^+$, on the other hand, led to specific peptide bond cleavage reactions adjacent to the matrix binding site and produced predominantly loss of an amino acid plus 1,5-DAN (Figure 12.12).

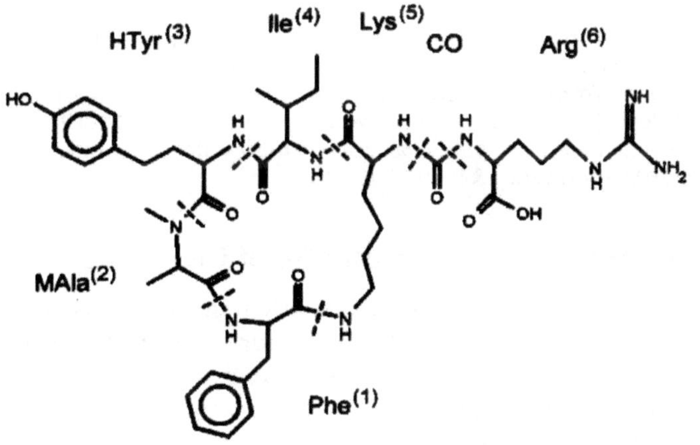

Peptide	$(M+H)^+$	(1)	(2)	(3)	(4)	(5)	(6)
Anabaenopeptolin A	844	Phe	MAla	HTyr	Val	Lys	Tyr
Oscillamide Y	858	Phe	MAla	HTyr	Ile	Lys	Tyr
Ferintoic acid A	867	Phe	MAla	HTyr	Val	Lys	Trp
Ferintoic acid B	881	Phe	MAla	HTyr	Ile	Lys	Trp
Anabaenopeptolin B	837	Phe	MAla	HTyr	Val	Lys	Arg
Anabaenopeptolin C	851	Phe	MAla	HTyr	Ile	Lys	Arg

Figure 12.11 Structure of the new anabaenopeptolin C. Anabaenopeptolin A and B were isolated from cyanobacterium *Anabaena flos-aquae*, oscillamide Y from cyanobacterium *Oscillatoria agardhii*, ferintoic acid A and B from cyanobacterium *Microcystis aeruginosa* and the new anabaenopeptolin C from cyanobacterium *Microcystis aeruginosa*. Reproduced by permission from Macmillan Publishers Ltd: *Nature Biotechnology* (ref. 86), copyright 1997.

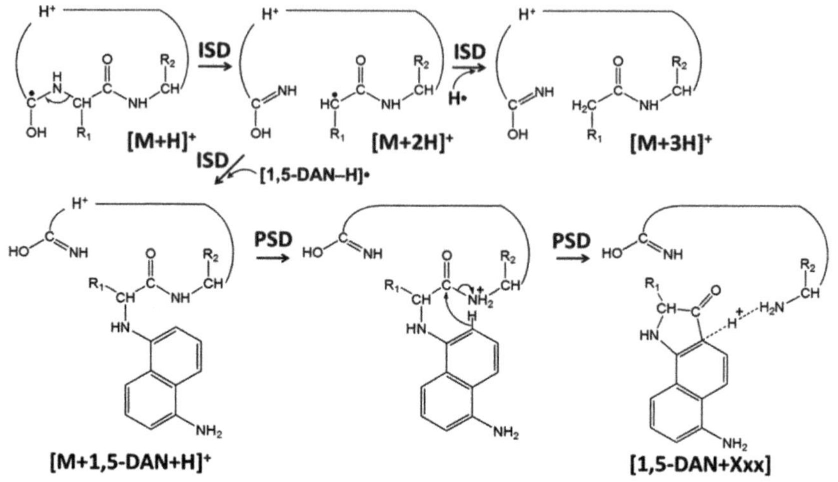

Figure 12.12 Potential fragmentation pathways of cyclic peptides by MALDI-ISD and PSD, when 1,5-DAN is used as the matrix. Reproduced with permission from D. Asakawa, N. Smargiasso and E. De Pauw, Estimation of Peptide N–Cα Bond Cleavage Efficiency During MALDI-ISD Using a Cyclic Peptide. *J. Mass Spectrom.*,[87] John Wiley and Sons. Copyright © 2016 John Wiley & Sons, Ltd.

12.6.5 Ion Mobility-mass Spectrometry of Cyclic Peptides

Currently, identification of cyclic peptides mostly relies on mass spectrometry, usually relying on liquid chromatography for pre-separation of topological isomers (*e.g.*, cyclic or linear peptides). However, this methodology requires numerous chromatographic runs and assignments of chromatographic retention times are often ambiguous. Recently, ion mobility spectrometry (IMS) was shown to be able to distinguish the small conformational differences of ions in the gas phase,[88] and it was also reported to be able to differentiate linear and cyclic peptides.[89] Hyphenation of IMS with MS provides an additional dimension of separation for determining the gas phase conformations of ions. In the ion mobility cell, ions travel in a buffer gas under the influence of an electric field; because of the unique collision cross-sections of different ions during the ion–neutral collisions, ions with different conformations can often be separated by IMS on a time scale of only a few milliseconds.

In 2004, Ruotolo *et al.* utilized IMS to study the structural differences of gramicidin S (cyclo(VOLFPVOLFP)) and five linear gramicidin S analogues.[89] Gramicidin S is a cyclic peptide with antimicrobial and hemolytic activities for both Gram-positive and Gram-negative bacteria.[90] Moreover, gramicidin S is an excellent model for studying the protein β-sheet formation due to its smaller size and lower degrees of freedom as compared to linear peptides. In their work, the authors measured the gas phase conformation of singly-charged, protonated and cationized (Cs, K, Li, Na, Rb) gramicidin

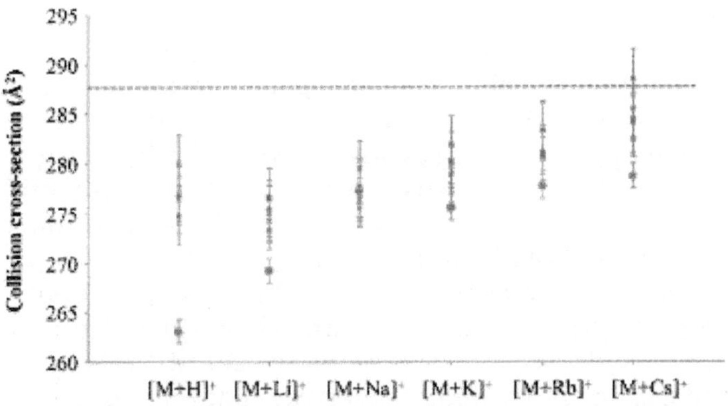

Figure 12.13 Plot of measured collision cross-section *versus* ion type (protonated, sodiated, *etc.*) for all peptides investigated in ref. 89. Data for cyclic GS shown in red (circles) and data for linear peptides in blue: G1 (filled diamonds), G2 (filled squares), G3 (filled triangles), G4 (x) and G5 (asterisk). Reproduced from *Journal of The American Society for Mass Spectrometry*, Ion Mobility Mass Spectrometry Applied to Cyclic Peptide Analysis: Conformational Preferences of Gramicidin S and Linear Analogs in the Gas Phase, **15**, 2004, 870–878, B. T. Ruotolo, C. C. Tate and D. H. Russell,[89] Copyright © 2004 American Society for Mass Spectrometry. With permission from Springer.

S and its linear analogues, and the collision cross-sections (Ω) as a function of the ionizing alkali metal ion (Figure 12.13). The results illustrated IMS to be a powerful complementary method to mass spectrometry, for rapid screening of peptide mixtures in the gas phase.

12.6.6 Quantification

Multiple reaction monitoring (MRM) remains the most widely applied method for quantitative LC-MS/MS analysis of target compounds in biological and environmental samples. However, in reality, the performance of MRM is often limited by the CID efficiency, due to the low selectivity and sensitivity seen for many compounds.[91] Usually, compound derivatization is the method of choice to overcome this problem. For example, for profiling vitamin D metabolites, which exhibit very low ionization efficiencies, detection limits by LC-MS/MS are not sensitive enough for the low abundant metabolites to be assessed at physiological concentrations. Therefore, Cookson-type reagents are often implemented for derivatization, given an increase in the response of the factor between 100- and 1000-fold over the non-derivatized compounds.[92–94] Unfortunately, for other molecules such as cyclic peptides, suitable derivatization reagents have not yet been described yet. Many cyclic peptides display poor CID efficiencies, which in turn make the intensities of product ions for the MRM experiments also too low for sensitive analysis.[95] Recently, Fu *et al.* demonstrated that differential ion mobility spectrometry

(DMS), which is a variation of IMS, can be successfully used together with LC-MS/MS for quantitative bioanalysis of cyclic peptides.[96] The DMS device is an add-on to conventional triple quadrupole or QqLIT mass spectrometers and is located between the ESI source and the MS orifice.[97] DMS separates different ion species by applying an asymmetric electric field and a compensation voltage resulting in increased assay specificity. Pasireotide, a therapeutic cyclic peptide for treatment of Cushing's disease, exhibited very poor CID efficiency in MRM mode. By selecting the most abundant precursor ion in both quadrupole 1 and quadrupole 3 of the triple quadrupole MS instrument with prior DMS separation, the matrix/chemical background noise was greatly reduced, and assay sensitivity was enhanced 5-fold as compared to measurements using the conventional MRM mode (Figure 12.14). This workflow can of course be expanded to general quantitative LC-MS analyses of molecules with poor CID fragmentation efficiency.

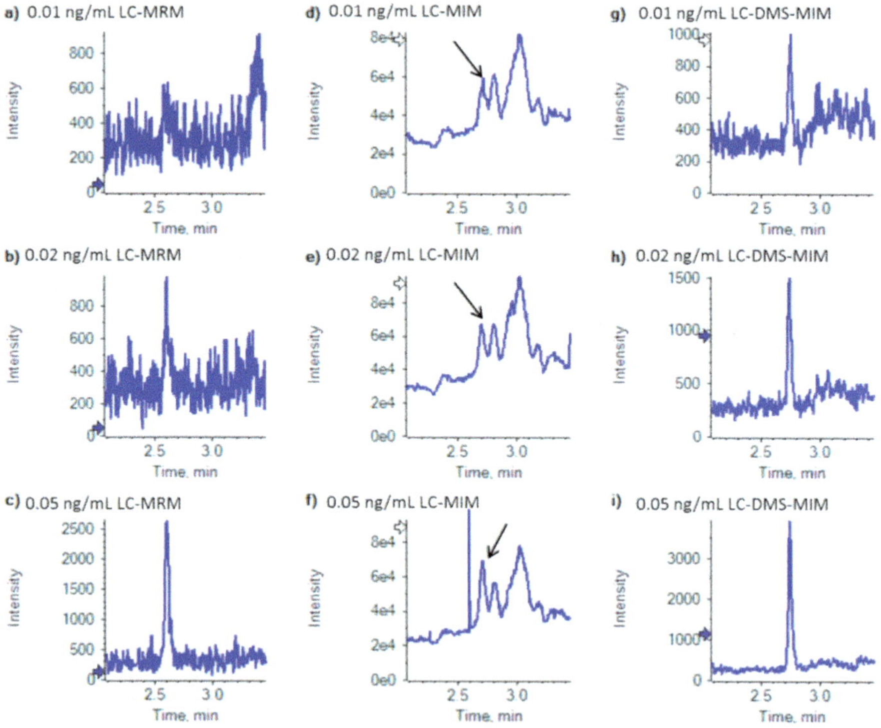

Figure 12.14 A comparison of assay sensitivity after injecting the same pasireotide human plasma extracts into a LC-quadrupole linear ion trap system. (Left) LC-MRM without differential ion mobility spectrometry (DMS) for concentrations of (a) 0.01, (b) 0.02, and (c) 0.05 ng mL^{-1}. (Center) LC-multiple ion monitoring (MIM) without DMS. (Right) LC-DMS-multiple ion monitoring. Reproduced with permission from ref. 96, Copyright (2016), American Chemical Society.

12.7 Conclusions

Structural characterization of cyclic peptides by mass spectrometry exhibits significant challenges. Compared to the other primary structural analytical technique, that is NMR, mass spectrometry offers direct sequencing of multiple mixture components with much higher sensitivity and speed than NMR. In this chapter, the principle MS techniques for structural identification of cyclic peptides were introduced and compared, and selected applications were shown. Fortunately, past serious problems during structural elucidation caused by the cyclic peptide structures have been overcome and structures can be readily identified by application of different mass spectrometry techniques. The considerable number of cyclic peptides that have been successfully characterized by mass spectrometry during the last 10 years is clear evidence for the impressive performance of mass spectrometry in cyclic peptide research.

References

1. D. J. Craik, Seamless proteins tie up their loose ends, *Science*, 2006, **311**, 1563–1564.
2. R. D. Hotchkiss, The chemical nature of gramicidin and tyrocidine, *J. Biol. Chem.*, 1941, **141**, 171–185.
3. G. W. Kenner and A. H. Laird, Resistance of a cyclic peptide to enzymic attack, *Chem. Commun.*, 1965, 305–306.
4. P. Roepstorff and J. Fohlman, Letter to the editors, *Biol. Mass Spectrom.*, 1984, **11**, 601.
5. K. Biemann, Contributions of mass spectrometry to peptide and protein structure, *Biol. Mass Spectrom.*, 1988, **16**, 99–111.
6. A. A. Tuinman and G. R. Pettit, Analysis of the amino acid sequence of peptides by mass spectrometry an ion notation proposal, *Int. J. Pept. Protein Res.*, 1990, **36**, 331–334.
7. L. C. M. Ngoka and M. L. Gross, A nomenclature system for labeling cyclic peptide fragments, *J. Am. Soc. Mass Spectrom.*, 1999, **10**, 360–363.
8. P. Edman and G. Begg, A protein sequenator, *Eur. J. Biochem.*, 1967, **1**, 80–91.
9. K. Eckart, Mass spectrometry of cyclic peptides, *Mass Spectrom. Rev.*, 1994, **13**, 23–55.
10. B. P. Roques, R. Rao and D. Marion, Use of nuclear overhauser effect in the study of peptides and proteins, *Biochimie*, 1980, **62**, 753–773.
11. K. F. Medzihradszky and R. J. Chalkley, Lessons in *de novo* peptide sequencing by tandem mass spectrometry, *Mass Spectrom. Rev.*, 2015, **34**, 43–63.
12. K. Biemann, [25] Sequencing of peptides by tandem mass spectrometry and high-energy collision-induced dissociation, *Methods Enzymol.*, 1990, **193**, 455–479.

13. M. Barber, R. S. Bordoli, R. D. Sedgwick and A. N. Tyler, Fast atom bombardment of solids as an ion source in mass spectrometry, *Nature*, 1981, **293**, 270–275.
14. M. Barber, R. S. Bordoli, R. D. Sedgwick and A. N. Tyler, Fast atom bombardment of solids (F.A.B.): a new ion source for mass spectrometry, *J. Chem. Soc., Chem. Commun.*, 1981, 325.
15. M. L. Gross, D. McCrery, F. Crow, K. B. Tomer, M. R. Pope and L. M. Ciuffetti, *et al.*, The structure of the toxin from helminthosporium carbonum, *Tetrahedron Lett.*, 1982, **23**, 5381–5384.
16. K. B. Tomer, F. W. Crow, M. L. Gross and K. D. Kopple, Fast-atom bombardment combined with tandem mass spectrometry for the determination of cyclic peptides, *Anal. Chem.*, 1984, **56**, 880–886.
17. K. Eckart, H. Schwarz, K. B. Tomer and M. L. Gross, Tandem mass spectrometry methodology for the sequence determination of cyclic peptides, *J. Am. Chem. Soc.*, 1985, **107**, 6765–6769.
18. J. B. Fenn, Electrospray wings for molecular elephants (Nobel Lecture), *Angew. Chem., Int. Ed.*, 2003, **42**, 3871–3894.
19. K. Markides and A. Gräslund, *The Nobel Prize in Chemistry 2002*, Nobelprize.org, 2002:13, http://www.nobelprize.org/nobel_prizes/chemistry/laureates/2002/, accessed July 31, 2016.
20. L. Rayleigh, XX. On the equilibrium of liquid conducting masses charged with electricity, *Philos. Mag. Ser. 5*, 1882, **14**, 184–186.
21. A. Gomez and K. Tang, Charge and fission of droplets in electrostatic sprays, *Phys. Fluids*, 1994, **6**, 404–414.
22. N. B. Cech and C. G. Enke, Practical implications of some recent studies in electrospray ionization fundamentals, *Mass Spectrom. Rev.*, 2001, **20**, 362–387.
23. J. B. Fenn, M. Mann, C. K. Meng, S. F. Wong and C. M. Whitehouse, Electrospray ionization–principles and practice, *Mass Spectrom. Rev.*, 1990, **9**, 37–70.
24. S. K. Chowdhury, V. Katta and B. T. Chait, An electrospray-ionization mass spectrometer with new features, *Rapid Commun. Mass Spectrom.*, 1990, **4**, 81–87.
25. M. W. F. Nielen and F. A. (Ab) Buijtenhuijs, Polymer analysis by liquid chromatography/electrospray ionization time-of-flight mass spectrometry, *Anal. Chem.*, 1999, **71**, 1809–1814.
26. R. Colton and J. C. Traeger, The application of electrospray mass spectrometry to ionic inorganic and organometallic systems, *Inorg. Chim. Acta*, 1992, **201**, 153–155.
27. J. A. Loo, Electrospray ionization mass spectrometry: a technology for studying noncovalent macromolecular complexes, *Int. J. Mass Spectrom.*, 2000, **200**, 175–186.
28. N. L. Anderson and N. G. Anderson, Proteome and proteomics: new technologies, new concepts, and new words, *Electrophoresis*, 1998, **19**, 1853–1861.

29. M. Karas, D. Bachmann and F. Hillenkamp, Influence of the wavelength in high-irradiance ultraviolet laser desorption mass spectrometry of organic molecules, *Anal. Chem.*, 1985, **57**, 2935–2939.
30. M. Karas, D. Bachmann, U. Bahr and F. Hillenkamp, Matrix-assisted ultraviolet laser desorption of non-volatile compounds, *Int. J. Mass Spectrom. Ion Processes*, 1987, **78**, 53–68.
31. K. Tanaka, H. Waki, Y. Ido, S. Akita, Y. Yoshida and T. Yoshida, *et al.*, Protein and polymer analyses up to m/z 100 000 by laser ionization time-of-flight mass spectrometry, *Rapid Commun. Mass Spectrom.*, 1988, **2**, 151–153.
32. J. Qin and B. T. Chait, Preferential fragmentation of protonated gas-phase peptide ions adjacent to acidic amino acid residues, *J. Am. Chem. Soc.*, 1995, **117**, 5411–5412.
33. J. Qin and B. T. Chait, Collision-induced dissociation of singly charged peptide ions in a matrix-assisted laser desorption ionization ion trap mass spectrometer2, *Int. J. Mass Spectrom.*, 1999, **190–191**, 313–320.
34. K. R. Jennings, Collision-induced decompositions of aromatic molecular ions, *Int. J. Mass Spectrom. Ion Phys.*, 1968, **1**, 227–235.
35. F. W. McLafferty, P. F. Bente, R. Kornfeld, S.-C. Tsai and I. Howe, Metastable ion characteristics. XXII. Collisional activation spectra of organic ions, *J. Am. Chem. Soc.*, 1973, **95**, 2120–2129.
36. J. W. Gauthier, T. R. Trautman and D. B. Jacobson, Sustained off-resonance irradiation for collision-activated dissociation involving Fourier transform mass spectrometry. Collision-activated dissociation technique that emulates infrared multiphoton dissociation, *Anal. Chim. Acta*, 1991, **246**, 211–225.
37. D. P. Little, J. P. Speir, M. W. Senko, P. B. O'Connor and F. W. McLafferty, Infrared multiphoton dissociation of large multiply charged ions for biomolecule sequencing, *Anal. Chem.*, 1994, **66**, 2809–2815.
38. L. Sleno and D. A. Volmer, Ion activation methods for tandem mass spectrometry, *J. Mass Spectrom.*, 2004, **39**, 1091–1112.
39. R. B. Cody and B. S. Freiser, Electron impact excitation of ions from organics: an alternative to collision induced dissociation, *Anal. Chem.*, 1979, **51**, 547–551.
40. Z. Guan, N. L. Kelleher, P. B. O'Connor, D. J. Aaserud, D. P. Little and F. W. McLafferty, 193 nm photodissociation of larger multiply-charged biomolecules, *Int. J. Mass Spectrom. Ion Processes*, 1996, **157**, 357–364.
41. R. A. Zubarev, N. L. Kelleher and F. W. McLafferty, Electron capture dissociation of multiply charged protein cations. A nonergodic process, *J. Am. Chem. Soc.*, 1998, **120**, 3265–3266.
42. Y. Qi and D. A. Volmer, Structural analysis of small to medium-sized molecules by mass spectrometry after electron-ion fragmentation (ExD) reactions, *Analyst*, 2016, **141**, 794–806.

43. Y. Qi and D. A. Volmer, Electron-based fragmentation methods in mass spectrometry: an overview, *Mass Spectrom. Rev.*, 2017, **36**, 4–15.
44. R. A. Zubarev, Reactions of polypeptide ions with electrons in the gas phase, *Mass Spectrom. Rev.*, 2003, **22**, 57–77.
45. H. J. Cooper, K. Håkansson and A. G. Marshall, The role of electron capture dissociation in biomolecular analysis, *Mass Spectrom. Rev.*, 2005, **24**, 201–222.
46. S. M. M. Sweet and H. J. Cooper, Electron capture dissociation in the analysis of protein phosphorylation, *Expert Rev. Proteomics*, 2007, **4**, 149–159.
47. N. L. Kelleher, Peer reviewed: top-down proteomics, *Anal. Chem.*, 2004, **76**, 196 A–203 A.
48. P. P. Hurtado and P. B. O'Connor, Differentiation of isomeric amino acid residues in proteins and peptides using mass spectrometry, *Mass Spectrom. Rev.*, 2012, **31**, 609–625.
49. H. Zhang, W. Cui, J. Wen, R. E. Blankenship and M. L. Gross, Native electrospray and electron-capture dissociation FTICR mass spectrometry for top-down studies of protein assemblies, *Anal. Chem.*, 2011, **83**, 5598–5606.
50. J. E. P. Syka, J. J. Coon, M. J. Schroeder, J. Shabanowitz and D. F. Hunt, Peptide and protein sequence analysis by electron transfer dissociation mass spectrometry, *Proc. Natl. Acad. Sci. U. S. A.*, 2004, **101**, 9528–9533.
51. J. P. Williams, A. J. Creese, D. R. Roper, B. N. Green and H. J. Cooper, Hot electron capture dissociation distinguishes leucine from isoleucine in a novel hemoglobin variant, Hb Askew, β54(D5)Val → Ile, *J. Am. Soc. Mass Spectrom.*, 2009, **20**, 1707–1713.
52. H. Lioe and R. A. J. O'Hair, Comparison of collision-induced dissociation and electron-induced dissociation of singly protonated aromatic amino acids, cystine and related simple peptides using a hybrid linear ion trap–FT-ICR mass spectrometer, *Anal. Bioanal. Chem.*, 2007, **389**, 1429–1437.
53. B. A. Budnik, K. F. Haselmann and R. A. Zubarev, Electron detachment dissociation of peptide di-anions: an electron–hole recombination phenomenon, *Chem. Phys. Lett.*, 2001, **342**, 299–302.
54. H. J. Yoo, N. Wang, S. Zhuang, H. Song and K. Håkansson, Negative-ion electron capture dissociation: radical-driven fragmentation of charge-increased gaseous peptide anions, *J. Am. Chem. Soc.*, 2011, **133**, 16790–16793.
55. B. Spengler, D. Kirsch and R. Kaufmann, Fundamental aspects of post-source decay in matrix-assisted laser desorption mass spectrometry. 1. Residual gas effects, *J. Phys. Chem.*, 1992, **96**, 9678–9684.
56. B. Spengler, Post-source decay analysis in matrix-assisted laser desorption/ionization mass spectrometry of biomolecules, *J. Mass Spectrom.*, 1997, **32**, 1019–1036.

57. R. S. Brown and J. J. Lennon, Sequence-specific fragmentation of matrix-assisted laser-desorbed protein/peptide ions, *Anal. Chem.*, 1995, **67**, 3990–3999.
58. T. Köcher, Å. Engström and R. A. Zubarev, Fragmentation of Peptides in MALDI in-source decay mediated by hydrogen radicals, *Anal. Chem.*, 2005, **77**, 172–177.
59. M. Takayama and A. Tsugita, Sequence information of peptides and proteins with in-source decay in matrix assisted laser desorption/ionization-time of flight-mass spectrometry, *Electrophoresis*, 2000, **21**, 1670–1677.
60. M. Takayama and A. Tsugita, Does in-source decay occur independent of the ionization process in matrix-assisted laser desorption?, *Int. J. Mass Spectrom.*, 1998, **181**, L1–L6.
61. S. Nicolardi, L. Switzar, A. M. Deelder, M. Palmblad and Y. E. M. van der Burgt, Top-down MALDI-in-source decay-FTICR mass spectrometry of isotopically resolved proteins, *Anal. Chem.*, 2015, **87**, 3429–3437.
62. R. Ait-Belkacem, M. Dilillo, D. Pellegrini, A. Yadav, E. L. de Graaf and L. A. McDonnell, In-source decay and pseudo-MS3 of peptide and protein ions using liquid AP-MALDI, *J. Am. Soc. Mass Spectrom.*, 2016, **27**, 2075–2079.
63. P. A. Jonsson, Mass spectrometry for protein and peptide characterisation, *Cell. Mol. Life Sci.*, 2001, **58**, 868–884.
64. L. C. M. Ngoka and M. L. Gross, Multistep tandem mass spectrometry for sequencing cyclic peptides in an ion-trap mass spectrometer, *J. Am. Soc. Mass Spectrom.*, 1999, **10**, 732–746.
65. K. Ishikawa, Y. Niwa, K. Oishi, S. Aoi, T. Takeuchi and S. Wakayama, Sequence determination of unknown cyclic peptide antibiotics by fast atom bombardment mass spectrometry, *Biol. Mass Spectrom.*, 1990, **19**, 395–399.
66. S. M. Williams and J. S. Brodbelt, MSn characterization of protonated cyclic peptides and metal complexes, *J. Am. Soc. Mass Spectrom.*, 2004, **15**, 1039–1054.
67. L. C. M. Ngoka and M. L. Gross, Location of alkali metal binding sites in endothelin A selective receptor antagonists, cyclo(D-Trp–D-Asp–Pro–D-Val–Leu) and cyclo(D-Trp–D-Asp–Pro–D-Ile–Leu), from multistep collisionally activated decompositions, *J. Mass Spectrom.*, 2000, **35**, 265–276.
68. L. C. M. Ngoka, M. L. Gross and P. L. Toogood, Sodium-directed selective cleavage of lactones: a method for structure determination of cyclodepsipeptides1, *Int. J. Mass Spectrom.*, 1999, **182–183**, 289–298.
69. S. Lin, S. Liehr, B. S. Cooperman and R. J. Cotter, Sequencing cyclic peptide inhibitors of mammalian ribonucleotide reductase by electrospray ionization mass spectrometry, *J. Mass Spectrom.*, 2001, **36**, 658–663.
70. N. Hue, L. Serani and O. Laprévote, Structural investigation of cyclic peptidolipids from *Bacillus subtilis* by high-energy tandem mass spectrometry, *Rapid Commun. Mass Spectrom.*, 2001, **15**, 203–209.

71. N. Leymarie, C. E. Costello and P. B. O'Connor, Electron capture dissociation initiates a free radical reaction cascade, *J. Am. Chem. Soc.*, 2003, **125**, 8949–8958.
72. M. E. van Apeldoorn, H. P. van Egmond, G. J. A. Speijers and G. J. I. Bakker, Toxins of cyanobacteria, *Mol. Nutr. Food Res.*, 2007, **51**, 7–60.
73. S. Bortoli and D. Volmer, Account: characterization and identification of microcystins by mass spectrometry, *Eur. J. Mass Spectrom.*, 2014, **20**, 1–19.
74. Y. Qi, L. Rosso, D. Sedan, L. Giannuzzi, D. Andrinolo and D. A. Volmer, Seven new microcystin variants discovered from a native *Microcystis aeruginosa* strain–unambiguous assignment of product ions by tandem mass spectrometry, *Rapid Commun. Mass Spectrom.*, 2015, **29**, 220–224.
75. F. F. del Campo and Y. Ouahid, Identification of microcystins from three collection strains of *Microcystis aeruginosa*, *Environ. Pollut.*, 2010, **158**, 2906–2914.
76. M. Lukač and R. Aegerter, Influence of trace metals on growth and toxin production of *Microcystis aeruginosa*, *Toxicon*, 1993, **31**, 293–305.
77. Y. Qi, S. Bortoli and D. A. Volmer, Detailed study of cyanobacterial microcystins using high performance tandem mass spectrometry, *J. Am. Soc. Mass Spectrom.*, 2014, **25**, 1253–1262.
78. M. A. Marahiel, Protein templates for the biosynthesis of peptide antibiotics, *Chem. Biol.*, 1997, **4**, 561–567.
79. M. Strieker, A. Tanović and M. A. Marahiel, Nonribosomal peptide synthetases: structures and dynamics, *Curr. Opin. Struct. Biol.*, 2010, **20**, 234–240.
80. R. Wills and P. O'Connor, Structural characterization of actinomycin D using multiple ion isolation and electron induced dissociation, *J. Am. Soc. Mass Spectrom.*, 2014, **25**, 186–195.
81. M. Barber, D. Bell, M. Morris, L. Tetler, M. Woods and B. W. Bycroft, *et al.*, Study of an actinomycin complex by mass spectrometry-mass spectrometry, *Talanta*, 1988, **35**, 605–611.
82. J. Liu, X. Duan, X. Chen and D. Zhong, Determination of eptifibatide concentration in human plasma utilizing the liquid chromatography–tandem mass spectrometry method, *J. Chromatogr. B*, 2009, **877**, 527–532.
83. S.-L. Wu, H. Jiang, Q. Lu, S. Dai, W. S. Hancock and B. L. Karger, Mass spectrometric determination of disulfide linkages in recombinant therapeutic proteins using online LC–MS with electron-transfer dissociation, *Anal. Chem.*, 2009, **81**, 112–122.
84. X. Duan, F. A. Engler and J. Qu, Electron transfer dissociation coupled to an orbitrap analyzer may promise a straightforward and accurate sequencing of disulfide-bridged cyclic peptides: a case study, *J. Mass Spectrom.*, 2010, **45**, 1477–1482.
85. B. Schilling, W. Wang, J. S. McMurray and K. F. Medzihradszky, Fragmentation and sequencing of cyclic peptides by matrix-assisted laser desorption/ionization post-source decay mass spectrometry, *Rapid Commun. Mass Spectrom.*, 1999, **13**, 2174–2179.

86. M. Erhard, H. von Dohren and P. Jungblut, Rapid typing and elucidation of new secondary metabolites of intact cyanobacteria using MALDI-TOF mass spectrometry, *Nat. Biotechol.*, 1997, **15**, 906–909.
87. D. Asakawa, N. Smargiasso and E. De Pauw, Estimation of peptide N–Cα bond cleavage efficiency during MALDI-ISD using a cyclic peptide, *J. Mass Spectrom.*, 2016, **51**, 323–327.
88. A. B. Kanu, P. Dwivedi, M. Tam, L. Matz and H. H. Hill, Ion mobility–mass spectrometry, *J. Mass Spectrom.*, 2008, **43**, 1–22.
89. B. T. Ruotolo, C. C. Tate and D. H. Russell, Ion mobility-mass spectrometry applied to cyclic peptide analysis: conformational preferences of gramicidin S and linear analogs in the gas phase, *J. Am. Soc. Mass Spectrom.*, 2004, **15**, 870–878.
90. L. H. Kondejewski, S. W. Farmer, D. S. Wishart, C. M. Kay, R. E. W. Hancock and R. S. Hodges, Modulation of structure and antibacterial and hemolytic activity by ring size in cyclic gramicidin S analogs, *J. Biol. Chem.*, 1996, **271**, 25261–25268.
91. P. Shipkova, D. M. Drexler, R. Langish, J. Smalley, M. E. Salyan and M. Sanders, Application of ion trap technology to liquid chromatography/mass spectrometry quantitation of large peptides, *Rapid Commun. Mass Spectrom.*, 2008, **22**, 1359–1366.
92. T. Higashi, K. Shimada and T. Toyo'oka, Advances in determination of vitamin D related compounds in biological samples using liquid chromatography–mass spectrometry: a review, *J. Chromatogr. B*, 2010, **878**, 1654–1661.
93. S. Ding, I. Schoenmakers, K. Jones, A. Koulman, A. Prentice and D. A. Volmer, Quantitative determination of vitamin D metabolites in plasma using UHPLC-MS/MS, *Anal. Bioanal. Chem.*, 2010, **398**, 779–789.
94. M. J. Müller, C. S. Stokes, F. Lammert and D. A. Volmer, Chemotyping the distribution of vitamin D metabolites in human serum, *Sci. Rep.*, 2016, **6**, 21080.
95. E. Ciccimaro, A. Ranasinghe, C. D'Arienzo, C. Xu, J. Onorato and D. M. Drexler, *et al.*, Strategy to improve the quantitative LC-MS analysis of molecular ions resistant to gas-phase collision induced dissociation: application to disulfide-rich cyclic peptides, *Anal. Chem.*, 2014, **86**, 11523–11527.
96. Y. Fu, Y.-Q. Xia, J. Flarakos, F. L. S. Tse, J. D. Miller and E. B. Jones, *et al.*, Differential mobility spectrometry coupled with multiple ion monitoring in regulated LC-MS/MS bioanalysis of a therapeutic cyclic peptide in human plasma, *Anal. Chem.*, 2016, **88**, 3655–3661.
97. C. Lapthorn, F. Pullen and B. Z. Chowdhry, Ion mobility spectrometry-mass spectrometry (IMS-MS) of small molecules: separating and assigning structures to ions, *Mass Spectrom. Rev.*, 2013, **32**, 43–71.

CHAPTER 13

Experimental and Computational Approaches to the Study of Macrocycle Conformations in Solution

JOSHUA SCHWOCHERT AND R. SCOTT LOKEY*

University of California Santa Cruz, USA
*E-mail: slokey@ucsc.edu

13.1 Introduction

Macrocycles – in particular, cyclic peptides – have shown remarkable versatility as ligands against challenging therapeutic targets such as protein–protein interactions (PPIs).[1] Cyclization is an established method for improving potency in peptides,[2–4] and cyclization can dramatically improve the proteolytic stability of peptides.[4,5] As the ability to understand and control physico-chemical properties in cyclic peptides has increased, so has interest in cyclic peptides as scaffolds for the design of next-generation therapeutics against undruggable targets. Designing drugs in this chemical space has some unique challenges. Properties such as cell permeability and water solubility can be highly dependent on the nature of a cyclic peptide's conformational ensemble in both aqueous and membrane-like environments. Therefore, innovations in experimental, computational, and hybrid approaches for the study of cyclic peptide solution conformations will have a significant impact on this emerging field.

Chemical Biology No. 6
Cyclic Peptides: From Bioorganic Synthesis to Applications
Edited by Jesko Koehnke, James Naismith and Wilfred A. van der Donk

Published by the Royal Society of Chemistry, www.rsc.org

Table 13.1 Overview of the different methods for structural elucidation, including pros, cons and the throughput-limiting steps.

Method	Pros	Cons	Limiting Step
Crystallography	Gives precise atomic coordinates	Provides only a conformational "snapshot". Crystal packing forces may bias the structure toward conformation(s) not present in solution	Growth of diffraction quality crystals
Computational	Does not require a physical sample	Limited confidence in results, sampling problems for larger molecules	Confidence in energy values and sampling coverage
NMR	Provides conformations relevant to bioactivity and permeability depending on the solvent. Can report directly on flexibility	Time-consuming, experimental and computational expertise required	Spectral acquisition and processing. Depending on the method, conformational sampling

This chapter provides a general review of current methods available for determining cyclic peptide conformations in solution, with a primary focus on NMR and the computational tools most commonly used to interpret the variety of phenomena that emerge from different NMR methods (Table 13.1).

De novo prediction of cyclic peptide structures remains a significant challenge. Therefore, to determine reliable solution conformations, some degree of experimental input is required. NMR is the most widely used method, and provides solution structures relevant to ADME properties (permeability and solubility prediction/verification), as well as target binding and biological activity. NMR provides experimental distance, torsional, and bond orientation information, which can be checked against independently generated conformers. Alternatively, NMR-derived information can be used as restraints in the conformer production process. Since experimentally derived solution structures require conformational sampling, there is an essential synergy between experiments and computation in determining the conformational states of macrocycles in solution.

13.2 Overview of Conformation Elucidation Techniques

13.2.1 X-ray Crystallography

X-ray crystallography is a widely-used technique for solving the structure of novel natural products, as well as providing atomic resolution protein structures. While X-ray crystallography provides unambiguous atomic coordinates, it can be technically challenging and can require a good deal of luck in obtaining diffractable crystals. Additionally, the conformation adopted in a crystal lattice may be only mildly relevant to either the solution phase conformation(s) or the conformation adopted by the macrocycle engaged with its target. Crystal packing forces, inter-molecular interactions, and solvent/salt co-crystals introduce interactions that are different from those in bulk solvent or when engaged with a binding partner.[6–9] This is especially true for larger, more flexible, peptidic or polyketide macrocycles, when compared to small molecules. However, in the rare cases where the target of the macrocycle is known and is crystallizable, the formation of co-crystals between the target and the ligand can be extremely informative and provide structure-guided optimization options. Nonetheless, crystallography is still used in macrocycle research, especially when crystals are easily or serendipitously formed.

13.2.2 Purely Computational Methods

While many methods exist for generating and ranking conformations of small molecules, some key limitations arise when they are applied to macrocycles. The large conformational search space and difficult-to-predict energetic contributions of long-range interactions and ring strain make both the adequate sampling of the relevant conformational landscape and subsequent energy ranking of macrocycle conformers a particularly challenging computational task. In particular, predicting the *cis–trans* equilibrium in *N*-methylated,[10] Pro-containing[11] and peptoid-containing[12] cyclic peptides can be challenging. There has been progress on both fronts, however, especially in recent years. Although computational methods alone will not be covered in depth in this chapter, they are often the starting point for experimentally derived structure generation and will be discussed in greater depth in that context.

13.2.3 Hybrid Methods

The most widely used techniques for macrocycle conformation analysis use a combination of NMR spectroscopy and computational techniques (Figure 13.1). NMR spectroscopy can provide experimental restraints that can limit

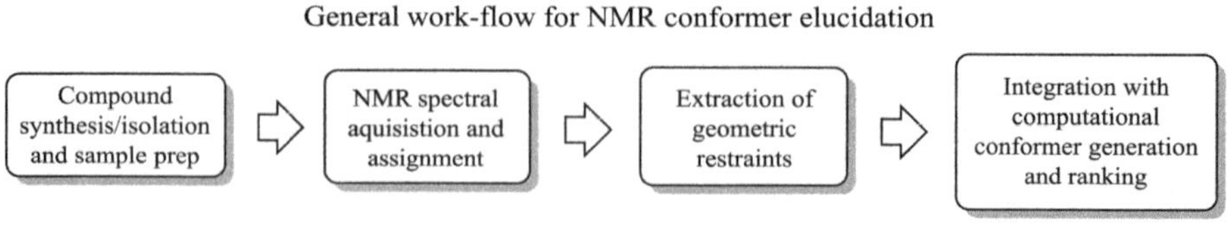

Figure 13.1 Flow chart showing the broad steps of NMR structure elucidation. From acquiring spectra to assignment to pulling out restraints/constraints.

the conformational space sampled by the computational methods or to prioritize candidate conformations from unrestrained sampling. Unlike crystallography, NMR does not provide a picture of a molecule's electron density; rather, NMR provides relative distance information, *via* Nuclear Overhauser Effect correlations (NOEs), relative bond vectors, *via* residual dipolar couplings (RDCs), and dihedral angles, *via* $3J$ correlations. When fed into a conformational search algorithm, these can exclude many potential conformations leading to more efficient sampling, as well as limiting the number of conformations that need to be ranked energetically. Additionally, emerging techniques such as chemical shift prediction have allowed for the generation of macrocycle conformations without the need for the above-mentioned restraints by relying solely on the chemical shifts of protons and carbons. These NMR-based techniques, especially those which include a limited number of restraints in conformational sampling, will be the focus of this section.

13.3 NMR Assignment and General Considerations

13.3.1 Introduction

Before 3D conformational analysis can take place, the NMR spectrum must be assigned, that is, peaks must be assigned to their corresponding atoms in the 2-dimensional structure. For natural products, this is particularly challenging because the molecular constitution and stereochemical configuration are often unknown. This review focuses on NMR assignments for cases in which the constitution and configuration are known. The reader is referred to a variety of useful books on the structural analysis of natural products.[13,14]

13.3.2 General Consideration: Solvent Systems

The first choice to be made in studying a conformation by NMR is the solvent. This choice is constrained both by technical issues, solubility *etc.*, and by the specific scientific question being asked. High-dielectric solvents such as DMSO, MeOH, and water are used to model cyclic peptides with respect to their behavior in a biological context, *e.g.*, in the study of potential binding modes with respect to their cognate targets. A variety of water suppression pulse sequences are available as part of most NMR acquisition routines, allowing the use of non-deuterated water or water–acetonitrile mixtures so that exchangeable protons can be observed.

In contrast to high-dielectric solvents, which model a compound's behavior in aqueous solution, low dielectric solvents have been used to model the behavior of cyclic peptides in the hydrophobic interior of the cell membrane.[15] Since one of the major limitations of cyclic peptides is their often-low cell permeability, understanding the interplay between conformation and permeability has inspired the study of cyclic peptide conformations in low-dielectric media.[16,17] Chloroform is commonly used because its dielectric constant of ~4 is similar to that proposed for the interior of a phospholipid bilayer, and chloroform is better at dissolving peptides compared to hydrocarbon solvents such as benzene or hexane. The chemical shift differences of NH groups in solvents of differing polarity have been used to identify their orientation with respect to the solvent, although this approach only works if the conformation is the same in the two solvents.[4]

13.3.3 Determining 2D Structure: Primary Sequence

Fortunately, there is a well-developed pipeline for the NMR assignment of peptides, which has benefited greatly from the field of structural biology. Along with ^{1}H and ^{13}C spectra, the assignment of side chain protons and carbons is generally achieved through 2D COSY, TOCSY, and ^{1}H–^{13}C HSQC spectra. Backbone carbonyls are generally assigned with the ^{1}H–^{13}C HMBC pulse sequence. Connectivities across amide bonds, which are essential for determining the relative position of amino acids, are also established with the HMBC pulse sequence.[18] In larger peptides and proteins, NOE's are often used to determine the relative sequence position of amino acids. This can be problematic for cyclic peptides, however, since short, through-space interactions can give rise to NOEs that do not represent the primary sequence, but instead are due to the 3-dimensional fold of the macrocycle. Thus, for cyclic peptides, NOEs are normally limited to conformational studies, and are not used in assigning spectra.

Spectral overlap can be a concern, especially with the aliphatic-rich hydrophobic peptides that dominate cyclic peptide natural products. Peptides that contain several repeat or similar residues, such as Ile and Leu, often have significant spectral overlap, especially in the aliphatic region of 0.75–2.5 ppm. Even if 2-dimensional spectra provide a complete assignment of chemical shifts, spectral overlap can preclude the use of these signals for conformational analysis. Diastereotopic protons, which can be particularly challenging in cases of spectral overlap, are normally assigned using a mixture of homo- and heteronuclear coupling constants. DQF COSY can provide *J*-couplings in a 2D NMR experiment and, when combined with rotamer libraries and the Karplus relation (see below), can simultaneously assign diastereotopic protons while giving insight into the relative population of rotamer states.[19] In the absence of diastereotopic assignments, pseudo-atoms representing the geometric average of diastereotopic protons can be used when calculating inter-proton distances.

Much of the tedium associated with making spectral assignments has been eased by modern software. Commercially available software packages such as MestreNova allow for the facile assignment of resonances across multiple NMR spectra to a graphical representation of a chemical structure, which has greatly increased the robustness and ease of spectral assignment (Figure 13.2). Additionally, databases such as the BMRB (Biological Magnetic Resonance Bank) provide chemical shift statistics for common amino acids, which can provide starting points for the assignment.

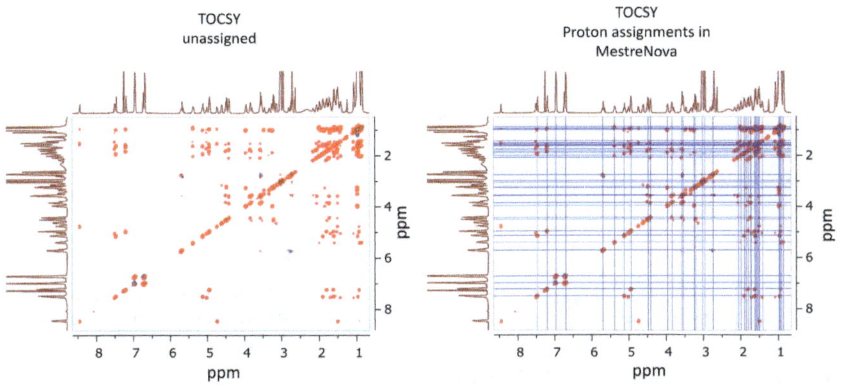

Figure 13.2 Example of the MestreNova assignment software, showing peak correlations across multiple spectra.

13.4 Conformational Information from NMR

13.4.1 Introduction

There are three primary types of information that can be gathered from NMR. First, *J*-couplings provide torsional information through the Karplus relation or derivatives thereof. Second, cross-peak volumes from NOESY and ROESY spectra provide information on through-space interproton distances. Third and most recently, residual dipolar couplings (RDCs) report on the relative orientation of bond vectors in anisotropic NMR samples. These techniques have all been widely used and validated, and their output can be fed into a variety of different computational approaches to help determine the solution conformation of a small molecule or peptide.

13.4.2 *3J* Correlations

The 3-*J* coupling, especially between the Cα and either NH or Cβ protons, provides torsional restraints, which can be particularly informative in a constrained system such as a cyclic peptide. The well-known Karplus relationship[20] relates a scalar ^3J coupling, often extracted from a sufficiently resolved ^{1}H spectrum or a ^{1}H–^{1}H DFQ COSY spectrum, to a dihedral angle. Of special interest is the vicinal coupling constant between Hα–HN protons, which is sensitive to the ϕ dihedral angle, for which the equation, as described by Wang and Bax, is shown in Figure 13.3.[21] When using the Karplus relationship to deduce dihedral angles from ^3J coupling constants, there are a number notational systems and care must be taken to ensure that the correct angle and equation form is being used. According to IUPAC, the correct atom orders are as follows: Phi: CO_{i-1}, N, Cα, CO; and Psi: N, Cα, CO, N_{i+1}.

It is important to mention that a given ^3J-coupling corresponds to multiple dihedral angle values, and when *J* is not at its extreme (high or low), the presence of other conformations will strongly affect the interpretation.

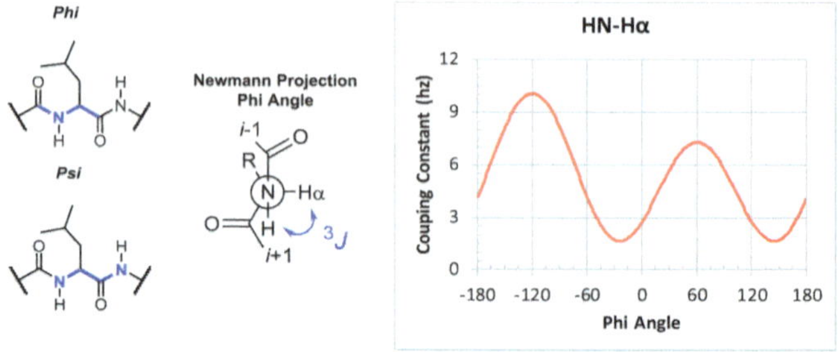

For L-amino acids: $J_{HNH\alpha} = 6.98\,cos^2(Phi - 60) - 1.38\cos(Phi - 60) + 1.72$

For D-amino acids: $J_{HNH\alpha} = 6.98\,cos^2(Phi + 60) - 1.38\cos(Phi + 60) + 1.72$:

Figure 13.3 The Karplus relationship.

Nonetheless, coupling constants have been used as restraints, for example, in the determination of the solution structure of cyclo[Pro–Pro–Ala–Ala–Ala] in DMSO using restrained MD.[22] This analysis, however, is dependent on a ^3J correlation between an alpha proton and the amide proton, and cannot be used for *N*- or α-alkylated residues, such as *N*-methylated residues or amino-isobutyric acid, respectively. Furthermore, free rotation about the peptide backbone will lead to a time-average Hα–HN ^3J coupling constant of ~6.5–7.5 Hz depending on the residue,[23] and so care must be taken in using ^3J values in this range as dihedral restraints. Like proton–proton ^3J correlations, heteronuclear proton–carbon coupling constants can yield conformational information, and even geminal homonuclear coupling constants can indicate the dihedral angle between the two protons and a neighboring carbonyl group.[24]

13.4.3 Through-space Couplings

Interatomic distances are the primary source of structural information in most cyclic peptide conformation-elucidation methods. ROESY and NOESY rely on the Nuclear Overhauser effect, which allows magnetization transfer between protons in close spatial proximity. The relative efficiency of this magnetization transfer is related to the distance between nuclei and can be easily extracted from the 2D NOESY or ROESY spectra as a volume of integration. The cross-peak volume of a reference proton pair, I_{ref}, and the corresponding reference distance, r_{ref}, can be used to calculate a proton–proton distance of interest, r_{ij}, from the following relationship:

$$r_{ij} = r_{ref}\sqrt[6]{\frac{I_{ref}}{I_{ij}}}$$

Acquisition of NOESY spectra can be problematic for many cyclic peptides of interest, as the NOESY intensity reaches zero in molecules between 0.5 and 2 kDa at room temperature. In some cases, lowering the temperature can give rise to larger NOE cross-peaks and more detailed structural information

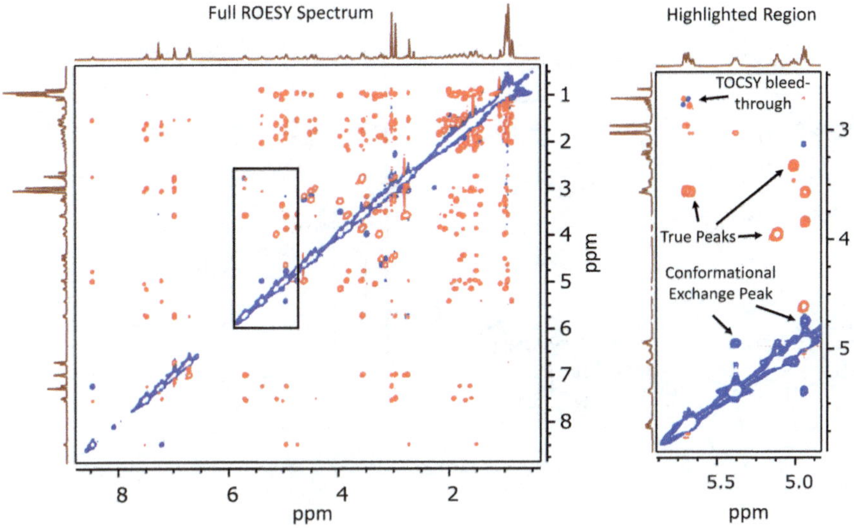

Figure 13.4 A ROESY spectrum of a cyclic hexapeptide containing two conformations in equilibrium (2 : 1). The highlighted region shows examples of peaks due to conformational exchange, TOCSY bleed-through, and true peaks (symmetric and the opposite sign of the diagonal).

in this regime.[25] ROESY spectra, which are independent of a molecule's correlation time, have robust NOE cross-peaks irrespective of molecular size. For this reason, ROESY is most often used for cyclic peptides. However, there are drawbacks to the use of ROESY in conformational analysis. First, since ROESY relies on the TOCSY spin lock sequence, there is the potential for TOCSY bleed-through for ^3J-coupled protons, resulting in an underestimated ROESY signal (Figure 13.4). Additionally, the efficiency of the spin lock sequence is dependent on the difference between the chemical shift of the proton being probed and the spin lock frequency. While TOCSY bleed-through must be avoided by looking for asymmetric peak shapes and focusing on inter-residue correlations, the spin lock effects can be subtracted mathematically.[26]

13.4.4 Establishing *Cis–Trans* Relationships for 3° Amides

Although the value of ω, the dihedral angle across the planar amide bond, is essentially restricted to two values (0° for *cis*, and 180° for *trans*), unlike ϕ, there is no proton–proton ^3J coupling available to report on the ω dihedral. This is not usually a problem associated with proteins, as this dihedral angle is always assumed to be *trans* for amino acids other than Pro. However, the ω dihedral is becoming increasingly important as *N*-methyl amino acids, peptoids, and Pro residues are being recognized as important backbone modifications for enhancing the permeability of cyclic peptides.[16,27,28] The exchange of an amide proton for an aliphatic carbon decreases the difference

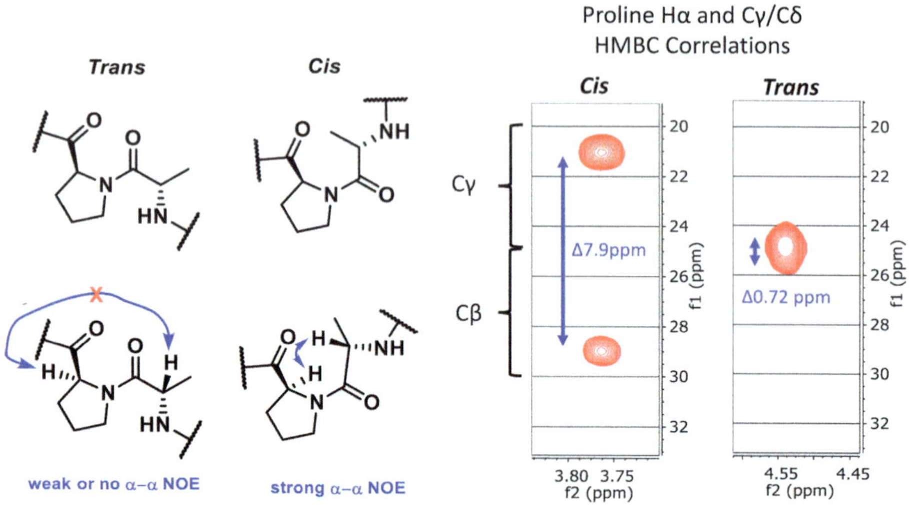

Figure 13.5 *Cis*/*trans* orientation determined by NOE cross-peak NMR chemical shift analysis. Notice that in the *trans* orientation, the Cβ and Cγ protons are so close in chemical shift as to overlap.

in steric strain between the two rotamers and reduces the dipole moment, which normally restricts amides to the *trans* geometry (a remarkable exception is the observation of an unexpected *cis*-peptide bond in the secondary amide of a cyclic peptide[32]).[29] Accordingly, while 1° amides are almost never found in the *cis* position, ~6% of Pro residues exist in the *cis* position in the PDB.[30,31] Establishing the conformation for Pro by NMR is a trivial process as a number of studies have shown that the difference in ^{13}C chemical shifts between β- and γ-carbons robustly reports on the *cis*/*trans* orientation of the Pro N-terminal amide bond. For *trans* and *cis* Pro amide bonds, Δβγ values of 4.5 ± 1.2 and 9.4 ± 1.3 ppm have been reported[4,30] (Figure 13.5). However, natural products such as CSA have inspired macrocycle chemists to include *N*-methylation as a backbone modification that can enhance the stability and permeability of peptides.[16,33] Unlike Pro, *N*-methylated residues do not have a robust chemical shift change associated with the *cis vs. trans* geometry, although this remains to be investigated in detail. Fortunately, when two amino acids are connected *via* a *cis*-amide, their Hα protons are normally in close proximity. Although not conclusive, a strong Hα–Hα correlation between adjacent residues often suggests the presence of a *cis*-amide bond between them.

In contrast to Pro, the *cis*–*trans* disposition of *N*-alkylated glycines, a.k.a., peptoids, are difficult to predict from NOEs as they contain two α-protons and are more conformationally flexible than Pro. Since the rate of interconversion between *cis* and *trans* amides is normally on the scale of milliseconds to minutes, there are two important considerations. First, MD simulations cannot efficiently sample the interconversion between *cis* and *trans* geometries at temperatures relevant to the laboratory. Because of this, *cis*/*trans* sampling must be done *via* replica exchange or parallel restrained MD runs (discussed in greater depth below). Second, many cyclic peptides, especially those containing multiple *N*-alkylated or Pro-residues, may appear as multiple conformations in intermediate or slow exchange on the NMR or chromatographic time scales. This is well documented for CSA, which adopts a single *cis–trans* conformation in a low dielectric environment, but many conformers in slow exchange in water or other polar solvents.[34–36]

13.4.5 Residual Dipolar Couplings (RDCs)

RDCs give information on the relative angles of atom pairs throughout a molecule. This relatively new NMR experiment relies on the anisotropic orientation of molecules in an NMR magnet to report on the orientation of N–H or C–H bonds relative to the external magnetic field. Originally relegated to paramagnetic proteins that spontaneously align in a strong external magnetic field,[37] and then generalized to biomolecules aligned in phospholipid bicelles,[38] more recently RDCs have been applied to proteins and small molecules[39] by performing the NMR experiment in an anisotropic sol–gel matrix embedded in the NMR tube.[40] RDCs provide unique information that is inaccessible through traditional NOE- and 3-*J*-based techniques. While NOEs and coupling constants report on local structure, RDCs provide relative vectors for bonds that are widely separated in space. Thus, RDCs are particularly useful for studying spatial relationships among structural domains in large proteins where local restraints are not sufficient to define global structure. RDCs have also proven useful in cyclic peptide systems, such as the natural product cyclosporine A (CSA). The incorporation of RDCs not only led to a revision in the overall shape of the backbone in the original NOE-based chloroform structure, but also provided insight into the orientation of the prenylated threonine (Bmt) side chain relative to the backbone conformation.[41] A combination of RDCs and NOEs was also used to obtain a solution structure of the somewhat rigid linear peptide efrapeptin C in dichloromethane.[42]

13.4.6 Measures of Intramolecular Hydrogen Bonding

Another simple yet powerful application of NMR in the analysis of cyclic peptide conformations derives from its ability to report on the degree of solvent exposure of exchangeable amide protons. Solvent occlusion of polar groups, either sterically or *via* intramolecular hydrogen bonds, can have a significant impact on physicochemical properties such as solubility and permeability. This is particularly true for orally bioavailable cyclic peptides such as CSA, in which the highly polar amide backbone must be sequestered in the membrane environment to achieve permeability. A striking example of this is CSA, in which all 4 of its non-methylated amide protons are engaged in transannular hydrogen bonds in both chloroform and in the crystal structure.[43] NMR can be used to probe hydrogen bonding in proteins, small molecules, and

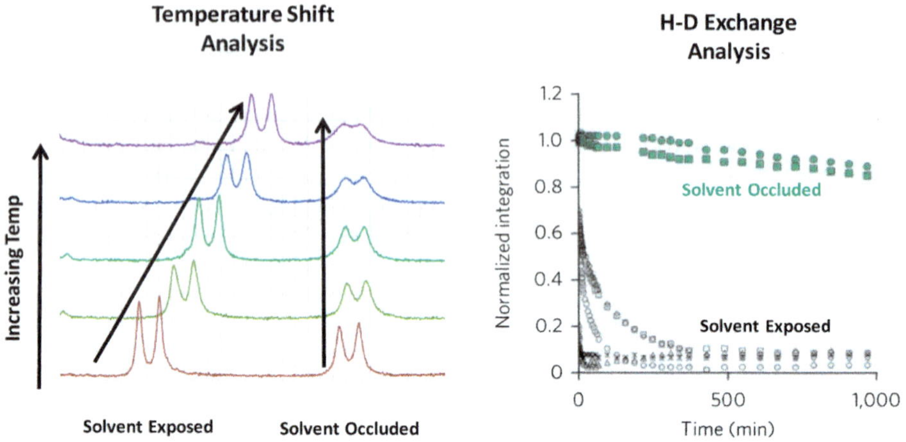

Figure 13.6 Examples of NMR temperature shift (left) and H–D exchange data (right), indicating solvent-exposed and solvent-occluded protons.

macrocycles. The two primary methods are temperature shift analysis, which correlates the temperature dependence of the amide proton's chemical shift with its solvent accessibility, and H-D exchange, measured by the titration of a protonated molecule with a deuterated solvent containing exchangeable protons.[4]

13.4.6.1 *Temperature Shift Analysis*

In water, a protein's amide NH chemical shifts display a temperature dependence that varies according to the degree of solvent exposure (Figure 13.6, left). This phenomenon has been attributed to the increased average distance between hydrogen bonded atoms at higher temperature, causing an upfield shift for hydrogen bond donors. Protons that are buried and/or involved in strong local 2° structural interactions show relatively small upfield (*i.e.*, negative) temperature shifts compared to amides that are solvent-exposed[4,44] (Figure 13.7). For hydrogen-bonding solvents, the temperature shift is greater for solvent-exposed protons than for internally hydrogen-bonded protons, because the ordering of the solvent about the exposed NH group is entropically less favorable than an intramolecular hydrogen bond.[4] The situation in non-hydrogen bonding solvents, such as chloroform, is more complex.[45] First, *inter*molecular hydrogen bonding can become a factor for peptides in low dielectric solvents, requiring that temperature shift experiments be performed at concentrations below which intermolecular interactions are observed.[46] In addition, a large temperature shift in chloroform indicates that a proton is weakly hydrogen bonded and becomes less hydrogen bonded at higher temperature. However, a small temperature shift could either indicate that the proton is involved in a strong intramolecular hydrogen bond or that it is solvent-exposed in a stable conformation that does not change

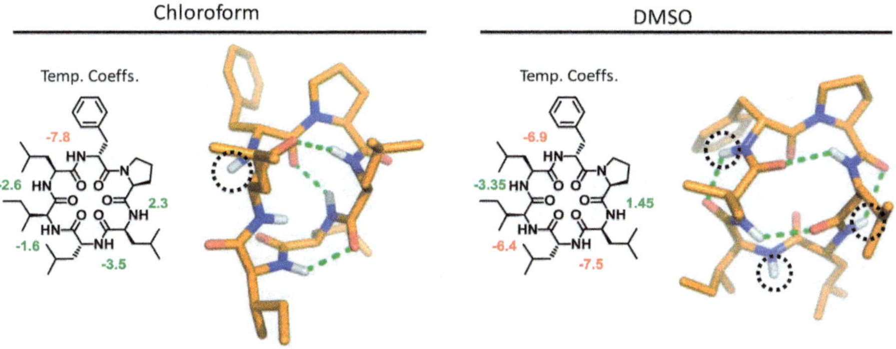

Figure 13.7 Example showing the solvent exposure of protons as measured by TC analysis, as well as their orientation in a 3D solution structure as elucidated by NMR and MD.

over the temperature range being examined. These data can be useful for corroborating NMR solution structures and for obtaining rough predictions of permeability for novel cyclic peptide scaffolds,[28] although care must be taken in their interpretation since the rate of cell penetration by carrier-mediated transport does not correlate with the degree of amide NH exposure.[47]

13.4.6.2 H–D Exchange

H–D exchange is historically one of the most commonly used measures of solvent exposure in proteins and peptides[48] (Figure 13.6, right). In a common version of this experiment, a fully protonated macrocycle is added to an NMR solvent of interest that does not contain exchangeable protons. After an initial spectrum is taken, a small amount, typically ~10 eq. or 1–5% v/v, of a reagent with exchangeable deuterons, such as CD_3OD, is added. NMR spectra (either simple ^1H 1D or a 2D method such as COSY or $^1H–^{15}N$ HSQC) are recorded as a function of time, and signals for solvent-exposed protons will exchange with deuterium (and therefore decrease in amplitude) as a function of time, while buried protons will exchange more slowly. H–D exchange rates of cyclic peptides in non-polar solvents have been shown to correlate roughly with temperature shift coefficients,[28] and both techniques provide a good qualitative picture of the degree of intramolecular hydrogen bonding and/or solvent exposure for a given proton.

13.5 Generation of NMR-informed Solution Conformations

The methods that combine computational and experimental data to generate a solution conformation of a macrocycle generally fit into two major categories: those that incorporate NMR data into the generation of conformations

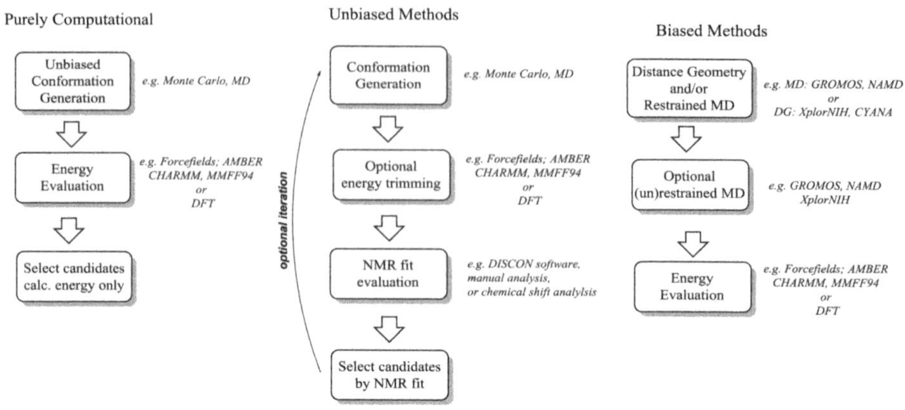

Figure 13.8 Chart describing various approaches to conformational analysis of cyclic peptides.

and those that generate conformers independently and use NMR data to select experimental structures among them (Figure 13.8). The latter methods rely on the computational efficiency of generating conformations and querying their deviation from experimental constraints; however, much like proteins whose conformational space is much too large to sample exhaustively, this method becomes significantly more challenging for larger macrocycles, such as cyclic peptides with greater than 6 amino acids. For macrocycles of sufficient complexity and flexibility, using NMR restraints to restrict the conformational space sampled is currently required, although as computational sampling and energy calculations become faster and more efficient, this will certainly change.

13.5.1 Unrestrained Conformation Generation and Sampling

13.5.1.1 Sampling

The goal of unrestrained conformational sampling is to provide a pool of diverse and energetically relevant conformations with which to compare NMR-derived distances, dihedral angles, and bond vectors. There are several computational methods whose goal is to efficiently search the conformational space in order select from the vast number of possible conformers a set of potential global energy minima with which to compare experimental data. Stochastic methods, such as the Monte-Carlo search algorithm[49] or variants thereof,[50,51] randomly sample a subset of dihedral angles at each step followed by an energy minimization step, with conformers being selected or rejected according to some energetic cutoff. These methods provide an efficient way of traversing large energy barriers (*e.g.*, Pro *cis–trans* isomerization) to avoid getting stuck in local minima.

In contrast to stochastic methods, molecular dynamics (MD) treats molecules as multi-body Newtonian objects where random initial velocities are applied to each atom and the system is set up to "evolve" deterministically through time using Newton's equations of motion. MD is often used to study conformational dynamics at experimentally relevant temperatures, but when run at "high temperature" or for a sufficient length of time, can provide a diverse conformer pool for comparison to NMR data. However, at high temperatures, instabilities can arise that cause simulations to fail, while at lower temperatures, MD kinetic trapping may lead to incomplete sampling of the conformational space. Methods that combine MD and stochastic approaches have emerged that combine the mechanistic relevance of MD with the sampling power of Monte-Carlo methods. These methods include replica exchange MD (REMD),[52,53] multi-canonical MD (MCMD)[54] and low-mode MD.[55]

A different approach to conformational sampling called "inverse kinematics" (IK) has been described,[56,57] which capitalizes on the fact that in macrocycles, cyclization fundamentally limits the degrees of freedom in backbone torsional space. This effectively reduces the domain of dihedral angles to be sampled, and allows complex ring topologies to be adequately sampled in cases where MD leads to kinetic traps. This method has recently been applied to exhaustively sample the backbone conformational landscape of several analogues of the large macrocyclic natural product thiocillin.[58]

13.5.1.2 Energetic Comparisons

The most common implementations of the above methods rely on parameterized "force fields",[59–61] which calculate energy based on molecular geometry to bias conformations towards energy minima. These force fields are computationally efficient but may be inaccurate, especially for macrocycles, and in some instances, can be replaced with much more accurate but computationally expensive methods that rely on Density Functional Theory (DFT). Additionally, most of these force fields have been parameterized on small molecules, proteins, and linear peptides, and may not accurately model peptide macrocycle energies, especially for macrocycles that contain unnatural amino acids. In one study, six commonly used force fields were found to over-stabilize α-helical and β-sheet backbone geometries in a model cyclic peptide, and under-weighted less populated regions of the Ramachandran plot.[62] In another example, the CHARMM force field was modified to more accurately represent energies in the conformational landscape of peptoids.[63]

13.5.2 Naïve Sampling and NMR-best Fit Selection

13.5.2.1 Single Conformer Fitting

The simplest method to generate NMR informed structures is to ask which conformation from a pool best fits the NMR data. This technique relies on the ability to compare NOE intensities, *3J* couplings, and RDCs obtained from NMR experiments to candidate conformations. Since RDC measurements are most often incorporated into other techniques that will be discussed further, this section will consider only NOE and *3J* restraints. This method is the most accessible as it requires only the software for generating and geometrically analyzing conformations. This analysis begins with a conformer pool generated without consideration for the NMR data. This collection of conformations can then optionally be dereplicated or energetically trimmed, although the computational ease of comparison between calculated and experimental values often doesn't limit the number of conformations used in subsequent analysis. The list of NMR restraints, which contains atom pair distances and dihedral angles, is first analyzed across the entire pool of conformations. Once the table of each conformation's distances/angles is obtained, it is compared to the list of values predicted by the NMR experiment. The least deviant conformation is then chosen as the NMR structure. It is often useful to inspect a number of the low error conformations to ensure that they have "converged" on a similar structure, which gives confidence that the NMR restraints have a single conformeric solution. This technique was recently used to determine the structure of cyclopropane-containing analogues of the model system 1NMe3,[64] as well as the natural product phepropeptin C.[65]

13.5.2.2 Fitting NMR Data to Conformational Ensembles

Distance geometry (DG) is one of the most commonly used computational methods for calculating protein structures using NMR data.[66–68] In general, DG is a method for determining the coordinates of a system of points (*i.e.*, atoms) based on a knowledge of a subset of the distances between those pairs. To calculate the molecular structure, DG uses upper and lower bounds for distances between atom pairs based on a combination of simple covalent connectivities (bond lengths and angles) and experimental distance restraints derived from NOESY/ROESY cross-peaks. Obtaining a conformational ensemble emerges from an iterative process of solving the DG matrix graphically followed by energy minimization of the resulting solutions. This routine is implemented in packages such as XPLOR and XPLOR-NIH and is well suited for calculating NMR structures of proteins in which there are many non-sequential NOE cross-peaks for generating useful distance restraints.

DISCON (Distribution of In Solution CONformations[69]) is a program that combines the sampling-first procedure mentioned above with the additional complexity of fitting conformational ensembles to the NMR data, rather than trying to fit the data to a single conformer. While short (<7 residue) peptides often adopt a relatively fixed backbone geometry, especially in low-dielectric solvent, this is often not the case for larger, more flexible structures. To address this problem, DISCON implements the NAMFIS (NMR Analysis of Molecular Flexibility In Solution)[70] algorithm, which allows ensembles of conformations to be fitted to the NMR data instead of a single conformation. DISCON is an open-source program with an intuitive graphical user interface that has functionality for labeling atoms and monitoring their geometric relationships.[71] These features make DISCON an accessible option for NMR solution structure solving, and the program has been used to help elucidate solution conformations of the macrolide spongistatin,[69] as well as a number of cyclic peptides.[27,72,73]

13.5.2.3 NMR Chemical Shift-based Methods

Chemical shift-based methods can be used to study the solution conformations of cyclic peptides and other macrocycles, providing a tool that is essentially orthogonal to the classic NOE- and 3J coupling-based restraints used for most NMR-based methods of structure prediction. Since the chemical shift of a nucleus is dependent on its local electronic environment, the chemical shifts of all nuclei in a molecule not only inform its 2D chemical structure, but also contain stereochemical and conformational information. For a given conformation, DFT methods can (at some computational expense) calculate the electron density around every nucleus and, after taking solvent effects into account, can be used to predict their chemical shifts.[74]

This technique has been used to correctly identify the correct carbon skeletons[75] and stereochemistries[76] of terpenoid natural products. In these highly rigid systems, torsional angles are relatively constrained by the carbon skeleton and stereochemistries of the pendant groups, making conformational sampling unnecessary.[77] However, for macrocycles in which the 2D chemical structure and stereochemistry is known *a priori*, DFT-derived chemical shifts can be used to provide conformational information. Recently this technique was used to confirm the solution conformations of two cyclic peptides.[78] In this study, an initial pool of hundreds of conformers was generated using Monte-Carlo search methods (using a relatively large energy window of 10 kcal mol^{-1}). This pool was further trimmed using QM energy calculations at successively higher levels of theory, keeping unique conformers within an increasingly narrow energy window and optimizing their geometries using DFT. Finally, chemical shifts were predicted for the ~20 conformers within 2.5 kcal mol^{-1} of the lowest energy structure, and the predictions were matched to the experimental NMR chemical shifts to arrive at a solution conformation.

References

1. E. M. Driggers, S. P. Hale, J. Lee and N. K. Terrett, The exploration of macrocycles for drug discovery–an underexploited structural class, *Nat. Rev. Drug Discovery*, 2008, **7**, 608–624.
2. Y. Gao and T. Kodadek, Direct comparison of linear and macrocyclic compound libraries as a source of protein ligands, *ACS Comb. Sci.*, 2015, **17**, 190–195.
3. S. K. Sia, P. A. Carr, A. G. Cochran, V. N. Malashkevich and P. S. Kim, Short constrained peptides that inhibit HIV-1 entry, *Proc. Natl. Acad. Sci. U. S. A.*, 2002, **99**, 14664–14669.
4. H. Kessler, Conformation and biological activity of cyclic peptides, *Angew. Chem., Int. Ed. Engl.*, 1982, **21**, 512–523.

5. L. Di, Strategic approaches to optimizing peptide ADME properties, *AAPS J.*, 2015, **17**, 134–143.
6. H. Kessler, G. Zimmermann, H. Förster, J. Engel, G. Oepen and W. S. Sheldrick, Does a molecule have the same conformation in the crystalline state and in solution? Comparison of NMR results for the solid state and solution with those of the X-ray structural determination, *Angew. Chem., Int. Ed.*, 1981, **20**, 1053–1055.
7. L. A. Bodack, T. B. Freedman, B. Z. Chowdhry and L. A. Nafie, Solution conformations of cyclosporins and magnesium-cyclosporin complexes determined by vibrational circular dichroism, *Biopolymers*, 2004, **73**, 163–177.
8. Z. W. Qu, H. Zhu and V. May, Unambiguous Assignment of vibrational spectra of cyclosporins A and H, *J. Phys. Chem. A*, 2010, **114**, 9768–9773.
9. U. K. Marelli, A. O. Frank, B. Wahl, V. La Pietra, E. Novellino, L. Marinelli, E. Herdtweck, M. Groll and H. Kessler, Receptor-bound conformation of cilengitide better represented by its solution-state structure than the solid-state structure, *Chemistry*, 2014, **20**, 14201–14206.
10. D. P. Slough, H. Yu, S. M. McHugh and Y.-S. Lin, Toward accurately modeling *N*-methylated cyclic peptides, *Phys. Chem. Chem. Phys.*, 2017, **19**, 5377–5388.
11. U. Doshi and D. Hamelberg, Reoptimization of the AMBER force field parameters for peptide bond (Omega) torsions using accelerated molecular dynamics, *J. Phys. Chem. B*, 2009, **113**, 16590–16595.
12. W. Brandt, T. Herberg and L. Wessjohann, Systematic conformational investigations of peptoids and peptoid-peptide chimeras, *Biopolymers*, 2011, **96**, 651–668.
13. R. G. Linington, P. G. Williams and J. B. MacMillan, *Problems in Organic Structure Determination: A Practical Approach to NMR Spectroscopy*, CRC Press, 2015.
14. P. Crews, J. Rodrigues and M. Jaspars, *Organic Structure Analysis*, Topics in Organic Chemistry, Oxford University Press, 2010.
15. W. Guba, R. Haessner, G. Breipohl, S. Henke, J. Knolle, V. Santagada and H. Kessler, Combined approach of NMR and molecular dynamics within a biphasic membrane mimetic: conformation and orientation of the bradykinin antagonist Hoe 140, *J. Am. Chem. Soc.*, 1994, **116**, 7532–7540.
16. T. R. White, C. M. Renzelman, A. C. Rand, T. Rezai, C. M. McEwen, V. M. Gelev, R. A. Turner, R. G. Linington, S. S. F. Leung, A. S. Kalgutkar, J. N. Bauman, Y. Zhang, S. Liras, D. A. Price, A. M. Mathiowetz, M. P. Jacobson and R. S. Lokey, On-resin *N*-methylation of cyclic peptides for discovery of orally bioavailable scaffolds, *Nat. Chem. Biol.*, 2011, **7**, 810–817.
17. T. Rezai, B. Yu, G. L. Millhauser, M. P. Jacobson and R. S. Lokey, Testing the conformational hypothesis of passive membrane permeability using synthetic cyclic peptide diastereomers, *J. Am. Chem. Soc.*, 2006, **128**, 2510–2511.

18. H. Kessler, C. Griesinger, J. Zarbock and H. R. Loosli, Assignment of carbonyl carbons and sequence analysis in peptides by heteronuclear shift correlation *via* small coupling constants with broadband decoupling in t1 (COLOC), *J. Magn. Reson.*, 1984, **57**, 331–336.
19. H. Kessler, C. Griesinger and K. Wagner, Peptide conformations. 42. Conformation of side chains in peptides using heteronuclear coupling constants obtained by two-dimensional NMR spectroscopy, *J. Am. Chem. Soc.*, 1987, **109**, 6927–6933.
20. M. J. Minch, Orientational dependence of vicinal proton-proton NMR coupling constants: The Karplus relationship, *Concepts Magn. Reson.*, 1994, **6**, 41–56.
21. A. C. Wang and A. Bax, Reparametrization of the Karplus relation for 3J(H.alpha.-N) and 3J(HN-C') in peptides from uniformly 13C/15N-enriched human ubiquitin, *J. Am. Chem. Soc.*, 1995, **117**, 1810–1813.
22. D. F. Mierke and H. Kessler, Combined use of homo- and heteronuclear coupling constants as restraints in molecular dynamics simulations, *Biopolymers*, 1992, **32**, 1277–1282.
23. C. J. Penkett, C. Redfield, I. Dodd, J. Hubbard, D. L. McBay, D. E. Mossakowska, R. A. G. Smith, C. M. Dobson and L. J. Smith, NMR analysis of main-chain conformational preferences in an unfolded fibronectin-binding protein, *J. Mol. Biol.*, 1997, **274**, 152–159.
24. D. F. Mierke, P. Schmieder, P. Karuso and H. Kessler, Conformational analysis of the *cis*- and *trans*-isomers of FK506 by NMR and molecular dynamics, *Helv. Chim. Acta*, 1991, **74**, 1027–1047.
25. H. Kessler, M. Köck, T. Wein and M. Gehrke, Reinvestigation of the Conformation of Cyclosporin A in Chloroform, *Helv. Chim. Acta*, 1990, **73**, 1818–1832.
26. J. G. Beck, A. O. Frank and H. Kessler, NMR of Peptides, in *NMR of Biomolecules*, Wiley-VCH Verlag GmbH & Co. KGaA, 2012, pp. 328–344.
27. J. Schwochert, R. Turner, M. Thang, R. F. Berkeley, A. R. Ponkey, K. M. Rodriguez, S. S. Leung, B. Khunte, G. Goetz, C. Limberakis, A. S. Kalgutkar, H. Eng, M. J. Shapiro, A. M. Mathiowetz, D. A. Price, S. Liras, M. P. Jacobson and R. S. Lokey, Peptide to peptoid substitutions increase cell permeability in cyclic hexapeptides, *Org. Lett.*, 2015, **17**, 2928–2931.
28. C. K. Wang, S. E. Northfield, B. Colless, S. Chaousis, I. Hamernig, R. J. Lohman, D. S. Nielsen, C. I. Schroeder, S. Liras, D. A. Price, D. P. Fairlie and D. J. Craik, Rational design and synthesis of an orally bioavailable peptide guided by NMR amide temperature coefficients, *Proc. Natl. Acad. Sci. U. S. A.*, 2014, **111**, 17504–17509.
29. S. S. Zimmerman and H. A. Scheraga, Stability of *cis*, *trans*, and nonplanar peptide groups, *Macromolecules*, 1976, **9**, 408–416.
30. Y. Shen and A. Bax, Prediction of Xaa-Pro peptide bond conformation from sequence and chemical shifts, *J. Biomol. NMR*, 2010, **46**, 199–204.
31. D. E. Stewart, A. Sarkar and J. E. Wampler, Occurrence and role of *cis* peptide bonds in protein structures, *J. Mol. Biol.*, 1990, **214**, 253–260.

32. H. Kessler, U. Anders and M. Schudok, An unexpected *cis* peptide bond in the minor conformation of a cyclic hexapeptide containing only secondary amide bonds, *J. Am. Chem. Soc.*, 1990, **112**, 5908–5916.
33. J. Chatterjee, F. Rechenmacher and H. Kessler, *N*-methylation of peptides and proteins: an important element for modulating biological functions, *Angew. Chem.*, 2013, **52**, 254–269.
34. P. F. Augustijns, S. C. Brown, D. H. Willard, T. G. Consler, P. P. Annaert, R. W. Hendren and T. P. Bradshaw, Hydration changes implicated in the remarkable temperature-dependent membrane permeation of cyclosporin A, *Biochemistry*, 2000, **39**, 7621–7630.
35. H. Kessler, M. Kock, T. Wein and M. Gehrke, Reinvestigation of the conformation of cyclosporin A in chloroform, *Helv. Chim. Acta*, 1990, **73**, 1818–1832.
36. H. R. Loosli, H. Kessler, H. Oschkinat, H. P. Weber, T. J. Petcher and A. Widmer, Peptide conformations. Part 31. The conformation of cyclosporin A in the crystal and in solution, *Helv. Chim. Acta*, 1985, **68**, 682–704.
37. J. R. Tolman, J. M. Flanagan, M. A. Kennedy and J. H. Prestegard, Nuclear magnetic dipole interactions in field-oriented proteins: information for structure determination in solution, *Proc. Natl. Acad. Sci. U. S. A.*, 1995, **92**, 9279–9283.
38. N. Tjandra and A. Bax, Direct measurement of distances and angles in biomolecules by NMR in a dilute liquid crystalline medium, *Science*, 1997, **278**, 1111–1114.
39. J. Yan, A. D. Kline, H. Mo, M. J. Shapiro and E. R. Zartler, A novel method for the determination of stereochemistry in six-membered chairlike rings using residual dipolar couplings, *J. Org. Chem.*, 2003, **68**, 1786–1795.
40. H. J. Sass, G. Musco, S. J. Stahl, P. T. Wingfield and S. Grzesiek, Solution NMR of proteins within polyacrylamide gels: diffusional properties and residual alignment by mechanical stress or embedding of oriented purple membranes, *J. Biomol. NMR*, 2000, **18**, 303–309.
41. J. Klages, C. Neubauer, M. Coles, H. Kessler and B. Luy, Structure refinement of cyclosporin a in chloroform by using RDCs measured in a stretched PDMS-gel, *Chembiochem*, 2005, **6**, 1672–1678.
42. S. Weigelt, T. Huber, F. Hofmann, M. Jost, M. Ritzefeld, B. Luy, C. Freudenberger, Z. Majer, E. Vass, J. C. Greie, L. Panella, B. Kaptein, Q. B. Broxterman, H. Kessler, K. Altendorf, M. Hollos and N. Sewald, Synthesis and conformational analysis of efrapeptins, *Chemistry*, 2012, **18**, 478–487.
43. H.-R. Loosli, H. Kessler, H. Oschkinat, H.-P. Weber, T. J. Petcher and A. Widmer, Peptide conformations. Part 31. The conformation of cyclosporin a in the crystal and in solution, *Helv. Chim. Acta*, 1985, **68**, 682–704.
44. N. J. Baxter and M. P. Williamson, Temperature dependence of 1H chemical shifts in proteins, *J. Biomol. NMR*, 1997, **9**, 359–369.
45. E. S. Stevens, N. Sugawara, G. M. Bonora and C. Toniolo, Conformational-analysis of linear peptides. 3. Temperature-dependence of NH chemical-shifts in chloroform, *J. Am. Chem. Soc.*, 1980, **102**, 7048–7050.

46. G. P. Dado and S. H. Gellman, Structural and thermodynamic characterization of temperature-dependent changes in the folding pattern of a synthetic triamide, *J. Am. Chem. Soc.*, 1993, **115**, 4228–4245.
47. U. K. Marelli, J. Bezencon, E. Puig, B. Ernst and H. Kessler, Enantiomeric cyclic peptides with different caco-2 permeability suggest carrier-mediated transport, *Chem.–Eur. J.*, 2015, **21**, 8023–8027.
48. C. E. Dempsey, Hydrogen exchange in peptides and proteins using NMR-spectroscopy, *Prog. Nucl. Magn. Reson. Spectrosc.*, 2001, **39**, 135–170.
49. Z. Li and H. A. Scheraga, Monte Carlo-minimization approach to the multiple-minima problem in protein folding, *Proc. Natl. Acad. Sci. U. S. A.*, 1987, **84**, 6611–6615.
50. S. B. Ozkan and H. Meirovitch, Efficient conformational search method for peptides and proteins: Monte Carlo minimization with an adaptive bias, *J. Phys. Chem. B*, 2003, **107**, 9128–9131.
51. S. B. Ozkan and H. Meirovitch, Conformational search of peptides and proteins: Monte Carlo minimization with an adaptive bias method applied to the heptapeptide deltorphin, *J. Comput. Chem.*, 2004, **25**, 565–572.
52. Y. Sugita and Y. Okamoto, Replica-exchange molecular dynamics method for protein folding, *Chem. Phys. Lett.*, 1999, **314**, 141–151.
53. H. T. Yu and Y. S. Lin, Toward structure prediction of cyclic peptides, *Phys. Chem. Chem. Phys.*, 2015, **17**, 4210–4219.
54. N. Nakajima, H. Nakamura and A. Kidera, Multicanonical ensemble generated by molecular dynamics simulation for enhanced conformational sampling of peptides, *J. Phys. Chem. B*, 1997, **101**, 817–824.
55. P. Labute, LowModeMD–implicit low-mode velocity filtering applied to conformational search of macrocycles and protein loops, *J. Chem. Inf. Model.*, 2010, **50**, 792–800.
56. E. A. Coutsias, K. W. Lexa, M. J. Wester, S. N. Pollock and M. P. Jacobson, Exhaustive conformational sampling of complex fused ring macrocycles using inverse kinematics, *J. Chem. Theory Comput.*, 2016, **12**, 4674–4687.
57. M. F. Thorpe and M. Lei, Macromolecular flexibility, *Philos. Mag.*, 2004, **84**, 1323–1331.
58. H. L. Tran, K. W. Lexa, O. Julien, T. S. Young, C. T. Walsh, M. P. Jacobson and J. A. Wells, Structure-activity relationship and molecular mechanics reveal the importance of ring entropy in the biosynthesis and activity of a natural product, *J. Am. Chem. Soc.*, 2017, **139**, 2541–2544.
59. H. Geng, F. Jiang and Y. D. Wu, Accurate structure prediction and conformational analysis of cyclic peptides with residue-specific force fields, *J. Phys. Chem. Lett.*, 2016, **7**, 1805–1810.
60. C. Keasar and R. Rosenfeld, Empirical modifications to the Amber/OPLS potential for predicting the solution conformations of cyclic peptides by vacuum calculations, *Folding Des.*, 1998, **3**, 379–388.

61. S. M. McHugh, J. R. Rogers, S. A. Solomon, H. Yu and Y. S. Lin, Computational methods to design cyclic peptides, *Curr. Opin. Chem. Biol.*, 2016, **34**, 95–102.
62. H. Yu and Y. S. Lin, Toward structure prediction of cyclic peptides, *Phys. Chem. Chem. Phys.*, 2015, **17**, 4210–4219.
63. D. T. Mirijanian, R. V. Mannige, R. N. Zuckermann and S. Whitelam, Development and use of an atomistic CHARMM-based forcefield for peptoid simulation, *J. Comput. Chem.*, 2014, **35**, 360–370.
64. K. Matsui, Y. Kido, R. Watari, Y. Kashima, Y. Yoshida and S. Shuto, Highly conformationally restricted cyclopropane tethers with three-dimensional structural diversity drastically enhance the cell permeability of cyclic peptides, *Chemistry*, 2017, **23**, 3034–3041.
65. J. Schwochert, Y. Lao, C. R. Pye, M. R. Naylor, P. V. Desai, I. C. Gonzalez Valcarcel, J. A. Barrett, G. Sawada, M. J. Blanco and R. S. Lokey, Stereochemistry balances cell permeability and solubility in the naturally derived phepropeptin cyclic peptides, *ACS Med. Chem. Lett.*, 2016, **7**, 757–761.
66. A. K. Ghose and G. M. Crippen, Quantitative structure activity relationship by distance geometry – quinazolines as dihydrofolate-reductase inhibitors, *J. Med. Chem.*, 1982, **25**, 892–899.
67. W. Jahnke, D. F. Mierke, L. Beress and H. Kessler, Structure of cobra cardiotoxin CTX I as derived from nuclear magnetic resonance spectroscopy and distance geometry calculations, *J. Mol. Biol.*, 1994, **240**, 445–458.
68. D. F. Mierke, A. Geyer and H. Kessler, Coupling constants and hydrogen bonds as experimental restraints in a distance geometry refinement protocol, *Int. J. Pept. Protein Res.*, 1994, **44**, 325–331.
69. O. Atasoylu, G. Furst, C. Risatti and A. B. Smith 3rd, The solution structure of (+)-spongistatin 1 in DMSO, *Org. Lett.*, 2010, **12**, 1788–1791.
70. D. O. Cicero, G. Barbato and R. Bazzo, NMR analysis of molecular flexibility in solution: a new method for the study of complex distributions of rapidly exchanging conformations. Application to a 13-residue peptide with an 8-residue loop, *J. Am. Chem. Soc.*, 1995, **117**, 1027–1033.
71. D. A. Evans, M. J. Bodkin, S. R. Baker and G. J. Sharman, Janocchio–a Java applet for viewing 3D structures and calculating NMR couplings and NOEs, *Magn. Reson. Chem.*, 2007, **45**, 595–600.
72. A. T. Bockus, K. W. Lexa, C. R. Pye, A. S. Kalgutkar, J. W. Gardner, K. C. Hund, W. M. Hewitt, J. A. Schwochert, E. Glassey, D. A. Price, A. M. Mathiowetz, S. Liras, M. P. Jacobson and R. S. Lokey, Probing the physicochemical boundaries of cell permeability and oral bioavailability in lipophilic macrocycles inspired by natural products, *J. Med. Chem.*, 2015, **58**, 4581–4589.
73. J. Schwochert, C. Pye, C. Ahlbach, Y. Abdollahian, K. Farley, B. Khunte, C. Limberakis, A. S. Kalgutkar, H. Eng, M. J. Shapiro, A. M. Mathiowetz, D. A. Price, S. Liras and R. S. Lokey, Revisiting N-to-O acyl shift for synthesis of natural product-like cyclic depsipeptides, *Org. Lett.*, 2014, **16**, 6088–6091.

74. A. Bagno, F. Rastrelli and G. Saielli, Toward the complete prediction of the 1H and 13C NMR spectra of complex organic molecules by DFT methods: application to natural substances, *Chemistry*, 2006, **12**, 5514–5525.
75. M. W. Lodewyk, C. Soldi, P. B. Jones, M. M. Olmstead, J. Rita, J. T. Shaw and D. J. Tantillo, The correct structure of aquatolide-experimental validation of a theoretically-predicted structural revision, *J. Am. Chem. Soc.*, 2012, **134**, 18550–18553.
76. G. Saielli, K. C. Nicolaou, A. Ortiz, H. Zhang and A. Bagno, Addressing the stereochemistry of complex organic molecules by density functional theory-NMR: vannusal B in retrospective, *J. Am. Chem. Soc.*, 2011, **133**, 6072–6077.
77. Q. N. Nguyen and D. J. Tantillo, Using quantum chemical computations of NMR chemical shifts to assign relative configurations of terpenes from an engineered streptomyces host, *J. Antibiot.*, 2016, **69**, 534–540.
78. S. Zaretsky, J. L. Hickey, M. A. St. Denis, C. C. G. Scully, A. L. Roughton, D. J. Tantillo, M. W. Lodewyk and A. K. Yudin, Predicting cyclic peptide chemical shifts using quantum mechanical calculations, *Tetrahedron*, 2014, **70**, 7655–7663.

CHAPTER 14

Trends in Cyclotide Research

MENG-WEI KAN AND DAVID J. CRAIK*

Institute for Molecular Bioscience, The University of Queensland, Brisbane QLD 4072, Australia
*E-mail: d.craik@imb.uq.edu.au

14.1 Introduction

Cyclotides[1] are disulfide-rich peptides from plants that have the unique features of a head-to-tail cyclic backbone and a knotted arrangement of three disulfide conserved bonds. This structural combination forms a motif that is referred to as the cyclic cystine knot (CCK). Cyclotides are the only known family of proteins that contain this motif and it makes them particularly stable against heat or proteolytic breakdown. Because of their stability, cyclotides have attracted much attention as scaffolds in drug design.[2–4] The term "cyclotide" was first introduced in 1999,[5] but the history of the field may be traced to earlier discoveries of a few macrocyclic peptides in the 1970s[6–10] and 1990s[5,11–16] that were later recognized as being part of this fascinating family of proteins characterized by a CCK motif.[5]

Figure 14.1 shows the structure of the prototypical cyclotide, kalata B1, originally discovered in 1970 as the bioactive ingredient in an African uterotonic medicine derived from the herb *Oldenlandia affinis*,[6] but not structurally characterized until 1995.[14] Subsequent studies of other macrocyclic cystine knotted peptides ultimately led to the definition of the cyclotide family.[5] It is estimated that the family might comprise more than 50 000 members,[17] although so far only around 300 cyclotides have been sequenced. Cyclotides

Chemical Biology No. 6
Cyclic Peptides: From Bioorganic Synthesis to Applications
Edited by Jesko Koehnke, James Naismith and Wilfred A. van der Donk

Published by the Royal Society of Chemistry, www.rsc.org

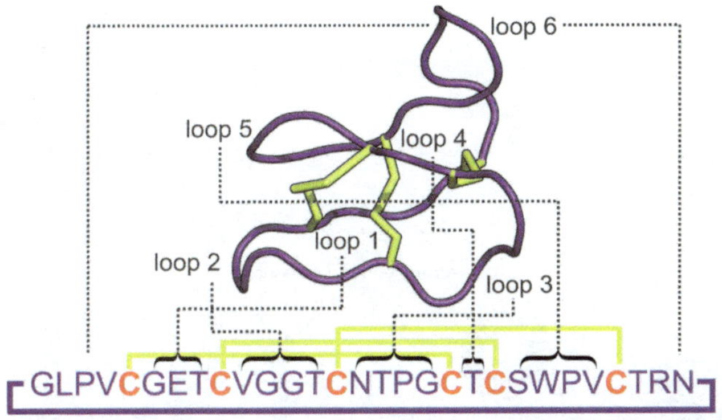

Figure 14.1 Sequence and structure of the prototypical cyclotide kalata B1 (PDBID: 1NB1), showing the labelling of backbone loops and the cystine knot motif.

range in size from 28–37 amino acids, with the variations occurring in the backbone loops (Figure 14.1) between the six conserved Cys residues characteristic of their structures. Extensive data on their sequences, structures and bioactivities are available in the online database, CyBase (www.cybase.org.au),[18–20] which also covers other classes of cyclic ribosomally synthesized polypeptides.[21,22] So far, cyclotides have been reported in species from five major plant families, namely the Rubiaceae, Violaceae, Cucurbitaceae, Fabaceae and Solanaceae families.

Cyclotides have been extensively reviewed and thus here we do not attempt another detailed review. Instead, the aim of this chapter is to provide a bibliographic analysis of the literature in the field. The articles identified and catalogued here are based on our in-house database of cyclotide literature, and on a 'Web of Science' search in December 2016 for books, chapters, original research papers and reviews in which cyclotides are the primary focus.

14.2 Trends in the Growth of the Cyclotide Field

Figure 14.2 shows the number of original research papers on cyclotides published each year since 1970, and highlights the growth of the field, which now comprises 279 original research papers. Key milestones are also indicated on Figure 14.2. For the purpose of analysis, we arbitrarily considered three periods of cyclotide research. The first comprises papers published prior to 1997, a period in which some of the initial important discoveries were made. The second period, covering the decade 1997 to 2006, was one of significant expansion, and the third period covers the most recent decade (2007–2016), when the field continued in a growth phase. The average number of cyclotide-related original research papers published per annum in the first period was two papers, in the second it was six, and the third period produced 21 papers per year on average over the decade, leading to a total

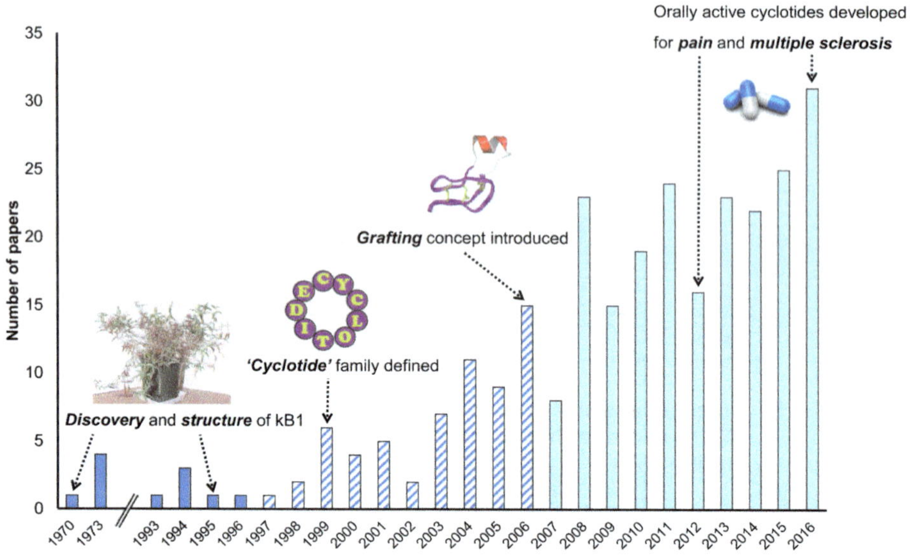

Figure 14.2 Cyclotide-related publications from 1970. Bars are shaded according to the decades 2007–2016 (light shading), 1997–2006 (hashed shading), and earlier than 1997 (dark shading). Key milestones are indicated. These include the discovery of kalata B1[7,8] and its structural characterization,[14] the definition of the cyclotide family, proof-of-concept of the ability to graft a sequence into a cyclotide framework, and finally the development of modified cyclotides as orally active drug leads for pain and multiple sclerosis, with efficacy demonstrated in animal models of these conditions. Grafted cyclotides have been made for a range of other diseases, but these two examples are those where oral activity was reported, compared with delivery *via* injection.

of 279 original research papers. Additionally, there are 102 reviews focused principally on cyclotides, leading to a total of 381 cyclotide references that we cite in this article. Cyclotides are mentioned incidentally in approximately another 40 papers or reviews, and are the subject of a recent book.[1]

14.3 Categories of Cyclotide Research: an Analysis

We analyzed the primary cyclotide publications, as well as some key reviews, and divided them into categories of discovery,[5–18,23–107] gene sequences,[30,38,44–49,64,65,71,80,86,92–95,108–123] analysis,[18–20,51,60,73,117,124–140] structure and folding,[5,14,29,33,41–45,56–60,108,124,131,134,135,141–210] bioactivity,[6,9,23,27,28,30,33,54,70,72,81–84,90,110,113,148,166,183,185,188,211–250] biosynthesis,[30,44,94,114–121,174,203,251–272] synthesis,[144,204,252,268,270–301] drug design,[2–4,165,187,191,201,210,234,253,282,302–346] membrane binding,[98,168,175,183,184,188,193,194,197,217,228,229,245,347–360] cell penetration,[360–366] and toxicity.[58,111,122,129,134,215,231,348] These categories are not mutually exclusive, and any given paper may have been categorized into more than one of these categories. For example, many of the papers categorized as 'discovery', also attract 'structure' or 'bioactivity' categorization. Our purpose in this analysis is to examine trends in the types of studies on cyclotides undertaken to date, to gain insight into the way in which the field is evolving and, perhaps, to predict future trends or areas in need of further study. Hopefully this analysis will also be useful to give new entrants to the field a comprehensive overview of progress since the initial discoveries of macrocyclic peptides in plants.

The first conclusion that can be drawn from Figure 14.3 is that most studies to date have focused on discovery, structure and bioactivities, and that this continues to be an active area of cyclotide research. Second, studies of the biology of cyclotides continue to be an active area of research, with much focus recently on their biosynthesis. Third, over the last decade, there has been a growing interest in applications of cyclotides. This trend has arisen because of the capacity to synthesize cyclotides that has developed over the last 15 years, thus alerting researchers to their potential for drug design purposes. Finally, there is a developing body of literature on studies of the membrane binding and cell penetrating properties of cyclotides. These studies have helped to understand the mechanisms underlying their bioactivity and have provided new opportunities for their applications as drug delivery scaffolds for intracellular targets. We now comment briefly on trends within the categories identified in Figure 14.3.

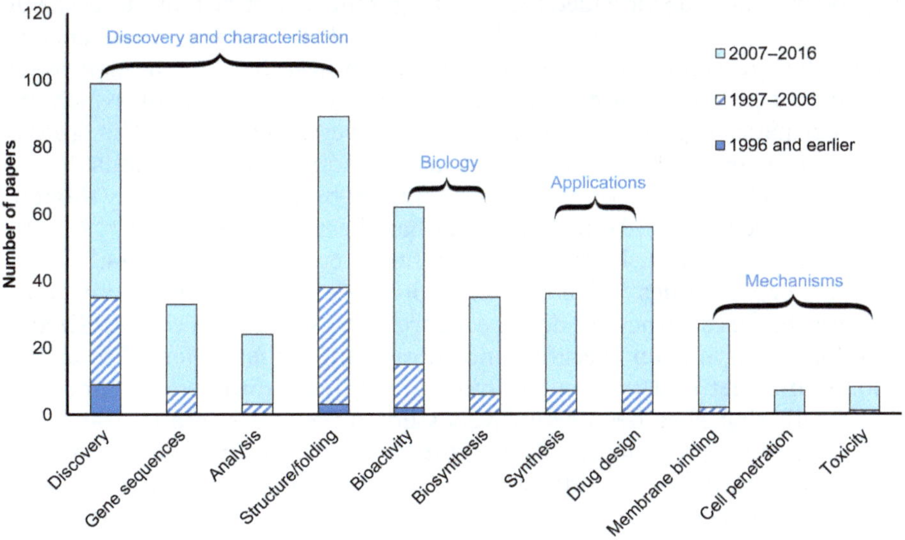

Figure 14.3 Analysis of cyclotide publications by topic category. Bars are shaded according to the decades 2007–2016 (light shading), 1997–2006 (hashed shading) and earlier than 1997 (dark shading). Papers were categorized by the main topic of the paper, but some papers were included in multiple categories if they addressed several topics in depth. The 'discovery' category (which includes 98 papers) refers to peptide-based discovery, with nucleic acid-based sequence discovery included in the 'gene sequences' category. Individual topics are grouped into the fields of 'discovery and characterization', 'biology', and 'applications and mechanisms', to give readers an idea of the relative publication records in these broad fields.

14.3.1 Peptide-based Discovery

'Discovery' accounts for the largest proportion of cyclotide papers (99/381 = 26% of published cyclotide research). Most of the early discoveries of cyclotides were made at the peptide level, by researchers isolating peptides from plants and characterizing them using a combination of mass spectrometry and NMR spectroscopy (for examples, see ref. 12, 14 and 16). Such studies led to the first structural characterizations of cyclotides and continue to play an important role in cyclotide research. Many of these discovery efforts were made as part of screening studies, but some were made based on ethnobotanical observations.[50,79,82,84,100] The latter include the detection of anti-HIV cyclotides in Chinese medicinal herbs, as well as discoveries of the uterotonic, antioxidant or antiviral activities of cyclotides. A noteworthy point that highlights the remarkable stability of cyclotides is that some cyclotides have been detected in herbarium dried plant specimens more than 50 years old.[86]

At the time cyclotides were discovered, it was not known whether they were ribosomally synthesized or whether they were biosynthesized by non-ribosomal routes, as was known to be the case for the macrocyclic peptide cyclosporin A, which is well-known for its clinical use as an immunosuppressive agent in transplant medicine. The discovery, in 2001, of genes encoding for cyclotide precursor proteins[30] not only provided the first insight into cyclotide biosynthesis, but also opened up new routes to their discovery based on nucleic acid sequencing, as described in Section 14.3.2. Briefly, their biosynthesis involves the expression of precursor proteins that contain one or more cyclotide domains, which are then enzymatically processed to produce mature cyclic peptide(s). The process appears to be remarkably efficient and most cyclotide-producer plants do not contain detectable traces of unprocessed (linear) precursors. Recently though, a range of acyclic cyclotide-like sequences have been reported.[42,65,71,76,80,94,104,117] They have been variously referred to as linear cyclotide derivatives,[44] uncyclotides,[71] or acyclotides[104,117] although we prefer the latter nomenclature. Violacin A was the first such acyclotide to be reported and structurally characterized.[44]

Cyclotides have been classified into three subfamilies, referred to as the Möbius, bracelet and trypsin inhibitor cyclotides. Prototypic members of each family that have been extensively studied include kalata B1 (Möbius), cycloviolacin O1 (bracelet) and MCoTI-II (trypsin inhibitor). Approximately two thirds of natural cyclotides fall into the bracelet category and one third into the Möbius category, with only a handful of trypsin inhibitor cyclotides known so far, and all of those occurring in a few species of the Cucurbitaceae plant family, most notably in the seeds of the tropical vine *Momordica cochinchinensis*. The three subfamilies differ subtly in their sequences and structures, but all comprise a CCK motif.

14.3.2 Gene-based Discovery and Cyclotide Gene Regulation

With the advent of advanced molecular biology techniques, it has become easier to rapidly sequence the genomes or transcriptomes of organisms or tissues and this capability has helped to accelerate the pace of cyclotide sequence discovery.[78] Table 14.1 lists the titles of studies involving nucleic

Table 14.1 Cyclotide nucleic acid-based discovery (with selected milestone papers highlighted).

Year	Title	◆ Milestone	Ref.
2001◆	Biosynthesis and insecticidal properties of plant cyclotides: the cyclic knotted proteins from *Oldenlandia affinis*	*The first paper to define cyclotide genes*	30
2004	Conserved structural and sequence elements implicated in the processing of gene-encoded circular proteins	–	108
2004	Variations in cyclotide expression in *Viola* species	–	109
2005	Processing of a 22 kDa precursor protein to produce the circular protein tricyclon A	–	38
2005◆	Dissecting defense-related and developmental transcriptional responses of maize during *Ustilago maydis* infection and subsequent tumor formation	*The first paper to examine the up-regulation of cyclotide-like genes in response to infection*	110
2006	Discovery of cyclotide-like protein sequences in graminaceous crop plants: ancestral precursors of circular proteins?	–	45
2006◆	Discovery and characterization of a linear cyclotide from *Viola odorata*: Implications for the processing of circular proteins	*The first paper to report the gene sequence of an 'acyclotide', providing insights into biosynthesis*	44
2008	The alpine violet, *Viola biflora*, is a rich source of cyclotides with potent cytotoxicity	–	46
2008	Molecular phylogenetic analysis of Violaceae (Malpighiales) based on plastid and nuclear DNA sequences	–	47
2009	Circular proteins from Melicytus (Violaceae) refine the conserved protein and gene architecture of cyclotides	–	48
2009	A transcriptional profile of metallophyte *Viola baoshanensis* involved in general and species-specific cadmium-defense mechanisms	–	111
2009	Two suites of cyclotide precursor genes from metallophyte *V. baoshanensis*: cDNA sequence variation, alternative RNA splicing and potential cyclotide diversity	–	112
2010◆	Cyclotide proteins and precursors from the genus *Gloeospermum*: Filling a blank spot in the cyclotide map of Violaceae	*A paper reporting cyclotide precursors and processing proteins in Violaceae plants*	49
2010	Cyclotides are a component of the innate defence of *Oldenlandia affinis*	–	113
2010	Identification of candidates for cyclotide biosynthesis and cyclisation by expressed sequence tag analysis of *Oldenlandia affinis*	–	114

(*continued*)

Table 14.1 (*continued*)

Year	Title	◆ Milestone	Ref.
2011◆	Discovery of an unusual biosynthetic origin for circular proteins in legumes	*Two papers reporting a novel precursor arrangement for cyclotide genes in butterfly pea*	115
2011◆	Discovery and characterization of novel cyclotides originated from chimeric precursors consisting of albumin-1 chain a and cyclotide domains in the Fabaceae family		64
2011	Discovery of a linear cyclotide from the bracelet subfamily and its disulfide mapping by top-down mass spectrometry	–	65
2012	Novel cyclotides and uncyclotides with highly shortened precursors from *Chassalia chartacea* and effects of methionine oxidation on bioactivities	–	71
2012◆	Cyclotides associate with leaf vasculature and are the products of a novel precursor in *Petunia* (Solanaceae)	*The first report of cyclotides in a Solanaceae plant*	117
2012	Cyclic peptides arising by evolutionary parallelism *via* asparaginyl-endopeptidase-mediated biosynthesis	–	116
2013	Discovery of linear cyclotides in monocot plant *Panicum laxum* of Poaceae family provides new insights into evolution and distribution of cyclotides in plants	–	80
2013	Cyclotide discovery in Gentianales revisited–identification and characterization of cyclic cystine-knot peptides and their phylogenetic distribution in Rubiaceae plants	–	78
2015◆	Distribution of circular proteins in plants: large-scale mapping of cyclotides in the Violaceae	*Paper reporting the detection of cyclotides in herbarium specimens*	86
2015	Molecular cloning and characterization of a cyclotide gene family in *Viola modesta*	–	92
2015	Exogenous plant hormones and cyclotide expression in *Viola uliginosa* (Violaceae)	–	120
2015	Micropropagation of *Viola uliginosa* (Violaceae) for endangered species conservation and for somaclonal variation-enhanced cyclotide biosynthesis	–	121
2015	Transcriptomic screening for cyclotides and other cysteine-rich proteins in the metallophyte *Viola baoshanensis*	–	122
2016◆	Gene coevolution and regulation lock cyclic plant defence peptides to their targets	*Paper making a link between cyclotide localization in plants and function*	123

acid-based discovery of cyclotides, and gives a flavor of the work done in this field. This includes discoveries of not only cyclotide precursor sequences[30,38,48,49,64,71,93,95,108,111,112,115,122] but the sequences of auxiliary processing enzymes involved in cyclotide production.[118,119,256] A significant proportion of papers reporting cyclotide-related nucleic acid sequences initially focused on *O. affinis* (Rubiaceae),[30,71,93,113,114] the plant in which kalata B1 was originally discovered,[8,9] and on several Violaceae species,[47–49,86,92,93,109,120–122] but recent studies have examined a wide range of species in other plant families, including the tropical vine *Momordica cochinchinensis*,[94] and the legume *Clitoria ternatea*,[64,115,118,123] as well as petunia plants.[117] It is interesting to note that although sequences encoding acyclotides have been found in species from both the monocot and dicot plant lineages, no cyclotides (*i.e.* cyclic peptides) have yet been found in monocots.

Perhaps the most significant finding to emerge from studies of nucleic acid sequences encoding cyclotides is that plants have used several alternative genetic blueprints for cyclotides and hence several strategies for their production. These include the evolution of dedicated genes specifically for making cyclotides, as well as alternatively co-opting other pre-existing genes, most notably albumin genes, to incorporate cyclotide nucleic acid sequences. The *Oak* genes from *O. affinis* provide an example of the former[30] and the cliotide genes from butterfly pea (*Clitoria ternatea*) provide an example of the latter.[64,115,123] This genetic diversity suggests that cyclotides arose in different species *via* convergent evolution, rather than having a common genetic or biosynthetic origin in all plant species.

14.3.3 Analysis

We populated this category with papers that involve experimental or computational approaches to detect or analyze cyclotides in some way (but not including 3D structure determination or membrane binding, which are included in their own sections below). Table 14.2 contains a list of articles classified as 'analysis' and demonstrates that a range of detection approaches, mass spectrometry analyses and other biophysical techniques have been applied to cyclotides, as have bioinformatics and database analyses. As far as the latter is concerned, as noted earlier, much information on cyclotides is contained in CyBase, a database dedicated to ribosomally synthesized cyclic peptides and circular proteins. CyBase is freely accessible and was first described in 2006,[18] but new features and analyses were reported in 2008[19] and 2010.[20] CyBase is curated and regularly updated in the Craik laboratory at the University of Queensland.

The key messages from Table 14.2 are that natural cyclotides are readily detectable in plants using methods ranging in sophistication from TLC to mass spectrometry, and that exogenously administered cyclotides can be detected in a range of biological and environmental samples. The importance of mass spectrometry in the sequencing and detection of cyclotides is paramount, and is perhaps the single most important technique for their analysis. NMR has played a major role in 3D structure determination (see

Table 14.2 Selection of analytical studies of cyclotides.

Technique	Title of published study	Ref.
A. Detection in plants and in the environment and SAR[a] studies		
TLC[b]	Application of TLC chemical method to detection of cyclotides in plants	51
Environment	Biomedicine in the environment: Sorption of the cyclotide kalata B2 to montmorillonite, goethite and humic acid	129
Quantification	Quantification of small cyclic disulfide-rich peptides	130
Extraction	Optimization of cyclotide extraction parameters	73
Isolation	A comparative study of extraction methods reveals preferred solvents for cystine knot peptide isolation from *Momordica cochinchinensis* seeds	133
QSAR[c]	Cyclotide structure–activity relationships: Qualitative and quantitative approaches linking cytotoxic and anthelmintic activity to the clustering of physicochemical forces	134
Localization	Immunolocalization of cyclotides in plant cells, tissues and organ supports their role in host defense	139
Pharmacokinetics	Biodistribution of the cyclotide MCoTI-II, a cyclic disulfide-rich peptide drug scaffold	140
B. Mass spectrometry		
MALDI	Peptide quantification by MALDI mass spectrometry: investigations of the cyclotide kalata B1 in biological fluids	125
MALDI	Quantitative analysis of backbone-cyclised peptides in plants	126
LC-MS	A new 'era' for cyclotide sequencing	127
MALDI	Elucidating the structure of two cyclotides of *Viola tianshanica* maxim by MALDI TOF/TOF MS analysis	60
LC–ESI-MS	A liquid chromatography-electrospray ionization-mass spectrometry method for quantification of cyclotides in plants avoiding sorption during sample preparation	128
MALDI imaging	Cyclotides associate with leaf vasculature and are the products of a novel precursor in *Petunia* (Solanaceae)	117
Pharmacokinetics	Improved method for quantitative analysis of the cyclotide kalata B1 in plasma and brain homogenate	138
C. Databases and bioinformatics		
Database	CyBase: a database of cyclic protein sequence and structure	18
Database	CyBase: a database of cyclic protein sequences and structures, with applications in protein discovery and engineering (an update)	19

Bioinformatics	Analysis and classification of circular proteins in CyBase	20
Bioinformatics	Prediction and characterization of cyclic proteins from sequences in three domains of life	132
Bioinformatics	Two Blast-independent tools, CyPerl and CyExcel, for harvesting hundreds of novel cyclotides and analogues from plant genomes and protein databases	136
Nomenclature	Nomenclature of homodetic cyclic peptides produced from ribosomal precursors: An IUPAC task group interim report	137
D. Biophysical methods[f]		
AUC[d]	A comparison of the self-association behaviour of the plant cyclotides kalata B1 and kalata B2 *via* analytical ultracentrifugation	132
Solution NMR	The self-association of the cyclotide kalata B2 in solution is guided by hydrophobic interactions	131
PFG[e] NMR	Translational diffusion of cyclic peptides measured using pulsed-field gradient NMR	135

[a]Structure–activity relationships.
[b]Thin layer chromatography.
[c]Quantitative structure–activity relationships.
[d]Analytical ultracentrifuge.
[e]Pulsed field gradient.
[f]Biophysical methods associated with 3D structure are covered in Table 14.3 and those associated with membrane binding are covered in Table 14.4.

Section 14.3.4), as well as in examining the self-association and translational diffusion of cyclotides.

14.3.4 Structures, Folding and Dynamics

As is clear from Figure 14.3, along with discovery, structure is a very active topic within the field of cyclotide research, with 89/381 = 23% of cyclotide papers fitting into this category. NMR is the predominant technique for determining the 3D structures of cyclotides, but molecular modelling,[33] neutron diffraction[197] and X-ray crystallography[184,205] have also been applied for studying either the structures or intermolecular interactions of cyclotides. The oxidative folding behavior of cyclotides has also been studied by NMR and HPLC or LCMS methods,[151,152,170,171] and stable two-disulfide intermediates have been identified. Interestingly, the folding pathways for Möbius and trypsin inhibitor cyclotides both involve a similar two-disulfide intermediate, but differ in their conversion to the final CCK product. In terms of dynamics, NMR diffusion measurements[168] have provided useful information on the molecular dynamics of cyclotides bound to micelles. There have also been studies on the dynamics of MCoTI-I and MCoTI-II bound to trypsin using NMR relaxation measurements.[186,198]

The PDB currently contains coordinates for 27 cyclotide-related structures, which are listed in Table 14.3. Of the PDB-deposited structures, most are of native cyclotides, but a number of synthetic derivatives have also been characterized, including some artificially linearized in the backbone and some containing point mutations aimed at exploring structure–activity relationships, along with some grafted cyclotide derivatives. One of the noteworthy features of cyclotide structures is that they are relatively rigid, which is one of the features that makes them useful as drug design scaffolds. Given the rigid CCK structure of cyclotides and the fact that several representative structures have been determined for each of the three subclasses of cyclotides, it can be anticipated that in the future, computer-based homology modelling may be all that is required for generating structural information on new cyclotides.

So far, there is no full 3D structure of a cyclotide precursor protein, but a recent study suggests that cyclotide precursors are likely to have an unstructured N-terminal Pro-domain upstream of well-structured cyclotide domains.[272] Some structural information is available on small fragments of these N-terminal domains, and interestingly, the two examples studied adopt a helical structure as isolated peptides in solution (see Table 14.3). It would certainly be of interest to gain more structural information on intact precursors, especially in the presence of their processing enzymes.

One key structural class missing from the PDB that is of much current interest in the field of cyclotide research is that of asparaginyl endopeptidase processing enzymes. The recent development of a recombinant method for expressing one of these proteins in *E. coli*[119] should facilitate such studies. Another sparsely studied area in terms of structure relates to complexes of cyclotides. The first study of a cyclotide complex was on metal ion binding

Table 14.3 PDB entries for cyclotide-related three-dimensional structures.[a]

Name of peptide	Organism	PDB ID	Comment	Year/Ref.
Plant-derived native cyclotides				
Kalata B1 (kB1)	*O. affinis*	1KAL[b]	First structure of a cyclotide	1995 [14]
"	"	1K48	Contains incorrect disulfide paring –not recommended for use[c]	2002 [149]
"	"	1JJZ	Un-optimized coordinates for 1K48	2002 [149]
"	"	1NB1◆	Best high resolution structure, the preferred kB1 structure for analysis[d]	2003 [154]
"	"	1ZNU	In DPC micelle solution	2006 [168]
Kalata B2	"	1PT4	In aqueous solution	2005 [161]
Kalata B2	"	2KCH	Micelle-bound structure	2009 [183]
Kalata B5	"	2KUX	In aqueous solution, pH study	2010 [57]
Kalata B7	"	2JWM	Mn^{2+} binding in DPC micelles	2008 [175]
Kalata B7	"	2M9O	Solution state structure	2013 [328]
Kalata B8	"	2B38	Solution state structure	2006 [41]
Kalata B12	"	2KVX	pH study to probe conserved Glu	2011 [194]
Varv F	*V. arvensis*	3E4H◆	First crystal structure of a cyclotide	2009 [184]
Varv F	"	2K7G	Solution state structure	2009 [184]
Vhr-1	*V. hederacea*	1VB8	In aqueous solution	2004 [251]
Vhl-1	"	1ZA8	In aqueous solution	2005 [37]
Vh1-2	"	2KUK	In aqueous solution	2010 [185]
Cycloviolacin O1	*V. odorata*	1NBJ◆	High resolution structure	2003 [154]
Cycloviolacin O2	"	2KNM	In aqueous solution	2009 [180]
[E6E^{Me}]CyO2[e]	"	2KNN	pH study to probe conserved Glu	2009 [180]
Cycloviolacin O14	"	2GJ0	In aqueous solution	2006 [43]
Violacin A	"	2FQA◆	First reported 'acyclotide'	2006 [44]
Tricyclon A	*V. tricolor*	1YP8	Extra-long loop 6 cyclotide	2005 [38]
Cter M	*C. ternatea*	2LAM	First structure of a Fabaceae cyclotide	2011 [115]
Palicourein	*P. condensata*	1R1F	Extra-long loop 6 cyclotide	2004 [155]
MCoTI-II	*M. cochinchinensis*	1IB9◆	First structure of a TI cyclotide	2001 [29]
"	"	1HA9◆	Comparison with acyclic analogues	2001 [147]
MCoTI-V	"	2LJS	Acyclic MCoTI peptide	2012 [116]
Synthetic cyclotides				
[Ala 1,15]kB1	Disulfide mutant	1N1U	Missing one disulfide bond	2003 [151]
[Δ23-28]kB1	Acyclic permutant	1ORX	Backbone opened in loop 6	2003 [213]

(*continued*)

Table 14.3 (*continued*)

Name of peptide	Organism	PDB ID	Comment	Year/ Ref.
[P20D,V21K] kB1	Grafted cyclotide	2F2I	Grafted kB1 analogue to test folding	2006 [165]
[W19K, P20D,V21K] kB1	Grafted cyclotide	2F2J	Grafted kB1 analogue to test folding	2006 [165]
All-D kalata B1	Synthetic cyclotide	2JUE◆	Mirror image synthetic form of kB1	2011 [193]
Linear kalata B1	Acyclic mutant	2KHB	Backbone opened in loop 6	2012 [197]
[W23WW]kB1	Synthetic mutant	2MN1	Extra Trp residue in loop 5	2014 [355]
[GGG] kB1[GGT]	Synthetic mutant	2MH1	kB1 cyclized using sortase A	2014 [268]
D/L Kalata B1	Racemic mixture	4TTM	Racemic kalata B1	2014 [205]
[G6A]kalata B1	Synthetic mutant	4TTN	Quasi racemic mixture	2014 [205]
[V25A]kalata B1	Synthetic mutant	4TTO	Quasi racemic mixture	2014 [205]
[Y15S]kalata B7	Synthetic mutant	2MW0	Used to probe cation binding	2016 [359]
[ΔSS]MCoTI-II	Folding intermediate	2PO8◆	Two-disulfide intermediate	2008 [171]
MCoTI-II linear	Acyclic mutant	2IT8	Backbone opened in loop 6	2011 [173]
MTabl13	Grafted cyclotide	2MT8◆	Grafted MCoTI-II analogue designed as a drug lead for CML	2015 [336]
kB1[ghrw; 23–28]	Grafted cyclotide	2LUR◆	Grafted kalata B1 analogue designed as a drug lead for obesity	2012 [322]
Cyclotide precursor fragments				
OaNTR	*O. affinis*	1WN8	NTR fragment of kalata B3 precursor	2004 [108]
VoNTR	*V. odorata*	1WN4	NTR fragment of kalata S precursor	2004 [108]
Cyclotide complexes				
Trypsin: MCoTI-II	*M. cochinchinensis*	4GUX◆	Crystal structure of complex	2013 [198]
Hdm2:MCoTI-I	Synthetic cyclotide	2M86◆	Hdm2 with anti-tumour cyclotide	2013 [327]
Kalata B7[f]	*O. affinis*	2JWM◆	In complex with Mn	2008 [175]

[a]Table includes entries found in the PDB (www.rcsb.org) using a search for the terms "cyclotide" or "kalata" or "knottin" or "MCoTI". Milestone or noteworthy structures are indicated with the ◆ symbol.

[b]1KAL, the prototypical structure of a cyclotide was determined before the term 'cyclotide' was introduced, so does not appear using the above search strings in the PDB, but is deposited there.

[c]A later study[154] showed that the 1K48 structure has an incorrect disulfide connectivity.

[d]This is the PDB file (1NB1) we recommend as the preferred structure of kalata B1.

[e]Cycloviolacin O2 with Glu6 methylated.

[f]The entry for kalata B7 is duplicated on this line (from earlier in the Table) to reflect that it can be considered as a complex (with manganese ions).

to the cyclotide kalata B1.[168] There have been two structures of complexes between cyclotides and other macromolecules reported, one of the MCoTI-II trypsin interaction and one of a grafted cyclotide interacting with the oncoprotein Hdm2.[327] There has also been a study on the dynamics of MCoTI-I with bound trypsin.[186,198] As noted above, NMR has been used to study the structure of cyclotides bound to DPC micelles, used as surrogates for membrane binding.[168,175]

14.3.5 Bioactivity

We can broadly divide studies on the biology of cyclotides into studies of bioactivity and studies of biosynthesis. Studies on the bioactivities of cyclotides comprise the third largest category by publication numbers (61/381 = 16%) of all categories in Figure 14.3. The presumed natural function of cyclotides is as host defense molecules and accordingly many of the bioactivity categorized papers have involved studies of the insecticidal,[30,104,176,217] nematocidal,[218,219,223,227,228,239] molluscicidal,[222] or other cytotoxic bioactivities of cyclotides.[33,148,166,188,215] Of course we cannot exclude the possibility that cyclotides might have additional or alternative natural functions and this is an area that warrants further investigation.

Recently, much of the focus of cyclotide bioactivity research has been not so much on the bioactivities of natural cyclotides, but on the artificial introduction, or grafting, of desired pharmaceutical bioactivities into a cyclotide framework, as discussed in more detail in Section 14.3.8. Introduced bioactivities include anti-tumor, analgesic, anti-HIV, anti-obesity and anti-angiogenic activities, amongst others. We make no further comment here on this topic, as it has been well covered in recent reviews.[232,233,244,333,367]

14.3.6 Biosynthesis

Biosynthesis is a very important topic in the field of cyclotide biology. As noted in Section 14.3.1, cyclotides are ribosomally synthesized by plants *via* enzymatic processing from a diverse range of precursor proteins. Some of these precursors encode just one cyclotide domain, whereas others encode up to six cyclotide domains, of either the same or different cyclotides. Recently, some precursors from the Cucurbitaceae plant family have been found to produce a series of tandem repeats of cyclic peptide domains terminated by a linear (acyclotide) domain,[94] demonstrating the versatility of cyclotide production and perhaps providing clues as to the evolution of cyclotides.

Research over the last few years has focused on the discovery of enzymes mediating the biosynthetic pathway by excising cyclotide domains from their precursors and cyclizing them. In particular, the enzymes butelase[118] and OAEP1[119] have been isolated and shown to cyclize cyclotide substrates *in vitro*. These asparaginyl endopeptidases recognize conserved Asn or Asp residues at the C-terminus of the cyclotide domain in the precursor sequences and ligate them with the N-terminal residue of the cyclotide domain to form loop 6 (Figure 14.1) of mature cyclotides. Precursors that lack the essential Asn or Asp residue are not able to be cyclized and this is the primary mechanism that leads to acyclotides.[44,65,66,117]

Together, the topics of bioactivity and biosynthesis have been the subjects of 96 papers[6,9,23,27,28,30,33,44,54,70,72,81–84,90,94,110,113–115,117–121,148,166,174,183,185,188,203,211–224,226–272] and 20 reviews[23,212,216,220,221,224,230,232,233,238,243,244,247,252,261,262,266,267,270,271] that fit into the topic of the natural 'biology' of cyclotides. In the future, we anticipate that this field will expand and lead to developments in the 'synthetic biology' of cyclotides. The beginnings of this field are already apparent in studies producing cyclotides in cells using intein-based approaches[266,282] and in the *in vitro* enzymatic cyclization of synthetic cyclotide precursors.[118,119] The stable nature of cyclotides, in particular, makes them amenable to adaptation of their biosynthesis using a range of chemical or biological approaches.

14.3.7 Synthesis

As with many peptides or proteins isolated from natural sources, it is useful to have methods of artificially making cyclotides so that their structures, functions and applications can be systematically studied, and/or they can be manufactured on a large scale. The main reason for wanting to synthesize cyclotides is not that are they are produced in limited amounts naturally, but that some can be difficult to separate and purify given that a single plant may express dozens to hundreds of similar cyclotides. Additionally, synthetic approaches allow very specific changes to be made to a peptide to explore structure–activity relationships. In the case of cyclotides, solid-phase approaches to their synthesis were developed in the 1990s[144,273–275] and have underpinned many applications since then.

Briefly, the main technique used for the solid phase peptide synthesis of cyclotides involves an adaptation of native chemical ligation, in which thioester-based chemistry is used to assemble and cyclize the cyclotide sequences, with disulfide bond formation being the second step in the process. As with bioactivity, we do not address this topic here in detail as it has been well covered in recent reviews.[285,287,301,368] We simply point out that synthesis and drug design of cyclotides go hand in hand, and we have grouped these topics in Figure 14.3 under the banner of 'applications'. Together they account for nearly 100 publications.

14.3.8 Drug Design and Protein Engineering Applications

Drug design is an increasingly important area of cyclotide research, with 56/381 = 15% of cyclotide papers fitting into this category, most of which were done in the last decade. Many of these studies involve cyclotide grafting. Figure 14.4 highlights the grafting concept, which basically involves the synthesis of a cyclotide in which a foreign bioactive epitope is inserted, with the aim of maintaining its biological activity while enhancing its

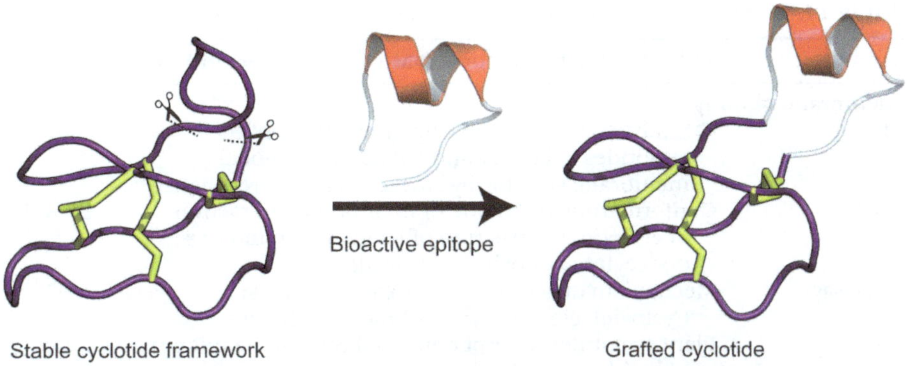

Figure 14.4 Schematic illustration of the grafting concept for cyclotides. Basically, a bioactive epitope that may have been discovered in one of a variety of ways (including screening, computational design, phage display or derived from a bioactive epitope of a larger protein) is inserted into one of the loops in the cyclotide framework. In the case illustrated, a bioactive helix is grafted into loop 6 of kalata B1.

stability. The targets of these 'designer' cyclotides include enzymes (both intracellular and extracellular), cell surface receptors and protein–protein interactions, highlighting the versatility of the cyclotide scaffold in drug design. Intracellular targets are currently of particular interest in the pharmaceutical industry and provide a point of difference between cyclotides and other protein-based modalities, such as antibodies, which cannot enter cells.

Although most studies involving grafting so far have focused on pharmaceutical applications, it can be anticipated that in future, the grafting concept is likely to be applied for agrichemical uses, including crop protection agents, as well as for diagnostic agents and imaging agents. The desirable features of cyclotides, including stability and plasticity to sequence substitution, by and large apply equally to these applications as they do to pharmaceuticals. One possible exception here is that for agrichemical applications, regulatory authorities do not favor extreme stability, and some level of digestibility may need to be engineered into agriculturally directed cyclotides. Overall, though, cyclotides can be regarded as useful molecular design and protein engineering frameworks.

14.3.9 Membrane Binding, Cell Penetration and Toxicity

Membrane binding is an increasingly active field of cyclotide research, with 27/381 = 7% of cyclotide papers fitting into this category, again with most activity in this field having occurred over the last decade. In contrast to the drug design aspects of cyclotide research described above, membrane binding studies have focused mainly on native cyclotide sequences. Table 14.4 provides a summary of the key papers in this field and highlights the

Table 14.4 Membrane binding and cell penetration studies of cyclotides.[a]

Technique	Title of published report (◆Milestone papers indicated)	Yr/Ref.
Membrane binding		
SPR[b]	◆Studies on the membrane interactions of the cyclotides kalata B1 and kalata B6 on model membrane systems by surface plasmon resonance	2005 [347]
NMR[c]	Conformation and mode of membrane interaction of cyclotides - structure of kalata B1 bound to a dodecylphosphocholine micelle	2006 [168]
Bioassay	Mechanism of action of cytotoxic cyclotides: Cycloviolacin O2 disrupts lipid membranes	2007 [348]
Bioassay	Plant cyclotides disrupt epithelial cells in the midgut of lepidopteran larvae	2008 [217]
NMR	Divalent cation coordination and mode of membrane interaction of cyclotides: NMR spatial structure of ternary complex kalata B7/Mn^{2+}/DPC micelle	2008 [175]
Electrophys[d]	◆The biological activity of the prototypic cyclotide kalata B1 is modulated by the formation of multimeric pores	2009 [349]
NMR	Despite a conserved cystine knot motif, different cyclotides have different membrane binding modes	2009 [183]
NMR/X-ray	Combined X-ray and NMR analysis of the stability of the cyclotide cystine knot fold that underpins its insecticidal activity and use as a drug scaffold	2009 [184]
Synthesis/ bioassay	Lysine-scanning mutagenesis reveals a previously unidentified amendable face of kalata B1 for optimisation of nematocidal activity	2010 [228]
Bioassay	The cyclotide cycloviolacin O2 from *Viola odorata* has potent bactericidal activity against Gram-negative bacteria	2010 [229]
SAR[e]	Cytotoxic potency of small macrocyclic knot proteins: Structure–activity studies of native and chemically modified cyclotides	2011 [188]
Vesicle leakage	Cyclotide-membrane interactions: Defining factors of membrane binding, depletion and disruption	2011 [351]
SPR	◆Decoding the membrane activity of the cyclotide kalata B1: The importance of PE phospholipids and lipid organization on hemolytic and anti-HIV activities	2011 [352]
SPR	A synthetic mirror image of kalata B1 reveals that cyclotide activity is independent of a protein receptor	2011 [193]
NMR	The role of conserved Glu residue on cyclotide stability and activity: A structural and functional study of kalata B12	2011 [194]
SPR/vesicle leakage	◆Phosphatidylethanolamine binding is a conserved feature of cyclotide-membrane interactions	2012 [354]
Neutron refl.[f] and ITC[g]	Cyclotides insert into lipid bilayers to form membrane pores and destabilize the membrane through hydrophobic and PE-specific interactions	2012 [197]
NMR/SPR/CM[h]	◆Anticancer and toxic properties of cyclotides are dependent on phosphatidylethanolamine phospholipid targeting	2014 [355]
MD[i]	Defining the membrane disruption mechanism of kalata B1 *via* coarse-grained molecular dynamics simulations	2014 [356]

Technique	Title of published report (◆Milestone papers indicated)	Yr/Ref.
MD	Dynamic scenario of membrane binding process of kalata B1	2014 [357]
CM	Emulsions stabilized by mini cyclic proteins for bioactive compound delivery	2014 [358]
SPR/bioassay	Lysine-rich cyclotides: A new subclass of circular knotted proteins from Violaceae	2015 [98]
Vesicle leakage	Selective membrane disruption by the cyclotide kalata B7: Complex ions and essential functional groups in the phosphatidylethanolamine binding pocket	2016 [359]
Cell penetration		
CM	◆The cyclic cystine knot mini-protein MCoTI-II is internalized into cells by macropinocytosis	2007 [361]
CM	◆Identification and characterization of a new family of cell-penetrating peptides. Cyclic cell-penetrating peptides	2011 [362]
CM	◆Cellular uptake of cyclotide MCoTI-I follows multiple endocytic pathways	2011 [363]
FC[j]/NMR	Structural parameters modulating the cellular uptake of disulfide-rich cyclic cell-penetrating peptides: MCoTI-II and SFTI-1	2014 [364]
SPR/FC/CM	◆The prototypic cyclotide kB1 has a unique mechanism of entering cells	2015 [360]
FC	Optimization of the cyclotide framework to improve cell penetration	2015 [365]
CM	Cellular uptake of a cystine-knot peptide and modulation of its intracellular trafficking	2016 [366]

[a]Table includes papers principally focused on membrane binding or cell penetration of cyclotides. Noteworthy milestones are indicated with the ◆ symbol.
[b]SPR, surface plasmon resonance.
[c]NMR, nuclear magnetic resonance.
[d]Electrophysiology.
[e]SAR, structure–activity relationship study.
[f]Neutron reflectrometry.
[g]ITC, isothermal calorimetry.
[h]CM, confocal microscopy.
[i]MD, molecular dynamics.
[j]FC, flow cytometry.

fact that the fundamental studies on understanding cyclotide membrane binding have helped to explain their natural functions as pesticidal agents. Furthermore, these studies have underpinned their potential uses as cell penetrating peptides.

The main finding to emerge from these cyclotide membrane binding studies is that one specific lipid, phosphatidylethanolamine, PE, is targeted by many cyclotides of the Möbius and bracelet, but not trypsin inhibitor subfamilies. In particular, a lack of PE binding for a given cyclotide correlates with a lack of biological activity. Furthermore, the fact that an all-D kalata B1 binds almost equally to biological membranes as does native kalata B1[193] supports the idea that there is no chiral protein recognition involved in the membrane binding process.

14.4 Reviews

We hope that the above analysis provides a useful summary of trends in the primary cyclotide literature. In addition to these original contributions, there have been a large number of articles that have reviewed various aspects of cyclotides. Overall, there have been 102 reviews on cyclotides[1–4,20–23,31,34–36,55,56,68,69,74–76,87,89,96,99,146,150,156,160,163,169,170,177,178,189,191,200–202,206,207,212,216,220,221,224,230,232,233,238,243,244,247,252,261,262,266,267,270,271,280,283–290,298,301,303,304,306,310,311,313,316,317,319,321,323,330,331,333,337,340,350,353,367–381] that have covered either general or specific aspects of cyclotide research, as summarized in Table 14.5. These include a book devoted to cyclotides,[1] nearly 20 book chapters, and around 80 reviews in journals. Several specialized areas of cyclotide research have been extensively reviewed, *e.g.* the discovery of cyclotides[23,31,34–36,55,56,68,69,74–76,87,89,96,99,220] and research on their bioactivities.[23,212,216,220,221,224,230,232,233,238,243,244,247] However, the analysis in Table 14.5 highlights that there has been little or no review activity associated with some aspects of cyclotide research, namely cyclotide gene sequences, cell penetration and toxicity. Cell penetration is a relatively new aspect of cyclotide research and is currently rapidly evolving.

Table 14.6 provides the titles of selected key reviews, both general and specific. We have highlighted some recommendations for reviews in specific topic areas, as well as identifying a few review articles that provide historical perspectives.

14.5 Conclusions

The number of papers devoted to cyclotide research is on an upward growth trend and cyclotides continue to be a fruitful area of investigation. They remain a structurally unique class of molecules, with no other proteins

Table 14.5 Summary of cyclotide review topics.

	Topic	References or comments
	General reviews	21,22,367–374,376–381
Discovery and Characterization	Discovery	23,31,34–36,55,56,68,69,74–76,87,89,96,99,220
	Gene sequences	Topic not yet reviewed
	Analysis	20
	Structure	56,146,150,156,160,163,169,170,177,178, 189,191,200–202,206,207
Biology	Bioactivity	23,212,216,220,221,224,230,232,233,238,243, 244,247
	Biosynthesis	252,261,262,266,267,270,271
Applications	Synthesis	252,270,271,280,283–290,298,301
	Drug design	2–4,191,201,303,304,306,310,311,313,316,317, 319,321,323,330,331,333,337,340
Mechanisms	Membrane binding	350,353,375
	Cell penetration	Topic not yet reviewed
	Toxicity	Topic not yet reviewed

Table 14.6 Selected key reviews on cyclotides. Noteworthy reviews are indicated with the * symbol.

Year[a]	Title of published report	*Noteworthy Reviews	Ref.
General reviews			
2001	Plant cyclotides: Circular, knotted peptide toxins	–	369
2001*	Circular proteins – no end in sight	*First review on circular proteins*	22
2006	The cyclotide family of circular mini-proteins: nature's combinatorial peptide template	–	304
2012*	Circular proteins from plants and fungi	*Comprehensive review on plant-derived cyclic peptides, including cyclotides*	69
2014	Chemistry and biology of cyclotides: Circular plant peptides outside the box	–	368
2016*	Advances in botanical research: plant cyclotides	*A book of 10 chapters covering all aspects of cyclotide research*	1
Historical overviews/accounts			
2000	*Oldenlandia affinis* (R&S) DC. A plant containing uteroactive peptides used in African traditional medicine	–	23
2012	The bountiful biological activities of cyclotides	–	367
2013	Joseph Rudinger Award memorial lecture: Discovery and applications of cyclotides	–	377
Reviews on cyclotide discovery and structures			
2004	Discovery, structure and biological activities of the cyclotides	–	34
2004	Squash inhibitors: From structural motifs to macrocyclic knottins	–	156
2007*	NMR as a tool for elucidating the structures of circular and knotted proteins	*Key review on structures of cyclotides*	169
2010	Naturally occurring circular proteins: Distribution, biosynthesis and evolution	–	374
2011	Structure and modelling of knottins, a promising molecular scaffold for drug discovery	–	191
2016*	Discovery, structure, function, and applications of cyclotides: Circular proteins from plants	*The most recent review on cyclotides*	380
Reviews on cyclotide bioactivity and biosynthesis			
2004*	Anti-HIV cyclotides	*An early comprehensive review on anti HIV activity of cyclotides*	36
2009	Circling the enemy: Cyclic proteins in plant defence	–	224
2011	Circular micro-proteins and mechanisms of cyclization	–	262
2013	Cyclotide biosynthesis	–	267

(*continued*)

Table 14.6 (*continued*)

Year[a]	Title of published report	*Noteworthy Reviews	Ref.
Reviews on cyclotide synthesis and drug design applications			
2010*	Cyclotide synthesis and supply: From plant to bioprocess	*A review on plant-cell-based production systems for cyclotides*	285
2010	Biological activities of natural and engineered cyclotides, a novel molecular scaffold for peptide-based therapeutics	–	286
2011*	The chemistry of cyclotides	*Comprehensive review on cyclotide chemistry*	287
2011	Cyclotides: A patent review	–	319
2013*	Cyclotides as grafting frameworks for protein engineering and drug design applications	*Overview of the principles and practise of cyclotide grafting*	4
2016	Synthesis and protein engineering applications of cyclotides	–	301
Reviews on the membrane binding and mechanisms of action of cyclotides			
2009	Membrane binding of cyclotides	–	350
2012*	Importance of the cell membrane on the mechanism of action of cyclotides	*Review on the relationship between membrane binding and biological activity*	353
2015	An increasing role for phosphatidylethanolamine as a lipid receptor in the action of host defence peptides	–	245

[a]Noteworthy reviews are indicated with the * symbol.

having *both* a cyclic backbone and a cystine knot. Discovery of cyclotide peptides accounts for a significant proportion of all cyclotide research, although we anticipate that in the future there will be a shift towards discovery at the nucleic acid level, followed by targeted proteomics, rather than front-line peptide-based discovery. Studies of the structures of cyclotides is a mature field and computer-based modelling is likely to be more common in the future for predictions of the 3D structures of newly discovered cyclotides. The main role for experimental structure determination is likely to be in studies of cyclotide complexes with their macromolecular targets and/or membranes, as there are relatively few such studies available at present. The role of cyclotides in plants and their mechanisms of host defense action are gradually being delineated, but much remains to be discovered in terms of understanding their biosynthesis. Also, it is important to note that we cannot exclude that cyclotides might have additional or alternative natural functions and we encourage further investigation in this area.

The number of papers categorized as 'applications' is increasing and in our opinion, the ultimate value of cyclotides derives from their stability and tolerance to sequence substitution, thus facilitating their use as scaffolds in drug design or in agriculture. Although still a small field (approximately 30 papers) there have been significant advances in understanding the mechanisms of action of cyclotides over the last decade and their potential applications as cell penetrating peptides is showing great promise. Overall, cyclotides have proven to be a topologically fascinating class of peptides but their full potential has probably not yet been realized.

Acknowledgements

Research on cyclotides in our laboratory is supported by grants from the Australian Research Council (ARC; grant IDs DP150100443 and LP130100550) and the National Health and Medical Research Council (NHMRC; grant IDs APP1084965 and APP1060225). DJC is an ARC Australian Laureate Fellow (FL150100146). We thank the many staff, students, collaborators and colleagues who have worked on cyclotides and whose names are listed amongst the references.

References

1. D. J. Craik, *Advances in Botanical Research: Plant Cyclotides*, ed. J.-P. Jacquot and P. Gadal, Academic Press, London, 2015, vol. 76.
2. D. J. Craik, S. Simonsen and N. L. Daly, *Curr. Opin. Drug Discovery Dev.*, 2002, **5**, 251–260.
3. K. Jagadish and J. A. Camarero, *Biopolymers*, 2010, **94**, 611–616.
4. A. G. Poth, L. Y. Chan and D. J. Craik, *Biopolymers*, 2013, **100**, 480–491.
5. D. J. Craik, N. L. Daly, T. Bond and C. Waine, *J. Mol. Biol.*, 1999, **294**, 1327–1336.
6. L. Gran, *Medd. Nor. Farm. Selsk.*, 1970, **12**, 173–180.
7. L. Gran, *Lloydia*, 1973, **36**, 174–178.
8. L. Gran, *Lloydia*, 1973, **36**, 207–208.
9. L. Gran, *Acta Pharmacol. Toxicol.*, 1973, **33**, 400–408.
10. K. Sletten and L. Gran, *Medd. Nor. Farm. Selsk.*, 1973, **7–8**, 69–82.
11. T. Schöpke, M. I. Hasan Agha, R. Kraft, A. Otto and K. Hiller, *Sci. Pharm.*, 1993, **61**, 145–153.
12. K. R. Gustafson, R. C. I. Sowder, L. E. Henderson, I. C. Parsons, Y. Kashman, J. H. I. Cardellina, J. B. McMahon, R. W. J. Buckheit, L. K. Pannell and M. R. Boyd, *J. Am. Chem. Soc.*, 1994, **116**, 9337–9338.
13. K. M. Witherup, M. J. Bogusky, P. S. Anderson, H. Ramjit, R. W. Ransom, T. Wood and M. Sardana, *J. Nat. Prod.*, 1994, **57**, 1619–1625.
14. O. Saether, D. J. Craik, I. D. Campbell, K. Sletten, J. Juul and D. G. Norman, *Biochemistry*, 1995, **34**, 4147–4158.
15. P. Claeson, U. Göransson, S. Johansson, T. Luijendijk and L. Bohlin, *J. Nat. Prod.*, 1998, **61**, 77–81.
16. U. Göransson, T. Luijendijk, S. Johansson, L. Bohlin and P. Claeson, *J. Nat. Prod.*, 1999, **62**, 283–286.
17. C. W. Gruber, A. G. Elliott, D. C. Ireland, P. G. Delprete, S. Dessein, U. Göransson, M. Trabi, C. K. Wang, A. B. Kinghorn, E. Robbrecht and D. J. Craik, *Plant Cell*, 2008, **20**, 2471–2483.

18. J. P. Mulvenna, C. Wang and D. J. Craik, *Nucleic Acids Res.*, 2006, **34**, D192–D194.
19. C. K. Wang, Q. Kaas, L. Chiche and D. J. Craik, *Nucleic Acids Res.*, 2008, **36**, D206–D210.
20. Q. Kaas and D. J. Craik, *Biopolymers*, 2010, **94**, 584–591.
21. D. J. Craik, *Science*, 2006, **311**, 1563–1564.
22. M. Trabi and D. J. Craik, *Trends Biochem. Sci.*, 2002, **27**, 132–138.
23. L. Gran, F. Sandberg and K. Sletten, *J. Ethnopharmacol.*, 2000, **70**, 197–203.
24. K. R. Gustafson, L. K. Walton, R. C. I. Sowder, D. G. Johnson, L. K. Pannell, J. H. I. Cardellina and M. R. Boyd, *J. Nat. Prod.*, 2000, **63**, 176–178.
25. Y. F. Hallock, R. C. I. Sowder, L. K. Pannell, C. B. Hughes, D. G. Johnson, R. Gulakowski, J. H. I. Cardellina and M. R. Boyd, *J. Org. Chem.*, 2000, **65**, 124–128.
26. J. F. Hernandez, J. Gagnon, L. Chiche, T. M. Nguyen, J. P. Andrieu, A. Heitz, T. Trinh Hong, T. T. Pham and D. Le Nguyen, *Biochemistry*, 2000, **39**, 5722–5730.
27. H. R. Bokesch, L. K. Pannell, P. K. Cochran, R. C. Sowder 2nd, T. C. McKee and M. R. Boyd, *J. Nat. Prod.*, 2001, **64**, 249–250.
28. A. M. Broussalis, U. Göransson, J. D. Coussio, G. Ferraro, V. Martino and P. Claeson, *Phytochemistry*, 2001, **58**, 47–51.
29. M. E. Felizmenio-Quimio, N. L. Daly and D. J. Craik, *J. Biol. Chem.*, 2001, **276**, 22875–22882.
30. C. Jennings, J. West, C. Waine, D. Craik and M. Anderson, *Proc. Natl. Acad. Sci. U. S. A.*, 2001, **98**, 10614–10619.
31. B. R. O'Keefe, *J. Nat. Prod.*, 2001, **64**, 1373–1381.
32. U. Göransson, A. M. Broussalis and P. Claeson, *Anal. Biochem.*, 2003, **318**, 107–117.
33. E. Svangård, U. Göransson, D. Smith, C. Verma, A. Backlund, L. Bohlin and P. Claeson, *Phytochemistry*, 2003, **64**, 135–142.
34. D. J. Craik, N. L. Daly, J. Mulvenna, M. R. Plan and M. Trabi, *Curr. Protein Pept. Sci.*, 2004, **5**, 297–315.
35. U. Göransson, E. Svangård, P. Claeson and L. Bohlin, *Curr. Protein Pept. Sci.*, 2004, **5**, 317–329.
36. K. R. Gustafson, T. C. McKee and H. R. Bokesch, *Curr. Protein Pept. Sci.*, 2004, **5**, 331–340.
37. B. Chen, M. L. Colgrave, N. L. Daly, K. J. Rosengren, K. R. Gustafson and D. J. Craik, *J. Biol. Chem.*, 2005, **280**, 22395–22405.
38. J. P. Mulvenna, L. Sando and D. J. Craik, *Structure*, 2005, **13**, 691–701.
39. S. M. Simonsen, L. Sando, D. C. Ireland, M. L. Colgrave, R. Bharathi, U. Göransson and D. J. Craik, *Plant Cell*, 2005, **17**, 3176–3189.
40. B. Chen, M. L. Colgrave, C. Wang and D. J. Craik, *J. Nat. Prod.*, 2006, **69**, 23–28.
41. N. L. Daly, R. J. Clark, M. R. Plan and D. J. Craik, *Biochem. J.*, 2006, **393**, 619–626.

42. M. R. R. Plan, U. Göransson, R. J. Clark, N. L. Daly, M. L. Colgrave and D. J. Craik, *ChemBioChem*, 2007, **8**, 1001–1011.
43. D. C. Ireland, M. L. Colgrave and D. J. Craik, *Biochem. J.*, 2006, **400**, 1–12.
44. D. C. Ireland, M. L. Colgrave, P. Nguyencong, N. L. Daly and D. J. Craik, *J. Mol. Biol.*, 2006, **357**, 1522–1535.
45. J. P. Mulvenna, J. S. Mylne, R. Bharathi, R. A. Burton, N. J. Shirley, G. B. Fincher, M. A. Anderson and D. J. Craik, *Plant Cell*, 2006, **18**, 2134–2144.
46. A. Herrmann, R. Burman, J. S. Mylne, G. Karlsson, J. Gullbo, D. J. Craik, R. J. Clark and U. Göransson, *Phytochemistry*, 2008, **69**, 939–952.
47. T. Tokuoka, *J. Plant Res.*, 2008, **121**, 253–260.
48. M. Trabi, J. S. Mylne, L. Sando and D. J. Craik, *Org. Biomol. Chem.*, 2009, **7**, 2378–2388.
49. R. Burman, C. W. Gruber, K. Rizzardi, A. Herrmann, D. J. Craik, M. P. Gupta and U. Göransson, *Phytochemistry*, 2010, **71**, 13–20.
50. C. K. Wang, M. L. Colgrave, K. R. Gustafson, D. C. Ireland, U. Göransson and D. J. Craik, *J. Nat. Prod.*, 2008, **71**, 47–52.
51. W. Y. Xu, J. Tang, C. J. Ji, W. J. He and N. H. Tan, *Chin. Sci. Bull.*, 2008, **53**, 1671–1674.
52. L. Y. Chan, C. K. Wang, J. M. Major, K. P. Greenwood, R. J. Lewis, D. J. Craik and N. L. Daly, *J. Nat. Prod.*, 2009, **72**, 1453–1458.
53. S. L. Gerlach, R. Burman, L. Bohlin, D. Mondal and U. Göransson, *J. Nat. Prod.*, 2010, **73**, 1207–1213.
54. S. L. Gerlach, R. Rathinakumar, G. Chakravarty, U. Göransson, W. C. Wimley, S. P. Darwin and D. Mondal, *Biopolymers*, 2010, **94**, 617–625.
55. C. W. Gruber, *Biopolymers*, 2010, **94**, 565–572.
56. D. C. Ireland, R. J. Clark, N. L. Daly and D. J. Craik, *J. Nat. Prod.*, 2010, **73**, 1610–1622.
57. M. R. Plan, K. J. Rosengren, L. Sando, N. L. Daly and D. J. Craik, *Biopolymers*, 2010, **94**, 647–658.
58. J. Tang, C. K. Wang, X. Pan, H. Yan, G. Zeng, W. Xu, W. He, N. L. Daly, D. J. Craik and N. Tan, *Peptides*, 2010, **31**, 1434–1440.
59. J. Tang, C. K. Wang, X. Pan, H. Yan, G. Zeng, W. Xu, W. He, N. L. Daly, D. J. Craik and N. Tan, *Helv. Chim. Acta*, 2010, **93**, 2287–2295.
60. B. Xiang, G. H. Du, X. C. Wang, S. X. Zhang, X. Y. Qin, J. Q. Kong, K. D. Cheng, Y. J. Li and W. Wang, *Acta Pharmacol. Sin.*, 2010, **45**, 1402–1409.
61. A. Ghassempour, M. Ghahramanzamaneh, H. Hashempour and K. Kargosha, *Acta Chromatogr.*, 2011, **23**, 641–651.
62. H. Hashempour, A. Ghassempour, N. L. Daly, B. Spengler and A. Rompp, *Protein Pept. Lett.*, 2011, **18**, 747–752.
63. W. He, L. Y. Chan, G. Zeng, N. L. Daly, D. J. Craik and N. Tan, *Peptides*, 2011, **32**, 1719–1723.
64. G. K. T. Nguyen, S. Zhang, T. K. N. Ngan, Q. T. N. Phuong, M. S. Chiu, A. Hardjojo and J. P. Tam, *J. Biol. Chem.*, 2011, **286**, 24275–24287.
65. G. K. T. Nguyen, S. Zhang, W. Wang, C. T. T. Wong, T. K. N. Ngan and J. P. Tam, *J. Biol. Chem.*, 2011, **286**, 44833–44844.

66. A. G. Poth, M. L. Colgrave, R. Philip, B. Kerenga, N. L. Daly, M. A. Anderson and D. J. Craik, *ACS Chem. Biol.*, 2011, **6**, 345–355.
67. M. Y. Yeshak, R. Burman, K. Asres and U. Göransson, *J. Nat. Prod.*, 2011, **74**, 727–731.
68. D. J. Craik, S. T. Henriques, J. S. Mylne and C. A. K. Wang, *Natural Product Biosynthesis by Microorganisms and Plants, Pt B*, 2012, vol. 516, pp. 37–62.
69. U. Göransson, R. Burman, S. Gunasekera, A. A. Strömstedt and K. J. Rosengren, *J. Biol. Chem.*, 2012, **287**, 27001–27006.
70. C. Gründemann, J. Koehbach, R. Huber and C. W. Gruber, *J. Nat. Prod.*, 2012, **75**, 167–174.
71. G. K. T. Nguyen, W. H. Lim, P. Q. T. Nguyen and J. P. Tam, *J. Biol. Chem.*, 2012, **287**, 17598–17607.
72. M. F. S. Pinto, I. C. M. Fensterseifer, L. Migliolo, D. A. Sousa, G. de Capdville, J. W. Arboleda-Valencia, M. L. Colgrave, D. J. Craik, B. S. Magalhaes, S. C. Dias and O. L. Franco, *J. Biol. Chem.*, 2012, **287**, 134–147.
73. M. Y. Yeshak, R. Burman, C. Eriksson and U. Göransson, *Phytochem. Lett.*, 2012, **5**, 776–781.
74. P. G. Arnison, M. J. Bibb and G. Bierbaum, *et al.*, *Nat. Prod. Rep.*, 2013, **30**, 108–160.
75. D. R. Demartini, G. Pasquali and C. R. Carlini, *J. Proteomics*, 2013, **93**, 224–233.
76. S. L. Gerlach, U. Göransson, Q. Kaas, D. J. Craik, D. Mondal and C. W. Gruber, *Biopolymers*, 2013, **100**, 433–437.
77. H. Hashempour, J. Koehbach, N. L. Daly, A. Ghassempour and C. W. Gruber, *Amino Acids*, 2013, **44**, 581–595.
78. J. Koehbach, A. F. Attah, A. Berger, R. Hellinger, T. M. Kutchan, E. J. Carpenter, M. Rolf, M. A. Sonibare, J. O. Moody, G. K.-S. Wong, S. Dessein, H. Greger and C. W. Gruber, *Biopolymers*, 2013, **100**, 438–452.
79. J. Koehbach and C. W. Gruber, *Commun. Integr. Biol.*, 2013, **6**, e27583.
80. G. K. Nguyen, Y. L. Lian, E. W. H. Pang, Q. T. N. Phuong, T. D. Tran and J. P. Tam, *J. Biol. Chem.*, 2013, **288**, 3370–3380.
81. M. Zarrabi, R. Dalirfardouei, Z. Sepehrizade and R. K. Kermanshahi, *J. Appl. Microbiol.*, 2013, **115**, 367–375.
82. R. K. Bachheti, S. Yousuf, R. G. Sharama, A. Joshi and A. Mathur, *Int. J. ChemTech Res.*, 2014, **6**, 2316–2322.
83. X. M. Ding, D. S. Bai and J. J. Qian, *Med. Chem. Res.*, 2014, **23**, 1406–1413.
84. M. Z. Liu, Y. Yang, S. X. Zhang, L. Tang, H. M. Wang, C. J. Chen, Z. F. Shen, K. D. Cheng, J. Q. Kong and W. Wang, *Acta Pharmacol. Sin.*, 2014, **49**, 905–912.
85. A. Berger, M. K. Kostyan, S. I. Klose, M. Gastegger, E. Lorbeer, L. Brecker and J. Schinnerl, *Phytochemistry*, 2015, **116**, 162–169.
86. R. Burman, M. Y. Yeshak, S. Larsson, D. J. Craik, K. J. Rosengren and U. Göransson, *Front. Plant Sci.*, 2015, **6**, 855.
87. D. J. Craik, *Advances in Botanical Research: Plant Cyclotides*, ed. D. J. Craik, Academic Press, 2015, vol. 76, ch. 1, pp. 1–13.

88. B. J. Flicker and H. E. Ballard, *Phytotaxa*, 2015, **230**, 39–53.
89. U. Göransson, S. Malik and B. Slazak, *Advances in Botanical Research: Plant Cyclotides*, ed. D. J. Craik, Academic Press, 2015, vol. 76, ch. 2, pp. 15–49.
90. S. T. Henriques and D. J. Craik, *Peptide Chemistry and Drug Design*, ed. B. M. Dunn, John Wiley & Sons, Inc, 2015, pp. 203–245.
91. E. Hu, D. Wang, J. Chen and X. Tao, *Int. J. Clin. Exp. Med.*, 2015, **8**, 4059–4065.
92. N. Kodari, B. Bahramnejad, J. Rostamzadeh, H. Maroofi and S. Torkaman, *J. Agric. Sci. Technol.*, 2015, **17**, 1637–1649.
93. N. B. Cunha, A. E. Barbosa, R. G. de Almeida, W. F. Porto, M. R. Maximiano, L. C. Alvares, C. B. Munhoz, C. U. Eugenio, A. A. Viana, O. L. Franco and S. C. Dias, *Biopolymers*, 2016, **106**, 784–795.
94. T. Mahatmanto, J. S. Mylne, A. G. Poth, J. E. Swedberg, Q. Kaas, H. Schaefer and D. J. Craik, *Mol. Biol. Evol.*, 2015, **32**, 392–405.
95. J. Koehbach and R. J. Clark, *Biopolymers*, 2016, **106**, 774–783.
96. J. Koehbach and C. W. Gruber, *Advances in Botanical Research: Plant Cyclotides*, ed. D. J. Craik, Academic Press, 2015, vol. 76, ch. 3, pp. 51–78.
97. P. Q. T. Nguyen, T. T. Luu, Y. Bai, G. K. T. Nguyen, K. Pervushin and J. P. Tam, *J. Nat. Prod.*, 2015, **78**, 695–704.
98. A. S. Ravipati, S. T. Henriques, A. G. Poth, Q. Kaas, C. K. Wang, M. L. Colgrave and D. J. Craik, *ACS Chem. Biol.*, 2015, **10**, 2491–2500.
99. N.-H. Tan and W.-J. He, *Advances in Botanical Research: Plant Cyclotides*, ed. D. J. Craik, Academic Press, 2015, vol. 76, ch. 4, pp. 79–111.
100. A. F. Attah, R. Hellinger, M. A. Sonibare, J. O. Moody, S. Arrowsmith, S. Wray and C. W. Gruber, *J. Ethnopharmacol.*, 2016, **179**, 83–91.
101. M. Farhadpour, H. Hashempour, Z. Talebpour, N. A-Bagheri, M. S. Shushtarian, C. W. Gruber and A. Ghassempour, *Anal. Biochem.*, 2016, **497**, 83–89.
102. H. N. Matsuura, A. G. Poth, A. C. A. Yendo, A. G. Fett-Neto and D. J. Craik, *J. Nat. Prod.*, 2016, **79**, 3006–3013.
103. P. Niyomploy, L. Y. Chan, A. G. Poth, M. L. Colgrave, P. Sangvanich and D. J. Craik, *Biopolymers*, 2016, **106**, 796–805.
104. M. F. Pinto, O. N. Silva, J. C. Viana, W. F. Porto, L. Migliolo, N. B. da Cunha, N. Gomes Jr, I. C. Fensterseifer, M. L. Colgrave, D. J. Craik, S. C. Dias and O. L. Franco, *J. Nat. Prod.*, 2016, **79**, 2767–2773.
105. W. F. Porto, V. J. Miranda, M. F. S. Pinto, S. M. Dohms and O. L. Franco, *Biopolymers*, 2016, **106**, 109–118.
106. A. Serra, X. Hemu, G. K. T. Nguyen, N. T. K. Nguyen, S. K. Sze and J. P. Tam, *Sci. Rep.*, 2016, **6**, 23005.
107. K. Thell, R. Hellinger, E. Sahin, P. Michenthaler, M. Gold-Binder, T. Haider, M. Kuttke, Z. Liutkevičiūtė, U. Göransson, C. Gründemann, G. Schabbauer and C. W. Gruber, *Proc. Natl. Acad. Sci. U. S. A.*, 2016, **113**, 3960–3965.
108. J. L. Dutton, R. F. Renda, C. Waine, R. J. Clark, N. L. Daly, C. V. Jennings, M. A. Anderson and D. J. Craik, *J. Biol. Chem.*, 2004, **279**, 46858–46867.

109. M. Trabi, E. Svangård, A. Herrmann, U. Göransson, P. Claeson, D. J. Craik and L. Bohlin, *J. Nat. Prod.*, 2004, **67**, 806–810.
110. C. W. Basse, *Plant Physiol.*, 2005, **138**, 1774–1784.
111. J. Zhang, M. Hu, J. T. Li, J. P. Guan, B. Yang, W. S. Shu and B. Liao, *J. Plant Physiol.*, 2009, **166**, 862–870.
112. J. Zhang, B. Liao, D. J. Craik, J. T. Li, M. Hu and W. S. Shu, *Gene*, 2009, **431**, 23–32.
113. J. S. Mylne, C. K. Wang, N. L. van der Weerden and D. J. Craik, *Biopolymers*, 2010, **94**, 635–646.
114. Q. Qin, E. J. McCallum, Q. Kaas, J. Suda, I. Saska, D. J. Craik and J. S. Mylne, *BMC Genomics*, 2010, **11**, 111.
115. A. G. Poth, M. L. Colgrave, R. E. Lyons, N. L. Daly and D. J. Craik, *Proc. Natl. Acad. Sci. U. S. A.*, 2011, **108**, 10127–10132.
116. J. S. Mylne, L. Y. Chan, A. H. Chanson, N. L. Daly, H. Schaefer, T. L. Bailey, P. Nguyencong, L. Cascales and D. J. Craik, *Plant Cell*, 2012, **24**, 2765–2778.
117. A. G. Poth, J. S. Mylne, J. Grassl, R. E. Lyons, A. H. Millar, M. L. Colgrave and D. J. Craik, *J. Biol. Chem.*, 2012, **287**, 27033–27046.
118. G. K. T. Nguyen, S. J. Wang, Y. B. Qiu, X. Hemu, Y. L. Lian and J. P. Tam, *Nat. Chem. Biol.*, 2014, **10**, 732–738.
119. K. S. Harris, T. Durek, Q. Kaas, A. G. Poth, E. K. Gilding, B. F. Conlan, I. Saska, N. L. Daly, N. L. van der Weerden, D. J. Craik and M. A. Anderson, *Nat. Commun.*, 2015, **6**, 10199.
120. B. Slazak, E. Jacobsson, E. Kuta and U. Göransson, *Phytochemistry*, 2015, **117**, 527–536.
121. B. Slazak, E. Sliwinska, M. Saługa, M. Ronikier, J. Bujak, A. Słomka, U. Göransson and E. Kuta, *Plant Cell, Tissue Organ Cult.*, 2015, **120**, 179–190.
122. J. Zhang, J. T. Li, Z. B. Huang, B. Yang, X. J. Zhang, D. H. Li, D. J. Craik, A. J. M. Baker, W. S. Shu and B. Liao, *J. Plant Physiol.*, 2015, **178**, 17–26.
123. E. K. Gilding, M. A. Jackson, A. G. Poth, S. T. Henriques, P. J. Prentis, T. Mahatmanto and D. J. Craik, *New Phytol.*, 2016, **210**, 717–730.
124. A. Nourse, M. Trabi, N. L. Daly and D. J. Craik, *J. Biol. Chem.*, 2004, **279**, 562–570.
125. M. L. Colgrave, A. Jones and D. J. Craik, *J. Chromatogr. A*, 2005, **1091**, 187–193.
126. I. Saska, M. L. Colgrave, A. Jones, M. A. Anderson and D. J. Craik, *J. Chromatogr. B: Anal. Technol. Biomed. Life Sci.*, 2008, **872**, 107–114.
127. M. L. Colgrave, A. G. Poth, Q. Kaas and D. J. Craik, *Biopolymers*, 2010, **94**, 592–601.
128. R. G. Ovesen, U. Göransson, S. H. Hansen, J. Nielsen and H. C. B. Hansen, *J. Chromatogr. A*, 2011, **1218**, 7964–7970.
129. R. G. Ovesen, J. Nielsen and H. C. Bruun Hansen, *Environ. Toxicol. Chem.*, 2011, **30**, 1785–1792.
130. A. C. Conibear, N. L. Daly and D. J. Craik, *Biopolymers*, 2012, **98**, 518–524.

131. K. J. Rosengren, N. L. Daly, P. J. Harvey and D. J. Craik, *Biopolymers*, 2013, **100**, 453–460.
132. P. Kedarisetti, M. J. Mizianty, Q. Kaas, D. J. Craik and L. Kurgan, *Biochim. Biophys. Acta, Proteins Proteomics*, 2014, **1844**, 181–190.
133. T. Mahatmanto, A. G. Poth, J. S. Mylne and D. J. Craik, *Fitoterapia*, 2014, **95**, 22–33.
134. S. Park, A. A. Strömstedt and U. Göransson, *PLoS One*, 2014, **9**, e91430.
135. C. K. Wang, S. E. Northfield, J. E. Swedberg, P. J. Harvey, A. M. Mathiowetz, D. A. Price, S. Liras and D. J. Craik, *J. Phys. Chem. B*, 2014, **118**, 11129–11136.
136. J. Zhang, Z. S. Hua, Z. B. Huang, Q. Z. Chen, Q. Y. Long, D. J. Craik, A. J. M. Baker, W. S. Shu and B. Liao, *Planta*, 2015, **241**, 929–940.
137. D. J. Craik, Y. Young Shim, U. Göransson, G. P. Moss, N. Tan, P. D. Jadhav, J. Shen and M. J. T. Reaney, *Biopolymers*, 2016, **106**, 917–924.
138. E. Melander, C. Eriksson, B. Jansson, U. Göransson and M. Hammarlund-Udenaes, *Biopolymers*, 2016, **106**, 910–916.
139. B. Slazak, M. Kapusta, S. Malik, J. Bohdanowicz, E. Kuta, P. Malec and U. Göransson, *Planta*, 2016, **244**, 1029–1040.
140. C. K. Wang, S. Stalmans, B. De Spiegeleer and D. J. Craik, *J. Pept. Sci.*, 2016, **22**, 305–310.
141. P. K. Pallaghy, K. J. Nielsen, D. J. Craik and R. S. Norton, *Protein Sci.*, 1994, **3**, 1833–1839.
142. R. Derua, K. R. Gustafson and L. K. Pannell, *Biochem. Biophys. Res. Commun.*, 1996, **228**, 632–638.
143. N. L. Daly, A. Koltay, K. R. Gustafson, M. R. Boyd, J. R. Casas-Finet and D. J. Craik, *J. Mol. Biol.*, 1999, **285**, 333–345.
144. N. L. Daly, S. Love, P. F. Alewood and D. J. Craik, *Biochemistry*, 1999, **38**, 10606–10614.
145. N. L. Daly and D. J. Craik, *J. Biol. Chem.*, 2000, **275**, 19068–19075.
146. D. J. Craik, N. L. Daly and C. Waine, *Toxicon*, 2001, **39**, 43–60.
147. A. Heitz, J. F. Hernandez, J. Gagnon, T. T. Hong, T. T. Pham, T. M. Nguyen, D. Le-Nguyen and L. Chiche, *Biochemistry*, 2001, **40**, 7973–7983.
148. P. Lindholm, U. Göransson, S. Johansson, P. Claeson, J. Gulbo, R. Larsson, L. Bohlin and A. Backlund, *Mol. Cancer Ther.*, 2002, **1**, 365–369.
149. L. Skjeldal, L. Gran, K. Sletten and B. F. Volkman, *Arch. Biochem. Biophys.*, 2002, **399**, 142–148.
150. D. J. Craik, D. G. Barry, R. J. Clark, N. L. Daly and L. Sando, *Toxin Rev.*, 2003, **22**, 555–576.
151. N. L. Daly, R. J. Clark and D. J. Craik, *J. Biol. Chem.*, 2003, **278**, 6314–6322.
152. N. L. Daly, R. J. Clark and U. Göransson, *Lett. Pept. Sci.*, 2003, **10**, 523–531.
153. U. Göransson and D. J. Craik, *J. Biol. Chem.*, 2003, **278**, 48188–48196.
154. K. J. Rosengren, N. L. Daly, M. R. Plan, C. Waine and D. J. Craik, *J. Biol. Chem.*, 2003, **278**, 8606–8616.
155. D. G. Barry, N. L. Daly, H. R. Bokesch, K. R. Gustafson and D. J. Craik, *Structure*, 2004, **12**, 85–94.

156. L. Chiche, A. Heitz, J. C. Gelly, J. Gracy, P. T. Chau, P. T. Ha, J. F. Hernandez and D. Le-Nguyen, *Curr. Protein Pept. Sci.*, 2004, **5**, 341–349.
157. M. L. Colgrave and D. J. Craik, *Biochemistry*, 2004, **43**, 5965–5975.
158. N. L. Daly, K. R. Gustafson and D. J. Craik, *FEBS Lett.*, 2004, **574**, 69–72.
159. S. M. Simonsen, N. L. Daly and D. J. Craik, *FEBS Lett.*, 2004, **577**, 399–402.
160. D. J. Craik and N. L. Daly, *Protein Pept. Lett.*, 2005, **12**, 147–152.
161. C. V. Jennings, K. J. Rosengren, N. L. Daly, M. Plan, J. Stevens, M. J. Scanlon, C. Waine, D. G. Norman, M. A. Anderson and D. J. Craik, *Biochemistry*, 2005, **44**, 851–860.
162. A. Koltay, N. L. Daly, K. R. Gustafson and D. J. Craik, *Int. J. Pept. Res. Ther.*, 2005, **11**, 99–106.
163. M. Cěmažar and D. J. Craik, *Int. J. Pept. Res. Ther.*, 2006, **12**, 253–260.
164. M. Cěmažar, N. L. Daly, S. Haggblad, K. P. Lo, E. Yulyaningsih and D. J. Craik, *J. Biol. Chem.*, 2006, **281**, 8224–8232.
165. R. J. Clark, N. L. Daly and D. J. Craik, *Biochem. J.*, 2006, **394**, 85–93.
166. A. Herrmann, E. Svangård, P. Claeson, J. Gullbo, L. Bohlin and U. Göransson, *Cell. Mol. Life Sci.*, 2006, **63**, 235–245.
167. S. S. Nair, J. Romanuka, M. Billeter, L. Skjeldal, M. R. Emmett, C. L. Nilsson and A. G. Marshall, *Biochim. Biophys. Acta*, 2006, **1764**, 1568–1576.
168. Z. O. Shenkarev, K. D. Nadezhdin, V. A. Sobol, A. G. Sobol, L. Skjeldal and A. S. Arseniev, *FEBS J.*, 2006, **273**, 2658–2672.
169. D. J. Craik and N. L. Daly, *Mol. BioSyst.*, 2007, **3**, 257–265.
170. M. Cěmažar, C. W. Gruber and D. J. Craik, *Antioxid. Redox Signaling*, 2008, **10**, 103–111.
171. M. Cěmažar, A. Joshi, N. L. Daly, A. E. Mark and D. J. Craik, *Structure*, 2008, **16**, 842–851.
172. C. Combelles, J. Gracy, A. Heitz, D. J. Craik and L. Chiche, *Proteins*, 2008, **73**, 87–103.
173. A. Heitz, O. Avrutina, D. Le-Nguyen, U. Diederichsen, J. F. Hernandez, J. Gracy, H. Kolmar and L. Chiche, *BMC Struct. Biol.*, 2008, **8**, 54.
174. J. S. Mylne and D. J. Craik, *Biopolymers*, 2008, **90**, 575–580.
175. Z. O. Shenkarev, K. D. Nadezhdin, E. N. Lyukmanova, V. A. Sobol, L. Skjeldal and A. S. Arseniev, *J. Inorg. Biochem.*, 2008, **102**, 1246–1256.
176. S. M. Simonsen, L. Sando, K. J. Rosengren, C. K. Wang, M. L. Colgrave, N. L. Daly and D. J. Craik, *J. Biol. Chem.*, 2008, **283**, 9805–9813.
177. N. L. Daly and D. J. Craik, *Oxidative Folding of Proteins and Peptides*, ed. J. Buchner and L. Moroder, RSC Publishing, London, 2009, pp. 318–344.
178. N. L. Daly, K. J. Rosengren and D. J. Craik, *Adv. Drug Delivery Rev.*, 2009, **61**, 918–930.
179. E. l. Dumont, A. l. D. Laurent, P.-F. o. Loos and X. Assfeld, *J. Chem. Theory Comput.*, 2009, **5**, 1700–1708.
180. U. Göransson, A. Herrmann, R. Burman, L. M. Haugaard-Jönsson and K. J. Rosengren, *ChemBioChem*, 2009, **10**, 2354–2360.
181. S. Gunasekera, N. Daly, R. Clark and D. J. Craik, *Antioxid. Redox Signaling*, 2009, **11**, 971–980.

182. S. K. Sze, W. Wang, W. Meng, R. Yuan, T. Guo, Y. Zhu and J. P. Tam, *Anal. Chem.*, 2009, **81**, 1079–1088.
183. C. K. Wang, M. L. Colgrave, D. C. Ireland, Q. Kaas and D. J. Craik, *Biophys. J.*, 2009, **97**, 1471–1481.
184. C. K. Wang, S. H. Hu, J. L. Martin, T. Sjogren, J. Hajdu, L. Bohlin, P. Claeson, U. Göransson, K. J. Rosengren, J. Tang, N. H. Tan and D. J. Craik, *J. Biol. Chem.*, 2009, **284**, 10672–10683.
185. N. L. Daly, B. Chen, P. Nguyencong and D. J. Craik, *Aust. J. Chem.*, 2010, **63**, 771–778.
186. S. S. Puttamadappa, K. Jagadish, A. Shekhtman and J. A. Camarero, *Angew. Chem., Int. Ed.*, 2010, **49**, 7030–7034.
187. C. P. Sommerhoff, O. Avrutina, H. U. Schmoldt, D. Gabrijelcic-Geiger, U. Diederichsen and H. Kolmar, *J. Mol. Biol.*, 2010, **395**, 167–175.
188. R. Burman, A. Herrmann, R. Tran, J. E. Kivela, A. Lomize, J. Gullbo and U. Göransson, *Org. Biomol. Chem.*, 2011, **9**, 4306–4314.
189. N. L. Daly, K. J. Rosengren, S. T. Henriques and D. J. Craik, *Eur. Biophys. J.*, 2011, **40**, 359–370.
190. E. Dumont, C. Michel and P. Sautet, *ChemPhysChem*, 2011, **12**, 2596–2603.
191. J. Gracy and L. Chiche, *Curr. Pharm. Des.*, 2011, **17**, 4337–4350.
192. S. S. Puttamadappa, K. Jagadish, A. Shekhtman and J. A. Camarero, *Angew. Chem., Int. Ed.*, 2011, **50**, 6948–6949.
193. L. Sando, S. T. Henriques, F. Foley, S. M. Simonsen, N. L. Daly, K. N. Hall, K. R. Gustafson, M. I. Aguilar and D. J. Craik, *ChemBioChem*, 2011, **12**, 2456–2462.
194. C. K. Wang, R. J. Clark, P. J. Harvey, K. J. Rosengren, M. Cěmažar and D. J. Craik, *Biochemistry*, 2011, **50**, 4077–4086.
195. C. T. Wong, M. Taichi, H. Nishio, Y. Nishiuchi and J. P. Tam, *Biochemistry*, 2011, **50**, 7275–7283.
196. K. Hall, T. H. Lee, N. L. Daly, D. J. Craik and M. I. Aguilar, *Biochim. Biophys. Acta, Biomembr.*, 2012, **1818**, 2354–2361.
197. C. K. Wang, H. P. Wacklin and D. J. Craik, *J. Biol. Chem.*, 2012, **287**, 43884–43898.
198. N. L. Daly, L. Thorstholm, K. P. Greenwood, G. J. King, K. J. Rosengren, B. Heras, J. L. Martin and D. J. Craik, *J. Biol. Chem.*, 2013, **288**, 36141–36148.
199. M. S. Goyder, F. Rebeaud, M. E. Pfeifer and F. Kálmán, *Expert Rev. Proteomics*, 2013, **10**, 489–501.
200. E. Schrank, G. E. Wagner and K. Zangger, *Molecules*, 2013, **18**, 7407–7435.
201. C. I. Schroeder, J. E. Swedberg and D. J. Craik, *Curr. Protein Pept. Sci.*, 2013, **14**, 532–542.
202. M. Góngora-Benítez, J. Tulla-Puche and F. Albericio, *Chem. Rev.*, 2014, **114**, 901–926.
203. K. Stanger, T. Maurer, H. Kaluarachchi, M. Coons, Y. Franke and R. N. Hannoush, *FEBS Lett.*, 2014, **588**, 4487–4496.

204. C. M. Taylor, S. E. Northfield, C. K. Wang and D. J. Craik, *Tetrahedron*, 2014, **70**, 7669–7674.
205. C. K. Wang, G. J. King, S. E. Northfield, P. G. Ojeda and D. J. Craik, *Angew. Chem., Int. Ed.*, 2014, **53**, 11236–11241.
206. M. L. Colgrave, *Advances in Botanical Research: Plant Cyclotides*, ed. D. J. Craik, Academic Press, 2015, vol. 76, ch. 5, pp. 113–154.
207. N. L. Daly and K. J. Rosengren, *Advances in Botanical Research: Plant Cyclotides*, ed. D. J. Craik, Academic Press, 2015, vol. 76, ch. 6, pp. 155–186.
208. H. Abdul Ghani, S. T. Henriques, Y. H. Huang, J. E. Swedberg, C. I. Schroeder and D. J. Craik, *Biopolymers*, 2017, **108**, DOI: 10.1002/bip.22927.
209. P. M. Jones and A. M. George, *Sci. Rep.*, 2016, **6**, 23174.
210. B. Senthilkumar, P. Kumar and R. Rajasekaran, *J. Cell. Biochem.*, 2016, **117**, 66–73.
211. J. P. Tam, Y. A. Lu, J. L. Yang and K. W. Chiu, *Proc. Natl. Acad. Sci. U. S. A.*, 1999, **96**, 8913–8918.
212. J. P. Tam, Y. A. Lu, J. L. Yang and Q. Yu, *Development of Novel Antimicrobial Agents: Emerging Strategies*, Horizon Scientific Press, Wymondham, UK, 2001, pp. 215–240.
213. D. G. Barry, N. L. Daly, R. J. Clark, L. Sando and D. J. Craik, *Biochemistry*, 2003, **42**, 6688–6695.
214. U. Göransson, M. Sjogren, E. Svangard, P. Claeson and L. Bohlin, *J. Nat. Prod.*, 2004, **67**, 1287–1290.
215. E. Svangård, U. Göransson, Z. Hocaoglu, J. Gullbo, R. Larsson, P. Claeson and L. Bohlin, *J. Nat. Prod.*, 2004, **67**, 144–147.
216. C. W. Gruber, M. Cemazar, M. A. Anderson and D. J. Craik, *Toxicon*, 2007, **49**, 561–575.
217. B. L. Barbeta, A. T. Marshall, A. D. Gillon, D. J. Craik and M. A. Anderson, *Proc. Natl. Acad. Sci. U. S. A.*, 2008, **105**, 1221–1225.
218. M. L. Colgrave, A. C. Kotze, Y. H. Huang, J. O'Grady, S. M. Simonsen and D. J. Craik, *Biochemistry*, 2008, **47**, 5581–5589.
219. M. L. Colgrave, A. C. Kotze, D. C. Ireland, C. K. Wang and D. J. Craik, *ChemBioChem*, 2008, **9**, 1939–1945.
220. L. Gran, K. Sletten and L. Skjeldal, *Chem. Biodiversity*, 2008, **5**, 2014–2022.
221. D. C. Ireland, C. K. Wang, J. A. Wilson, K. R. Gustafson and D. J. Craik, *Biopolymers*, 2008, **90**, 51–60.
222. M. R. R. Plan, I. Saska, A. G. Cagauan and D. J. Craik, *J. Agric. Food Chem.*, 2008, **56**, 5237–5241.
223. M. L. Colgrave, A. C. Kotze, S. Kopp, J. S. McCarthy, G. T. Coleman and D. J. Craik, *Acta Trop.*, 2009, **109**, 163–166.
224. D. J. Craik, *Trends Plant Sci.*, 2009, **14**, 328–335.
225. A. M. Broussalis, S. Clemente and G. E. Ferraro, *Crop Prot.*, 2010, **29**, 953–956.
226. R. Burman, E. Svedlund, J. Felth, S. Hassan, A. Herrmann, R. J. Clark, D. J. Craik, L. Bohlin, P. Claeson, U. Göransson and J. Gullbo, *Biopolymers*, 2010, **94**, 626–634.

227. M. L. Colgrave, Y. H. Huang, D. J. Craik and A. C. Kotze, *Antimicrob. Agents Chemother.*, 2010, **54**, 2160–2166.
228. Y. H. Huang, M. L. Colgrave, R. J. Clark, A. C. Kotze and D. J. Craik, *J. Biol. Chem.*, 2010, **285**, 10797–10805.
229. M. Pränting, C. Loov, R. Burman, U. Göransson and D. I. Andersson, *J. Antimicrob. Chemother.*, 2010, **65**, 1964–1971.
230. C. W. Gruber and M. O'Brien, *Planta Med.*, 2011, **77**, 207–220.
231. R. G. Ovesen, K. K. Brandt, U. Göransson, J. Nielsen, H. C. Hansen and N. Cedergreen, *Environ. Toxicol. Chem.*, 2011, **30**, 1190–1196.
232. D. J. Craik, *Toxins*, 2012, **4**, 139–156.
233. M. F. S. Pinto, I. C. M. Fensterseifer and O. L. Franco, *Prog Biol Control*, ed. J. M. Mérillon and K. G. Ramawat, Springer, Netherlands, 2012, vol. 12, pp. 333–344.
234. C. T. T. Wong, D. K. Rowlands, C. H. Wong, T. W. C. Lo, G. K. T. Nguyen, H. Y. Li and J. P. Tam, *Angew. Chem., Int. Ed.*, 2012, **51**, 5620–5624.
235. X. Duan, J. Xu, E. Ling and P. Zhang, *Plant Mol. Biol.*, 2013, **83**, 131–141.
236. S. L. Gerlach, M. Yeshak, U. Göransson, U. Roy, R. Izadpanah and D. Mondal, *Biopolymers*, 2013, **100**, 471–479.
237. C. Gründemann, K. Thell, K. Lengen, M. Garcia-Käufer, Y.-H. Huang, R. Huber, D. J. Craik, G. Schabbauer and C. W. Gruber, *PLoS One*, 2013, **8**, e68016.
238. F. Harris, S. R. Dennison, J. Singh and D. A. Phoenix, *Med. Res. Rev.*, 2013, **33**, 190–234.
239. D. Malagón, B. Botterill, D. J. Gray, E. Lovas, M. Duke, C. Gray, S. R. Kopp, L. M. Knott, D. P. McManus, N. L. Daly, J. Mulvenna, D. J. Craik and M. K. Jones, *Biopolymers*, 2013, **100**, 461–470.
240. Z. Sen, X. K. Zhan, J. Jing, Z. Yi and Z. Wanqi, *Oncol. Lett.*, 2013, **5**, 641–644.
241. R. Hellinger, J. Koehbach, H. Fedchuk, B. Sauer, R. Huber, C. W. Gruber and C. Gründemann, *J. Ethnopharmacol.*, 2014, **151**, 299–306.
242. I. C. M. Fensterseifer, O. N. Silva, U. Malik, A. S. Ravipati, N. R. F. Novaes, P. R. R. Miranda, E. A. Rodrigues, S. E. Moreno, D. J. Craik and O. L. Franco, *Peptides*, 2015, **63**, 38–42.
243. J. J. Guzman-Rodriguez, A. Ochoa-Zarzosa, R. Lopez-Gomez and J. E. Lopez-Meza, *BioMed Res. Int.*, 2015.
244. G. K. Oguis, M.-W. Kan and D. J. Craik, *Advances in Botanical Research: Plant Cyclotides*, ed. D. J. Craik, Academic Press, 2015, vol. 76, ch. 7, pp. 187–226.
245. D. A. Phoenix, F. Harris, M. Mura and S. R. Dennison, *Prog. Lipid Res.*, 2015, **59**, 26–37.
246. M. A. Esmaeili, N. Abagheri-Mahabadi, H. Hashempour, M. Farhadpour, C. W. Gruber and A. Ghassempour, *Fitoterapia*, 2016, **109**, 162–168.
247. F. Harris, S. Prabhu, S. R. Dennison, T. J. Snape, R. Lea, M. Mura and D. A. Phoenix, *Protein Pept. Lett.*, 2016, **23**, 676–687.
248. K. N. T. Nguyen, G. K. T. Nguyen, P. Q. T. Nguyen, K. H. Ang, P. C. Dedon and J. P. Tam, *FEBS J.*, 2016, **283**, 2067–2090.

249. C. K. Wang, S. E. Northfield, Y. H. Huang, M. C. Ramos and D. J. Craik, *Eur. J. Med. Chem.*, 2016, **109**, 342–349.
250. D. G. Wang, J. Y. Chen, J. F. Zhu and Y. H. Mou, *Int. J. Clin. Exp. Med.*, 2016, **9**, 9521–9526.
251. M. Trabi and D. J. Craik, *Plant Cell*, 2004, **16**, 2204–2216.
252. S. Gunasekera, N. L. Daly, M. A. Anderson and D. J. Craik, *IUBMB Life*, 2006, **58**, 515–524.
253. R. H. Kimura, A.-T. Tran and J. A. Camarero, *Angew. Chem., Int. Ed.*, 2006, **118**, 987–990.
254. P. Seydel and H. Dörnenburg, *Plant Cell, Tissue Organ Cult.*, 2006, **85**, 247–255.
255. J. A. Camarero, R. H. Kimura, Y.-H. Woo, A. Shekhtman and J. Cantor, *ChemBioChem*, 2007, **8**, 1363–1366.
256. C. W. Gruber, M. Cemazar, R. J. Clark, T. Horibe, R. F. Renda, M. A. Anderson and D. J. Craik, *J. Biol. Chem.*, 2007, **282**, 20435–20446.
257. I. Saska, A. D. Gillon, N. Hatsugai, R. G. Dietzgen, I. Hara-Nishimura, M. A. Anderson and D. J. Craik, *J. Biol. Chem.*, 2007, **282**, 29721–29728.
258. P. Seydel, C. W. Gruber, D. J. Craik and H. Dornenburg, *Appl. Microbiol. Biotechnol.*, 2007, **77**, 275–284.
259. A. D. Gillon, I. Saska, C. V. Jennings, R. F. Guarino, D. J. Craik and M. A. Anderson, *Plant J.*, 2008, **53**, 505–515.
260. C. W. Gruber, M. Čemažar, A. Mechler, L. L. Martin and D. J. Craik, *Biopolymers*, 2009, **92**, 35–43.
261. B. F. Conlan, A. D. Gillon, D. J. Craik and M. A. Anderson, *Biopolymers*, 2010, **94**, 573–583.
262. B. F. Conlan and M. A. Anderson, *Curr. Pharm. Des.*, 2011, **17**, 4318–4328.
263. B. F. Conlan, A. D. Gillon, B. L. Barbeta and M. A. Anderson, *Am. J. Bot.*, 2011, **98**, 2018–2026.
264. B. F. Conlan, M. L. Colgrave, A. D. Gillon, R. Guarino, D. J. Craik and M. A. Anderson, *J. Biol. Chem.*, 2012, **287**, 28037–28046.
265. C. J. S. Barber, P. T. Pujara, D. W. Reed, S. Chiwocha, H. Zhang and P. S. Covello, *J. Biol. Chem.*, 2013, **288**, 12500–12510.
266. R. Borra and J. A. Camarero, *Biopolymers*, 2013, **100**, 502–509.
267. D. J. Craik and U. Malik, *Curr. Opin. Chem. Biol.*, 2013, **17**, 546–554.
268. X. Jia, S. Kwon, C. I. Wang, Y. H. Huang, L. Y. Chan, C. C. Tan, K. J. Rosengren, J. P. Mulvenna, C. I. Schroeder and D. J. Craik, *J. Biol. Chem.*, 2014, **289**, 6627–6638.
269. R. Hellinger, J. Koehbach, A. Puigpinos, R. J. Clark, T. Tarrago, E. Giralt and C. W. Gruber, *J. Nat. Prod.*, 2015, **78**, 1073–1082.
270. Y. Li, T. Bi and J. A. Camarero, *Advances in Botanical Research: Plant Cyclotides*, ed. D. J. Craik, Academic Press, 2015, vol. 76, ch. 9, pp. 271–303.
271. T. Shafee, K. Harris and M. Anderson, *Advances in Botanical Research: Plant Cyclotides*, ed. D. J. Craik, Academic Press, 2015, vol. 76, ch. 8, pp. 227–269.

272. N. L. Daly, S. Gunasekera, R. J. Clark, F. Lin, J. D. Wade, M. A. Anderson and D. J. Craik, *Biopolymers*, 2016, **106**, 825–833.
273. J. P. Tam and Y.-A. Lu, *Tetrahedron Lett.*, 1997, **38**, 5599–5602.
274. J. P. Tam and Y.-A. Lu, *Protein Sci.*, 1998, **7**, 1583–1592.
275. J. P. Tam, Y.-A. Lu and Q. Yu, *J. Am. Chem. Soc.*, 1999, **121**, 4316–4324.
276. O. Avrutina, H.-U. Schmoldt, H. Kolmar and U. Diederichsen, *Eur. J. Org. Chem.*, 2004, **2004**, 4931–4935.
277. P. Thongyoo, E. W. Tate and R. J. Leatherbarrow, *Chem. Commun. (Camb.)*, 2006, 2848–2850.
278. P. Thongyoo, A. M. Jaulent, E. W. Tate and R. J. Leatherbarrow, *ChemBioChem*, 2007, **8**, 1107–1109.
279. M. Cěmažar and D. J. Craik, *J. Pept. Sci.*, 2008, **14**, 683–689.
280. H. Dörnenburg, *Biotechnol. Lett.*, 2008, **30**, 1311–1321.
281. P. Thongyoo, N. Roque-Rosell, R. J. Leatherbarrow and E. W. Tate, *Org. Biomol. Chem.*, 2008, **6**, 1462–1470.
282. J. Austin, W. Wang, S. Puttamadappa, A. Shekhtman and J. A. Camarero, *ChemBioChem*, 2009, **10**, 2663–2670.
283. H. Dörnenburg, *Biotechnol. J.*, 2009, **4**, 632–645.
284. R. J. Clark and D. J. Craik, *Biopolymers*, 2010, **94**, 414–422.
285. H. Dörnenburg, *Biopolymers*, 2010, **94**, 602–610.
286. A. E. Garcia and J. A. Camarero, *Curr. Mol. Pharmacol.*, 2010, **3**, 153–163.
287. D. J. Craik and A. C. Conibear, *J. Org. Chem.*, 2011, **76**, 4805–4817.
288. T. Katoh, Y. Goto, M. S. Reza and H. Suga, *Chem. Commun.*, 2011, **47**, 9946–9958.
289. T. L. Aboye and J. A. Camarero, *J. Biol. Chem.*, 2012, **287**, 27026–27032.
290. M. Reinwarth, D. Nasu, H. Kolmar and O. Avrutina, *Molecules*, 2012, **17**, 12533–12552.
291. J. S. Zheng, H. N. Chang, J. Shi and L. Liu, *Sci. China: Chem.*, 2012, **55**, 64–69.
292. J. S. Zheng, S. Tang, Y. Guo, H. N. Chang and L. Liu, *ChemBioChem*, 2012, **13**, 542–546.
293. M. Akcan and D. J. Craik, in *Peptide Synthesis and Applications*, ed. K. J. Jensen, P. Tofteng Shelton and S. L. Pedersen, Humana Press, 2013, vol. 1047, pp. 89–101.
294. B. Cowper, D. J. Craik and D. Macmillan, *ChemBioChem*, 2013, **14**, 809–812.
295. S. Gunasekera, T. Aboye, W. Madian, H. El-Seedi and U. Göransson, *Int. J. Pept. Res. Ther.*, 2013, **19**, 43–54.
296. O. Cheneval, C. I. Schroeder, T. Durek, P. Walsh, Y.-H. Huang, S. Liras, D. A. Price and D. J. Craik, *J. Org. Chem.*, 2014, **79**, 5538–5544.
297. X. Hemu, Y. B. Qiu and J. P. Tam, *Tetrahedron*, 2014, **70**, 7707–7713.
298. S. E. Northfield, C. K. Wang, C. I. Schroeder, T. Durek, M. W. Kan, J. E. Swedberg and D. J. Craik, *Eur. J. Med. Chem.*, 2014, **77**, 248–257.
299. T. Aboye, Y. Kuang, N. Neamati and J. A. Camarero, *ChemBioChem*, 2015, **16**, 827–833.

300. X. Hemu, Y. Qiu, G. K. T. Nguyen and J. P. Tam, *J. Am. Chem. Soc.*, 2016, **138**, 6968–6971.
301. H. Qu, B. J. Smithies, T. Durek and D. J. Craik, *Aust. J. Chem.*, 2017, **70**, 152–161.
302. O. Avrutina, H. U. Schmoldt, D. Gabrijelcic-Geiger, D. Le Nguyen, C. P. Sommerhoff, U. Diederichsen and H. Kolmar, *Biol. Chem.*, 2005, **386**, 1301–1306.
303. D. J. Craik, M. Cěmažar and N. L. Daly, *Curr. Opin. Drug Discovery Dev.*, 2006, **9**, 251–260.
304. D. J. Craik, M. Cěmažar, C. K. Wang and N. L. Daly, *Biopolymers*, 2006, **84**, 250–266.
305. M. Werle, T. Schmitz, H.-L. Haung, A. Wentzel, H. Kolmar and A. Bernkop-Schnurch, *J. Drug Targeting*, 2006, **14**, 137–146.
306. D. J. Craik, R. J. Clark and N. L. Daly, *Expert Opin. Invest. Drugs*, 2007, **16**, 595–604.
307. T. L. Aboye, R. J. Clark, D. J. Craik and U. Göransson, *ChemBioChem*, 2008, **9**, 103–113.
308. O. Avrutina, H. U. Schmoldt, D. Gabrijelcic-Geiger, A. Wentzel, H. Fraudendorf, C. P. Sommerhoff, U. Diederichsen and H. Kolmar, *ChemBioChem*, 2008, **9**, 33–37.
309. S. Gunasekera, F. M. Foley, R. J. Clark, L. Sando, L. J. Fabri, D. J. Craik and N. L. Daly, *J. Med. Chem.*, 2008, **51**, 7697–7704.
310. H. Kolmar, *FEBS J.*, 2008, **275**, 2684–2690.
311. N. L. Daly and D. J. Craik, *Future Med. Chem.*, 2009, **1**, 1613–1622.
312. P. Thongyoo, C. Bonomelli, R. J. Leatherbarrow and E. W. Tate, *J. Med. Chem.*, 2009, **52**, 6197–6200.
313. D. J. Craik, J. S. Mylne and N. L. Daly, *Cell. Mol. Life Sci.*, 2010, **67**, 9–16.
314. Y. Gao, T. Cui and Y. Lam, *Bioorg. Med. Chem.*, 2010, **18**, 1331–1336.
315. L. Y. Chan, S. Gunasekera, S. T. Henriques, N. F. Worth, S. J. Le, R. J. Clark, J. H. Campbell, D. J. Craik and N. L. Daly, *Blood*, 2011, **118**, 6709–6717.
316. N. L. Daly and D. J. Craik, *Curr. Opin. Chem. Biol.*, 2011, **15**, 362–368.
317. N. L. Daly, C. W. Gruber, U. Göransson and D. J. Craik, *Folding of Disulfide Proteins*, ed. S. Ventura and R. W. Chang, Springer, New York, USA, 2011, ch. 3, pp. 43–61.
318. J. A. Getz, J. J. Rice and P. S. Daugherty, *ACS Chem. Biol.*, 2011, **6**, 837–844.
319. A. B. Smith, N. L. Daly and D. J. Craik, *Expert Opin. Ther. Pat.*, 2011, **21**, 1657–1672.
320. T. L. Aboye, H. Ha, S. Majumder, F. Christ, Z. Debyser, A. Shekhtman, N. Neamati and J. A. Camarero, *J. Med. Chem.*, 2012, **55**, 10729–10734.
321. D. J. Craik, J. E. Swedberg, J. S. Mylne and M. Čemažar, *Expert Opin. Drug Discovery*, 2012, **7**, 179–194.
322. R. Eliasen, N. L. Daly, B. S. Wulff, T. L. Andresen, K. W. Conde-Frieboes and D. J. Craik, *J. Biol. Chem.*, 2012, **287**, 40493–40501.
323. V. Baeriswyl and C. Heinis, *ChemMedChem*, 2013, **8**, 377–384.

324. J. A. Getz, O. Cheneval, D. J. Craik and P. S. Daugherty, *ACS Chem. Biol.*, 2013, **8**, 1147–1154.
325. B. Glotzbach, M. Reinwarth, N. Weber, S. Fabritz, M. Tomaszowski, H. Fittler, A. Christmann, O. Avrutina and H. Kolmar, *PLoS One*, 2013, **8**, e76956.
326. K. Jagadish, R. Borra, V. Lacey, S. Majumder, A. Shekhtman, L. Wang and J. A. Camarero, *Angew. Chem., Int. Ed.*, 2013, **52**, 3126–3131.
327. Y. Ji, S. Majumder, M. Millard, R. Borra, T. Bi, A. Y. Elnagar, N. Neamati, A. Shekhtman and J. A. Camarero, *J. Am. Chem. Soc.*, 2013, **135**, 11623–11633.
328. J. Koehbach, M. O'Brien, M. Muttenthaler, M. Miazzo, M. Akcan, A. G. Elliott, N. L. Daly, P. J. Harvey, S. Arrowsmith, S. Gunasekera, T. J. Smith, S. Wray, U. Göransson, P. E. Dawson, D. J. Craik, M. Freissmuth and C. W. Gruber, *Proc. Natl. Acad. Sci. U. S. A.*, 2013, **110**, 21183–21188.
329. P. Quimbar, U. Malik, C. P. Sommerhoff, Q. Kaas, L. Y. Chan, Y.-H. Huang, M. Grundhuber, K. Dunse, D. J. Craik, M. A. Anderson and N. L. Daly, *J. Biol. Chem.*, 2013, **288**, 13885–13896.
330. K. Rolka, A. Lesner, A. Łęgowska and M. Wysocka, *Antitumor Potential and Other Emerging Medicinal Properties of Natural Compounds*, ed. E. F. Fang and T. B. Ng, Springer, Netherlands, 2013, pp. 187–204.
331. S. E. Ackerman, N. V. Currier, J. M. Bergen and J. R. Cochran, *Expert Rev. Proteomics*, 2014, **11**, 561–572.
332. K. Gray, S. Elghadban, P. Thongyoo, K. A. Owen, R. Szabo, T. H. Bugge, E. W. Tate, R. J. Leatherbarrow and V. Ellis, *Thromb. Haemostasis*, 2014, **112**, 402–411.
333. K. Thell, R. Hellinger, G. Schabbauer and C. W. Gruber, *Drug Discovery Today*, 2014, **19**, 645–653.
334. C. K. Wang, C. W. Gruber, M. Cěmažar, C. Siatskas, P. Tagore, N. Payne, G. Sun, S. Wang, C. C. Bernard and D. J. Craik, *ACS Chem. Biol.*, 2014, **9**, 156–163.
335. L. Y. Chan, D. J. Craik and N. L. Daly, *Biosci. Rep.*, 2015, **35**, e00270.
336. Y.-H. Huang, S. T. Henriques, C. K. Wang, L. Thorstholm, N. L. Daly, Q. Kaas and D. J. Craik, *Sci. Rep.*, 2015, **5**, 12974.
337. M. A. Jackson and E. K. Gilding, *Advances in Botanical Research: Plant Cyclotides*, ed. D. J. Craik, Academic Press, 2015, vol. 76, ch. 10, pp. 305–333.
338. K. Jagadish, A. Gould, R. Borra, S. Majumder, Z. Mushtaq, A. Shekhtman and J. A. Camarero, *Angew. Chem., Int. Ed.*, 2015, **54**, 8390–8394.
339. F. Maass, J. Wustehube-Lausch, S. Dickgiessr, B. Valldorf, M. Reinwarth, H. U. Schmoldt, M. Daneschdar, O. Avrutina, U. Sahin and H. Kolmar, *J. Pept. Sci.*, 2015, **21**, 651–660.
340. A. Mollica, R. Costante, A. Stefanucci and E. Novellino, *J. Enzyme Inhib. Med. Chem.*, 2015, **30**, 575–580.
341. A. Sangphukieo, W. Nawae, T. Laomettachit, U. Supasitthimethee and M. Ruengjitchatchawalya, *PLoS One*, 2015, 10.

342. T. Aboye, C. Meeks, S. Majumder, A. Shekhtman, K. Rodgers and J. Camarero, *Molecules*, 2016, **21**, 152.
343. L. Y. Chan, D. J. Craik and N. L. Daly, *Sci. Rep.*, 2016, **6**, 35347.
344. A. C. Conibear, S. Chaousis, T. Durek, K. J. Rosengren, D. J. Craik and C. I. Schroeder, *Biopolymers*, 2016, **106**, 89–100.
345. C. D'Souza, S. T. Henriques, C. K. Wang, O. Cheneval, L. Y. Chan, N. J. Bokil, M. J. Sweet and D. J. Craik, *Biochemistry*, 2016, **55**, 396–405.
346. J. E. Swedberg, T. Mahatmanto, H. Abdul Ghani, S. J. de Veer, C. I. Schroeder, J. M. Harris and D. J. Craik, *J. Med. Chem.*, 2016, **59**, 7287–7292.
347. H. Kamimori, K. Hall, D. J. Craik and M. I. Aguilar, *Anal. Biochem.*, 2005, **337**, 149–153.
348. E. Svangård, R. Burman, S. Gunasekera, H. Lovborg, J. Gullbo and U. Göransson, *J. Nat. Prod.*, 2007, **70**, 643–647.
349. Y.-H. Huang, M. L. Colgrave, N. L. Daly, A. Keleshian, B. Martinac and D. J. Craik, *J. Biol. Chem.*, 2009, **284**, 20699–20707.
350. C. K. Wang, Y.-H. Huang, K. Greenwood and D. J. Craik, *Membrane-active Peptides: Methods and Results on Structure and Function*, ed. M. Castanho, IUL Publishers, La Jolla, USA, 2009, pp. 597–672.
351. R. Burman, A. A. Stromstedt, M. Malmsten and U. Göransson, *Biochim. Biophys. Acta, Biomembr.*, 2011, **1808**, 2665–2673.
352. S. T. Henriques, Y. H. Huang, K. J. Rosengren, H. G. Franquelim, F. A. Carvalho, A. Johnson, S. Sonza, G. Tachedjian, M. A. R. B. Castanho, N. L. Daly and D. J. Craik, *J. Biol. Chem.*, 2011, **286**, 24231–24241.
353. S. T. Henriques and D. J. Craik, *ACS Chem. Biol.*, 2012, **7**, 626–636.
354. S. T. Henriques, Y. H. Huang, M. A. R. B. Castanho, L. A. Bagatolli, S. Sonza, G. Tachedjian, N. L. Daly and D. J. Craik, *J. Biol. Chem.*, 2012, **287**, 33629–33643.
355. S. T. Henriques, Y.-H. Huang, S. Chaousis, C. K. Wang and D. J. Craik, *ChemBioChem*, 2014, **15**, 1956–1965.
356. W. Nawae, S. Hannongbua and M. Ruengjitchatchawalya, *Sci. Rep.*, 2014, **4**, 3933.
357. W. Nawae, S. Hannongbua and M. Ruengjitchatchawalya, *PLoS One*, 2014, **9**, e114473.
358. W. Xu, B. Wang, Y. Lin, Y. Li, Z. Su, W. He, N. Tan and Q. Wang, *RSC Adv.*, 2014, **4**, 48000–48003.
359. A. A. Strömstedt, P. E. Kristiansen, S. Gunasekera, N. Grob, L. Skjeldal and U. Göransson, *Biochim. Biophys. Acta, Biomembr.*, 2016, **1858**, 1317–1327.
360. S. T. Henriques, Y.-H. Huang, S. Chaousis, M.-A. Sani, A. G. Poth, F. Separovic and D. J. Craik, *Chem. Biol.*, 2015, **22**, 1087–1097.
361. K. P. Greenwood, N. L. Daly, D. L. Brown, J. L. Stow and D. J. Craik, *Int. J. Biochem. Cell Biol.*, 2007, **39**, 2252–2264.
362. L. Cascales, S. T. Henriques, M. C. Kerr, Y. H. Huang, M. J. Sweet, N. L. Daly and D. J. Craik, *J. Biol. Chem.*, 2011, **286**, 36932–36943.
363. J. Contreras, A. Y. O. Elnagar, S. F. Hamm-Alvarez and J. A. Camarero, *J. Controlled Release*, 2011, **155**, 134–143.

364. C. D'Souza, S. T. Henriques, C. K. Wang and D. J. Craik, *Eur. J. Med. Chem.*, 2014, **88**, 10–18.
365. Y. H. Huang, S. Chaousis, O. Cheneval, D. J. Craik and S. T. Henriques, *Front. Pharmacol.*, 2015, **6**, 17.
366. X. Gao, K. Stanger, H. Kaluarachchi, T. Maurer, P. Ciepla, C. Chalouni, Y. Franke and R. N. Hannoush, *Sci. Rep.*, 2016, **6**, 35179.
367. S. L. Gerlach and D. Mondal, *Chron. Young Sci.*, 2012, **3**, 169–177.
368. R. Burman, S. Gunasekera, A. A. Stromstedt and U. Göransson, *J. Nat. Prod.*, 2014, **77**, 724–736.
369. D. J. Craik, *Toxicon*, 2001, **39**, 1809–1813.
370. D. J. Craik, M. A. Anderson, D. G. Barry, R. J. Clark, N. L. Daly, C. V. Jennings and J. Mulvenna, *Lett. Pept. Sci.*, 2002, **8**, 119–128.
371. N. H. Tan and J. Zhou, *Chem. Rev.*, 2006, **106**, 840–895.
372. D. J. Craik, M. Cĕmažar and N. L. Daly, *Curr. Opin. Drug Discovery Dev.*, 2007, **10**, 176–184.
373. P. B. Pelegrini, B. F. Quirino and O. L. Franco, *Peptides*, 2007, **28**, 1475–1481.
374. L. Cascales and D. J. Craik, *Org. Biomol. Chem.*, 2010, **8**, 5035–5047.
375. S. T. Henriques and D. J. Craik, *Drug Discovery Today*, 2010, **15**, 57–64.
376. M. Cĕmažar, S. Kwon, T. Mahatmanto, A. S. Ravipati and D. J. Craik, *Curr. Top. Med. Chem.*, 2012, **12**, 1534–1545.
377. D. J. Craik, *J. Pept. Sci.*, 2013, **19**, 393–407.
378. D. J. Craik, D. P. Fairlie, S. Liras and D. Price, *Chem. Biol. Drug Des.*, 2013, **81**, 136–147.
379. S. C. da Silva Lima, A. M. Benko-Iseppon, J. P. Bezerra-Neto, L. L. Amorim, J. R. Neto, S. Crovella and V. Pandolfi, *Curr. Protein Pept. Sci.*, 2016, **18**, 375–390.
380. J. Weidmann and D. J. Craik, *J. Exp. Bot.*, 2016, **67**, 4801–4812.
381. A. M. White and D. J. Craik, *Expert Opin. Drug Discovery*, 2016, **11**, 1151–1163.

CHAPTER 15

Cyclic Peptides – A Look to the Future

CRISTINA N. ALEXANDRU-CRIVAC[a,b], LUCA DALPONTE[a,b], WAEL E. HOUSSEN*[a,b,d], MOHANNAD IDRESS[a,b], MARCEL JASPARS*[b], KIRSTIE A. RICKABY[c] AND LAURENT TREMBLEAU†*[c]

[a]Institute of Medical Sciences, University of Aberdeen, Aberdeen AB25 2ZD, UK; [b]Marine Biodiscovery Centre, Department of Chemistry, University of Aberdeen, Meston Walk, Aberdeen AB24 3UE, UK; [c]Department of Chemistry, University of Aberdeen, Meston Walk, Aberdeen AB24 3UE, UK; [d]Pharmacognosy Department, Faculty of Pharmacy, Mansoura University, Mansoura 35516, Egypt
*E-mail: m.jaspars@abdn.ac.uk, l.trembleau@abdn.ac.uk, w.houssen@abdn.ac.uk

15.1 Introduction

This chapter sums up the different approaches to harness the promising therapeutic potential of cyclic peptides in drug discovery and proposes a number of areas which need improvement in order to make cyclic peptides live up to their full potential as drug candidates. It is clear that improved methods to rapidly and efficiently synthesize cyclic peptides and modify them are essential to explore this region of chemical space. A better understanding of what governs the physicochemical characteristics of this compound class is

† All authors contributed equally and their names are listed in alphabetical order.

Chemical Biology No. 6
Cyclic Peptides: From Bioorganic Synthesis to Applications
Edited by Jesko Koehnke, James Naismith and Wilfred A. van der Donk

Published by the Royal Society of Chemistry, www.rsc.org

essential to allow the better prediction of the properties of designed cyclic peptides. With this goes the ability to accurately and reliably predict the solution and binding conformations of cyclic peptides, as well as theoretical approaches for determining which peptides are likely to cyclize easily, and which are not. How such compounds bind to their target proteins is just beginning to be understood and improvements are necessary to allow the design of cyclic peptides that bind specifically to extended binding sites. Finally, the ability to utilize biosynthetic machineries from diverse pathways to create hybrid molecules with desirable characteristics is proposed as a major target for future investigation.

15.1.1 Advantages of Cyclic Peptides

Cyclic peptides offer great potential as therapeutic agents because of their target selectivity and activity in nanomolar to picomolar concentrations. However, issues such as their bioavailability and/or size falling outside of the desired therapeutic range are often encountered. Cyclic peptides are robust alternatives to linear peptides that can help overcome several hurdles in therapeutic development.[1] Eliminating the charges on the N- and C-termini through cyclization can improve stability against metabolic enzymes and confer higher biological membrane permeability, as well as increase bioavailability.[2,3] Compared to their linear counterparts, cyclic peptides can engage targets through numerous and spatially distributed binding interactions, thereby having increased binding affinity and selectivity.[4] What is more, cyclization considerably reduces the conformational flexibility of the peptide backbone, allowing control over the size and shape of the ring and providing an entropic advantage on binding.[3,5–7]

There are already over 100 cyclic peptide-containing drugs in the clinic or in late stage clinical trials including the immunosuppressant cyclosporine, the anti-tumor agent plitidepsin and the antibiotics daptomycin, vancomycin and gramicidin.[3,5–7] These naturally-derived or synthetically produced cyclic peptides are very attractive scaffolds for the pharmaceutical industry due to their diverse activities, which include, amongst others, antibacterial potential and the ability to re-sensitize bacteria to previously-used antibiotics, as well as anti-tumor and anti-inflammatory properties.[3,5–7] There is evidence that cyclic peptides can modulate therapeutic targets that involve flat and extended binding sites, such as class B GPCRs and protein–protein interactions (PPIs) (Figure 15.1).[2,3]

Having the pharmacokinetic qualities of small molecules and the target-binding characteristics of biologics, cyclic peptides have become widely recognized as a promising class of molecules for novel therapeutics (Figure 15.1).[2,8]

However, challenges in the synthesis of these compounds on a large scale and with a sufficient level of purity have significantly contributed to their under-exploitation. In order to maximize the therapeutic potential of these cyclic structures and to favor their commercial use, alternative approaches

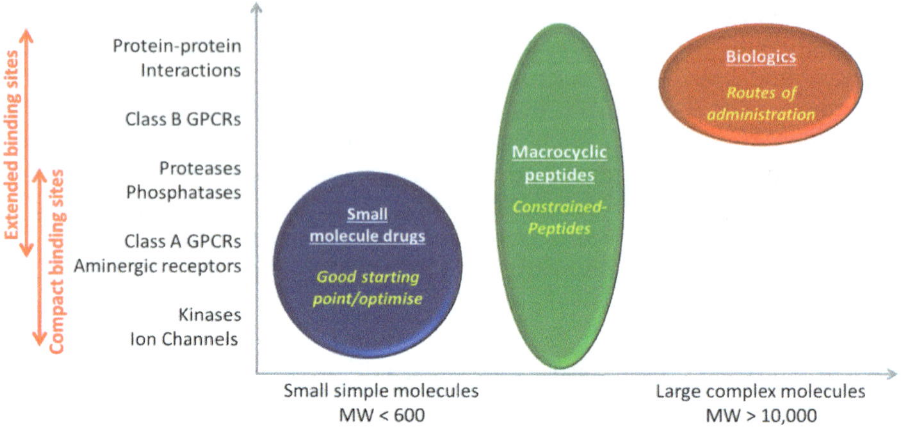

Figure 15.1 Cyclic peptides can modulate targets not amenable to small molecules, while still being cheap and offering an oral route of administration (*c.f.* biological drugs).

are needed to address and satisfy certain key aspects. Ideal strategies would involve the use of small reaction volumes, fast reactions with high yields, and a low risk of intermolecular reactions. Moreover, the final product should be easy to purify upon completion and obtained in a ready-to-use form, while the overall processes should be reproducible and easy to scale up when necessary. While most of these requirements can be attained and successfully completed through many current chemical and/or biochemical approaches, method-specific advantages and drawbacks remain with regard to high-throughput production, compatibility with specific amino acid sequences and overall cost.[7]

15.2 Synthetic and Biosynthetic Approaches to Cyclic Peptides

15.2.1 Synthetic Methods for Cyclization

Various bioactive natural cyclic peptides and analogues, including modified peptide-based macrocyclic derivatives (*e.g.* heterocycle-containing cyclic peptides) have been synthetically prepared, with different cyclization methods highlighted in various reviews.[7,9–12] Depending on its functional groups, a peptide can be cyclized in four different ways: head-to-tail, head-to-side chain, side chain-to-tail or side chain-to-side chain (Figure 15.2). In this section, we give an overview of the key problems associated with peptide cyclization, focusing on head-to-tail cyclization of homodetic (all-amide linked) peptides, as similar problems can be encountered with the other modes of cyclization. Recent advances in the field will also be highlighted, as well as aspects to be addressed towards improving the availability and utility of cyclic peptides as drugs.

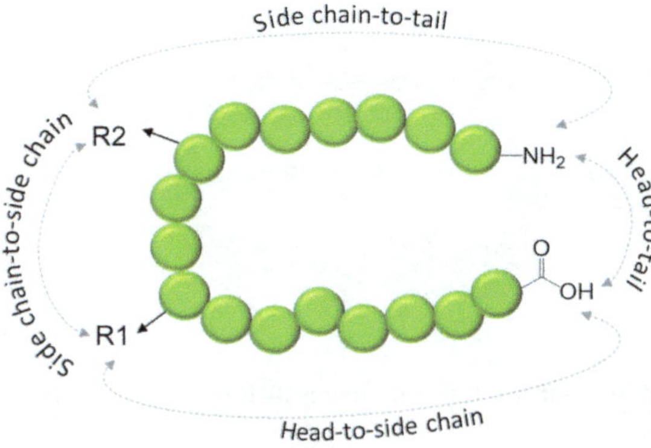

Figure 15.2 The four possible ways a peptide can be constrained in a macrocycle.

The chemical methods for cyclic peptide synthesis generally involve preparation of the required peptide sequence using proteogenic or non-proteogenic amino acids (AA), which is often performed on resin, typically followed by activation of the carboxyl terminal group and subsequent macrocyclization with a free amino group of the peptide. This last step can be performed in solution, or on resin *via* attachment of an AA side chain such as Asp, Glu, Ser, Thr, Cys, Lys, Tyr, His or Arg.

There are several issues associated with peptide macrocyclization. Firstly, the carboxyl group needs to be activated by a coupling reagent. Contrary to the case of Fmoc or Boc-protected α-AAs,[13] the activation of peptide acids and cyclization in the presence of a base will be associated with a high risk of epimerization, as typically observed in peptide fragment condensations.[14] Secondly, the activated peptide must adopt an entropically disfavored pre-cyclization conformation before being able to react. The energy of this conformation depends on the peptide length and sequence, and α-AA configuration.[9] A too high pre-cyclization conformation energy will lead to slow cyclization and will result in the generation of a significant number of side-products, as well as low yields. The side-products are often linear and cyclic oligomers, which result from intermolecular reactions. This has been often observed even when performing on-resin cyclization reactions in high dilution conditions (typically 10^{-3}–10^{-5} M) or with pseudodilution.[15,16] Moreover, slow cyclizations are commonly accompanied by significant quantities of epimerized by-products and impurities associated with the solvents and coupling reagents used.

The effects of parameters such as the length and sequence of the peptide precursor, the configuration of α-AAs and the method of activation on the yield of the desired cyclic peptides have been extensively investigated (Figure 15.3). Constrained small-sized rings (2 to 4 AA residues) and medium-sized rings (5 to 8 AA residues) are generally difficult to prepare in good yields, unless their sequences include AAs that help induce a β-turn in the peptide precursor. For instance, the formation of cyclodipeptides (or

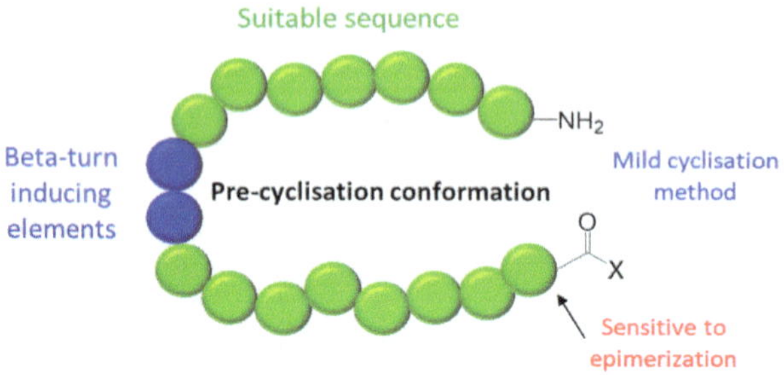

Figure 15.3 The key parameters for successful macrocyclization.

diketopiperazines) can be greatly facilitated by the presence of D-amino acids or AAs that can adopt a *cis*-conformation such as glycine, proline and *N*-alkylated AAs. Cyclic tripeptides such as cyclotriproline have been successfully prepared by on-resin cyclization, but extended formation of cyclic oligomers is frequently observed with other AAs.[17,18] Recalcitrant serine- and threonine-containing cyclic tripeptides and tetrapeptides have also been obtained in good yields using pseudoproline derivatives of these residues, enabling the required β-turn conformations to be attained for cyclization.[19,20] What is more, novel approaches have led to significant advances in the synthesis of natural-like cyclotetrapeptides, as reviewed by De Leon Rodriguez *et al.*[21] The methods involve the use of pseudoproline residues, microwaves or solid-phase and ring contraction reactions by rearrangement of amino ester fragments.

The cyclization of natural and synthetic pentapeptides, hexapeptides and heptapeptides has been reported to lead to extensive epimerization of the C-terminal activated residue and/or to the formation of cyclic dimers.[22–25] Alkali ions such as sodium and caesium ions were shown to effectively assist the cyclization of pentapeptides and heptapeptides, respectively, supposedly by chelating the amide carbonyl oxygens and therefore templating the intramolecular reactions.[26,27] Meutermans used a ring contraction reaction involving a removable amino ester linker to produce cyclopentapeptide cyclo(Ala–Ala–Phe–Leu–Pro), albeit in low overall yield.[28] In a more recent study, a dehydrophenylalanine residue was used in the peptide chain as a traceless turn-inducer to help cyclization, leading to an impressive 74% yield under non-diluted conditions. An L-Phe AA residue was then generated by Rh-catalyzed hydrogenation, leading to natural product dichotomine E.[29] Cyclization of hexapeptides can also be problematic as demonstrated by the low yield (2%) obtained for H-Phe–Pro–Phe–Pro–Phe–Pro-OH. Interestingly, replacing the first L-Phe residue with D-Phe gave a significantly higher yield of 57% of cyclohexapeptide.[25] In another study, the group of Yudin exploited the relative weakness of the amide bond at an aziridine residue in a cyclotetrapeptide to incorporate a dipeptide fragment leading to the generation of a cyclohexapeptide.[30] Moreover, in a systematic study that compared the cyclization yields for all the possible heptapeptide precursors of cyclo(Tyr–Gly–Gly–Pro–Phe–Pro–Gly), it was found that small or large residues at both termini of the peptides gave low yields of cyclic product. However, an excellent 88% yield was achieved after 24 h of cyclization using H-Gly–Tyr–Gly–Gly–Pro–Phe–Pro-OH.[24]

Although the use of an on-resin pseudodilution strategy has proven successful, there have been cases where this technique has led to unsatisfactory yields. For example, the synthesis of the cytostatic cycloheptapeptide phakellistatin-5 *via* resin attachment to the β-carboxyl group of an Asp residue gave only 15% yield of cyclic product.[31] An elegant on-resin peptide cyclization was developed by means of the Kenner's 'safety-catch' sulfonamide linker.[32] The technique is particularly useful, as it removes the need to anchor the side chain of a residue to perform on-resin cyclizations. Impressively, it has enabled a 192-member library of cyclic decapeptides to be prepared and tested for antibacterial activity.[33] The yield of the macrocyclizations, however, remains relatively modest.

Another important parameter for successful macrocyclization is the choice of the coupling reagent. For instance, cyclization of the thymopentin analogue H-Val–Arg–Lys(Ac)–Ala–Val–D-Tyr-OH using benzotriazole-based coupling reagents gave <25% of product with extensive epimerization of the tyrosine residue, while HAPyU gave 55% yield of the product within 30 minutes with negligible epimerization. Interestingly, in the case of the cyclization of a thymopentin analogue from the linear precursor H-Arg–Lys–Asp–Val–Tyr-OH, HAPyU was also very effective with no epimerization or oligomerization, even at a high peptide concentration (0.1–0.2 M).[34] Other studies have highlighted the importance of this aspect, and parameters such as the solvent and base used, the presence of racemization suppressants, and the temperature of cyclization.[24,35–37]

Other strategies have also been investigated, such as the use of thioester C-terminal groups, in which the cyclization is performed by intramolecular ligation,[38,39] or assisted by a metal such as silver(I).[40] Although these approaches require the installation of a thioester group in the peptide, they have the advantage of being performed under non-diluted conditions. For example, a recent study reported the efficient synthesis of a small cyclic RGD peptide based on a combination of micro-flow technology, triphosgene-mediated peptide chain elongation and micro-flow photochemical macrolactamization. This enabled a more rapid (<5 min) and clean synthesis of the cyclic peptide.[41]

Many larger peptides, usually containing 8 to 15 AA residues, have been cyclized in modest to good yields. Examples include the synthesis of cyclic nonapeptide chevalierin C under high dilution,[42] the antibiotic cyclic decapeptide gramicidin S obtained *via* cyclodimerization,[43] the immunosuppressant cyclic undecapeptide cyclosporine A and its analogues generated by solid-phase synthesis[44] and a library of cyclic pentadecapeptides as β-hairpin mimetics synthesized in solution.[45] Compared to the smaller peptides described above, cyclization of large peptides (*i.e.* up to about 20 AA residues) may be facilitated through the increased flexibility and stability of the required pre-cyclization conformation.

In summary, in the past 20 years, major advances have been recorded in the field of linear peptide synthesis and macrocyclization. These have enabled chemists to cyclize peptide sequences of various sizes in modest to excellent yields. One of the major advances has been the development of coupling reagents and mild coupling methods that minimize the risk of epimerization of the C-terminal residue over the course of the cyclization. Nevertheless, effective peptide cyclization is largely contingent on the existence of peptide sequence features, and reagents that promote it.[9] As highlighted in Section 15.4, computer modelling methodologies should also help researchers to select the optimal method of peptide macrocyclization. All these areas of peptide research are expected to remain very active in the years to come.

15.2.2 Biochemical Methods for Cyclization

In contrast to synthetic methodologies, biochemical approaches do not require high dilution conditions, specific catalysts or larger scale preparation of the starting peptide material.[7,46] Biochemical strategies include sortase-mediated ligation,[47] split-intein circular ligation of peptides and proteins (SICLOPPS)[48] and enzymatic macrocyclization by biosynthetic enzymes from both non-ribosomal peptide (NRP) and ribosomally synthesized and post-translationally modified peptide (RiPP) pathways.[49] These enzyme-catalyzed macrocyclizations are alternative systems that present several advantages over other approaches. They do not require high dilution conditions or large quantities of organic solvent and are unlikely to lead to epimerization and oligomerizations, including cyclic oligomer side-products. Therefore, these biochemical methods can be considered "greener" technologies than purely chemical synthesis of cyclic peptides.

An LPXTG motif at the C-terminus, along with an oligo-G motif at the N-terminus, is needed in a peptide sequence for sortase-mediated ligation, where X is a variable residue. This motif is then incorporated into the resulting product.[47] SICLOPPS is another very powerful method to produce a large diverse library *in vivo* that suits high-throughput screening systems.[48,50] It involves rearranging the order of the elements of the intein, yielding an active *cis*-intein (IC:target peptide:IN) and resulting in cyclization of the target protein/peptide sequence upon splicing (Figure 15.4). The main advantage of this method is that any target can be incorporated into the SICLOPPS vector, without limitations on sequence identity by engineering restriction sites into the C-terminal (IC) and N-terminal inteins (IN).[48,51] A disadvantage is that only the 20 canonical amino acids can be included, which limits the chemical diversity accessible. Although this has been partially addressed by the use of amber stop-codon suppression, it is still limited to one non-natural amino acid per peptide.[52]

Alternative biosynthetic approaches involve the use of cyclization systems employed in the synthesis of NRPs and RiPPs. Many bioactive cyclic peptides are produced by template-directed syntheses on large and multi-modular non-ribosomal peptide synthetases or hybrid polyketide (PKS/NRPS)-based systems. Cyclization through NRPS-based enzymology is often obtained through a thioesterase domain and associated modifications include D-amino

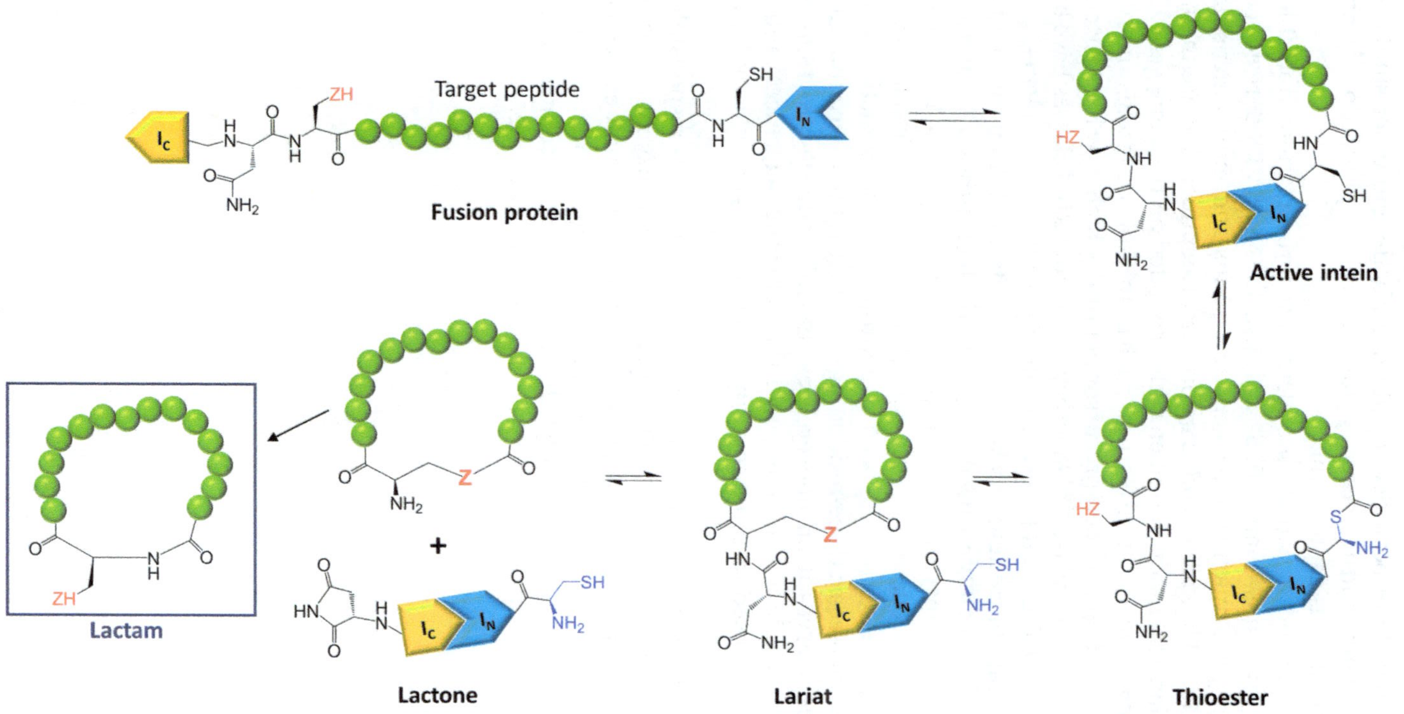

Figure 15.4 The expressed fusion protein folds to form an active intein. An N-to-S acyl shift at the target N-terminal intein junction produces a thioester, which undergoes transesterification with a side chain nucleophile (serine or cysteine, Z = O or S) at the C-terminal intein junction to form a lariat intermediate. An asparagine side chain liberates the cyclic product as a lactone, and a Z-to-N acyl shift generates the thermodynamically favored lactam product *in vivo*.

acids, β-hydroxy amino acids and non-proteinogenic amino acids. Its engineering principles have been developed on surfactin and other similar systems and efficiently and consistently applied for the production of many cyclic peptides, such as daptomycin.[53]

Ribosomally encoded natural products have proven more tractable to genetic manipulations than NRPS and hybrid NRPS/PKS-derived compounds, as they are much simpler genetic systems and more amenable to bioengineering.[53] They are encoded as precursor peptides that are matured by post-translational modification enzymes encoded on the same gene cluster. These enzymes incorporate features such as thiazoles, oxazoles and D-stereocentres to give the modified precursor peptide, which may be cyclized by means of a ligase or a cyclase.[49,53] Macrocyclases from RiPP classes have been recently identified and used for cyclization of peptides, such as $PatG_{mac}$ produced by the cyanobacterium *Prochloron* sp.,[54] butelase-1 from the plant *Clitoria ternatea*,[55] GmPOPB involved in α-amanitin biosynthesis[56,57] and PCY1 responsible for the biosynthesis of the plant orbitide segetalin A.[58]

$PatG_{mac}$, a macrocyclase from the patellamide biosynthetic pathway, is a subtilisin-like protease that has been structurally and mechanistically characterized. It recognizes a three-residue signal (AYD) at the C-terminus of the core sequence, cleaves it off and cyclizes the substrate (Figure 15.5). The enzyme is highly tolerant to changes in the substrate sequence[59] and has been successfully used to produce a library of patellamide-like cyclic peptides *in vitro*.[60]

Butelase-1 is an asparagine/aspartate peptide ligase that is responsible for the formation of plant cyclic peptide cyclotides.[55] This enzyme is currently isolated from the medicinal plant *Clitoria ternatea*, and was only expressed in an insoluble form in *E. coli*, a problem that has halted its wider application in biotechnology. GmPOPB is a prolyl oligopeptidase involved in α-amanitin biosynthesis. This enzyme requires the substrate to have a 25-mer at the

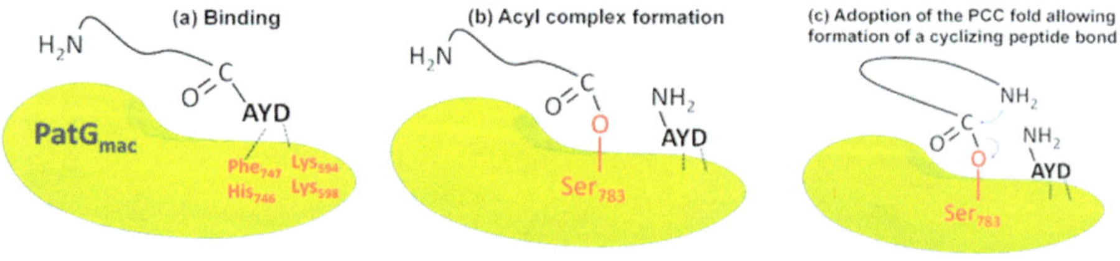

Figure 15.5 Cartoon showing a substrate binding with $PatG_{mac}$ and the steps of enzyme-mediated cyclization. (a) The peptide binds to the enzyme through the recognition signal. (b) $PatG_{mac}$ cleaves the recognition signal and forms an acyl complex with the residue at position P1. (c) An amide bond is formed and macrocyclization of the peptide is completed. Reproduced from J. Booth, C.-N. Alexandru-Crivac, K. A. Rickaby *et al.*, *J. Phys. Chem. Lett.*, 2017, 8, 2310–2315, DOI: 10.1021/acs.jpclett.7b00848.[61] © 2017 American Chemical Society. Published under the terms of the CC BY 4.0 licence, https://creativecommons.org/licenses/by/4.0/.

C-terminus that is cleaved off before cyclization and is not highly tolerant to variations in substrate sequence.[56,57] PCY1 is a member of the S9A family of serine proteases involved in the biosynthesis of the orbitide segetalin A. The enzyme requires and cleaves off a 13-mer at the C-terminus of the sequence to be cyclized.[58]

Cyclic peptides are attractive scaffolds with the potential to incorporate a large number of protein domains of medicinal value, in order to allow for more efficient delivery. However, despite the success of current synthetic and biosynthetic approaches used for generating cyclic compounds, there are still method-specific limitations that need to be addressed. Future developments, both chemical and biological, should aim for 100% conversion to the final cyclic products along with reduced costs and time of production. Furthermore, in order to emphasize an improved utility, novel systems would need to be able to use traceless macrocyclases and process a broader range of sequences, including both canonical and non-natural amino acids. Therefore, the development of practical protocols that mimic the power of Nature's strategies remains paramount for the advancement of novel peptide-based drugs.

15.3 PK/ADMET Properties of Cyclic Peptides

15.3.1 Introduction

The use of bioinformatics tools and the creation of a reliable model is a priority in the design of new membrane-permeable cyclic peptides. A recent study conducted by Hewitt *et al.*, based on a library of 1152 cyclic peptides with diverse backbone geometries inspired by natural products, describes the role of the side chain orientation and steric factors in determining membrane permeability. It also highlights how important the combination of bioinformatics with experimental work is in designing and predicting the potential of synthesized cyclic compounds.[62]

Cyclic peptides have surface areas suitable for interacting with so-called "difficult" drug targets, which typically possess large, flat or groove-shaped binding sites.[63–65] This is highly significant, as it is thought that approximately half of the exploitable drug targets belong to this class.[66] However, the difficulty in the design of cyclopeptide pharmaceuticals resides in the fact that these compounds belong to a chemical space of molecules that do not satisfy the currently accepted Lipinski's rule-of-5 (Ro5) for orally absorbed drugs.[67,68] Large compounds often suffer from poor pharmacokinetics, such as low solubility and cell permeability,[69] increased efflux and metabolism.

15.3.2 Prediction of PK/ADMET Properties of Cyclic Peptides: The New 'Beyond Rule of 5' Guidelines

A recent study by Villar *et al.* on a small number of natural macrocycle–protein complexes was able to define key characteristics of macrocycles.[64] Based on this dataset, physicochemical guidelines comparable to Lipinski's rules[70] are proposed for both orally available and non-orally available macrocycles (Table 15.1). It is clear that the macrocycle ranges are far larger than those for conventional drugs, and despite their higher polar surface areas, some are already orally available. These guidelines need to be verified against larger datasets and computational studies, which are usually hampered by the difficulty in reliably predicting the solution structure of macrocycles.

To facilitate the design of macrocycles with drug-like properties, Villar *et al.* described several key points that should be tested to verify their assertions regarding the physicochemical properties (Table 15.2).[64] The authors concluded that drug-like macrocycles typically have one or two large substituents comprising 20–30 heavy atoms, as well as multiple smaller substituents that bind to the protein. Roughly 1 in 3 ring atoms interact with the protein, which contributes to *ca.* 25% of the contact area, suggesting that all regions of the macrocycle may be of importance to achieve good binding. Secondly, critical binding interactions are mediated by a single polar heavy atom substituent, meaning that great attention should be paid to the role of such atoms in the macrocycle design process. Furthermore, the balance of polar and non-polar atoms in macrocycles is similar to that for conventional drugs

Table 15.1 Physicochemical guidelines relevant to the design of macrocycles based on macrocyclic drugs in current clinical use.[64]

Property	Conventional drugs	Oral MC drugs	Non-oral MC drugs
MW	≤500	600–1200	600–1300
clog*P*	≤5	−2 to 6	−7 to 2
PSA	≤140 Å^2	180–320 Å^2	150–500 Å^2
Number of HBDs	≤5	≤12	≤17
Number of HBAs	≤10	12–16	9–20
Number of rotatable bonds	≤10	≤15	≤30

Table 15.2 Structural guidelines relevant to the design of macrocycles based on macrocyclic drugs in current clinical use.[64]

Property	Observed range[a]
Ring size (R)	14–38
Number of substituents	4.4 (3–8)
Large substituents (≥5 HA)[b]	1.9 (1–3)
Small substituents (2–4 HA)[b]	2.4 (1–6)
Proportion of HA that are in substituents	47% (40–59%)
Number of peripheral groups[c]	5–12
Polar/nonpolar balance, substituents	~30/70
Polar/nonpolar balance, peripheral groups	~60/40
Degrees of unsaturation in ring	~0.4*R*–4 (±3)
N : O ratio	0.25 : 1 (0–0.4 : 1)
Chiral centers	15 (9–18)

[a]Mean (10–90% range).
[b]HA (non-hydrogen atoms).
[c]Peripheral groups are groups connected to the MC ring that contain only a single HA.

and the *c*log*P* values remain in the same range, whilst the polar surface area increases with molecular size (Table 15.1).

A second study generated a family of 200 *de novo* designed non-peptidic macrocycles to try and understand the structural and conformational determinants of macrocycle cell permeability.[71] This showed that certain functional groups, substituents and molecular properties overwhelmingly affect the cell permeability. Molecular simulations were used to try and understand how the structures of these macrocycles affect their physicochemical properties. It was found that the incorporation of functionalities like primary and secondary amines, triazoles, urea and sulfonamides lowered the cell permeability and/or rapid efflux, whereas groups like phenyl, pyridine, tertiary amines and isoxazole had a positive effect on the macrocycles' permeability. What is more, a combination of some of these groups, like phenyl and tertiary amines, gave a synergistic increase in permeability, while others were antagonistic (*e.g.* numbers of oxygens *vs.* nitrogens in the molecule). It was also concluded that cellular efflux and passive permeability is inversely proportional to the ring size.

The low diversity of the test set limits the apparent applicability of these conclusions, one of which is that the extension of the rule-of-5 is possible for macrocycle space. However, this study emphasized the need for larger test sets with consistently-measured physicochemical properties and experimentally-derived solution conformations, in order to derive a reliable set of design parameters for orally and non-orally available macrocycles. This may suggest that specific design criteria would need to be developed for different classes of compounds, such as cyclic peptides and other types of macrocycles.

There are several factors involved in the cell permeability of cyclic peptides such as peptide size, cyclization, net charge, lipophilicity, intermolecular hydrogen bonding potential, internal hydrogen bonds, formulation, solvent accessibility of amide hydrogens and *N*-methylation. These can have an effect as a single modification, but can also act in combination. More than that, orally available cyclic peptides need to overcome additional challenges, such as acid and enzymatic degradation in the gastrointestinal tract, the loss of efficacy before reaching the target and entering the cells *via* passive diffusion and/or active transport.[72]

Currently, the physicochemical properties and cell permeability of drug candidates (quantitative-structure–property-relationship or QSPR) can be calculated *via* training sets based on properties of known drugs.[73,74] This method is therefore limited by the lack of data for macrocycles and cyclic peptides.[64,75] An alternative strategy with promising results is computational modelling of cyclic peptides' conformations in low and high dielectric environments.[76] As described above, cyclization can improve a drug's permeability and stability against enzymatic degradation and increase its oral availability.[77] However, further research is required to establish more comprehensive and inclusive guidelines towards the design of macrocycles. Extensive experimental datasets and bioinformatics tools will help to fill the current gaps and facilitate the prediction and design of highly specific drugs to target the 'undruggable' space.

15.3.3 Backbone Modifications Affecting PK and ADMET

A critical step involved in passive diffusion is the ability to form intramolecular hydrogen bonds leading to a reduction of exposed hydrogen bonds and the polar surface area. Several strategies aiming at masking or burying polar groups of cyclic peptides have successfully led to increased membrane permeability and bioavailability (Figure 15.6). These include the *N*-methylation of solvent-exposed amide groups,[7,78,79] while preserving intramolecularly hydrogen-bonded amide NH groups,[10,76,80,81] as well as the shielding of polar groups by large hydrophobic groups.[62,80] There are several other modifications leading to favorable ADMET properties, such as manipulating the stereochemistry of the cyclic peptide scaffold in certain positions, for example through incorporation of γ-amino acids as in didemnin B,[82] or D-amino acids.[83,84] The use of an exocyclic control element (ECE) contributes to the overall rigidification of the macrocyclic structure and has been shown to improve the permeability of cyclic peptides with polar side chains.[85]

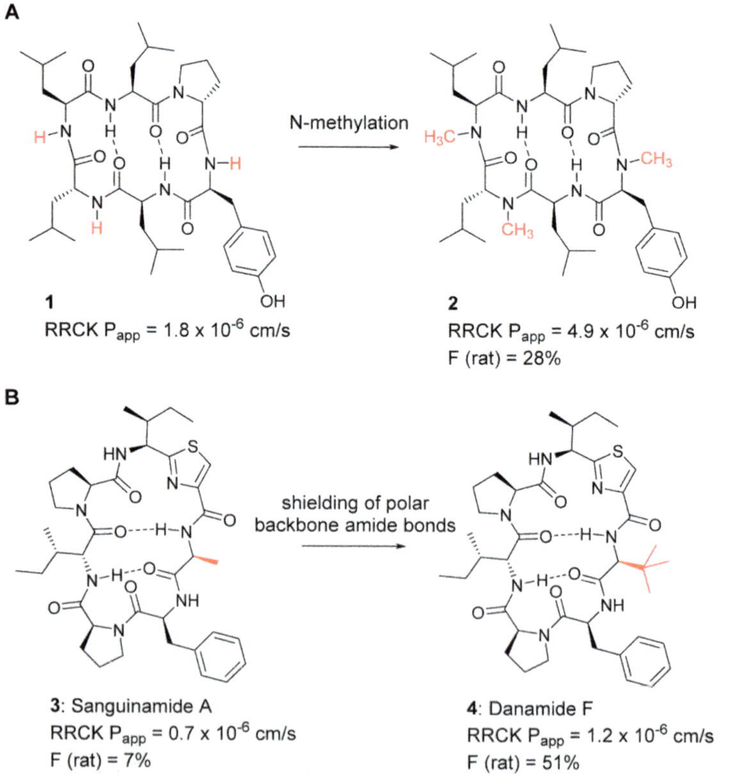

Figure 15.6 Effect of *N*-methylation[7] (A) and shielding[80] (B) on RRCK permeability, and bioavailability (*F*). RRCK P_{app}: Ralph–Russ canine kidney permeability. Intramolecular hydrogen-bonding is shown as presumed in a lipophilic environment.

A study based on somatostatin analogues showed improved permeability and bioavailability in rats after multiple *N*-methylations of exposed amides.[83] However, a separate systematic study in which cyclohexaalanine was *N*-methylated in different combinations recorded contradictory results.[86] In certain cases, *N*-methylation of solvent exposed amides nullified the permeability, while in others *N*-methylation increased it. It was concluded that while *N*-methylation of solvent-exposed NHs can be beneficial, the number and position of *N*-methyl groups is not directly correlated with the permeability of polar as well as non-polar peptides.[86]

The main drawback of these methods is the generation of cyclopeptides that are significantly more lipophilic, but less soluble, which could impact on their oral bioavailability. Hydrophobic compounds, while being more membrane permeable, are also more rapidly metabolized by cytochrome P450 enzymes in the liver, leading to rapid blood clearance.[77] In contrast, compounds that are too hydrophilic, such as multiply charged antibacterial cyclopeptides, will most probably not be orally absorbed and will not reach blood circulation.[78,83,85]

Remarkable environment-dependent conformational effects have been observed for thiazole-containing cycloheptapeptide **6**, which unexpectedly displayed both high permeability and high solubility (Figure 15.7).[87] NMR spectroscopy in $CDCl_3$ (non-polar, membrane-like) and DMSO-d_6 (polar, water-like) combined with computer-aided conformational analysis of **6** revealed that it exists in a single conformation in $CDCl_3$, where the NH amide groups are shielded, but in multiple conformations in DMSO-d_6, where the NH amide groups are more exposed to the solvent. On the contrary, poorly soluble analogue **5** was described to exist in similar conformations in non-polar and polar solvents. The limited number of permeable cyclic peptides with no *N*-methylation or intramolecular hydrogen bonds[88] strongly supports the key role of the conformation in the bioavailability and bioactivity.[89]

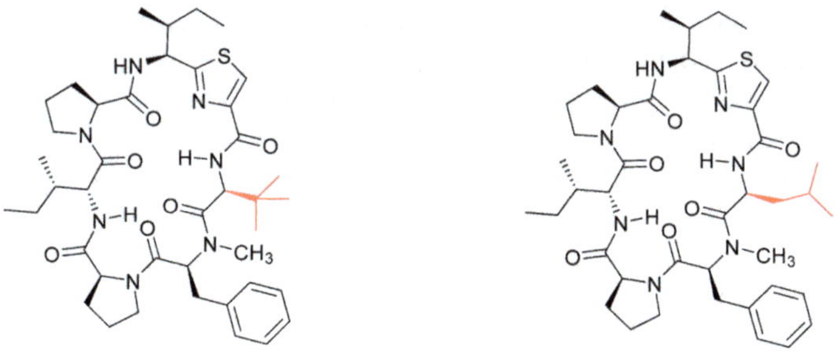

5: Danamide D
PAMPA $P_e = 2.1 \times 10^{-6}$ cm/s
Thermodynamic solubility = 8 μM (PBS, pH 7.4)

6: Danamide D analogue
PAMPA $P_e = 17 \times 10^{-6}$ cm/s
Thermodynamic solubility = 111 μM (PBS, pH 7.4)

Figure 15.7 Effect of environment-dependent conformations on membrane permeability and solubility. PAMPA P_e: permeability (P_e) in the parallel artificial membrane permeability assay.

By analyzing pairs of polar enantiomers using PAMPA (passive cell permeability on artificial membrane) and Caco-2 cell membrane assays, Marelli *et al.* could predict human intestinal permeability and investigate the drug efflux.[86] Other examples of peptides using active transport are lysine- or arginine-rich cell-penetrating peptides (CPPs). The guanidinium cationic groups of Arg seem to be crucial for efficient cell penetration, a role attributed to their capacity to interact with the polar head of the phospholipids in the membrane or sulfate groups of the glycosaminoglycans. The CPP peptides are usually described as entering the cells *via* the endosome. A recent study, however, demonstrated the influence of the position of the cyclization and stereochemistry on the cellular uptake of octa- and nona-arginine derivatives.[83,90]

An alternative approach to improve membrane permeability is the use of stapled α-helices, in which the side chains that are cyclized are placed on the same face of the helix in order to stabilize the secondary structure. A number of these stapled peptides have been reported to have cellular penetration, likely through an active uptake mechanism.[83] Another strategy is the hydrogen-bond surrogate (HBS) approach that differs in the attachment of the olefins directly to the backbone. More HBS-stabilized helices were reported with improved metabolic stability compared to that of non-stabilized controls, reduced cytotoxicity, effective cell penetration, and/or potent cellular activity consistent with selective inhibition of their intended targets.[89]

Overall, significant progress has been made in the understanding and design of cyclic peptides with enhanced solubility and membrane passive diffusion, which will, in turn, lead to an increase in the number of orally absorbed cyclic peptide drugs on the market. Most notably, the discovery of the properties of chameleon-type macrocycles constitutes a major advancement in the development of drug discovery tools for medicinal chemists.[91] One could expect the systematic computer-aided design of cyclic peptides possessing such properties, whether intrinsically present or intentionally introduced *via* customized side chains to create interactions that are beneficial for solubility and cell permeability. Progress in the development of permeability and clearance prediction models, whether based on training sets or new *in silico* approaches, will undoubtedly help the development of cyclic peptide drugs. Finally, understanding the mechanisms of active transporter proteins such as the recently revealed structure of P-glycoprotein[92] and the study of solute carrier membrane transport proteins[93] are among the key aspects for further tackling the medicinally promising realm of "difficult" drug targets using macrocycles and cyclic peptides.

15.4 Prediction of Structures of Cyclic Peptides

The three-dimensional structure of a given cyclic peptide directly influences its physical properties, thus making this a key goal in drug discovery. However, despite extensive research into how exactly the 3D structure can affect the properties of macrocycles, the elucidation of 3D structures remains non-trivial. NMR can be used to elucidate which atoms are involved in intramolecular H-bonding, and circular dichroism is an indispensable technique for exploring aspects of protein secondary structure, such as α-helices and β-sheets, which can be embedded into macrocyclic structures;[10] however, neither of these gives an accurate depiction of the global structure of a given cyclic peptide. While this information can be obtained *via* the use of X-ray crystallography, it must be remembered that the conformation elucidated using this technique is representative only of a single conformation that a cyclic peptide can adopt. In reality, the conformational behavior exhibited by the large majority of cyclic peptides is extremely solvent-dependent and changing this variable may lead to drastic changes in the overall 3D structure.[94]

In addition, the aforementioned techniques can only be employed when there is sufficient physical material with which to work. In the context of the design process for a given cyclic peptide being targeted for a functional application, this is not ideal due to the multitude of features which must be considered.[95] For example, it may be advantageous to investigate different types of cyclization (*i.e.* head-to-tail, head-to-side chain, disulfide, side chain-to-side chain) or the incorporation of non-canonical, *N*-methylated or D-amino acids; however, experimentally generating, testing, analyzing and comparing large libraries of what are often synthetically challenging compounds would be extremely demanding in terms of time and resources. For that reason, when knowledge of sequence–structure relationships is the intended goal, computational methods tend to be favored.

Great efforts have been made in the development and validation of computer-based structural prediction techniques.[76,96–101] These methods normally evaluate the (free) energy of a certain structure using an effective energy function (which can be physics-based or knowledge-based) with implicit solvation, in addition to a conformational search (sometimes with clustering) algorithm to locate the low-energy minima in a similar fashion to the *de novo* methods used for proteins and peptides, such as ROSETTA,[102] I-TASSER,[103] QUARK and PepLook[104] and Pep-Fold.[105] It must be noted that, despite the progress made regarding the refinement of these techniques, they are still limited. They often struggle with consistently predicting the experimental structures of CPs as in the case of α-conotoxin, for which ROSETTA cannot produce a structure resembling that observed experimentally.[106] This can be attributed to the CPs themselves, which often have distinct conformations separated by minute differences in free energy.[107] This puts a very high demand on the accuracy of the energy function.[108] Furthermore, the bioactivity of a CP may be related to its conformational dynamics, information that is difficult to obtain without physically realistic conformational sampling.

15.4.1 Conformational Search Algorithms

There are a vast array of different conformational search algorithms from which to choose, the main difference between each being the mechanism by which they reduce the dimensionality of the conformational space.[95]

Distance geometry methods describe a molecule using distance restraints. Stochastic search algorithms, such as those utilized by popular software such as MacroModel and MOE,[101] operate by defining variables that characterize the system (such as dihedral angles or bond lengths) and then randomly assigning discrete values to each of these variables, after which the resulting conformations are scored. Methods such as CAESAR[109] and Omega[110] employ systematic search algorithms. Here, a variety of likely conformations of individual fragments are amassed to build the final structures.

The techniques discussed above are all popular choices for search algorithms used in the generation of cyclic peptide libraries for virtual screening. However, if low-energy structures are desired, it is often necessary to couple these with additional software, for example combining kinematic search algorithms with additional ROSETTA functions.[95]

Fragment based algorithms for the structure prediction of cyclic peptides have largely been adapted from those used to predict structures in linear peptide systems.[111,112] They function by identifying conformations that individual residues or specific fragments are likely to adopt before assessing them for the structural prediction of the sequence in question.[95] Some popular pieces of software that use fragment based algorithms include PepLook,[105] PEPstrMOD,[113] PEP-FOLD,[114] and I-TASSER.[115] The efficacy with which each of these programs can model cyclic peptides is highly dependent on the nature of the cyclic peptide. In order to make the most appropriate selection, factors such as size, method of cyclization, and the nature of the included residues must be considered.[95] As observed in Table 15.3, there are options available for the modelling of CPs, which include various different cyclization methods, varying lengths of CP and non-canonical residues. The reliability with which these structures can be predicted depends on the capability of the conformational search algorithm to generate all plausible structures that could be adopted, as well as the accuracy of the energetic function.[95] Research into both of these aspects is still being actively conducted.

15.4.2 Molecular Dynamics Simulations

A molecular dynamics simulation involves the numerical integration of Newton's equation of motion to simulate the dynamics of the system. Provided the run is of sufficient length, it is possible to obtain a Boltzmann-weighted ensemble. Unfortunately, conventional MD simulations are often limited due to the tendency to remain kinetically trapped in local free energy minima, which often remain separated from the global minimum by large free energy barriers. This is particularly problematic in small cyclic peptides where coherent changes in dihedral angles over multiple residues are required to sample a new conformation.

Molecular dynamics has the advantage that water molecules can be dealt with explicitly.[116,117] It is, however, limited by both the sampling efficiency and the efficiency of the force field used. Additional limitations of conventional MD simulations arise due the structures of the cyclic molecules: the

Table 15.3 List of the aforementioned methods and their compatibility in relation to the stated CP properties.

Method	Description	MC Size (no. of residues)	Methods of cyclization	Non-canonical residues
PepLook[105]	Iterative Boltzmann-Stochastic algorithm which utilizes ϕ and ψ; angles derived from the structural alphabet of Etchebest *et al.* Generates a series of random peptide conformations before selecting the lowest energy model	Up to 30	Disulfide, head-to-tail (*via* distance restraints)	Yes
PEP-FOLD-fold[114]	Uses a structural alphabet of 27 motifs derived from the Hidden Markov Model. First determines the letters from the structural alphabet which make up the sequence and then uses a greedy algorithm and coarse grained forcefield to build these into a model	9–25	Disulfide (*via* the inclusion of disulfide potentials)	No
I-TASSER[115]	A combination of both threading and *ab initio* methods. Sequence is first threaded through a structural library from the PDB using LOMETS. A global structure is built up by assembling fragments. Unaligned regions are generated using an *ab initio* approach	10–1500	Disulfide, head-to-tail (*via* distance restraints)	No
PEPstr-MOD[113]	An update on PEPstr that allows a user to introduce modifications/insertions at chosen positions throughout the structure. The secondary structure is predicted using PSPIRED and BetaTurns. An initial structure is generated using the tleap module of AMBER11, which is then minimized and subjected to a very short MD simulation using AMBER11 or GROMACS to yield the final 3D structure prediction	7–25	Disulfide, head-to-tail (*via* specification of covalent bonds)	Yes

inherent ring strain present in small macrocycles creates large free energy barriers between the different conformations and larger macrocycles can have incredibly vast structural ensembles.[117,118] Both of these issues mean that sampling the full conformational ensembles can be extremely difficult. Because of this, there has recently been a great deal of focus on the development of a number of advanced sampling techniques which employ parallelization algorithms.[27,119–125] Thanks to these, it is now possible to conduct MD simulations in the microsecond time range with some reports of simulations being run on a millisecond timescale even being reported.[126–128] One of the most popular of these methods is replica-exchange molecular dynamics (REMD), or parallel tampering as it is also known.[119,125]

REMD explores conformational space by using different replicas which are each simulated at a different temperature in parallel. At regular time intervals, the configurations of these replicas are exchanged using Metropolis acceptance criteria. Conformational sampling is more efficient at high temperatures. REMD exploits this fact to enhance the sampling of the replicas in all general degrees of freedom. The technique has previously been used to model structural ensembles of some cyclic peptides and peptoids.[129–131] It should be noted that optimal performance of REMD calculations can only be achieved provided that the overlap in energy space between replicas is sufficient so as to allow adequate exchanges to be accepted. In addition, although REMD simulations enhance the sampling of conformations, they do so for all degrees of freedom. In instances where there are large numbers of conformations separated by large free energy barriers, it may be preferable to utilize a method in which the slow degrees of freedom are specifically targeted.

Metadynamics is another well-used method of enhanced sampling that operates by accelerating rare events.[122] A time-dependent external potential is applied to the space of a small number of degrees of freedom, known as collective variables (CVs), to discourage the system from revisiting configurations which have already been sampled and making it possible for the system to quickly escape low energy space and explore conformational space more rapidly.[95] Bias exchange metadynamics, BE-MetaD, is a variant of metadynamics. BE-MetaD runs several replicas in parallel, in a similar fashion to REMD, each biased on one of the CVs. Regular exchanges are attempted and accepted or rejected by a probability generated by a metropolis criterion.

REMD and BE-MetaD have been evaluated and both shown to be able to reproduce the same conformational distribution of a small cyclic peptide observed in standard MD simulations but over a much shorter time (1/12 and 1/175 of the time for the respective techniques).[129]

15.4.3 Force Fields

Ideally, a good force field should strike a good balance in how it handles different aspects of secondary structure and should demonstrate the ability to accurately describe thermodynamic properties in addition to kinetic rates pertaining to relevant biological processes such as protein folding.[129]

This is a non-trivial matter as the parameterization of the majority of force fields is carried out by fitting quantum chemistry calculations to empirical data and, as a result, even well-established force fields often have to be re-parameterized occasionally.[128,132–137] A number of force fields have been developed for linear peptide and protein systems, including various iterations of AMBER,[128,136–139] GROMOS,[140,141] OPLS[142,143] and CHARMM,[134,144] all of which remain popular. However, cyclic peptides cannot be simply treated in the same manner as linear peptides. For example, the increased ring strain in smaller cyclic peptides can force them to assume non-canonical dihedrals which are almost never encountered in the standard systems for which these force fields were designed and prevents them from adopting regular secondary structures.[95] In addition, great care should be taken to ensure correct description of the interactions between cyclic peptides and the solvent. The intramolecular H-bonds induced by the ring strain present in small CPs can force them to adopt conformations in which the CONH groups in the backbone are exposed to the solvent. This means that the solvent can play a pivotal role in the resulting CP structure.

Yu *et al.* have evaluated how effectively several force fields, which were designed for linear peptides, could be extended to model cyclic peptide systems.[145] In this NMR based study, it was possible to experimentally identify only one highly populated conformation, however, computationally, each of the force fields tested returned numerous conformations with significant populations. This led the authors to conclude that re-parameterization is vital in order to achieve accurate modelling of cyclic peptides.

In addition, there has recently been some effort directed towards the development of force fields capable of handling CP systems,[108] including residue specific force fields, such as RSFF1 and RSFF2 [146,147], which have proven efficient at both folding proteins and at modelling cyclic peptides in molecular dynamics simulations. Interestingly, RSFF2 has also been shown to be capable of predicting structures of *N*-methylated CPs (provided there was experimental information regarding the presence of *cis*/*trans* isomers).[148] It is hoped that this force field could be developed to have the capacity to describe isomerization, and thus it could be utilized for reliable *de novo* structure prediction for cyclic peptides.

15.4.4 Predicting Whether Peptides Will Cyclize

Peptide trajectories have been successfully studied through accelerated dynamics (AXD), through which reaction coordinates are constrained within two "boxes" containing the transition states of interest. This simulates an accelerated rate of reaction, which then requires multiplication by correction factors. This has often proven to be difficult. Boxed molecular dynamics (BXD) complements AXD as the reaction coordinate is more efficiently explored by being split into several "boxes", instead of just two, including regions that would otherwise be rarely considered. Trajectories are therefore run consecutively, generating kinetic rate coefficients for exchange between neighboring boxes and efficiently mapping the free energy in each state.[149]

The BXD strategy has been successfully applied to simulate the loop formation dynamics in small peptides, which is the fold adopted to favor non-enzymatic peptide cyclization.[150] Therefore, this method was also implemented in predicting cyclizable peptide sequences, as reported by Booth *et al.* A computational method using BXD was compared with experimental data obtained from macrocyclization reactions using $PatG_{mac}$ macrocyclase and an 84% prediction accuracy was reported. A pre-cyclic conformation (PCC) was observed to be obtained after the macrocyclization signal is cleaved by the enzyme, which was used to indicate the propensity of a given peptide to bend onto itself. The BXD system provides both thermodynamic and kinetic information and its simplicity and speed make it suitable for broad use in obtaining reliable information on the least and most favorable sequences for peptide macrocyclization[61] (Figure 15.8).

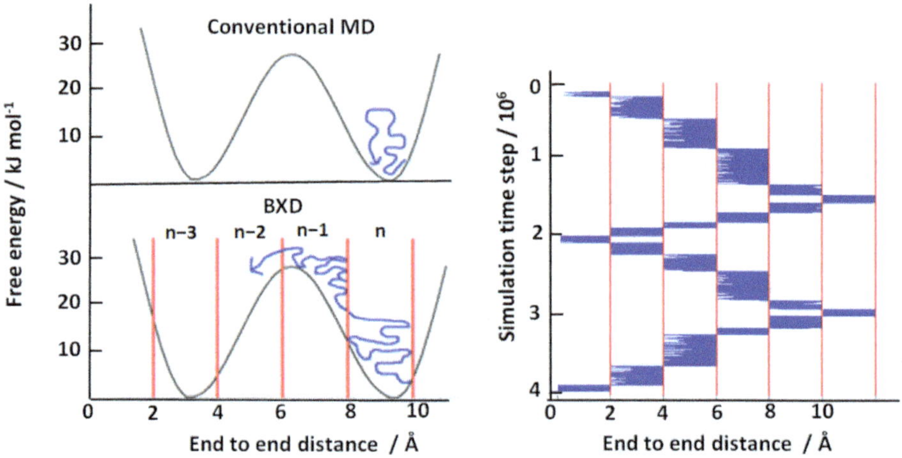

Figure 15.8 A sketch of the BXD method is shown on the left. With conventional MD, a simulated trajectory (blue line) will not be able to cross large free energy barriers, so sampling is poor. With BXD, reflecting boundaries are placed along the reaction coordinate chosen to describe the process, splitting the phase space into boxes (n, $n+1$, $n+2\cdots$). By restricting the trajectory within a box for a length of time, and then allowing it to pass into the next box and restricting it there, the boxes act as a thermal ratchet and allow free energy barriers to be crossed. On the right, a plot of the reaction coordinate value against simulation time from a BXD simulation shows how the trajectory (blue) moves through the boxes and samples the space. The boxes' boundaries are shown by vertical lines. Reproduced from J. Booth, C.-N. Alexandru-Crivac, K. A. Rickaby *et al.*, *J. Phys. Chem. Lett.*, 2017, 8, 2310–2315, DOI: 10.1021/acs.jpclett.7b00848.[61] © 2017 American Chemical Society. Published under the terms of the CC BY 4.0 licence, https://creativecommons.org/licenses/by/4.0/.

15.4.5 Conclusions

In conclusion, there has been much research effort expended on the development of computational techniques for the accurate structure prediction of cyclic peptides. While great improvements have been, and are continuing to be made, there is still some way to go to reach a stage where *de novo* rational design of cyclic peptides *via* a computational route can be accomplished, and it remains a very active field of study. This will require the development of force fields that can parametrize amino acid residues in the unusual conformations sometimes present in cyclic peptides, rapidly and reliably search the conformational space taking into account solvent effects, and predict whether any given linear peptide will fold to allow chemical or biosynthetic macrocyclization.

15.5 Binding of Cyclic Peptides to Targets

Antibodies are well-known for their highly-specific binding to billions of targets using different sequences and spatial arrangements of the 20 canonical amino acids that make up their complementarity determining regions (CDRs). These CDRs are assemblies of conformationally defined loops that interact with their cognate targets. Similarly, β-hairpins are considered important recognition motifs for many therapeutically relevant receptors.[151–153] Large libraries of small molecule ligands have been synthesized to simulate these structural motifs and interact with the relevant targets.[154] Cyclic peptides are considered the prototypical β-turn mimetics which have the ability constrain pharmacophore elements in different conformations, giving an additional dimension to the chemical diversity of cyclic peptides. Therefore, macrocycles may have antibody-like binding specificities, as well as 'drug-like' characteristics. Binding of macrocycles to target proteins can be categorized into three major binding geometries (Figure 15.9).[64] In the first of these, the structure sits in the receptor binding groove perpendicularly with other side chains interacting near to the active site, with the rest of the structure exposed to the solvent. It has been found that most large macrocycles adopt this mode of binding. The remainder of the large macrocycles favor a second type of binding geometry where the macrocycle lies horizontally (face-on) on the protein surface with one or more big substituents anchored in adjacent binding pockets. The third category is mainly adopted by small macrocycles, which tend to have a compact globular conformation, fitting inside the binding cleft of the target protein. Macrocycle binding, in physicochemical terms, tends to account for the overall balance between hydrophobic/hydrophilic groups in the molecule. This was evident by analyzing the ratio of polar/non-polar groups interacting with a hydrophobic binding site (73% non-polar and 27% polar) and it was found to be identical to the ratio of non-polar/polar atoms constituting the macrocycle, a characteristic observed in some natural products that can help guide chemists in designing good ligands. Various atoms are involved in ligand–receptor

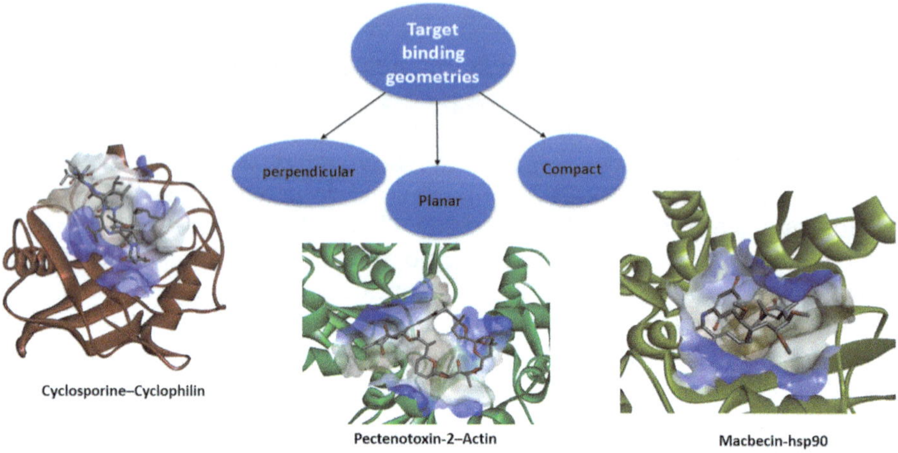

Figure 15.9 Binding modes of macrocyclic drugs. This figure illustrates 3 major binding modes: perpendicular to the protein surface (Cyclosporine-cyclophilin) and face on to the protein (Pectenotoxin-2-Actin) for large macrocycles, and a compact globular conformation for smaller macrocycles (Macbecin-hsp90).[64]

contacts including σ-bonded backbone atoms, peripheral atoms (*e.g.* methyl, carbonyl and hydroxyl groups) and larger ring substituents, such as amino acid side chains. The latter was found to represent 60% of ligand atoms involved in direct contact with the target protein, while the macrocycle backbone and peripheral atoms constitute 38% and 15% of the molecule interacting with the receptor, respectively. This makes cyclic peptides ideal as high affinity binders given the variety of natural and non-natural amino acid side groups available and the possible conformations that can be adopted by peptide macrocycles.[64]

The ability of cyclic peptides to interfere with protein–protein interactions (PPI) can be exploited effectively in rational design by first identifying the most critical protein–protein contact points or hotspots. Computational tools such as ROSETTA can be used to find these by using alanine scans. This has been very successful with 79% agreement with experimental data.[155,156] LoopFinder is a powerful tool in searching structural databases to identify peptide loops involved in PPIs.[64]

What is critical is the improved finding of hotspots in the extended binding sites of targets and rational strategies to design macrocycles with complementary characteristics and desirable drug-like properties. This is hampered by the difficulty in accurately predicting the solution conformation of the cyclic peptides (see Section 15.4) and when bound to the target protein. Computational methods have been utilized effectively to predict the PPI hotspots as a starting point for drug discovery. One of the powerful tools in this regard is alanine scanning based on the ROSETTA molecular modelling package, in which all residues at the interface of each of the interacting proteins is substituted with alanine followed by binding energy calculations to map the critical loops for PPI.[155] Having determined the active site and the corresponding interacting loop sequences, the design process then starts by building analogues of these loops and more importantly their solution conformations. Many search algorithms are available to predict the conformation of such peptides with Monte Carlo conformational searching being the most appropriate option available.[157] The binding of generated conformations can be scored using various types of docking software, such as GLIDE, CLICK DOCKING, AADS and AUTODOCK. Although these computational approaches are good starting points, they need to be tailored to cyclic peptides and supported with experimental data available from natural products to improve their efficiency.

15.6 Hybrid Systems to Generate Diversity in Cyclic Peptides

Recent advances in our knowledge of how complex natural products are biosynthesized has allowed the recruitment of biosynthetic machineries to synthesize these compounds. Genome and metagenome mining approaches have identified several biosynthetic gene clusters that remain intractable or cryptic in their native hosts.[158,159] The expression of these clusters in heterologous systems has generated a wide array of novel structures.[160] Combinatorial biosynthesis is an emerging field that allows the reprogramming of biosynthetic pathways by mixing and matching genes from known biosynthetic clusters to produce unnatural compounds or to introduce structural modifications to natural scaffolds.[161] Given the interest from pharmaceutical companies in exploring the therapeutic potential of cyclic peptides,[10,162] and the challenges encountered in their chemical synthesis, there are concerted efforts to harness these biosynthetic machineries. Cyclic peptides are biosynthesized either *via* ribosomal or non-ribosomal routes. Although ribosomal peptides do not incorporate amino acids beyond the canonical 20 proteinogenic amino acids and thus have, in theory, limited structural diversity, they can be extensively post-translationally modified, and these modifications lead to products with many features resembling the non-ribosomal peptides.[163] It is now evident that many natural macrocyclic peptides incorporating diverse modifications are produced by a ribosomal route and are collectively known as ribosomally synthesized and post-translationally modified peptides (RiPPs).[49] The precursor of these compounds is usually a longer, typically ~20–110 residues in length, linear precursor peptide encoded by a structural gene that contains the sequence to be processed into the final product (the core peptide), as well as sequence(s) for recognition and for directing the processing enzymes. RiPPs offer many advantages when compared to non-ribosomal peptides that make them very attractive for bioengineering. These include: (1) the simplicity of the biosynthetic machinery; (2) the co-linearity between the structure of the final product and the sequence of the core peptide and (3) the distant separation of the sequence to be processed from the enzyme recognition sequence. These factors make it possible to predict the final product structure from the sequence of the gene cluster[164] and allow the application of site-directed mutagenesis to alter the core peptide and thus the final product without jeopardizing the enzyme–substrate binding. The latter approach has been successfully used to produce unnatural cyanobactins,[60,165] lantibiotics,[166–168] thiopeptides,[169,170] lasso peptides[171] and linear azol(in)e-containing peptides (LAPs).[172] Generating diversity by introducing variations in the core peptide sequence was one of the features that enabled the combinatorial biosynthesis of RiPPs. Other enabling features include the presence of a specific portable short recognition signal for each processing enzyme, the promiscuous nature of the processing enzymes and the pathway modularity.[173] Examples of how these features enabled combinatorial biosynthesis include the recent development of engineered post-translational modifying enzymes that are fused to their recognition signals and thus could successfully modify stand-alone core peptides.[174,175] Interestingly, some processing enzymes that don't require recognition signals outside of the core peptide have also been reported.[176–178] The promiscuous nature of the enzymes allowed the generation of cyclic peptides containing unnatural amino acids[178–181] and non-amino acid chemical building blocks, *e.g.* benzyl rings, polyethers and alkyl chains.[182,183] The modular nature of the pathways allowed the exchange of genes between different pathways. However, this exchange is still restricted to pathways that share high sequence similarity and encode compounds from the same class of RiPPs, *e.g.* the patellamide and the trunkamide cyanobactin pathways. Current efforts are focused on exchanging genes from pathways that encode compounds from different classes of RiPPs. These unnatural compounds will have hybrid structural features from these different classes. Critical steps are to engineer a precursor peptide that has the recognition determinants of these enzymes and to control the timing of processing by each enzyme. The recent development of engineered enzymes that are fused with their recognition signals[174,175] is a great advance and should enable the *in vitro* generation of hybrid RiPPs. Another challenge arises from the fact that some of these post-translational modifications involve the same residue, *e.g.* cysteine is involved in the formation of the cyanobactin thiazoline heterocycles and the lantibiotic lanthionine bridges. An understanding of the kinetics and directionality of processing by the modifying enzymes could help overcome this problem. For example, the cyanobactin heterocyclase from the trunkamide pathway requires a 37-amino acid leader sequence in order to process all cysteine residues in the core sequence. In the absence of this leader sequence, the enzyme can only process the C-terminal cysteine[182] leaving the internal cysteines available for processing by enzymes from other classes of RiPPs. In summary, the use of biosynthetic enzymes from different classes of RiPPs could generate novel compounds with hybrid structural features from these classes and this represents a new direction to generate diversity in cyclic peptides (Figure 15.10).

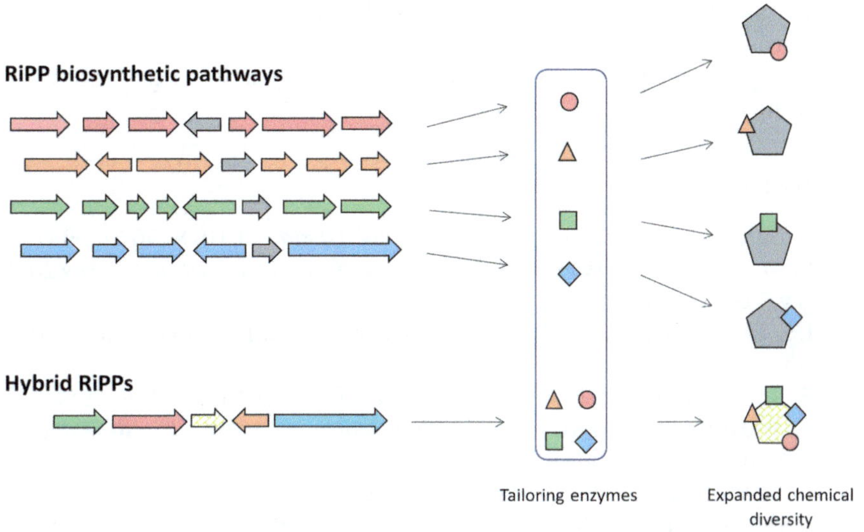

Figure 15.10 Expanding the chemical diversity of RiPPs *via* incorporation of hybrid structural features from different classes. RiPP biosynthetic pathways contain gene(s) encoding the precursor peptide (grey) and others encoding the tailoring enzymes. Each set of tailoring enzymes incorporates structural features characteristic to one class of RiPPs. Making hybrid RiPPs involves the construction of an engineered precursor peptide (striated yellow) that can be recognized by a cocktail of enzymes from different biosynthetic pathways.

15.7 Conclusions

This chapter has provided insight into which areas of cyclic peptide research need further effort to make this compound class of greater utility in the drug discovery process. It is clear that they show much promise in that they combine the properties of biologics in addressing extended binding sites with the desirable drug-like characteristics of small molecules. The synthesis of highly diverse cyclic peptide libraries will need development of both synthetic and biosynthetic strategies, with the optimal strategy likely to adopt the best elements of both. A clearer understanding of how cyclic peptides bind to their targets will assist in the design of future generations of cyclic peptides, which have greater affinity and specificity. Coupled to this is the prediction of the solution conformation and the structural parameters of cyclic peptides that allow them to cross membranes and adopt biologically relevant conformations. Understanding which modifications improve or worsen desirable physicochemical properties will permit better design of drug-like cyclic peptides and lead to the development of a useful set of guidelines for the design of such compounds. Predicting whether designed peptides can be cyclized will also prevent much wasted effort. A great deal of research effort is being extended in all of these areas with solutions to these problems arising regularly as a result. Cyclic peptide research is likely to thrive in the next few years heralding an era in which we can design and rapidly synthesize a cyclic peptide that binds with high affinity and specificity to its target and has efficacy in the clinic.

References

1. L. Cascales and D. J. Craik, *Org. Biomol. Chem.*, 2010, **8**, 5035–5047.
2. E. M. Driggers, S. P. Hale, J. Lee and N. K. Terrett, *Nat. Rev. Drug Discovery*, 2008, **7**, 608–624.
3. J. I. Levin, in *Macrocycles in Drug Discovery*, ed. J. I. Levin, 2015, vol. 40, pp. 1–499.
4. F. Giordanetto and J. Kihlberg, *J. Med. Chem.*, 2014, **57**, 278–295.
5. W. E. Houssen and M. Jaspars, *ChemBioChem*, 2010, **11**, 1803–1815.
6. F. Kopp and M. A. Marahiel, *Nat. Prod. Rep.*, 2007, **24**, 735–749.
7. C. J. White and A. K. Yudin, *Nat. Chem.*, 2011, **3**, 509–524.
8. I. Sainis, D. Fokas, K. Vareli, A. G. Tzakos, V. Kounnis and E. Briasoulis, *Mar. Drugs*, 2010, **8**, 629–657.
9. V. Marti-Centelles, M. D. Pandey, M. I. Burguete and S. V. Luis, *Chem. Rev.*, 2015, **115**, 8736–8834.
10. A. K. Yudin, *Chem. Sci.*, 2015, **6**, 30–49.
11. G. Cheng, A. Montero, P. Gastaminza, C. Whitten-Bauer, S. F. Wieland, M. Isogawa, B. Fredericksen, S. Selvarajah, P. A. Gallay, M. R. Ghadiri and F. V. Chisari, *Proc. Natl. Acad. Sci. U. S. A.*, 2008, **105**, 3088–3093.
12. J. S. Davies, *J. Pept. Sci.*, 2003, **9**, 471–501.
13. N. L. Benoiton, *Chemistry of Peptide Synthesis*, CRC Press, 2005.
14. H. Benz, *Synthesis*, 1994, 337–358.
15. M. Fridkin, A. Patchorn and E. Katchals, *J. Am. Chem. Soc.*, 1965, **87**, 4646–4648.
16. L. T. Scott, J. Rebek, L. Ovsyanko and C. L. Sims, *J. Am. Chem. Soc.*, 1977, **99**, 625–626.
17. M. Rothe, K. D. Steffen and I. Rothe, *Angew. Chem., Int. Ed.*, 1965, **4**, 356.
18. M. Rothe, M. Lohmuller, W. Fischer, W. Taiber and U. Breuksch, *Multiple Cyclo-Oligomerisations on Polymeric Supports in Solid Phase Synthesis*, ed. R. Epton, SPCC (UK) Ltd, Birmingham, UK, 1990, pp. 551–558.
19. P. Dumy, M. Keller, D. E. Ryan, B. Rohwedder, T. Wohr and M. Mutter, *J. Am. Chem. Soc.*, 1997, **119**, 918–925.
20. D. Skropeta, K. A. Jolliffe and P. Turner, *J. Org. Chem.*, 2004, **69**, 8804–8809.
21. L. M. D. Rodriguez, A. J. Weidkamp and M. A. Brimble, *Org. Biomol. Chem.*, 2015, **13**, 6906–6921.
22. A. Ehrlich, H. U. Heyne, R. Winter, M. Beyermann, H. Haber, L. A. Carpino and M. Bienert, *J. Org. Chem.*, 1996, **61**, 8831–8838.
23. X. M. Gao, Y. H. Ye, M. Bernd and B. Kutscher, *J. Pept. Sci.*, 2002, **8**, 418–430.

24. Y. C. Tang, H. B. Xie, G. L. Tian and Y. H. Ye, *J. Pept. Sci.*, 2002, **60**, 95–103.
25. S. F. Brady, S. L. Varga, R. M. Freidinger, D. A. Schwenk, M. Mendlowski, F. W. Holly and D. F. Veber, *J. Org. Chem.*, 1979, **44**, 3101–3105.
26. Y. H. Ye, X. M. Gao, M. Liu, Y. C. Tang and G. L. Tian, *Lett. Pept. Sci.*, 2003, **10**, 571–579.
27. P. Liu, B. Kim, R. A. Friesner and B. J. Berne, *Proc. Natl. Acad. Sci. U. S. A.*, 2005, **102**, 13749–13754.
28. W. D. F. Meutermans, S. W. Golding, G. T. Bourne, L. P. Miranda, M. J. Dooley, P. F. Alewood and M. L. Smythe, *J. Am. Chem. Soc.*, 1999, **121**, 9790–9796.
29. D. N. Le, J. Riedel, N. Kozlyuk, R. W. Martin and V. M. Dong, *Org. Lett.*, 2017, **19**, 114–117.
30. C. J. White, J. L. Hickey, C. C. G. Scully and A. K. Yudin, *J. Am. Chem. Soc.*, 2014, **136**, 3728–3731.
31. G. R. Pettit, B. E. Toki, J. P. Xu and D. C. Brune, *J. Nat. Prod.*, 2000, **63**, 22–28.
32. L. H. Yang and G. Morriello, *Tetrahedron Lett.*, 1999, **40**, 8197–8200.
33. C. G. Qin, X. Z. Bu, X. F. Zhong, N. L. J. Ng and Z. H. Guo, *J. Comb. Chem.*, 2004, **6**, 398–406.
34. A. Ehrlich, S. Rothemund, M. Brudel, M. Beyermann, L. A. Carpino and M. Bienert, *Tetrahedron Lett.*, 1993, **34**, 4781–4784.
35. T. Jeremic, A. Linden and H. Heimgartner, *Helv. Chim. Acta*, 2004, **87**, 3056–3079.
36. H. Kaur, A. M. Heapy and M. A. Brimble, *Synlett*, 2012, 2284–2288.
37. Y. H. Ye, H. T. Li and X. H. Jiang, *Biopolymers*, 2005, **80**, 172–178.
38. Y. Shao, W. Y. Lu and S. B. H. Kent, *Tetrahedron Lett.*, 1998, **39**, 3911–3914.
39. L. Z. Yan and P. E. Dawson, *J. Am. Chem. Soc.*, 2001, **123**, 526–533.
40. L. S. Zhang and J. P. Tam, *J. Am. Chem. Soc.*, 1999, **121**, 3311–3320.
41. Y. Mifune, H. Nakamura and S. Fuse, *Org. Biomol. Chem.*, 2016, **14**, 11244–11249.
42. C. Baraguey, C. Auvin-Guette, A. Blond, F. Cavelier, F. Lezenven, J. L. Pousset and B. Bodo, *J. Chem. Soc., Perkin Trans. 1*, 1998, 3033–3039.
43. M. Tamaki, S. Akabori and I. Muramatsu, *J. Am. Chem. Soc.*, 1993, **115**, 10492–10496.
44. P. Raman, S. S. Stokes, Y. M. Angell, G. R. Flentke and D. H. Rich, *J. Org. Chem.*, 1998, **63**, 5734–5735.
45. L. J. Jiang, K. Moehle, B. Dhanapal, D. Obrecht and J. A. Robinson, *Helv. Chim. Acta*, 2000, **83**, 3097–3112.
46. P. Thapa, M. J. Espiritu, C. Cabalteja and J.-P. Bingham, *Int. J. Pept. Res. Ther.*, 2014, **20**, 545–551.
47. T. Katoh, Y. Goto, M. S. Rezab and H. Suga, *Chem. Commun.*, 2011, **47**, 9946–9958.
48. A. Tavassoli and S. J. Benkovic, *Nat. Protoc.*, 2007, **2**, 1126–1133.
49. P. G. Arnison, M. J. Bibb, G. Bierbaum, A. A. Bowers, T. S. Bugni, G. Bulaj, J. A. Camarero, D. J. Campopiano, G. L. Challis, J. Clardy, P. D. Cotter, D. J. Craik, M. Dawson, E. Dittmann, S. Donadio, P. C. Dorrestein,

K. D. Entian, M. A. Fischbach, J. S. Garavelli, U. Goransson, C. W. Gruber, D. H. Haft, T. K. Hemscheidt, C. Hertweck, C. Hill, A. R. Horswill, M. Jaspars, W. L. Kelly, J. P. Klinman, O. P. Kuipers, A. J. Link, W. Liu, M. A. Marahiel, D. A. Mitchell, G. N. Moll, B. S. Moore, R. Muller, S. K. Nair, I. F. Nes, G. E. Norris, B. M. Olivera, H. Onaka, M. L. Patchett, J. Piel, M. J. T. Reaney, S. Rebuffat, R. P. Ross, H. G. Sahl, E. W. Schmidt, M. E. Selsted, K. Severinov, B. Shen, K. Sivonen, L. Smith, T. Stein, R. D. Sussmuth, J. R. Tagg, G. L. Tang, A. W. Truman, J. C. Vederas, C. T. Walsh, J. D. Walton, S. C. Wenzel, J. M. Willey and W. A. van der Donk, *Nat. Prod. Rep.*, 2013, **30**, 108–160.
50. J. E. Townend and A. Tavassoli, *ACS Chem. Biol.*, 2016, **11**, 1624–1630.
51. T. Passioura, T. Katoh, Y. Goto and H. Suga, in*Ann Rev Biochem*, ed. R. D. Kornberg, 2014, vol. 83, pp. 727–752.
52. K. R. Lennard and A. Tavassoli, *Chem.–Eur. J.*, 2014, **20**, 10608–10614.
53. A. A. Bowers, *MedChemComm*, 2012, **3**, 905–915.
54. J. Koehnke, A. Bent, W. E. Houssen, D. Zollman, F. Morawitz, S. Shirran, J. Vendome, A. F. Nneoyiegbe, L. Trembleau, C. H. Botting, M. C. M. Smith, M. Jaspars and J. H. Naismith, *Nat. Struct. Mol. Biol.*, 2012, **19**, 767–772.
55. G. K. T. Nguyen, S. Wang, Y. Qiu, X. Hemu, Y. Lian and J. P. Tam, *Nat. Chem. Biol.*, 2014, **10**, 732–738.
56. H. Luo, S. Y. Hong, R. M. Sgambelluri, E. Angelos, X. Li and J. D. Walton, *Chem. Biol.*, 2014, **21**, 1610–1617.
57. J. D. Walton, H. E. Hallen-Adams and H. Luo, *Biopolymers*, 2010, **94**, 659–664.
58. C. J. S. Barber, P. T. Pujara, D. W. Reed, S. Chiwocha, H. Zhang and P. S. Covello,*J. Biol. Chem.*, 2013, **288**, 12500–12510.
59. M. S. Donia, B. J. Hathaway, S. Sudek, M. G. Haygood, M. J. Rosovitz, J. Ravel and E. W. Schmidt, *Nat. Chem. Biol.*, 2006, **2**, 729–735.
60. W. E. Houssen, A. F. Bent, A. R. McEwan, N. Pieiller, J. Tabudravu, J. Koehnke, G. Mann, R. I. Adaba, L. Thomas, U. W. Hawas, H. T. Liu, U. Schwarz-Linek, M. C. M. Smith, J. H. Naismith and M. Jaspars, *Angew. Chem., Int. Ed.*, 2014, **53**, 14171–14174.
61. J. Booth, C.-N. Alexandru-Crivac, K. A. Rickaby, A. F. Nneoyiegbe, U. Umeobika, A. R. McEwan, L. Trembleau, M. Jaspars, W. E. Houssen and D. V. Shalashilin,*J. Phys. Chem. Lett.*, 2017, **8**, 2310.
62. W. M. Hewitt, S. S. Leung, C. R. Pye, A. R. Ponkey, M. Bednarek, M. P. Jacobson and R. S. Lokey,*J. Am. Chem. Soc.*, 2015, **137**, 715–721.
63. D. E. Scott, A. R. Bayly, C. Abell and J. Skidmore, *Nat. Rev. Drug Discovery*, 2016, **15**, 533–550.
64. E. A. Villar, D. Beglov, S. Chennamadhavuni, J. A. Porc Jr, D. Kozakov, S. Vajda and A. Whitty, *Nat. Chem. Biol.*, 2014, **10**, 723–731.
65. B. C. Doak, J. Zheng, D. Dobritzsch and J. Kihlberg,*J. Med. Chem.*, 2016, **59**, 2312–2327.
66. S. Surade and T. L. Blundell, *Chem. Biol.*, 2012, **19**, 42–50.

67. C. A. Lipinski, F. Lombardo, B. W. Dominy and P. J. Feeney, *Adv. Drug Delivery Rev.*, 1997, **23**, 3–25.
68. P. Matsson, B. C. Doak, B. Over and J. Kihlberg, *Adv. Drug Delivery Rev.*, 2016, **101**, 42–61.
69. M. M. Hann, *MedChemComm*, 2011, **2**, 349.
70. C. A. Lipinski, *Adv. Drug Delivery Rev.*, 2016, **101**, 34–41.
71. B. Over, P. Matsson, C. Tyrchan, P. Artursson, B. C. Doak, M. A. Foley, C. Hilgendorf, S. E. Johnston, M. D. Lee, R. J. Lewis, P. McCarren, G. Muncipinto, U. Norinder, M. W. Perry, J. R. Duvall and J. Kihlberg, *Nat. Chem. Biol.*, 2016, **12**, 1065–1074.
72. K. Fosgerau and T. Hoffmann, *Drug Discovery Today*, 2015, **20**, 122–128.
73. K. Palm, K. Luthman, A. L. Ungell, G. Strandlund and P. Artursson, *J. Pharm. Sci.*, 1996, **85**, 32–39.
74. H. vandeWaterbeemd, G. Camenisch, G. Folkers and O. A. Raevsky, *Quant. Struct.-Act. Relat.*, 1996, **15**, 480–490.
75. A. M. Mathiowetz, S. S. F. Leung and M. P. Jacobson, in *Marcrocycles in Drug Discovery*, ed. J. I. Levin, RSC, 2015, vol. 40, pp. 367–397.
76. T. Rezai, J. E. Bock, M. V. Zhou, C. Kalyanaraman, R. S. Lokey and M. P. Jacobson, *J. Am. Chem. Soc.*, 2006, **128**, 14073–14080.
77. C. K. Wang, S. E. Northfield, J. E. Swedberg, B. Colless, S. Chaousis, D. A. Price, S. Liras and D. J. Craik, *Eur. J. Med. Chem.*, 2015, **97**, 202–213.
78. P. Thansandote, R. M. Harris, H. L. Dexter, G. L. Simpson, S. Pal, R. J. Upton and K. Valko, *Bioorg. Med. Chem.*, 2015, **23**, 322–327.
79. I. Lewis, M. Schaefer, T. Wagner, L. Oberer, E. Sager, P. Wipfli and T. Vorherr, *Int. J. Pept. Res. Ther.*, 2015, **21**, 205–221.
80. D. S. Nielsen, H. N. Hoang, R. J. Lohman, T. A. Hill, A. J. Lucke, D. J. Craik, D. J. Edmonds, D. A. Griffith, C. J. Rotter, R. B. Ruggeri, D. A. Price, S. Liras and D. P. Fairlie, *Angew. Chem., Int. Ed. Engl.*, 2014, **53**, 12059–12063.
81. J. G. Beck, J. Chatterjee, B. Laufer, M. U. Kiran, A. O. Frank, S. Neubauer, O. Ovadia, S. Greenberg, C. Gilon, A. Hoffman and H. Kessler, *J. Am. Chem. Soc.*, 2012, **134**, 12125–12133.
82. A. T. Bockus, K. W. Lexa, C. R. Pye, A. S. Kalgutkar, J. W. Gardner, K. C. Hund, W. M. Hewitt, J. A. Schwochert, E. Glassey, D. A. Price, A. M. Mathiowetz, S. Liras, M. P. Jacobson and R. S. Lokey, *J. Med. Chem.*, 2015, **58**, 4581–4589.
83. A. Bhat, L. R. Roberts and J. J. Dwyer, *Eur. J. Med. Chem.*, 2015, **94**, 471–479.
84. U. K. Marelli, O. Ovadia, A. O. Frank, J. Chatterjee, C. Gilon, A. Hoffman and H. Kessler, *Chem.–Eur. J.*, 2015, **21**, 15148–15152.
85. J. L. Hickey, S. Zaretsky, M. A. S. Denis, S. K. Chakka, M. M. Morshed, C. C. G. Scully, A. L. Roughton and A. K. Yudin, *J. Med. Chem.*, 2016, **59**, 5368–5376.
86. U. K. Marelli, J. Bezencon, E. Puig, B. Ernst and H. Kessler, *Chem.–Eur. J.*, 2015, **21**, 8023–8027.

87. A. T. Bockus, J. A. Schwochert, C. R. Pye, C. E. Townsend, V. Sok, M. A. Bednarek and R. S. Lokey, *J. Med. Chem.*, 2015, **58**, 7409–7418.
88. T. A. Hill, R. J. Lohman, H. N. Hoang, D. S. Nielsen, C. C. G. Scully, W. M. Kok, L. G. Liu, A. J. Lucke, M. J. Stoermer, C. I. Schroeder, S. Chaousis, B. Colless, P. V. Bernhardt, D. J. Edmonds, D. A. Griffith, C. J. Rotter, R. B. Ruggeri, D. A. Price, S. Liras, D. J. Craik and D. P. Fairlie, *ACS Med. Chem. Lett.*, 2014, **5**, 1148–1151.
89. J. E. Bock, J. Gavenonis and J. A. Kritzer, *ACS Chem. Biol.*, 2013, **8**, 488–499.
90. H. Traboulsi, H. Larkin, M. A. Bonin, L. Volkov, C. L. Lavoie and E. Marsault, *Bioconjugate Chem.*, 2015, **26**, 405–411.
91. A. Whitty, M. Zhong, L. Viarengo, D. Beglov, D. R. Hall and S. Vajda, *Drug Discovery Today*, 2016, **21**, 712–717.
92. S. G. Aller, J. Yu, A. Ward, Y. Weng, S. Chittaboina, R. Zhuo, P. M. Harrell, Y. T. Trinh, Q. Zhang, I. L. Urbatsch and G. Chang, *Science*, 2009, **323**, 1718–1722.
93. A. Cesar-Razquin, B. Snijder, T. Frappier-Brinton, R. Isserlin, G. Gyimesi, X. Bai, R. A. Reithmeier, D. Hepworth, M. A. Hediger, A. M. Edwards and G. Superti-Furga, *Cell*, 2015, **162**, 478–487.
94. S. Zaretsky, C. C. G. Scully, A. J. Lough and A. K. Yudin, *Chem.–Eur. J.*, 2013, **19**, 17668–17672.
95. S. M. McHugh, J. R. Rogers, S. A. Solomon, H. T. Yu and Y. S. Lin, *Curr. Opin. Chem. Biol.*, 2016, **34**, 95–102.
96. C. Baysal and H. Meirovitch, *Biopolymers*, 2000, **53**, 423–433.
97. J. Beaufays, L. Lins, A. Thomas and R. Brasseur, *J. Pept. Sci.*, 2012, **18**, 17–24.
98. Y. Goldtzvik, M. Goldstein and R. B. Gerber, *Chem. Phys.*, 2013, **415**, 168–172.
99. G. V. Nikiforovich, K. E. Kover, W. J. Zhang and G. R. Marshall, *J. Am. Chem. Soc.*, 2000, **122**, 3262–3273.
100. M. T. Oakley and R. L. Johnston, *J. Chem. Theory Comput.*, 2014, **10**, 1810–1816.
101. A. Rayan, H. Senderowitz and A. Goldblum, *J. Mol. Graphics Modell.*, 2004, **22**, 319–333.
102. P. Bradley, K. M. S. Misura and D. Baker, *Science*, 2005, **309**, 1868–1871.
103. S. T. Wu, J. Skolnick and Y. Zhang, *BMC Biol.*, 2007, **5**.
104. D. Xu and Y. Zhang, *Proteins: Struct., Funct., Bioinf.*, 2012, **80**, 1715–1735.
105. A. Thomas, S. Deshayes, M. Decaffmeyer, M. H. Van Eyck, B. Charloteaux and R. Brasseur, *Proteins: Struct., Funct., Bioinf.*, 2006, **65**, 889–897.
106. R. Das, *PLoS One*, 2011, **6**.
107. P. Cuniasse, I. Raynal, A. Yiotakis and V. Dive, *J. Am. Chem. Soc.*, 1997, **119**, 5239–5248.
108. H. Geng, F. Jiang and Y. D. Wu, *J. Phys. Chem. Lett.*, 2016, **7**, 1805–1810.
109. J. Li, T. Ehlers, J. Sutter, S. Varma-O'Brien and J. Kirchmair, *J. Chem. Inf. Model.*, 2007, **47**, 1923–1932.
110. P. C. D. Hawkins, A. G. Skillman, G. L. Warren, B. A. Ellingson and M. T. Stahl, *J. Chem. Inf. Model.*, 2010, **50**, 572–584.

111. I. J. Chen and N. Foloppe, *Bioorg. Med. Chem.*, 2013, **21**, 7898–7920.
112. K. S. Watts, P. Dalal, A. J. Tebben, D. L. Cheney and J. C. Shelley, *J. Chem. Inf. Model.*, 2014, **54**, 2680–2696.
113. S. Singh, H. Singh, A. Tuknait, K. Chaudhary, B. Singh, S. Kumaran and G. P. S. Raghava, *Biol. Direct*, 2015, **10**.
114. P. Thevenet, Y. M. Shen, J. Maupetit, F. Guyon, P. Derreumaux and P. Tuffery, *Nucleic Acids Res.*, 2012, **40**, W288–W293.
115. Y. Zhang, *BMC Bioinf.*, 2008, **9**.
116. A. M. Razavi, W. M. Wuest and V. A. Voelz, *J. Chem. Inf. Model.*, 2014, **54**, 1425–1432.
117. R. N. Riemann and M. Zacharias, *J. Pept. Sci.*, 2004, **63**, 354–364.
118. C. Ramakrishnan, P. K. C. Paul and K. Ramnarayan, *J. Biosci.*, 1985, **8**, 239–251.
119. D. J. Earl and M. W. Deem, *Phys. Chem. Chem. Phys.*, 2005, **7**, 3910–3916.
120. R. Faller, Q. L. Yan and J. J. de Pablo, *J. Chem. Phys.*, 2002, **116**, 5419–5423.
121. M. K. Fenwick and F. A. Escobedo, *J. Chem. Phys.*, 2003, **119**, 11998–12010.
122. A. Laio and M. Parrinello, *Proc. Natl. Acad. Sci. U. S. A.*, 2002, **99**, 12562–12566.
123. A. Mitsutake, Y. Sugita and Y. Okamoto, *Biopolymers*, 2001, **60**, 96–123.
124. S. W. Rick, *J. Chem. Phys.*, 2007, **126**.
125. Y. Sugita and Y. Okamoto, *Chem. Phys. Lett.*, 1999, **314**, 141–151.
126. R. Day, D. Paschek and A. E. Garcia, *Proteins: Struct., Funct., Bioinf.*, 2010, **78**, 1889–1899.
127. T. J. Lane, D. Shukla, K. A. Beauchamp and V. S. Pande, *Curr. Opin. Struct. Biol.*, 2013, **23**, 58–65.
128. K. Lindorff-Larsen, S. Piana, R. O. Dror and D. E. Shaw, *Science*, 2011, **334**, 517–520.
129. J. M. Damas, L. C. S. Filipe, S. R. R. Campos, D. Lousa, B. L. Victor, A. M. Baptista and C. M. Soares, *Eur. Biophys. J. Biophys. Lett.*, 2013, **42**, S90.
130. V. A. Voelz, K. A. Dill and I. Chorny, *Biopolymers*, 2011, **96**, 639–650.
131. E. Yedvabny, P. S. Nerenberg, C. So and T. Head-Gordon, *J. Phys. Chem. B*, 2015, **119**, 896–905.
132. N. Schmid, A. P. Eichenberger, A. Choutko, S. Riniker, M. Winger, A. E. Mark and W. F. van Gunsteren, *Eur. Biophys. J. Biophys. Lett.*, 2011, **40**, 843–856.
133. M. M. Reif, P. H. Hunenberger and C. Oostenbrink, *J. Chem. Theory Comput.*, 2012, **8**, 3705–3723.
134. A. D. MacKerell, M. Feig and C. L. Brooks, *J. Am. Chem. Soc.*, 2004, **126**, 698–699.
135. D. W. Li and R. Bruschweiler, *Angew. Chem., Int. Ed.*, 2010, **49**, 6778–6780.
136. V. Hornak, R. Abel, A. Okur, B. Strockbine, A. Roitberg and C. Simmerling, *Proteins: Struct., Funct., Bioinf.*, 2006, **65**, 712–725.
137. R. B. Best and G. Hummer, *J. Phys. Chem. B*, 2009, **113**, 9004–9015.
138. W. D. Cornell, P. Cieplak, C. I. Bayly, I. R. Gould, K. M. Merz, D. M. Ferguson, D. C. Spellmeyer, T. Fox, J. W. Caldwell and P. A. Kollman, *J. Am. Chem. Soc.*, 1996, **118**, 2309.

139. J. M. Wang, P. Cieplak and P. A. Kollman, *J. Comput. Chem.*, 2000, **21**, 1049–1074.
140. L. D. Schuler, X. Daura and W. F. Van Gunsteren, *J. Comput. Chem.*, 2001, **22**, 1205–1218.
141. C. Oostenbrink, A. Villa, A. E. Mark and W. F. Van Gunsteren, *J. Comput. Chem.*, 2004, **25**, 1656–1676.
142. W. L. Jorgensen, D. S. Maxwell and J. TiradoRives, *J. Am. Chem. Soc.*, 1996, **118**, 11225–11236.
143. G. A. Kaminski, R. A. Friesner, J. Tirado-Rives and W. L. Jorgensen, *J. Phys. Chem. B*, 2001, **105**, 6474–6487.
144. A. D. MacKerell, D. Bashford, M. Bellott, R. L. Dunbrack, J. D. Evanseck, M. J. Field, S. Fischer, J. Gao, H. Guo, S. Ha, D. Joseph-McCarthy, L. Kuchnir, K. Kuczera, F. T. K. Lau, C. Mattos, S. Michnick, T. Ngo, D. T. Nguyen, B. Prodhom, W. E. Reiher, B. Roux, M. Schlenkrich, J. C. Smith, R. Stote, J. Straub, M. Watanabe, J. Wiorkiewicz-Kuczera, D. Yin and M. Karplus, *J. Phys. Chem. B*, 1998, **102**, 3586–3616.
145. H. T. Yu and Y. S. Lin, *Phys. Chem. Chem. Phys.*, 2015, **17**, 4210–4219.
146. F. Jiang, C. Y. Zhou and Y. D. Wu, *J. Phys. Chem. B*, 2014, **118**, 6983–6998.
147. C. Y. Zhou, F. Jiang and Y. D. Wu, *J. Phys. Chem. B*, 2015, **119**, 1035–1047.
148. D. P. Slough, H. T. Yu, S. M. McHugh and Y. S. Lin, *Phys. Chem. Chem. Phys.*, 2017, **19**, 5377–5388.
149. D. R. Glowacki, E. Paci and D. V. Shalashilin, *J. Phys. Chem. B*, 2009, **113**, 16603–16611.
150. D. V. Shalashilin, G. S. Beddard, E. Paci and D. R. Glowacki, *J. Chem. Phys.*, 2012, **137**, 165102.
151. J. B. Ball, R. A. Hughes, P. F. Alewood and P. R. Andrews, *Tetrahedron*, 1993, **49**, 3467–3478.
152. S. Z. Li, J. H. Lee, W. Lee, C. J. Yoon, J. H. Baik and S. K. Lim, *FEBS J.*, 1999, **265**, 430–440.
153. G. D. Rose, L. M. Gierasch and J. A. Smith, *Adv. Protein Chem.*, 1985, **37**, 1–109.
154. H. J. Bohm and G. Klebe, *Angew. Chem., Int. Ed. Engl.*, 1996, **35**, 2589–2614.
155. T. Kortemme, D. E. Kim and D. Baker, *Sci. STKE*, 2004, **2004**, pl2.
156. S. Chaudhury, S. Lyskov and J. J. Gray, *Bioinformatics*, 2010, **26**, 689–691.
157. S. B. Ozkan and H. Meirovitch, *J. Phys. Chem. B*, 2003, **107**, 9128–9131.
158. S. F. Brady, L. Simmons, J. H. Kim and E. W. Schmidt, *Nat. Prod. Rep.*, 2009, **26**, 1488–1503.
159. B. Wilkinson and J. Micklefield, *Nat. Chem. Biol.*, 2007, **3**, 379–386.
160. G. L. Challis, *Microbiology*, 2008, **154**, 1555–1569.
161. E. Kim, B. S. Moore and Y. J. Yoon, *Nat. Chem. Biol.*, 2015, **11**, 649–659.
162. SciBx, *Innovation in Drug Discovery and Development Summit on Macrocycles and Constrained Peptides*, 2012.
163. J. A. McIntosh, M. S. Donia and E. W. Schmidt, *Nat. Prod. Rep.*, 2009, **26**, 537–559.

164. W. E. Houssen, J. Koehnke, D. Zollman, J. Vendome, A. Raab, M. C. M. Smith, J. H. Naismith and M. Jaspars, *ChemBioChem*, 2012, **13**, 2683–2689.
165. M. D. B. Tianero, M. S. Donia, T. S. Young, P. G. Schultz and E. W. Schmidt, *J. Am. Chem. Soc.*, 2012, **134**, 418–425.
166. A. N. Appleyard, S. Choi, D. M. Read, A. Lightfoot, S. Boakes, A. Hoffmann, I. Chopra, G. Bierbaum, B. A. M. Rudd, M. J. Dawson and J. Cortes, *Chem. Biol.*, 2009, **16**, 490–498.
167. P. D. Cotter, L. H. Deegan, E. M. Lawton, L. A. Draper, P. M. O'Connor, C. Hill and R. P. Ross, *Mol. Microbiol.*, 2006, **62**, 735–747.
168. E. M. Molloy, D. Field, P. M. O'Connor, P. D. Cotter, C. Hill and R. P. Ross, *PLoS One*, 2013, **8**.
169. A. A. Bowers, M. G. Acker, A. Koglin and C. T. Walsh, *J. Am. Chem. Soc.*, 2010, **132**, 7519–7527.
170. T. S. Young, P. C. Dorrestein and C. T. Walsh, *Chem. Biol.*, 2012, **19**, 1600–1610.
171. S. J. Pan and A. J. Link, *J. Am. Chem. Soc.*, 2011, **133**, 5016–5023.
172. C. D. Deane, J. O. Melby, K. J. Molohon, A. R. Susarrey and D. A. Mitchell, *ACS Chem. Biol.*, 2013, **8**, 1998–2008.
173. D. Sardar and E. W. Schmidt, *Curr. Opin. Chem. Biol.*, 2016, **31**, 15–21.
174. J. Koehnke, G. Mann, A. F. Bent, H. Ludewig, S. Shirran, C. Botting, T. Lebl, W. E. Houssen, M. Jaspars and J. H. Naismith, *Nat. Chem. Biol.*, 2015, **11**, 548–558.
175. T. J. Oman, P. J. Knerr, N. A. Bindman, J. E. Velasquez and W. A. van der Donk, *J. Am. Chem. Soc.*, 2012, **134**, 6952–6955.
176. A. F. Bent, G. Mann, W. E. Houssen, V. Mykhaylyk, R. Duman, L. Thomas, M. Jaspars, A. Wagner and J. H. Naismith, *Acta Crystallogr., Sect. D: Struct. Biol.*, 2016, **72**, 1174–1180.
177. A. Parajuli, D. H. Kwak, L. Dalponte, N. Leikoski, T. Galica, U. Umeobika, L. Trembleau, A. Bent, K. Sivonen, M. Wahlsten, H. Wang, E. Rizzi, G. De Bellis, J. Naismith, M. Jaspars, X. Y. Liu, W. Houssen and D. P. Fewer, *Angew. Chem., Int. Ed.*, 2016, **55**, 3596–3599.
178. X. Yang and W. A. van der Donk, *J. Am. Chem. Soc.*, 2015, **137**, 12426–12429.
179. A. Kuthning, P. Durkin, S. Oehm, M. G. Hoesl, N. Budisa and R. D. Sussmuth, *Sci. Rep.*, 2016, **6**.
180. E. Oueis, C. Adamson, G. Mann, H. Ludewig, P. Redpath, M. Migaud, N. J. Westwood and J. H. Naismith, *ChemBioChem*, 2015, **16**, 2646–2650.
181. E. Oueis, M. Jaspars, N. J. Westwood and J. H. Naismith, *Angew. Chem., Int. Ed. Engl.*, 2016, **55**, 5842–5845.
182. J. Koehnke, A. F. Bent, D. Zollman, K. Smith, W. E. Houssen, X. F. Zhu, G. Mann, T. Lebl, R. Scharff, S. Shirran, C. H. Botting, M. Jaspars, U. Schwarz-Linek and J. H. Naismith, *Angew. Chem., Int. Ed. Engl.*, 2013, **52**, 13991–13996.
183. E. Oueis, B. Nardone, M. Jaspars, N. J. Westwood and J. H. Naismith, *ChemistryOpen*, 2017, **6**, 11–14.

Subject Index